19th Argentinean Bioengineering Society Congress

(SABI 2013)

Journal of Physics: Conference Series Volume 477

Tucuman, Argentina
4-6 September 2013

ISBN: 978-1-62993-939-1
ISSN: 1742-6588

Printed from e-media with permission by:

Curran Associates, Inc.
57 Morehouse Lane
Red Hook, NY 12571

proceedings.com

Some format issues inherent in the e-media version may also appear in this print version.

Printed by Curran Associates, Inc. (2014)

Institute of Physics
Dirac House, Temple Back
Bristol BS1 6BE UK

Phone: 44 1 17 929 7481
Fax: 44 1 17 920 0979

techtracking@iop.org

Additional copies of this publication are available from:

Curran Associates, Inc.
57 Morehouse Lane
Red Hook, NY 12571 USA
Phone: 845-758-0400
Fax: 845-758-2634
Email: curran@proceedings.com
Web: www.proceedings.com

TABLE OF CONTENTS

BIOMECHANICS

EQUIPMENT, INSTRUMENTATION, SENSORS AND MEASUREMENTS

MEDICAL INFORMATICS

CLINICAL ENGINEERING AND MANAGEMENT OF MEDICAL TECHNOLOGY

MODELING AND SIMULATION OF BIOMEDICAL SYSTEMS

BIOMEDICAL IMAGE PROCESSING

BIOMEDICAL SIGNAL PROCESSING

SABI 2013 – MEMORIES

From the 4–6 September 2013, the XIX Argentine Congress of Bioengineering and VIII Clinical Engineering Conference, SABI 2013, took place in the Catalinas Park Hotel in the city of San Miguel de Tucumán, Tucumán, Argentina. The Congress was held in a privileged central area of the city, opposite a splendid park, which provided a wonderful show of color and joy with their flowered Lapachos.

The present Congress was organized by the NOA Regional of the Argentine Bioengineering Society and the Bioengineering Department of the Faculty of Exact Sciences and Technology of the National University of Tucuman (UNT). SABI'2013 achieved a multidisciplinary convergence of all areas of bioengineering at the same academic/scientific regional, national and international forum. The conference was attended by researchers, teachers and students, engineers, physicists, mathematicians, biologists, doctors and health professionals working in the field of bioengineering.

The general objectives of the Argentine Congress of Bioengineering SABI 2013 were:

- To generate an adequate academic and scientific forum to present the work and experiences of researchers/teachers/students, contributing this way, to the production of new knowledge in the field of Bioengineering.
- Provide the approach to the main problems facing bioengineering and clinical engineering in Latin America and especially in Argentina, creating spaces that encourage discussion and joint development of research, technology development, transfer and innovation.
- Establish specific relationships between students majoring in bioengineering/biomedical engineering in the country and Latin America, with activities and/or workshops independently and free.
- To inform the scientific community and other social sectors, the topics addressed in this Congress, as well as progress and proposals, so that the social impact of bioengineering becomes a reality.

The subject areas in which papers were presented were:

- bioinformatics
- medical informatics
- biomaterials
- biomechanics
- education and bioengineering topics
- equipment, instrumentation, sensors and measurements
- clinical engineering and management of medical technology
- modelling and simulation of biomedical systems
- processing of biomedical images
- biomedical signal processing
- simulation systems and neural engineering
- neuromuscular systems and rehabilitation engineering

- high value bioengineering products and exportables

Five Conferences took place during the three days. The invited speakers were:

ScD Anthony Guiseppi-Elie
Center for Bioelectronics, Biosensors and Biochips (C3B), Clemson University
Advanced Materials Center, South Carolina, USA
Department of Chemical and Biomolecular Engineering
Department of Bioengineering
Department of Electrical and Computer Engineering, Clemson University, Clemson, USA
Implantable bioanalytical microsystems for the management of trauma

PhD Max E Valentinuzzi
Professor, Biomedical Entineering Institute, Buenos Aires University, Argentina
Professor Emeritus, National University of Tucumán, Tucumán, Argentina
Editor, Retrospectroscope section, IEEE Pulse, USA
Bioengineering in Latin America: looking back, but mostly ahead

Dra Élida Beatriz Hermida
National University of San Martín, Buenos Aires, Argentina
National Atomic Energy Commission
Biomatter: a path from applied science to technology transfer

Dr Renato García Ojeda
Coordinator of the Biomedical Entineering Institute
Federal University of Santa Catarina – Brasil
CORAL President
Technology Assessment for the incorporation and acquisition of medical equipment

Dr Carmelo José Felice
Media and Interfaces Laboratory (LAMEIN),
Bioengineering Departament, Faculty of Sciences and Technology of the National University of Tucuman
INSIBIO – CONICET
Argentina's dream: science for explosive growth of wealth

Six panel sessions took place, entitled: Young Professionals, Professional Associations, Clinical Engineering: Standards and Regulations, Rehabilitation Engineering, Education in Bioengineering and Tissue Engineering. They were an enabling environment for discussion and exchange of ideas, during which topics such as the future of the Biomedical Engineering graduates, professional performance, the national reality of medical technologies and their influence on the economies of the country and of Latin America prevailed. An important issue was the discussion of the Responsibilities of Health Technology Services under the new Law 'Traceability and Verification of Medical Products in use in Health' recently adopted in our country.

Both conferences and panel sessions were attended by many conference attendees, allowing closer relationships, both academic, as well as scientific and personal. The Congress was sponsored by major academic institutions in the country related to Bioengineering. It also received monetary and/or organizational support of the following organizations:

- Ministry of Health of the Province of Tucumán
- The National Agency for Scientific and Technological Promotion
- IEEE
- Tourism Secretary of Tucumán
- State Secretary for Innovation and Technological Development
- Science and Technology Council, National University of Tucumán

191 papers were received, of which 165 were approved, after revision, for presentation at the Congress. The Congress was organized in 19 oral sessions in different areas and 3 poster sessions, one per day.

Summaries of all of the activities of the Congress were published in a book and the complete papers on a CD. It is important to highlight that the authors were given the opportunity to publish their articles in the *Journal of Physics: Conference Series*, taking this option, 42 of the papers approved in Congress.

The conference wass attended by 470 delegates from different provinces of the country and neighboring countries like Brazil, Colombia, Peru, Uruguay and Chile, and was attended by five companies, who sponsored the event in different ways.

Within the framework of SABI 2013, the 5th Student SABI took place the previous day, 3 September, in the 'Eugenio Flavio Virla Cultural Center' of the National University of Tucumán. This event included 120 students, many of whom also participated in the General Congress, and were from all the country, as well as students from Brazil, Colombia and Uruguay.

In this context, 29 papers were presented in three categories: advances in graduation projects, initiating research projects and academic projects. All papers had a very good level, and oral and posters mentions were given for the best works.

As in the previous 18 meetings of the Society, SABI 2013 was again a space for undisputed knowledge integration and shared of peer experiences within the specific area of bioengineering. The cordiality and good treatment of the attendees promoted a climate of friendship, which we could enjoy during all the development of the congress.

facet

Hemodynamic differences in intracranial anerusysm blebs due to blood rheology

Marcelo A. Castro[1]
CONICET. Universidad Tecnológica Nacional, Facultad Regional Buenos Aires
Grupo de Investigación y Desarrollo en Bioingeniería
Medrano 951, Ciudad Autónoma de Buenos Aires, CP 1179, Argentina

E-mail: marcelo.a.castro@gmail.com

María C. Ahumada Olivares
Universidad Favaloro, Facultad de Ingeniería, Ciencias Exactas y Naturales
Solís 453, Ciudad Autónoma de Buenos Aires, CP 1078, Argentina

Juan R. Cebral
George Mason University. College of Sciences
4400 University Drive, Fairfax, VA, 22030, USA

Abstract. Computational blood flow simulations over patient-specific image-based three dimensional domains under personalized flow conditions have been extensively used to investigate associations between flow characteristics and aneurysm initiation, progress and rupture. Although it is widely accepted that the wall shear stress is associated to those processes, there is still no agreement about what stress characteristics are responsible for triggering those biomechanical processes. Although the incorporation of the blood rheology in large arterial systems containing aneurysms resulted in similar hemodynamic characterizations for most aneurysms, large aneurysms, especially those containing blebs, are expected to have flow rates in the range where Newtonian and non-Newtonian models largely differ. However, there is no consent among authors about the impact of blood rheology on the intraaneurysmal wall shear stress magnitude. In this work we used high resolution models reconstructed from rotational angiography images to perform unsteady finite element blood flow simulations to investigate the differences in WSS distribution and alignment for Newtonian and non-Newtonian rheologies. Unstructured finite element meshes were generated using an advancing front technique. Personalized flow conditions were imposed at the inlets of the models. The Casson model was incorporated as a velocity-dependent apparent viscosity and the results were compared to those using the Newtonian rheology. Associations between the localization of regions with large differences in wall shear stress magnitude and orientation, and the regions of differentiated wall shear stress magnitude were studied in a cohort of patients.

[1] To whom any correspondence should be addressed.

Introduction

During the last years some research groups have investigated the correlation between hemodynamic characteristic and initiation, development and rupture of cerebral aneurysms using numerical simulations. A previous study showed a possible association between high maximum WSS at the systolic peak and rupture in a cohort of AComA aneurysms [7]. A large study including 210 cerebral aneurysms at different locations it was found a statistical significant association between rupture and WSS distributions with elevated maximum WSS, high flow concentration and small impingement size [8]. Cebral et al. also showed a connection between location of aneurysm blebs and regions of high WSS in models where blebs were virtually removed [9]. In another study, it was presented a relationship between rupture, and coexistence of high WSS and high positive spatial WSS gradient, observed in three-patient scanned before and after aneurysm formation [10]. Although other investigators reported possible associations between low shear stress and either rupture [11,12] or blister formation [13], most experimental, clinical and numerical evidence suggest a connection between high WSS values and aneurysm formation. Particularly, the 20 vascular models harboring aneurysms in the middle cerebral artery reconstructed in [12], considered a limited portion of the parent artery, neglecting important features of the vascular geometry that resulted in a simplified simulated blood flow. Later, it was demonstrated that such simplifications significantly alter intra aneurysmal flow patterns [4].

Although most previous computational studies assumed that blood behaves as a Newtonian fluid, this assumption may not be completely accurate under some hemodynamic conditions. WSS in idealized models of saccular aneurysms was studied using Casson rheology under different flow conditions [14-16]. Particularly, Khanafer et al. showed that non-Newtonian wall shear stress is greater during the peak systole in aortic aneurysm models [15]. A limited number of patient-specific studies have been presented. Rayz et al. found no significant differences between low wall shear stress regions that may be associated with risk of thrombus formation using Newtonian and non-Newtonian computational fluid dynamic simulations in three patients [17]. However, accounting for non-Newtonian behavior improved the agreement with observations using longitudinal MRI studies. Xiang et al. showed that Newtonian viscosity model could overestimate normalized wall shear stress and consequently underestimate the risk of rupture of intracranial aneurysm in three internal carotid artery saccular aneurysms [18]. These results would suggest a dependence of WSS distributions on the real vascular geometry.

The purpose of this work is to compare the WSS values and orientation in patient-specific cerebral aneurysm models containing blebs where low flow is expected, using both Newtonian and Casson rheology in order to investigate regions of underestimation and overestimation of WSS in realistic geometries.

Materials and methods

1.1. Imaging and modeling

Five patients with cerebral aneurysms containing blebs were selected from our data base. Three dimensional rotational angiography (3DRA) images were acquired in four patients with aneurysms in the internal carotid artery (Cases #1, #2, #3 and #4) using a Philips Integris System (128^3 voxels covering a field of view of 54.02 mm, with a voxel resolution of 0.422 mm), while the other patient with an aneurysm in the anterior communicating artery (Case #5) was imaged using dynamic computerized angiographic tomography at 19 time points along the cardiac cycle (512 x 512 x 205 pixels with a spatial resolution of 0.3125 x 0.3125 x 0.6125 mm covering a field of view of 16.0 x 16.0 x 12.81 cm.). Vascular modelling was performed using a previously presented methodology [3,7,19]. For the 4DCTA case, each three dimensional volume was cropped and averaged to reduce noise and computational cost. Masks were applied to differentiate internal carotid arteries from extravascular structures to properly reconstruct parent arteries, which is needed for a realistic blood flow simulation. The vascular model was reconstructed from the average image. High-quality volumetric finite element grids composed of tetrahedral elements with an advancing front technique were generated for each case [20-22]. Element size was adjusted in order to approximately maintain the same number of elements in both large and small arteries, which resulted in grids containing up to 2.43 million tetrahedra (see Figure 1).

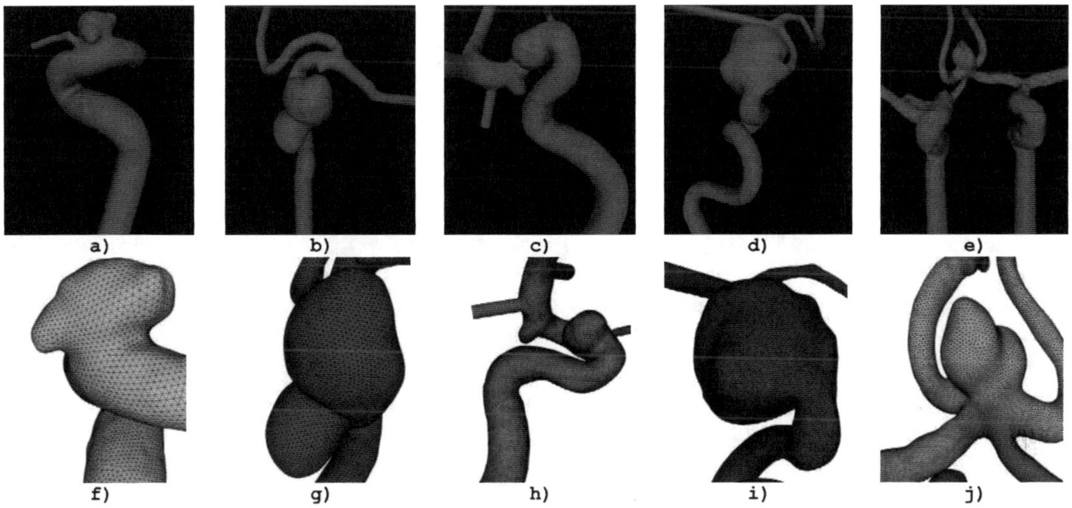

Figure 1. Vascular models for cases #1 (a, f), #2 (b, g), #3 (c, h), #4 (d, i) and #5 (e, j). First row corresponds to vascular models. Second and third rows correspond to unstructured grid of the computational domain from different views. Element size is adjusted to keep roughly the same number of elements across a cross-sectional area.

1.2. Numerical simulations

Finite element blood flow numerical simulations were performed using both Newtonian and non-Newtonian Casson rheologies. For the Newtonian case, blood was modelled as an incompressible fluid with attenuation 1.0 g/cm^3 and viscosity 0.04 Poise. For the Casson model a velocity dependent apparent viscosity derived from that non-Newtonian model is computed at every mesh node (1)

$$\mu_{app} = \left(\sqrt{\tau_0 \frac{1 - e^{-m\dot{\gamma}}}{\dot{\gamma}}} + \sqrt{\mu} \right)^2 \qquad (1)$$

where τ_0 is the yield stress, which was assumed as 0.09 gr (cm sec)$^{-1}$, μ is the Newtonian viscosity and $\dot{\gamma}$ is the shear strain rate. The exponent m, which is assumed greater that 10, is included in order for the apparent viscosity to be bonded for any range of shear strain rates [23].

The governing equations were the unsteady Navier-Stokes equations in 3D [24]. Vessel walls were assumed rigid, and no slip boundary conditions were applied at the walls. Pulsatile flow conditions derived from PCMR measurement in healthy subjects were imposed at the inlet of the models. Flow waveforms were scaled with the inlet area to achieve a mean WSS of 15 dyne/cm^2 at the inflow boundary of each model. This choice is consistent with studies relating vessel area and flow rates in internal carotid and vertebral arteries [25], as well as with the principle of minimal work expressed by Murray's law [26]. Fully developed pulsatile velocity profiles were prescribed with use of the Womersley solution [1,27]. Assuming that all distal vascular beds have similar total resistance to flow, traction-free boundary conditions with the same pressure level were applied at outlet boundaries. The Navier-Stokes equations were numerically integrated by using a fully implicit finite-element formulation [3].

Maps of WSS magnitude were created to visualize the distribution of shear forces on the aneurysm wall for both Newtonian and Casson rhoeologies. The percent relative difference between both distributions was computed. Additionally, the angular difference of WSS vector orientations was also calculated. Above mentioned maps were compared in order to establish possible correlations and determine the impact of blood rheology on the underestimation or overestimation of WSS. Intraaneurysmal WSS was not normalized with respect to the parent artery values. Instead, normalization was achieved by scaling the mean flow rate inflow waveform using the cross-sectional area according to the principle of minimal work expressed by Murray's law [26].

Results

Five vascular images exhibiting large aneurysms were selected from our data base. Cases #1 to #4 have aneurysms in the internal carotid artery. Particularly, cases #1 and #3 have the aneurysm in the bifurcation with the ophthalmic artery, while cases #2 and #4 are internal carotid siphon aneurysms. Cases #2 and #5 correspond to patients with a two-bleb aneurysm, while the other ones have multiple blebs on the aneurysm wall. The number of elements in the unstructured grids of tetrahedra depends on both the model size and the number of narrow vessels included in the model. Particularly, the AComA model (case #5) contains 2.43 million elements and 449 thousand points because it contains both left and right vasculature. However, ICA models reached up to 2.34 million elements and 4.25 thousand points because they include some small arteries where element size is automatically adjusted by means of sources that locally modify the element size from the background grid [28].

Cross sectional areas of the internal carotid arteries were used to scale the flow rates previously acquired using phase-contrast magnetic resonance imaging in a normal volunteer [29] in order to meet the criterion of minimal biological work for a typical mean WSS of 15 dyn/cm^2 at that artery [25].

For the wall shear stress analysis, distribution of the WSS magnitude at the systolic peak within the aneurysm wall was investigated for each case. Figures #2 through #6 correspond to cases #1 through #5. In all figures first and second row correspond to different views of the same aneurysm. First two columns show the maps of absolute value of WSS for both Newtonian rheology and Casson rheology, respectively. The relative difference between absolute values of WSS for Newtonian and Casson flows, is shown in the third column. Color maps range between blue (where the Newtonian WSS is as much as twice the Casson WSS) to magenta (where the Casson WSS is as much as twice the Newtonian WSS). Dark regions correspond to similar WSS values. Finally, the angle between the coplanar WSS vectors at each mesh node was computed and shown in the fourth column. Angles range from 0° to 20° for all cases but cases #4 where the angle ranges from 0° to 12°.

The first observation is that WSS distributions have similar characteristics for both rheologies. That result is in agreement with a previous sensitivity study [3]. The second observation is that large arteries, where greater flow rates are expected, have relatively larger WSS values for Casson model, which appears as dark red regions in the third column of Figures 2 through 6. This fact is in line with theoretical and computational results in simplified configurations and flow conditions [24,28]. The third observation is that not only WSS values differ between Newtonian and non-Newtonian rheologies, but also their vector orientation does. Those differences systematically take place in regions of low WSS. However, there is no clear correlation between regions with different WSS values, and regions where WSS vectors exhibit different orientations. In one of the blebs in case #1 large angular differences are associated to regions where any of the rheologies predict larger WSS values than the other one (compare Figure 3g to Figure 3h), however, that is not the case for case #4 (compare Figure 5c to Figure 5d, and Figure 5g to Figure 5h) where little WSS vectors have similar orientations despite their different magnitude values. Additionally, in the neck of one of the blebs of case #2, WSS vector orientations significantly differ in regions where WSS magnitude values do not.

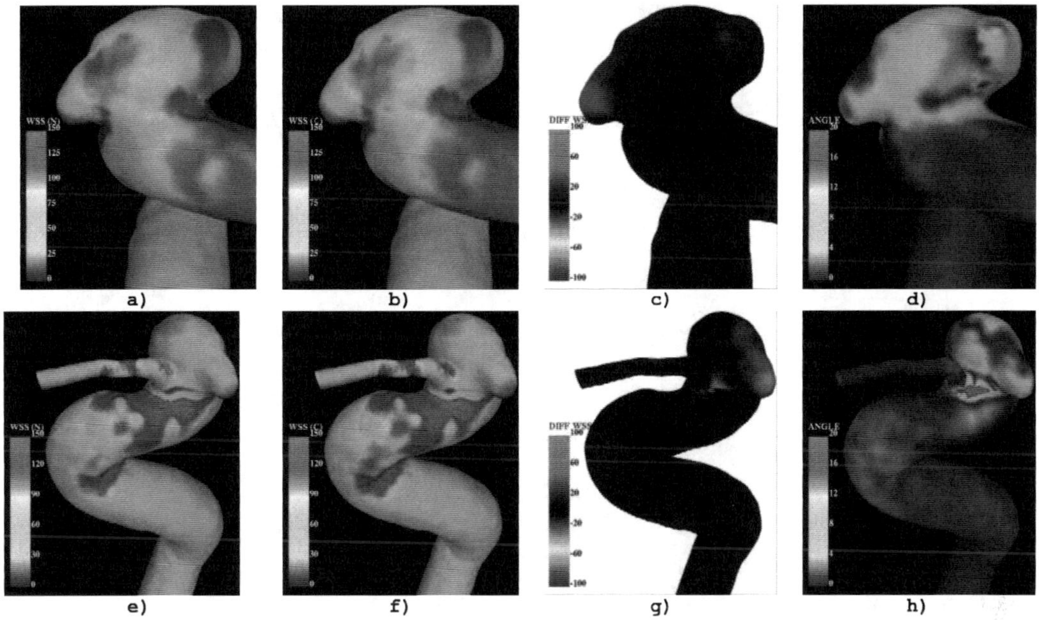

Figure 2. Wall shear stress analysis for Case #1. First column (a, e): WSS maps for Newtonian rheology; Second column (b, f): WSS maps for Casson rheology; Third column (c, g): relative difference between the Newtonian and Casson WSS, where magenta indicates that Casson WSS is as much as twice the Newtonian WSS, blue indicates that Newtonian WSS is as much as twice the Casson WSS, and dark regions correspond to similar WSS values; Fourth column (d, h): difference in WSS vector orientation, where magenta represents a maximum of 20°.

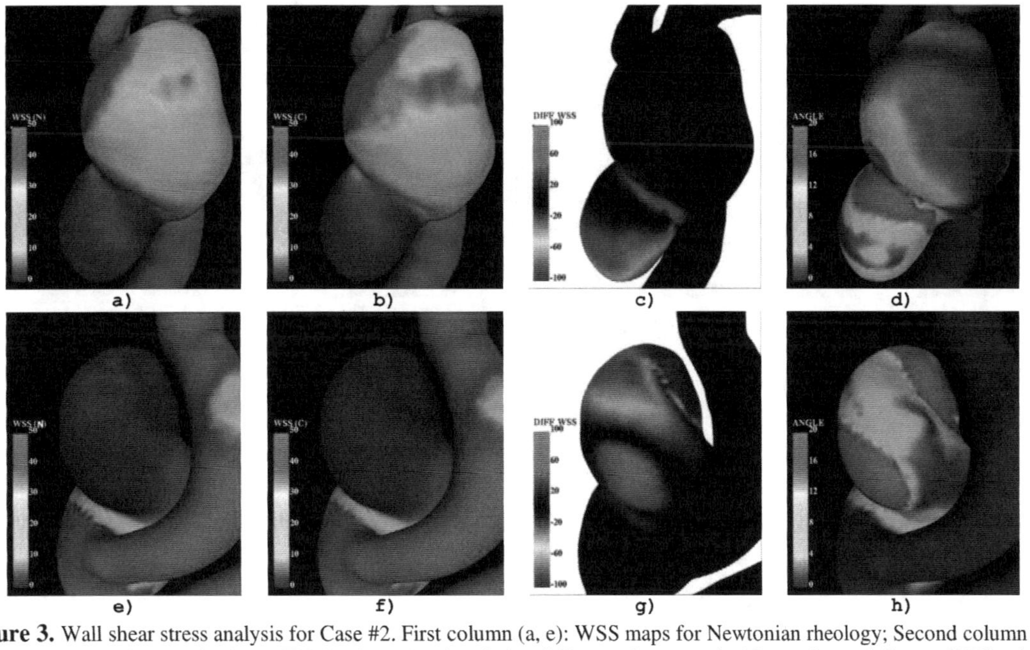

Figure 3. Wall shear stress analysis for Case #2. First column (a, e): WSS maps for Newtonian rheology; Second column (b, f): WSS maps for Casson rheology; Third column (c, g): relative difference between the Newtonian and Casson WSS, where magenta indicates that Casson WSS is as much as twice the Newtonian WSS, blue indicates that Newtonian WSS is as much as twice the Casson WSS, and dark regions correspond to similar WSS values; Fourth column (d, h): difference in WSS vector orientation, where magenta represents a maximum of 20°.

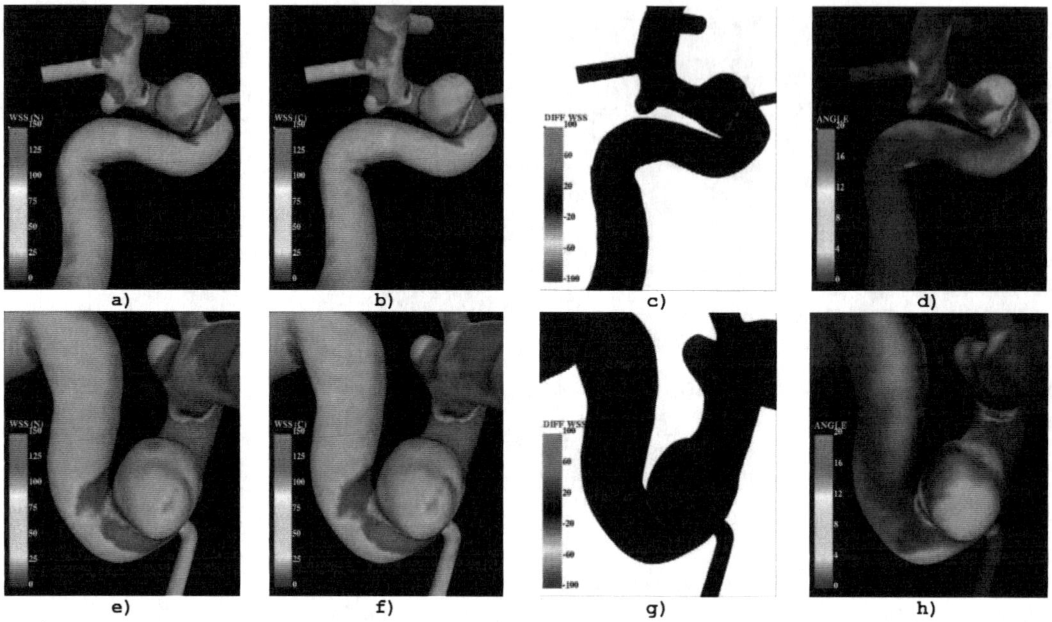

Figure 4. Wall shear stress analysis for Case #3. First column (a, e): WSS maps for Newtonian rheology; Second column (b, f): WSS maps for Casson rheology; Third column (c, g): relative difference between the Newtonian and Casson WSS, where magenta indicates that Casson WSS is as much as twice the Newtonian WSS, blue indicates that Newtonian WSS is as much as twice the Casson WSS, and dark regions correspond to similar WSS values; Fourth column (d, h): difference in WSS vector orientation, where magenta represents a maximum of 20°.

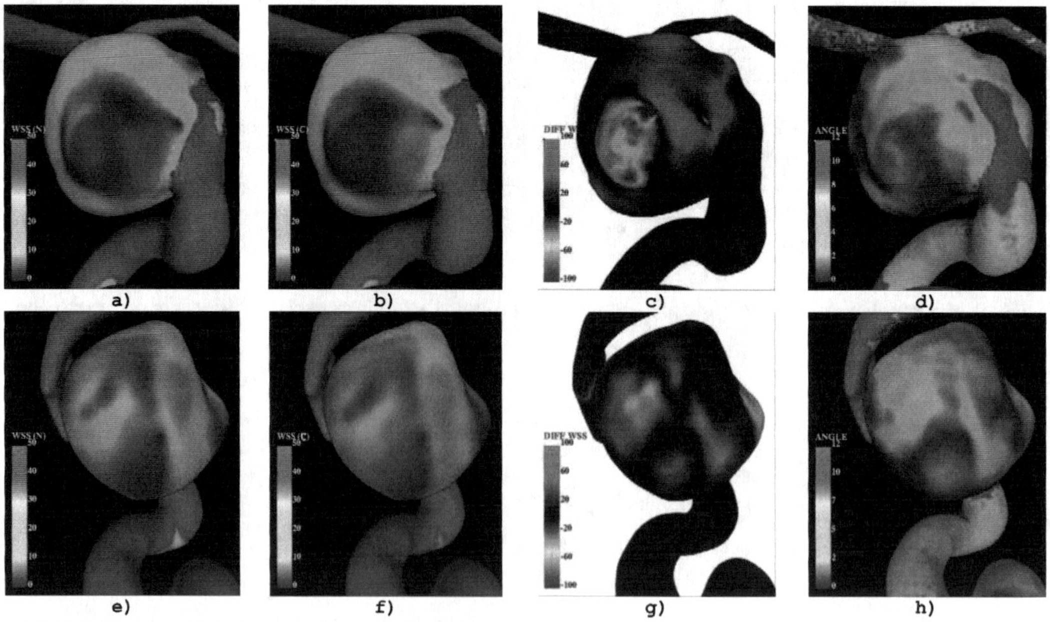

Figure 5. Wall shear stress analysis for Case #4. First column (a, e): WSS maps for Newtonian rheology; Second column (b, f): WSS maps for Casson rheology; Third column (c, g): relative difference between the Newtonian and Casson WSS, where magenta indicates that Casson WSS is as much as twice the Newtonian WSS, blue indicates that Newtonian WSS is as much as twice the Casson WSS, and dark regions correspond to similar WSS values; Fourth column (d, h): difference in WSS vector orientation, where magenta represents a maximum of 12°.

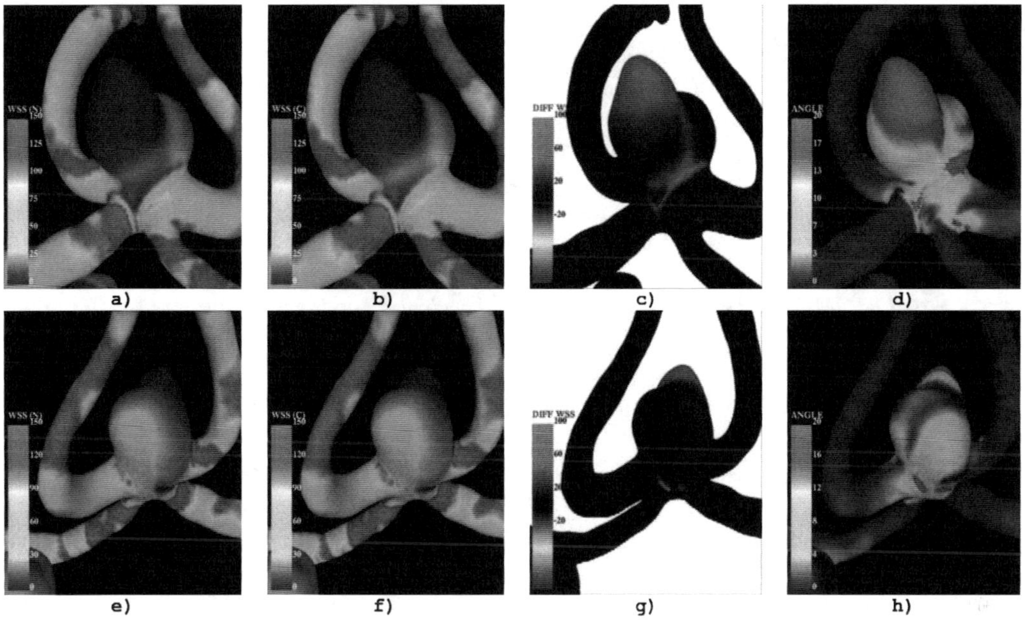

Figure 6. Wall shear stress analysis for Case #5. First column (a, e): WSS maps for Newtonian rheology; Second column (b, f): WSS maps for Casson rheology; Third column (c, g): relative difference between the Newtonian and Casson WSS, where magenta indicates that Casson WSS is as much as twice the Newtonian WSS, blue indicates that Newtonian WSS is as much as twice the Casson WSS, and dark regions correspond to similar WSS values; Fourth column (d, h): difference in WSS vector orientation, where magenta represents a maximum of 20°.

Finally, for cases #1 through #4 there is no clear correlation either between low WSS regions and regions where any of the rheologies predict larger WSS values than the other one. Particularly, that is clearly observed in case #2 where Casson WSS is significantly greater or less than Newtonian WSS in regions of low WSS (compare Figure 3a to Figure 3c, and Figure 3e to Figure 3g). The same fact is also observed in case #4, when comparing Figure 5a to Figure 5c, and Figure 5e to Figure 5g. For case #5, Casson WSS is greater that Newtonian WSS in all arterial and aneurysmal wall. This difference is greater in regions of low shear stress, particularly, in the bleb that is not directly impacted by the inflow jet. On the other hand, in the impaction zone, which occurs in the other bleb, WSS values are roughly the same for both rheologies (see Figures 6 a-c and e-g).

Additionally, WSS was space-averaged over the parent artery (under relatively high WSS) and one of the blebs (under relatively low WSS), and the time evolution for case #2 and case #3 for both Casson and Newtonian rheologies was computed (see Figure 7). In arteries, Casson WSS is systematically higher than Newtonian WSS over the entire cardiac cycle (less than 10%), as shown in Figures 7a and 7c. However, for low WSS blebs findings differ. While in case #2 the same behaviour as in the artery is observed even for WSS values lower than 5 dyn/cm^2 (Figure 7b), for case #3 Casson WSS is higher than Newtonian WSS only around the systolic peak, where the WSS is higher. However, no threshold could be found, therefore for the same WSS value, Casson WSS may be higher than Newtonian WSS, or viceversa, depending on what part of the cardiac cycle is analysed.

Discussion

The purpose of this work is to investigate the differences between Newtonian and Casson WSS characteristics under a pulsatile flow condition at image-based aneurysm models. Our results corroborate that the smallest differences appear in high flow rate regions [24]. Besides, the WSS distributions do not exhibit significant differences among rheologies. This result is in line with a previous sensitivity study [3].

The characteristics of a non-Newtonian flow for idealized vascular models harboring aneurysms have been investigated in the past [14-16]. When a stationary flow in a rigid and straight pipe is considered, WSS values are larger for a Casson flow than for a Newtonian rheology. This difference is more significant for low Reynold's numbers [28]. However, when realistic geometries are considered under pulsatile regimes the differences must be investigated in a patient-specific basis. Rayz et al. found no significant differences between low wall shear stress regions that may be associated with risk of thrombus formation using Newtonian and non-Newtonian computational fluid dynamic simulations in three patients [17]. However, accounting for non-Newtonian behavior improved the agreement with observations using longitudinal MRI studies. Xiang et al. showed that Newtonian viscosity model could overestimate normalized wall shear stress and consequently underestimate the risk of rupture of intracranial aneurysm in three internal carotid artery saccular aneurysms [18]. These results would suggest a dependence of WSS distributions on the real vascular geometry. Fisher et al. studied a given number of rheologies in different idealized aneurysm models. It was found that spatially averaged Newtonian WSS was higher than the corresponding Casson WSS at the aneurysm dome during the whole cardiac cycle. The relative difference was higher during the diastole rather than the systole [16].

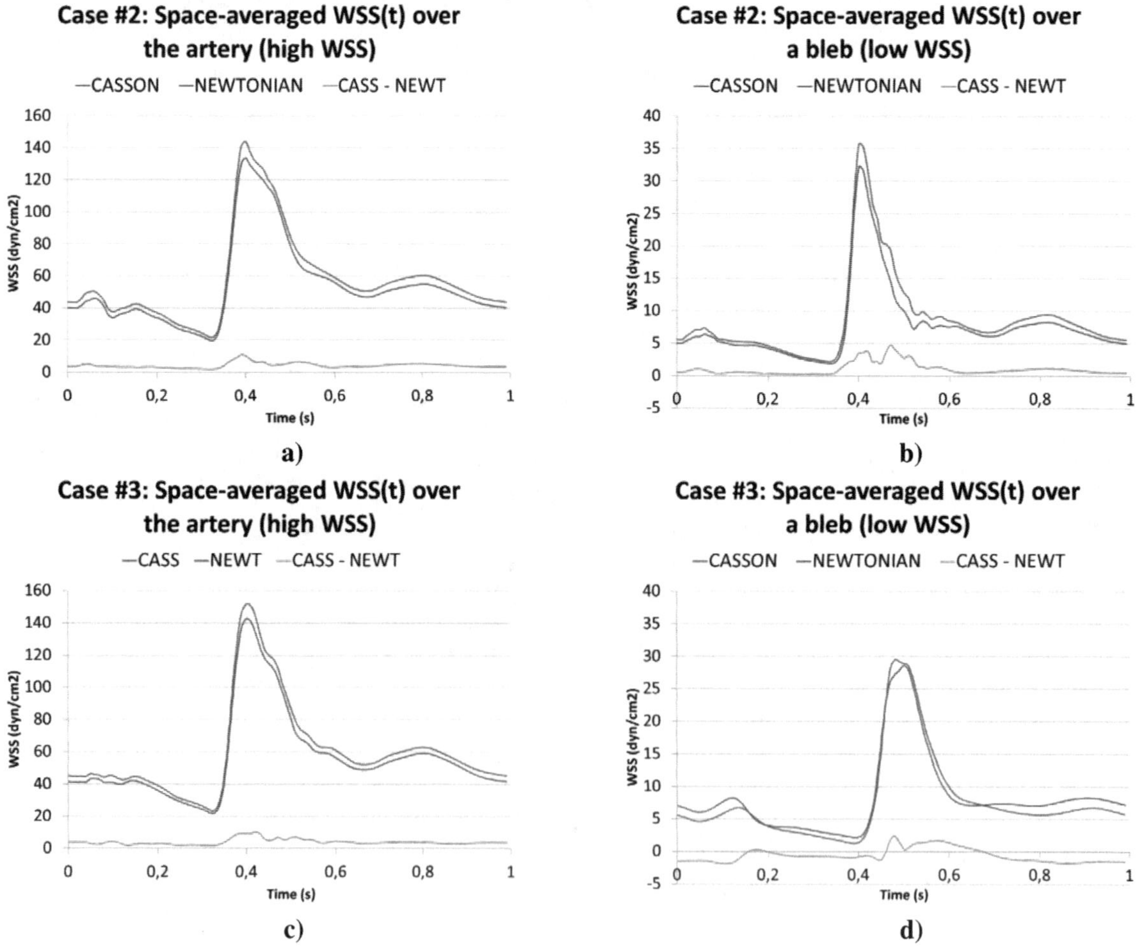

Figure 7. Space-averaged for case #2 (a and c) and case #3 (b and d), in the artery under a relatively high WSS (a and b) and a bleb (c and d). Casson WSS is plotted in blue, Newtonian WSS is plotted in red, and the difference between the Casson WSS and the Newtonian WSS is plotted in red.

The main finding is that there is no clear correlation between low WSS regions and regions where any of the rheologies predict larger WSS values. Particularly, large relative differences between Casson and Newtonian WSS may appear in a low WSS region (<10 dyn/cm^2). Additionally, flow in the internal carotid artery produces little greater WSS values for the non-Newtonian rheology with similar vector orientations. Finally, a non negligible change in WSS vector orientation is observed: as large as 20° in regions of low WSS. The location of those regions is not necessarily associated with regions where large differences in WSS occur when considering different rheologies. These observations may indicate that the realistic geometry plays an important role in the intra-aneurysmal hemodynamic characteristics, resulting in low flow regions where either the Newtonian or non-Newtonian WSS may be larger than the other one, exhibiting also differences in the vector orientation. However, considering that high WSS values are responsible for aneurysm initiation, growth and rupture, differences of WSS in regions of low WSS may not have a clinical impact [7,9,10,30]. Additionally, the rheology does not change the main hemodynamic characterization whose association with aneurysm initiation to rupture is investigated [28].

Given the small number of cases presented in this work, no statistical implications can be derived. However, these findings show some discrepancy with the results presented by Xiang et al. It iw worth mentioning that the Casson model is only a possible representation that not necessarily depicts in an accurate manner the intensity and distribution of blood internal forces [16]. Finally, the same mean WSS was imposed at the inlets of the models based on the principle of minimal work expressed by Murray's law [26]. Although that assumption is an approximation, it allows comparability when searching for associations between hemodynamic features and aneurysm risk factors [7,25,29]. Instead, Xiang et al. analyzed the WSS normalized with respect to the typical WSS values at the near parent artery [18], which may affect comparability.

Acknowledgements

Marcelo Castro wants to acknowledge CONICET (Consejo Nacional de Investigaciones Científicas y Técnicas) and MINCyT (Ministerio de Ciencia y Tecnología, PICT #279) for financial support.

References

[1] Taylor CA, Hughes TJR, Zarins CK, "Finite element modeling of blood flow in arteries. Computational Methods in Applied Mechanical Engineering 1998:158:155–196
[2] Steinman DA, Milner JS, Norley CJ, Lownie SP, Holdsworth DW. Image-based computational simulation of flow dynamics in a giant intracranial aneurysm. Am J Neurorad 2003:24(4):553-554
[3] Cebral JR, Castro MA, Appanaboyina S, Putman CM, Millán D, Frangi A. Efficient Pipeline for Image-Based Patient-Specific Analysis of Cerebral Aneurysms Hemodynamics: Technique and Sensitivity. IEEE - Transactions on Medical Imaging - Special Issue on Vascular Imaging 2005:24(4):457-467
[4] Castro MA, Putman CM, Cebral JR. Computational Fluid Dynamics Modeling of Intracranial Aneurysms: Effects of Parent Artery Segmentation on Intraaneurysmal Hemodynamics. Am J Neuroradiol 2006:27:1703-1709
[5] Nakatani H, Hashimoto N, Kang H, Yamazoe N, Kikuchi H, et al. Cerebral blood flow patterns at major vessel bifurcations and aneurysms in rats. J Neurosurg 1991:74:258-262
[6] Crompton M. Mechanisms of growth and rupture in cerebral berry aneurysms. Br J Med 1996:1: 1138-1142
[7] Castro MA, Putman CM, Cebral JR. Hemodynamic Patterns of Anterior Communicating Artery Aneurysms: A Possible Association with Rupture. Am J Neuroradiol 2009:30(2):297-302

[8] Cebral JR, Mut F, Weir J, Putman CM. Quantitative characterization of hemodynamic environment in ruptured and unruptured brain aneurysms. Am J Neuroradiol 2011:32:145-151
[9] Cebral JR, Sheridan M, Putman CM. Hemodynamics and Bleb Formation in Intracranial Aneurysms. Am J Neuroradiol 2010: 31:304-310
[10] Kulcsar Z, Ugron A, Marosfo M, Berentei Z, Paal G, Szikora I. Hemodynamics of Cerebral Aneurysm Initiation: The Role of Wall Shear Stress and Spatial Wall Shear Stress Gradient. Am J Neuroradiol 2011: 32(3):587-594
[11] Jou LD, Lee DH, Morsi H, Mawad ME. Wall Shear Stress on Ruptured and Unruptured Intracranial Aneurysms at the Internal Carotid Artery. Am J Neuroradiol 2008:29:1761-1767
[12] Shojima M, et al. Magnitude and role of wall shear stress on cerebral aneurysm: Computational fluid dynamic study of 20 middle cerebral aneurysms. Stroke 2004:35:2500-2505
[13] Shojima M, Nemoto S, Morita A, Oshima M, Watanabe E, Saito N. Role of Shear Stress in the Blister Formation of Cerebral Aneurysms. Neurosurg 2010:67(5):1268-1275
[14] Low M, Perktold K, Raunig R. Hemodynamics in rigid and distensible saccular aneurysms: A numerical study of pulsatile flow characteristics. Biorheol 1993:30:287-298
[15] Khanafer KM, Gadhoke P, Berguer R, Bull JL. Modeling pulsatile flow in aortic aneurysms: Effect on non-Newtonian properties of blood. Biorheol 2006:43:661-679
[16] Fisher C, Stroud Rossmann J. Effects of non-Newtonian behavior on hemodynamics of cerebral aneurysms. J Biomech Eng 2009:31:1-9
[17] Rayz VL, Boussel L, Lawton MT, Acevedo-Bolton G, Ge L, Young WL, Higashida RT, Saloner D. Numerical modeling of the flow in intracranial aneurysms: Prediction of regions prone to thrombus formation. Ann Biomed Eng 2011:36:1793-1804
[18] Xiang J, Tremmel M, Kolega J, Levy E, Natarajan S, Meng H. Newtonian viscosity model could overestimate wall shear stress in intracranial aneurysm domes and underestimated rupture risk. J Neurointerv Surg 2011:4(5):351-357
[19] Yim P, Vasbinder GB, Ho VB, Choyke PL. Isosurfaces as deformable models for magnetic resonance angiography. IEEE – Trans Med Imag 2003: 22(7):875-881
[20] Löhner R. Extensions and improvements of the advancing front grid generation technique. Comput Meth Appl Mech Eng 1996:5:119–132
[21] Löhner R. Regridding surface triangulations. J Comput Phys 1996:126:1–10
[22] Löhner R. Automatic unstructured grid generators. Finite Elem Analysis Design 1997:25:111–134
[23] Pham TV, Mitsoulis E. Entry and exit flows of Casson fluids. Canad J Biomech Eng 1994:72: 1080-1084
[24] Mazumdar JN. Biofluid Mechanics 1992. World Scientific. Singapore
[25] Cebral JR, Castro MA, Putman CM, Alperin N. Flow-area relationship in internal carotid and vertebral arteries. Physiol Meas 2008:29(10):585-594
[26] Sherman TF. On connecting large vessels to small. The meaning of Murray's law. J Gen Physiol 1981:78:431–453
[27] Womersley JR. Method for the calculation of velocity, rate of flow and viscous drag in arteries when the pressure gradient is known. J Physiol 1995:127:553–563
[28] Castro MA, Putman CM, Cebral JR. Computational Hemodynamics of cerebral aneurysms: Assessing the risk of rupture from hemodynamic patterns. VDM Verlag 2008. ISBN 9783639094411
[29] Cebral JR, Castro MA, Soto O, Löhner R, Alperin N. Blood flow models of the circle of Willis from magnetic resonance data. J Eng Math 2003:47(3/4):369-386
[30] Cebral JR, Mut F, Weir J, Putman CM. Association of hemodynamic characteristics and cerebral aneurysm rupture. Am J Neuroradiol 2011:32:264-270

Spacers in the treatment of hip joint infections: numerical analysis of their durability

Brenda Weiss, Sebastián Vanrell, Marcelo Berli, Sebastián Ubal y José Di Paolo

Grupo Biomecánica Computacional, Facultad de Ingeniería, Universidad Nacional de Entre Ríos. Ruta Provincial 11, km 10, 3100 Oro Verde, Entre Ríos, Argentina

E-mail: bweiss@bioingenieria.edu.ar, srvanrell@conicet.gov.ar

Abstract. Hip spacers are temporary implants having a geometry similar to the femoral component of a hip prosthesis, and they are manufactured with antibiotic-impregnated bone cement. The use of spacers in two stage revisions is the most effective treatment to eradicate infections and to avoid limb shortening. The most frequent complication associated with spacers is fatigue failure, for which doctors recommend patients to stay at rest. In this work, several spacer designs are analyzed in order to determine the feasibility of doing activities like walking, standing up or sitting down while performing the antibiotics treatment. Designs combine both different neck diameters and the presence/absence of an internal, stainless steel reinforcement. By means of computational simulations based on the finite element method, stress fields are calculated for various hip spacer designs under several load and fixing conditions. For this purpose, a 3D model of human femur is generated by processing tomographic images with segmentation techniques and inverse engineering. The results allow us to estimate the life expectancy of each design, by considering the fatigue behavior of the bone cement. Only the introduction of a reinforcement with a proper diameter into the bone cement matrix could assure the integrity of the spacer along the treatment period.

1. Introduction

Total hip replacement is a successful solution to relieve pain and restore hip movement [1]. Despite the success achieved with hip prosthesis, hip arthroplasty could lead to several complications, among which infections are the most feared because available treatments are complex, expensive and not always effective [2, 3].

Currently, the most effective treatment to eradicate the infection involves the extraction of the implant, followed by the administration of antibiotics and then the placement of a new implant after 2-6 months since the extraction surgery. Therefore, the patient must remain at rest and practically incapable to move the affected limb, which leads to leg shortening. In order to overcome these drawbacks, hip spacers are now usually implanted during the time interval between the two surgeries, showing infection healing in more than 93% of the cases [2].

Hip spacers have a geometry similar to the femoral component of metallic prosthesis and are manufactured from antibiotic-impregnated bone cement. While handmade spacers (crafted at the moment of implantation) are sometimes implanted, preformed spacers show less complications [4] and are the only ones to be analyzed in this work. In fact, mechanical failures are considered the most important complication among those associated with spacers.

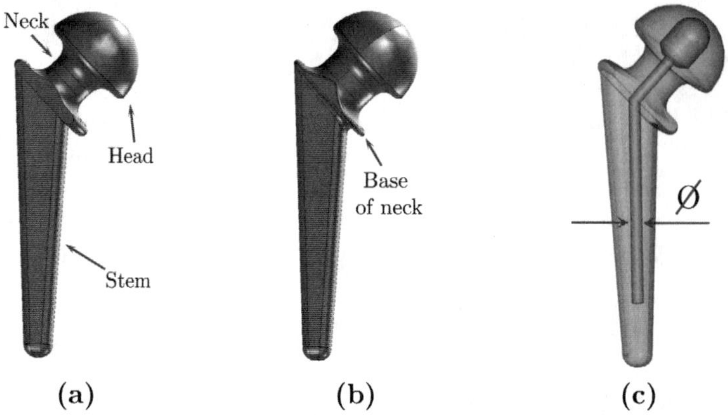

Figure 1. (a) Standard Spacer. (b) Thick Neck Spacer. (c) Standard Spacer reinforced with 316L stainless steel endoskeleton. The bone cement matrix is shown in pink and the metallic endoskeleton is shown in blue.

Therefore, most doctors recommend to patients having an implanted spacer to remain at rest until the following replacement.

The aim of this work is to provide numerical assessment about the basic activities (e.g. walking, standing up or sitting down) that a patient with a hip spacer can do without compromising the integrity of the implant. This objective is driven by the need to improve the quality of life of these patients. Our finite element–based study allow us to evaluate the stresses which develop for several spacer designs and thus predict their lifespan for each of the activities analyzed .

2. Materials and methods

Since bone cement can be considered as a ductile material we choose a failure criterion based on Von Mises stress. The software COMSOL Multiphysics, which is based on finite element analysis, was used to study the stress distribution on spacers.

2.1. Description of spacers

At first, a spacer's geometry design (called *Standard Spacer*, SS_{nR}, see Fig. 1a) was proposed by the *Grupo de Biomateriales para Prótesis* (GBP) from *Facultad de Ingeniería, Universidad de Buenos Aires*; later, as a result of a collaborative work, this design was improved based on our numerical predictions. The first design has a 24 mm–diameter–neck, is entirely built with bone cement and nowadays is implanted in Argentina.

The neck of the spacer presents the highest stresses (as will be seen later) hence a design with a larger (10%) neck diameter, called *Thick Neck Spacer* (TNS_{nR} – Fig. 1b), is proposed. In addition, the incorporation of a 316L stainless steel reinforcement is proposed (Fig. 1c). The diameter of the reinforcement, being a design parameter of this study, is modified throughout the simulations. Due to this reason, the spacer designs simulated are: SS_6, $SS_{8.4}$ and $SS_{10.2}$ (*Standard Spacer with Reinforcement* of 6 mm, 8.4 mm and 10.2 mm of diameter, respectively), and TNS_6, $TNS_{8.4}$ and $TNS_{10.2}$ (*Thick Neck Spacer with Reinforcement* of 6 mm, 8.4 mm and 10.2 mm of diameter, respectively)

2.2. Human femur geometric reconstruction

The tomographic images (CT images) employed in this work correspond to the right femur of an adult male (subset of the dataset called *Obelix*, freely available in the *OsiriX-Viewer* repository).

Segmentation of CT images was achieved by combining three techniques implemented in the software *itk-SNAP* [5]: threshold filtering, active contour method (region competition) and manual segmentation. This last step consist of manually filling the space occupied by the medulla (to assure an intimate contact with the spacer) and tweak the sectors with geometric complexity, such as the joint surfaces.

The result is a collection of binary images where the intensity of each pixel codifies whether the corresponding voxel belongs or not to the space occupied by femur. This kind of information is enough to delimit the femur geometry; however, it still is a set of stacked 2D images that can not be entered (as a valid geometry) into the simulation software. Therefore, the outer surface of femur was extracted and represented by an interconnected mesh of triangles that results from employing the marching cubes algorithm, implemented in *itk-SNAP* [5].

Given the irregularity of the resulting surface, a Laplacian smoothing process (low pass filter) was applied to recover the curvature of the actual anatomical surface. This filter does not change the mesh topology (number of nodes and their interconnections), but instead alters the nodes position which define the triangles so that there are no abrupt changes between the orientation of neighboring facets.

While generation of a volumetric model is not required, it is desirable given the advantages that arise at later stages. To obtain that model, which is suitable for being incorporated in a finite element software, we have used a discrete set of irregular parameterized surfaces (like patches), as NURBS surfaces (*non-uniform rational basis spline*, see [6]), to generate a volumetric model by interpolation of a smoothed surface mesh. This procedure was carried out using CAD tools.

The volumetric model of the femur is incorporated into COMSOL Multiphysics as a geometric domain. The aim of this work is to study the stresses that develop on the spacer, which is located in the proximal femur. Hence, in order to reduce computational requirements and simplify the model geometry in a region that is not of interest (condyles), the femur was transversely sectioned at the distal end of the diaphysis.

2.3. Model characteristics

The following assumptions were considered to formulate the physical model:

- Materials are modeled as linear elastic homogeneous solids.
- During a hip arthroplasty the proximal epiphysis is sectioned, thus leaving out almost all cancellous bone. Therefore, the femur model considered in this work is made of cortical bone only. Distal epiphysis is not modeled.
- The human femur can be modeled as an orthotropic solid, whose properties are shown in Table 1. Ashman measured them using ultrasound techniques [7].
- The materials employed to manufacture the hip spacer (both stainless steel and bone cement) can be considered isotropic solids; their mechanical properties were taken from literature: 316L stainless steel ($E = 193\ GPa$ y $\nu = 0.25$ [8]), bone cement ($E = 1.47\ GPa$ y $\nu = 0.35$ [9]).
- Materials undergo small deformations, hence these and displacements are linearly related.
- Metal-polymer and polymer-bone interfaces are intimate and perfect, hence there is no sliding, verifying continuity of displacements and stresses there.

2.4. Fixing conditions

Two fixing conditions were compared by applying identical loads to spacers of equal characteristics. The aim of this study was to evaluate advantages or disadvantages of employing a

Table 1. Mechanical properties of cortical femur bone [7], where directions correspond to the coordinate system that is shown in figures 2c and 2d.

Young Modulus		Shear modulus		Poisson ratio	
E_x	12,9 GPa	G_{xy}	4,53 GPa	ν_{xy}	0,376
E_y	13,4 GPa	G_{xz}	5,61 GPa	ν_{xz}	0,222
E_z	20,0 GPa	G_{xy}	6,23 GPa	ν_{yz}	0,235

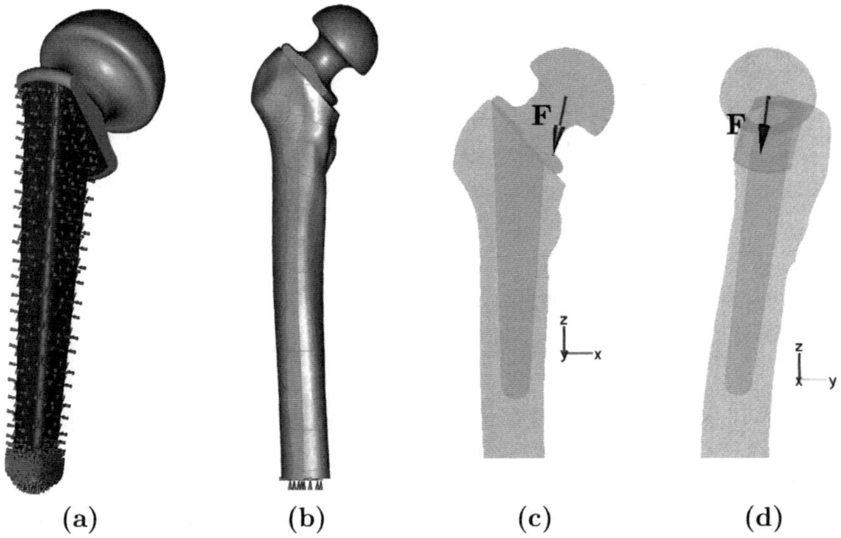

Figure 2. Fixing conditions (null displacement). (a) Fixed-stem. (b) Fixed-femur (at bottom section). Surfaces affected by the condition are marked with red arrows. (c) and (d) Application point of equivalent punctual load **F** and its direction.

more complex model (spacer inserted in the femur), which requires more computational resources than simulations of an isolated spacer.

- *Fixed-stem spacer (FS):* If femur is assumed as a perfectly rigid solid, we can avoid modeling it, by fixing the surface of the spacer that is in contact with the femur —which correspond to the stem—. There, null displacement boundary conditions are considered (Fig. 2a).
- *Fixed-femur spacer (FF):* Null displacements condition is applied to the distal section of the femur (Fig. 2b), which corresponds to mechanical tests that have been done in other research works [10–12].

2.5. Loading conditions
Although the applied load, as on a healthy femur, is distributed over the spacer's head surface, we consider an equivalent point load **F** which is applied on the center of spacer's head (Fig. 2c and 2d). Both the magnitude and direction of **F** were determined by Bergmann [13]. For each activity, the value of **F** represents the maximum force that appears once per gait cycle.

Simulated activities in this work are normal walking (NW), slow walking (SW), fast walking (FW), up stairs (US), down stairs (DS), stand up (SU) and sit down (SD).

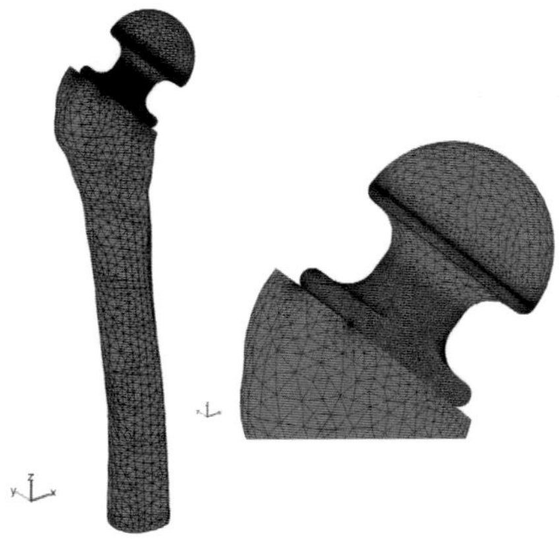

Figure 3. Characteristic element mesh distribution.

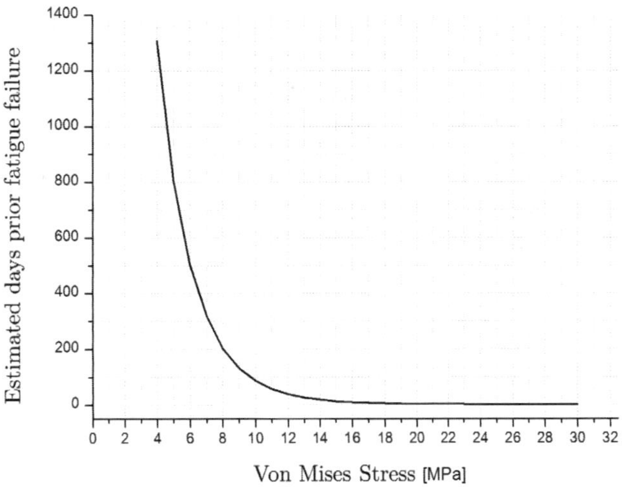

Figure 4. Estimated days prior to fatigue of bone cement as a function of maximum Von Mises stress on this material.

2.6. Solving methodology

The greatest stresses appear on the spacer's neck, so it is important to analyze the mesh influence in this particular zone, specifically on the maximum Von Mises stress developed there. Therefore, a convergence analysis has been carried out by manually modifying parameter h —maximum size of any mesh element— on neck's surfaces. The convergence criterion adopted in this work is that a mesh if fine enough if the change in the maximum Von Mises stress is less than or equal to 2% when halving h in the neck region. This resulted in a maximum element size of 0.001 m.

Meshes are automatically generated by COMSOL Multiphysics (Fig. 3) according to the parameters manually established. These meshes have quadratic Lagrange elements —tetrahedral on volumes and triangular on surfaces— with high resolution (higher quantity of elements per unit volume) in regions where surfaces present significant curvature, and also in those areas where variables present steep gradients. For the solution of the linear system, we use PARDISO (direct) and *Conjugated Gradients* (iterative) solvers.

The maximum values of Von Mises stresses are observed at qualitatively the same regions where fatigue failures are initiated [14]. In addition, the stresses are concentrated on the spacer's neck. As a consequence of this observations, we are interested in the values of the maximum Von Mises stress developed there, as well as in the positions where they appear, since these are the most likely points to result in the spacer's failure. On this stage, results are processed using COMSOL Multiphysics.

The main flaw of spacers is fatigue failure, therefore, considering the results obtained by Lewis [15] and the estimation that a prosthesis implanted in a person of normal activity is subjected to about 1 million annual cycles [10], we built a curve of estimated days prior to cement's fatigue as a function of Von Mises maximum stresses to which this material is subjected (Fig. 4). Then, to estimate the duration of each design, the maximum stresses were *translated* into estimated days prior to cement's fatigue.

3. Results

The maximum Von Mises stresses obtained in all analyzed cases are placed on the neck of spacers. This suggests this region is potentially prone to fatigue failures; indeed this is consistent with

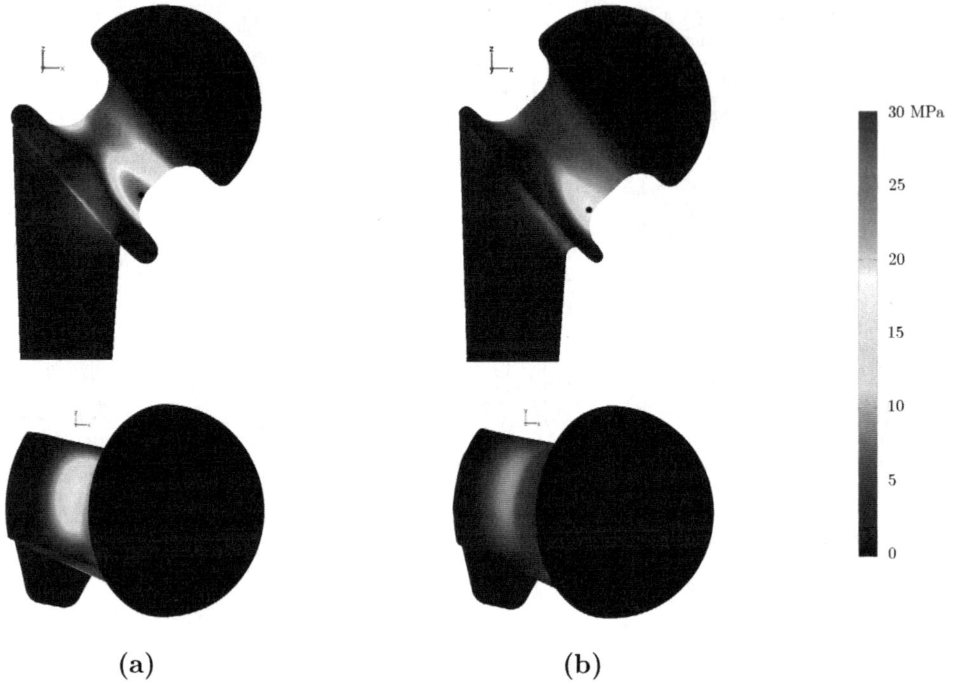

Figure 5. Von Mises stress distribution. A black dot indicates where the maximum Von Mises stress is located. (a) Stantard spacer (SS_{nR}). (b) Thick neck spacer (TNS_{nR}).

Table 2. Maximum Von Mises stresses on the neck, cycles prior fatigue and estimated lifetime of non-reinforced spacers. Applied load: 100 kg–patient, normal walking. Fixing condition: fixed-femur.

Spacer	Maximum stress on neck [MPa]	Estimated cycles prior fatigue	Estimated lifetime [days]
SS_{nR}	29,24	1.044	<1 day
TNS_{nR}	18,13	13.151	∼5 days

observations from clinical practice [16]. Comparing results for each spacer design and every fixing condition, under identical load direction and magnitude, it is seen that the stress field maintains a similar spatial distribution. As an example, figure 5 shows the stress field for SS_{nR} and TNS_{nR}, under a normal walking of a 100 kg–patient.

3.1. Influence of the diameter of the spacer neck

Table 2 shows the results obtained for implants SS_{nR} and TNS_{nR}, the load corresponding to normal walking of a 100 kg–patient and fixed-femur condition. As can be seen, the maximum Von Mises stress on the neck of both models are significantly different. Under these conditions, the TNS_{nR} presents stresses 38% lower than the SS_{nR}. A similar behavior is observed when comparing SS and TNS reinforced designs under the same loading and fixing conditions. These are reasonable observations since the neck is under bending, thus an increment of neck diameter increases the moment of inertia of the neck's section; therefore this impacts on a stress reduction in this area.

Figure 6a shows the variation of the maximum Von Mises stress on SS_{nR} and TNS_{nR} under

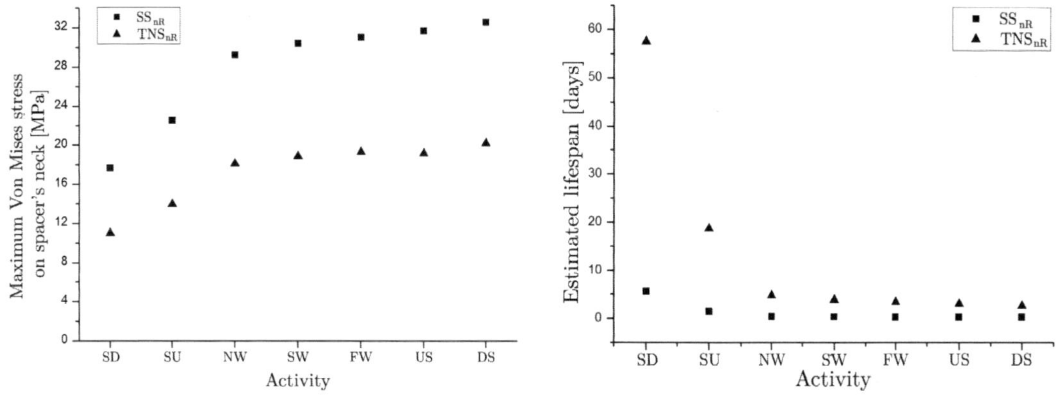

(a) (b)

Figure 6. (a) Von Mises maximum stress and (b) estimated lifetime of non-reinforced spacers (SS_{nR} and TNS_{nR}) depending on simulated activities.

routine activities loading conditions: SD, SU, NW, SW, FW, US and DS. Regardless of the activity, the maximum stresses are higher in standard spacers. The highest maximum stresses are observed in FW, US and DS, while the lowest are seen when sitting down or standing up, two essential activities to improve the patient life quality during infection treatment.

Figure 6b is obtained by *translating* stresses shown in figure 6a into estimated duration days. In figure 6b can be seen that most of the activities determine a similar spacer lifetime (<7 days), except when sitting or standing with a TNS_{nR} implanted. The difference observed in the latter cases is accentuated by the non-linearity (exponential) of the curve used to estimate the durability of the spacer (Fig. 4). Thus, it can be noticed that a small stress reduction from values around 14 MPa causes a considerable increase in the estimated days prior fatigue. In addition, these estimations may be affected by the uncertainty of the experimental curve obtained by Lewis [15].

Consequently, we can estimate that a TNS_{nR} could be under normal walking for 5 days, and that SS_{nR} could not even withstand one day (Table 2). Then, none of the designs (SS_{nR} and TNS_{nR}) presents stress values on the bone cement small enough to ensure the spacer integrity during the treatment, when carrying out the analyzed activities.

3.2. Influence of fixing condition

When a TNS_{nR} is implanted in a 100 kg–patient and the two fixation conditions (Fixed-Stem and Fixed-Femur) are compared, figure 7 is obtained. It can be noted that the fixed-femur condition shows stresses ∼5% higher than the fixed-stem condition. This trend in the results suggests that the simplest model (FS) could be used to estimate the spacer lifetime, after applying a safety factor to the computed maximum stress. This would allow to employ the FS model, which is considerably less expensive computationally, due to the smaller degrees freedom to be solved for in the model.

3.3. Influence of reinforcement diameter

A metallic reinforcement incorporation significantly decreases maximum Von Mises stresses on the necks of SS_{nR} and TNS_{nR} (Fig. 8). This is reasonable since the stresses that the spacer must withstand are absorbed by the higher stiffness zone: the reinforcement.

For example, incorporating a reinforcement of 6 mm diameter into an TNS (TNS_6) represents

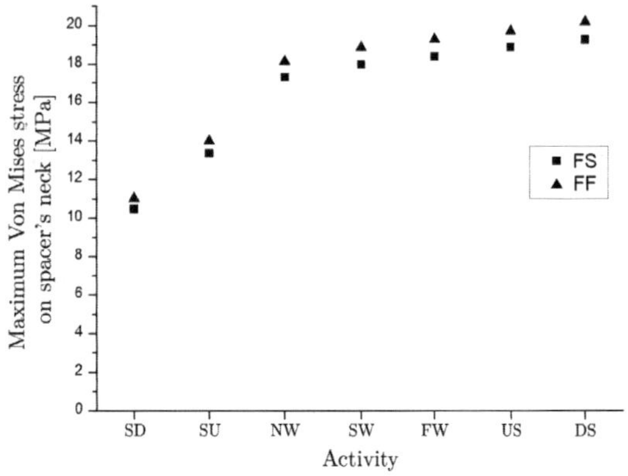

Figure 7. Maximum Von Mises stress on the neck of TNS_{nR} for each activity. FS: Fixed stem. FF: Fixed femur.

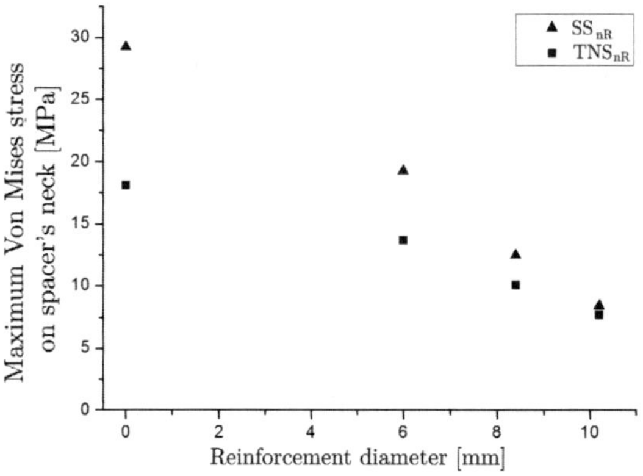

Figure 8. Maximum Von Mises stress on SS and TNS spacer's neck depending on the reinforcement diameter (Non-reinforced spacers are represented with null diameter). Applied load: 100 kg–patient, normal walking.

a 25% reduction of maximum stresses in comparison to TNS_{nR} (non-reinforced). If the reinforcement diameter is increased up to 8.4 mm ($TNS_{8.4}$) the reduction in stresses is now 45%, whereas it is 58% if the diameter is 10.2 mm ($TNS_{10.2}$). In terms of lifetime, the incorporation of a 10.2 mm diameter endoskeleton would increase the TNS_{nR} durability —under NW condition— up to approximately 230 days. In the same way, the SS_{nR} lifetime would be 170 days longer when a 10.2 mm diameter reinforcement is incorporated (Table 3).

It should be noted that the influence of neck spacer diameter is significant for non-reinforced spacers —maximum Von Mises stresses for TNS_{nR} are 38% lower than for SS_{nR}— being this not the case for those with a 10.2 mm diameter reinforcement —maximum Von Mises stresses for TNS_{nR} are 9% lower than for SS_{nR}— (Fig. 8).

4. Discussion

Table 3 summarizes the results obtained for 100 kg–patients, showing estimated duration (in days) until failure by fatigue, for different spacer designs and activities. In all the cases considered there, the fixed–femur (FF) condition is employed since the results obtained are more conservatives, given the larger maximum stresses observed under this type of fixation.

Even though some of the simulated designs —those in yellow in Table 3— have an estimated lifetime between 2 and 6 months (i.e. the typical medical treatment interval, see [10]), a durability equal to or larger than 6 months is desirable —results in green in Table 3— in order to assure the integrity of the implant during the treatment.

As indicated in Table 3, spacer $TNS_{10.2}$ is the most promising, since it is the only one that allows the patient to develop all the activities considered in this work, during the whole treatment interval. Designs $TNS_{8.4}$ and $SS_{10.2}$ would also allow performing most of the activities, though for a reduced period.

For simplicity, we considered in this study that the implanted patient possesses a level of activity comparable to that of a normal person (same number of cycles per unit time, same force pattern transferred to the affected hip). Therefore, even though the remaining designs are

Table 3. Estimated durability (in days) of the different spacer designs, subjected to several daily activities (SW, NW, FW, US, DS, SU, SD), when implanted in a 100 kg–patient. Fixed–femur condition. Yellow cells: durability between 2 and 6 months. Green cells: durability equal to or larger than 6 months.

Spacer	SD	SU	NW	SW	FW	US	DS
SS_{nR}	6	1	0	0	0	0	0
SS_6	47	14	3	3	2	2	2
$SS_{8.4}$	262	107	33	27	24	21	19
$SS_{10.2}$	765	399	168	141	130	111	108
TNS_{nR}	58	19	5	4	3	3	3
TNS_6	189	73	21	17	15	13	12
$TNS_{8.4}$	493	230	85	71	65	55	52
$TNS_{10.2}$	969	526	232	199	182	161	155

not expected to attain the desired duration when subjected to normal walk, a reduced set of activities (sitting down, standing up, taking a few steps), would be probably possible without compromising the integrity of the spacer during the full treatment interval.

Some other activities (not simulated here), like walking with reduced support of the body weight on the affected limb (e.g. because of the use of a cane or a walker) would surely benefit from lower maximum stresses and consequently and extended lifetime.

Besides, in order to estimate the life expectancy of the implant, we considered that the whole of the loading cycles correspond to a unique activity. As is well known, activities like going up or down the stairs do not represent a substantial fraction of the daily cycles. A more realistic approach should take into account the distribution of daily activities when estimating the duration of the hip spacer.

5. Conclusions

The results shown in this work, obtained from a model with several simplifications, allowed us to estimate the lifetime of different spacer designs under several loading conditions. A particular design ($TNS_{10.2}$) meets the strength and durability requirements that enable patients both to complete the antibiotics treatment and to carry out basic daily activities that notoriously improve their quality of life. These results could also provide a guideline for the recommendations given to patients regarding the activities they can perform.

The methodology to reconstruct the geometry of the femur could be easily adapted to the reconstruction of the geometry of many other anatomical structures. Furthermore, additional processing of the tomographic images should allow the extraction of material properties of the tissues, leading to more realistic, complete, patient-specific models.

Acknowledgments

The authors would like to thank CIN (*Consejo Interuniversitario Nacional*) and UNER (*Universidad Nacional de Entre Ríos*) for support provided by the *Stimulus to scientific vocation scolarship* and the *Research initiation scolarship under PID UNER 6103*, respectively.

References

[1] Pagès E, Iborra J and Cuxart A 2007 *Rehabilitación* **41** 280–9
[2] García-Sánchez I 2001 *Gaceta Médica de Bilbao* **98** 66–67 ISSN 03044858

[3] Militz M and Bühren V 2010 *Der Chirurg; Zeitschrift Für Alle Gebiete Der Operativen Medizen* **81** 310–320 ISSN 1433-0385

[4] Grassi F, D'Angelo F, Pietri M D and Cherubino P 2005 *Journal of Bone & Joint Surgery, British Volume* **87-B** 256–256 ISSN 1358-992X, 2049-4416

[5] Yushkevich P A, Piven J, Hazlett H C, Smith R G, Ho S, Gee J C and Gerig G 2006 *NeuroImage* **31** 1116–1128 ISSN 1053-8119

[6] Piegl L 1991 *IEEE Computer Graphics and Applications* **11** 55–71 ISSN 0272-1716

[7] Ashman R, Cowin S, Van Buskirk W and Rice J 1984 *Journal of Biomechanics* **17** 349–361 ISSN 0021-9290

[8] Aalco Metals Ltd 2011 Stainless steel 1.4404 - 316L Product data sheet Aalco Metals Ltd Cobham, Surrey

[9] Kurtz S, Villarraga M, Zhao K and Edidin A 2005 *Biomaterials* **26** 3699–3712 ISSN 0142-9612

[10] Thielen T, Maas S, Zuerbes A, Waldmann D, Anagnostakos K and Kelm J 2009 *International Journal of Medical Sciences* **6** 280–286 ISSN 1449-1907

[11] Schileo E, Taddei F, Malandrino A, Cristofolini L and Viceconti M 2007 *Journal of Biomechanics* **40** 2982–2989 ISSN 0021-9290

[12] Taddei F, Cristofolini L, Martelli S, Gill H and Viceconti M 2006 *Journal of Biomechanics* **39** 2457–2467 ISSN 0021-9290

[13] Bergmann G, Deuretzbacher G, Heller M, Graichen F, Rohlmann A, Strauss J and Duda G 2001 *Journal of Biomechanics* **34** 859–871 ISSN 0021-9290

[14] Hung J P, Chen J H, Chiang H L and Shih-Shyn Wu J 2004 *Computer Methods and Programs in Biomedicine* **76** 103–113 ISSN 0169-2607

[15] Lewis G and Ed Austin G 1994 *Journal of Applied Biomaterials* **5** 307–314 ISSN 1549-9316

[16] Jung J, Schmid N V, Kelm J, Schmitt E and Anagnostakos K 2009 *International Journal of Medical Sciences* **6** 265–273 ISSN 1449-1907

Cerebrovascular systems with concomitant pathologies:A computational hemodynamics study

Marcelo A. Castro[1]
CONICET. Universidad Tecnológica Nacional, Facultad Regional Buenos Aires
Grupo de Investigación y Desarrollo en Bioingeniería
Medrano 951, Ciudad Autónoma de Buenos Aires, CP 1179, Argentina

E-mail: marcelo.a.castro@gmail.com

Nora L. Peloc
Universidad Favaloro, Facultad de Ingeniería, Ciencias Exactas y Naturales
Solís 453, Ciudad Autónoma de Buenos Aires, CP 1078, Argentina

Juan R. Cebral
George Mason University. College of Sciences
4400 University Drive, Fairfax, VA, 22030, USA

Abstract. Although unusual, cerebral aneurysms may coexist with a proximal artery stenosis. In that small percent of patients, such coexistence poses a challenge for interventional neuroradiologists and neurosurgeons to make the best treatment decision. According to previous studies, the incidence of cerebral aneurysms in patients with internal carotid artery stenosis is not greater than 5%, where the aneurysm is usually incidentally detected, being 2% for aneurysms and stenoses in the same cerebral circulation. Those cases pose a difficult management decision for the physician. Case reports showed patients who died due to aneurysm rupture months after endarterectomy but before aneurysm clipping, while others did not show any change in the aneurysm after plaque removal, having optimum outcome after aneurysm coiling. The aim of this study is to investigate the intraaneurysmal hemodynamic changes before and after treatment of stenotic plaque. Idealized models were constructed with different stenotic grade, distance and relative position to the aneurysm. Digital removal of the stenotic plaque was performed in the reconstructed model of a patient with both pathologies. Computational fluid dynamic simulations were performed using a finite element method approach. Blood velocity field and hemodynamic forces were recorded and analyzed. Changes in the flow patterns and wall shear stress values and distributions were observed in both ideal and image-based models. Detailed investigation of wall shear stress distributions in patients with both pathologies is required to make the best management decision.

1 To whom any correspondence should be addressed.

Introduction
Intracranial aneurysms and artery stenosis are vascular diseases with different characteristics. According to previous studies, the incidence of cerebral aneurysms in patients with internal carotid artery stenosis is 2.3-4.9 %, where the aneurysm is usually incidentally detected [1-3]. A retrospective study of 853 patients with stenosis in the carotid bifurcation showed that 46 patients (5.4%) had a cerebral aneurysm, while only 2% had the aneurysm in the same arterial circulation [4]. A case report showed that a patient with a stenosis in the carotid artery and an asymptomatic unruptured aneurysm (5.0x10.0 mm) in the same artery at the posterior communicating artery bifurcation underwent a successful endarterectomy. The patient died after a subarachnoid hemorrhage five months after the intervention. Autopsy revealed that the aneurysm had growth up to 14.0x10.0 mm. before rupture [5]. Another case report describes the treatment option for a patient with similar characteristics. A stenosis in the carotid artery was located proximal to an unruptured aneurysm (14.0x8.0 mm.) at the ophthalmic artery bifurcation. A stent was successfully deployed to remove the stenosis. A few months after the intervention, no change was observed in the aneurysm size and the aneurysm was treated with a coil embolization procedure [1]. The treatment option has risk and benefits that must be studied by the physician in a patient-specific basis. Removing the stenosis may result in aneurysm rupture due to flow changes. In addition, a brain surgery to treat the aneurysm may cause a stroke because of flow reduction. Simultaneous interventions were also reported [6]. In this work we analyze the hemodynamic characteristics at cerebral aneurysms before and after the virtual removal of a proximal stenotic plaque. Idealized models were created to investigate changes of velocity and wall shear stress at the aneurysm after intervention, depending on stenosis grade and relative position to the aneurysm. A patient-specific model with a proximal stenosis was included. The plaque was digitally removed using a Laplacian filter and the aneurysm hemodynamic characteristics were compared.

Materials and methods

1.1. Idealized models
Three patients with cerebral aneurysms in the AComA were selected from our data base. In order to minimize the dependency of the results on the reconstruction of the normal vascular models, all images presenting an extension of the disease towards further regions beyond the aneurysm neck were rejected for this study. Bilateral rotational scans were obtained to visualize both avenues of flow into the anterior communicating artery (AcomA) aneurysm using a Philips Integris System [8]. Images were obtained during a 180° rotation and imaging at 15 frames per second for 8 seconds. The 120 projection images were reconstructed into a 3D dataset of 128 128 128 voxels covering a field of view of 54.02 mm on a dedicated Philips Integris workstation. The voxel resolution was therefore 0.422 mm. These data were exported into a PC for mathematic vascular modelling using a previously presented methodology [19,20,10]. Models were edited and aneurysms were virtually removed.

1.2. Patients and images
Rotational cerebral angiography images are the preferred modality for reconstructing vascular models harboring aneurysms due to their highest spatial resolution and contrast. For the selected patient, a rotational scan was obtained to visualize the pathologic vasculature. An 8-second acquisition consisted in 120 projections obtained during a 180° rotation, which were reconstructed into a 3D dataset of 128^3 voxels covering a field of view of 54.02 mm on a dedicated workstation. The voxel resolution was therefore 0.422 mm. These data were exported into a PC for mathematic vascular modeling using a previously presented methodology [8,9,10,11]. The imaging protocol was approved by the institutional review board and informed consent was obtained from the subject.

1.3. Vascular modeling
Finite element blood flow numerical simulations were performed for all models under the same flow conditions. Blood was modelled as an incompressible Newtonian fluid with attenuation 1.0 g/cm^3 and

viscosity 0.04 Poise. The governing equations were the unsteady Navier-Stokes equations in 3D [15]. Vessel walls were assumed rigid, and no slip boundary conditions were applied at the walls. Pulsatile flow conditions derived from phase-contrast magnetic resonance measurement in a healthy subject were imposed at the inlet of the models. Flow waveforms were scaled with the inlet area in order to achieve a mean WSS of 15 dyn/cm^2 at the inflow boundary of each model. This choice is consistent with studies relating vessel area and flow rates in internal carotid and vertebral arteries [16], as well as with the principle of minimal work expressed by Murray's law [17]. Fully developed pulsatile velocity profiles were prescribed with the use of the Womersley solution [18,19]. Assuming that all distal vascular beds have similar total resistance to flow, traction-free boundary conditions with the same pressure level were applied at outlet boundaries. The Navier-Stokes equations were numerically integrated by using a fully implicit finite element formulation [9]. We computed 2 cardiac cycles using 100 time-steps per cycle, and all of the reported results correspond to the second cardiac cycle.

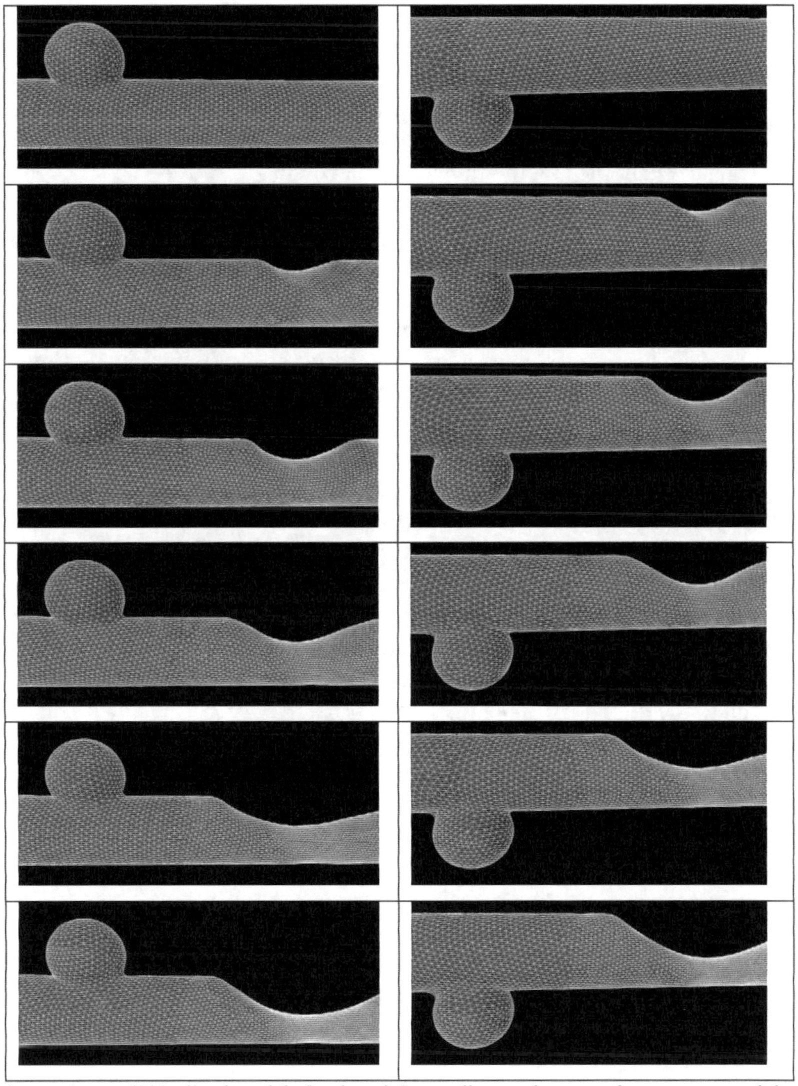

Figure 1. This picture shows all idealized models for the minimum distance between the aneurysm and the stenosis (1.60 cm). Left column: aneurysm ipsilateral to the stenosis. Right column: aneurysm contralateral to the stenosis. Each row corresponds to a different grade of stenosis. From top to bottom: 0, 20, 40, 55, 65 and 75.

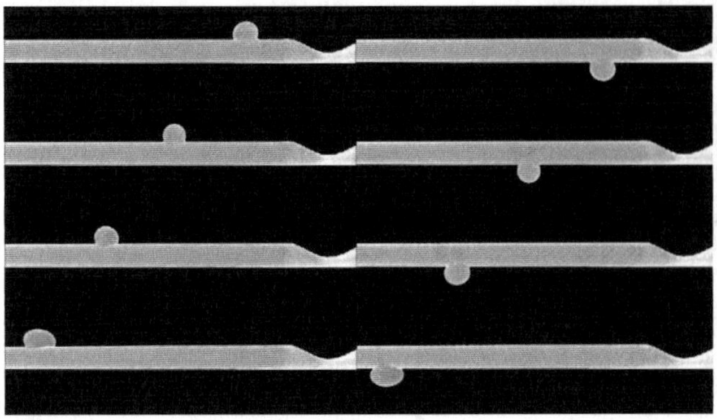

Figure 2. This picture shows all idealized models for the 75-degree stenosis. Left panel: aneurysm ipsilateral to the stenosis. Right panel: aneurysm contralateral to the stenosis. Each row corresponds to a different distance between the aneurysm and the stenosis. From top to bottom: 1.60 cm, 3.20 cm, 4.80 cm and 6.40 cm.

1.4. Data analysis

Maps of WSS magnitude were created to visualize the distribution of shear forces on the aneurysm wall and the vessel. Magnitude of the velocity and vector field was computed at cutplane containing the vessel axis, the aneurysm dome and the maximum narrowing of the artery.

Results

In order to investigate how changes in the flow pattern due to the stenosis grade may affect the intraaneurysmal hemodynamics, magnitude of the velocity and vector field at a cutplane containing the vessel axis, the aneurysm dome and the maximum narrowing at the systolic peak were displayed (Figures 3a and 3b). Wall shear stress maps were constructed to study how its distribution depends on the stenosis grade (Figures 4a and 4b). The same flow rate wave form was applied to all models. Severe stenoses were not included in this study, therefore no inflow correction was imposed at the inlet of the model after virtual intervention with respect to the original diseased arteries.

Eccentric stenoses generate a recirculation at distal regions adjacent and ipsilateral to the stenosis along with a main jet narrowing and shift towards the opposite side. That produces lower velocity and wall shear stress in the post stenotic vortex adjacent to the stenosis, and higher values in the opposite side. The extension of the recirculation zone is minimal for low stenosis grades but significantly increases with the severity of the disease. That region is associated with lower wall shear stress values compared to the vessel without stenotic plaque. For higher stenosis grades, the recirculation pattern results in higher velocity and wall shear stress values over a larger area. For distal aneurysms ipsilateral to the high degree stenoses, the vortex may reach the aneurysm depending on the stenosis grade and the distance between the aneurysm and the stenosis. Under this condition, the part of the neck closer to the aneurysm witnesses high velocities (see Figures 3a) resulting in higher wall shear stress values (see Figure 4a).

It is also observed that the higher the stenosis grade, the higher the jet narrowing and the higher the wall shear stress values right downstream the vessel narrowing. However, because of jet concentration, those values become lower than the average in the normal vessel when the jet heads towards the vessel axis (see Figures 4a and 4b). Those values increase again when the jet develops towards its ideal Womersley profile in a cylindrical vessel.

For aneurysms contralateral to the stenosis the intraaneurysmal flow pattern is not affected by the recirculation. However, for high stenosis grades the jet is deflected towards the aneurysm resulting in higher WSS values in the distal part of the neck (see Figure 4b). The part of the vessel under flow recirculation experiments higher WSS values for mild stenoses compared to a normal vessel. But now that effect takes place diametrically opposite to the aneurysm, which is contralateral to the stenosis.

The typical lateral aneurysm in a straight vessel exhibits a higher WSS region in the distal part of the neck. However, when a proximal aneurysm is present changes in the WSS values both in the distal and proximal part of the aneurysm neck appear. Therefore, if the stenosis is removed, forces over the aneurysm neck will change depending on the WSS distributions before and after the intervention. For aneurysms ipsilateral to the stenosis, the WSS at the distal part of the neck is less affected by the plaque. However, for higher grades, that region starts to feel the disturbance caused by the recirculation vortex resulting in lower WSS values. For the distance between the aneurysm and the stenosis and the stenosis grades considered here, the distal part of the neck remains outside that vortex. On the other hand, the proximal part is under lower velocities and WSS values when the vortex reaches that region, which is observed for 55% and 65% degree stenosis, and suddenly increases for higher grades (e.g. 75% degree stenosis) because the vortex region is large enough to enter the aneurysm resulting in higher velocities upstream in the proximal part of the neck and WSS values even higher than those observed when no stenosis is present (see Figure 5). Note that for all other configurations the WSS in the aneurysm is lower with than without stenosis because the stenosis produces a concentration of the flow with higher velocities and flow rate confined to a main jet far from the walls in the post-stenotic region.

It is also observed that for higher stenosis grades the WSS at the distal part of the neck is much lower in aneurysms ipsilateral to the stenosis than for contralateral ones. That is due to the fact that ipsilateral aneurysms are affected by the recirculation vortex for higher grades.

Considering that an endarterectomy would affect the intra-aneurysmal hemodynamics, the pre- and post-intervention WSS maximum values in the aneurysm at the systolic peak were compared for the 65% degree stenosis case for both ipsilateral and contralateral configurations. It was observed that for ipsilateral configurations the maximum WSS, which occurs at the distal region of the neck, would increase from 39.0 dyn/cm^2 to 65.3 dyn/cm^2 (67.6%), while for contralateral configurations it would increase from 51.5 dyn/cm^2 to 65.3 dyn/cm^2 (26.7%). Note that 65.3 dyn/cm^2 corresponds to the WSS value at the systolic peak in the distal region of the aneurysm neck when no proximal stenosis is present. That observation shows that hemodynamics of lateral aneurysms contralateral to a proximal eccentric artery stenosis is considerably less affected compared to ipsilateral configurations.

For the patient-specific model, the same inflow rate waveform was used, which was scaled in order to impose a mean wall shear stress of 15 dyn/cm^2 at the internal carotid artery [16,17]. The inflow waveform had been previously acquired from a normal subject using phase-contrast magnetic resonance angiography imaging [8,11]. Three regions were selected to compare the WSS values before and after the intervention, ordered according to the distance to the stenosis: I) proximal part of the first lobulation neck close to the stenosis; II) peak value at the neck of the second lobulation; III) peak value at the dome of the second lobulation. High WSS values at region (I) dropped from 720 dyn/cm^2 to 270 dyn/cm^2. The region (II), which is located downstream, exhibits almost no difference in peak WSS: 235 dyn/cm^2 (before) compared to 219 dyn/cm^2 (after). That is due to the fact that although the jet is less concentrated, it impacts straight on the neck of the second lobulation. Finally, for region (III) the jet is much less concentrated after the intervention resulting in a low WSS of 20 dyn/cm^2 at the dome after the intervention compared to 150 dyn/cm^2 before the intervention (see Figure 6). In this patient-specific case it can be observed that some high WSS regions reduce their values after intervention while others remain approximately the same.

Although it is not expected flow alterations after the removal of a 65% grade stenotic plaque, a virtual experiment was performed by decreasing the pre-intervention flow in 30% according to reported data for higher stenosis grades. In this experiment it was observed a lower WSS in the dome before the intervention, because of a lower flow. However, it is still higher that that after the plaque removal, due to jet dissipation. On the other hand, lower flow before the intervention results in lower WSS values at the neck of the first lobulations. Those values significantly increase after the intervention because of a higher flow. This virtual experiment suggest that removing a stenotic plaque may relocate regions of high and low WSS, which may eventually change the biological response of endothelial cells to flow forces (see Figure 7).

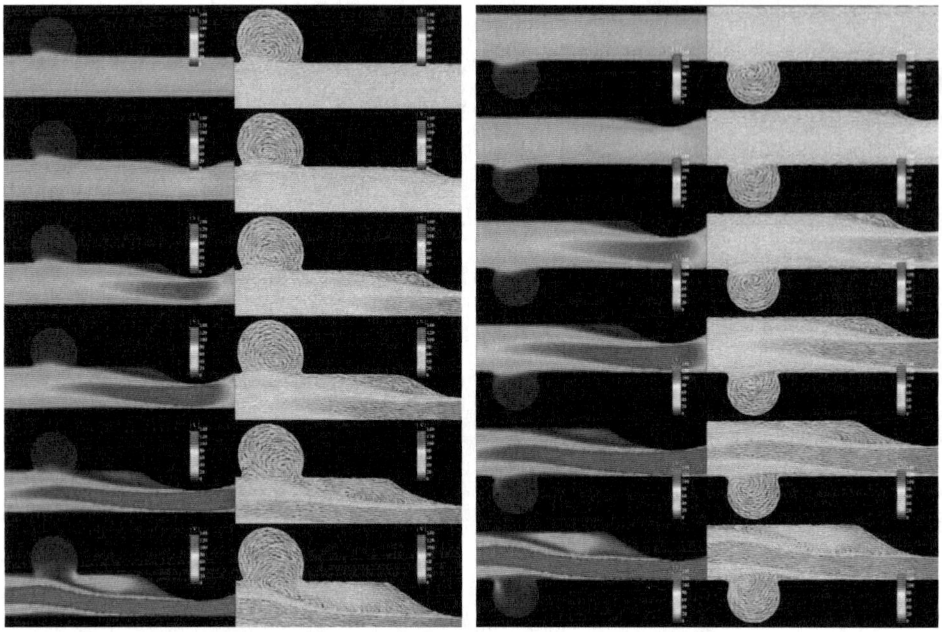

Figure 3. a) LEFT PANEL: ipsilateral cases ; b) RIGHT PANEL: contralateral cases. Magnitude of the velocity (left columns) and vector field (right columns) at the systolic peak at a cutplane containing the vessel axis, the aneurysm dome and the maximum narrowing, for aneurysms ipsilateral to the stenosis. Each row corresponds to a different grade of stenosis. From top to bottom: 0, 20, 40, 55, 65 and 75. Color maps range between the minimum velocity value of 0 cm/sec (blue) and >140 cm/sec (magenta).

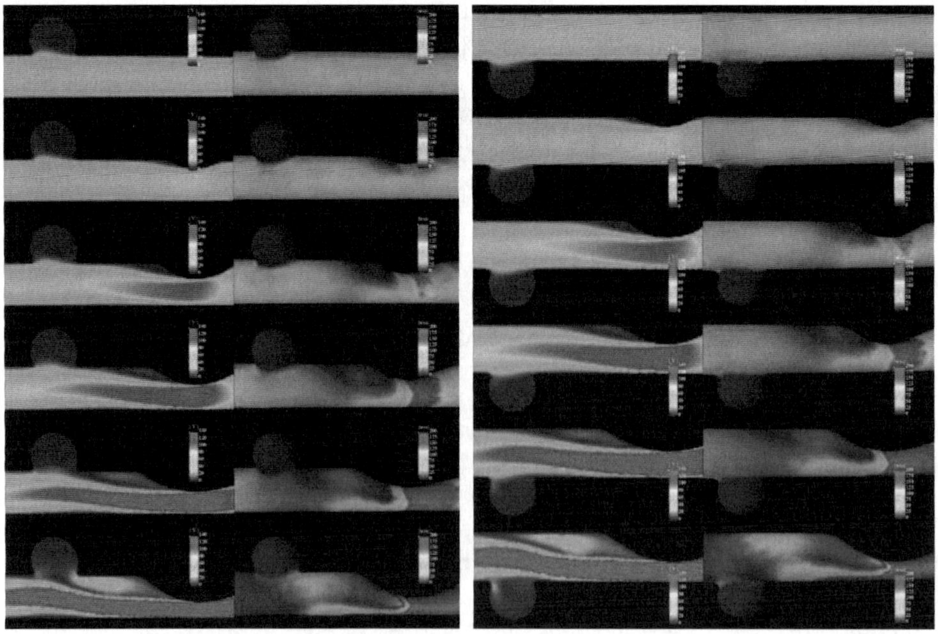

Figure 4. a) LEFT PANEL: ipsilateral cases ; b) RIGHT PANEL: contralateral cases. Magnitude of the velocity (left columns) and wall shear stress distribution (right columns) at the systolic peak at a cutplane for aneurysms ipsilateral to the stenosis. Each row corresponds to a different grade of stenosis. From top to bottom: 0, 20, 40, 55, 65 and 75. Color maps range for the velocity range between 0 cm/sec (blue) and >140 cm/sec (magenta), while for the wall shear stress range between 0 dyn/cm^2 and >200 dyn/cm^2.

Journal of Physics: Conference Series **477** (2013) 012003 doi:10.1088/1742-6596/477/1/012003

Discussion

In this paper we applied a previous developed methodology for patient-specific image-based computational hemodynamics studies of vascular cerebral networks harboring intracranial aneurysms [4,10,14,19] to study the intra-aneurysmal hemodynamics in patients having a cerebral aneurysm distal to a artery stenosis in the same circulation. Patients with concomitant carotid artery stenosis and unruptured intracranial aneurysms pose a difficult management decision for the physicians. Case reports showed patients who died due to aneurysm rupture months after endarterectomy but before aneurysm clipping, while others did not show any change in the aneurysm after plaque removal, having optimum outcome after aneurysm coiling [1,5].

The aim of this preliminary work is to investigate the main characteristics of the intra-aneurysmal hemodynamics affected by a proximal artery stenosis, and the expected changes after a plaque removal. For that purpose, a set of forty eight idealized CFD models of lateral aneurysms located at different positions from a proximal eccentric stenosis was constructed. Vector velocity field and wall shear stress distributions were examined for different distances between the aneurysm and the stenosis, for aneurysms both ipsilateral and contralateral to the stenosis, and for six different stenosis grades (low and mild). It could be observed that the recirculation distal to the stenosis may significantly affect the intraaneurysmal hemodynamics of ipsilateral aneurysms increasing the wall shear stress values in the proximal part of the neck. On the other hand, contralateral aneurysms may experience an increment of the wall shear stress in the distal part of the neck due to the deflection and narrowing of the main jet in the post-stenotic region. Those effects are relevant in the vascular zone closer to the aneurysm. However, for high stenosis grades, the recirculation extends further downstream resulting in higher wall shear stress gradients. Comparison between the intraaneurysmal hemodynamics before and after a plaque removal may help to estimate the changes in the hemodynamic quantities and their impact on the risk of rupture.

Flow alterations after the removal of a 65% grade stenotic plaque are not expected. However, a virtual experiment was performed by reducing the pre-intervention flow in 30% according to reported data for higher stenosis grades. It was observed a lower WSS in the dome before the intervention, because of a lower flow. However, it is still higher that that after the plaque removal, due to jet dissipation. On the other hand, lower flow before the intervention results in lower WSS values at the neck of the first lobulations. Those values significantly increase after the intervention because of a higher flow. This virtual experiment suggest that removing a stenotic plaque may relocate regions of high and low WSS, which may eventually change the biological response of endothelial cells to flow forces.

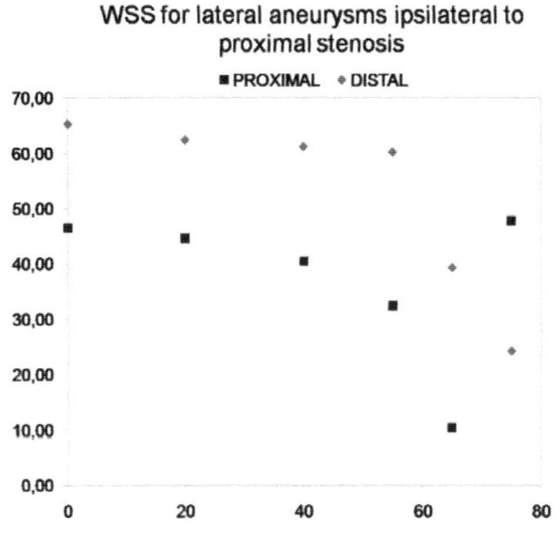

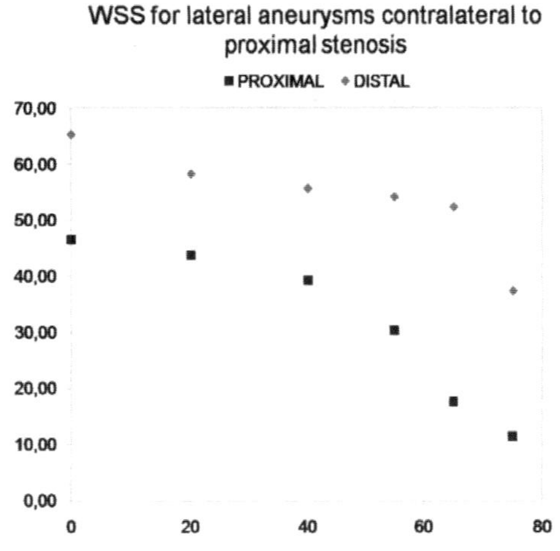

Figure 5. Magnitude of the absolute value of the wall shear stress distribution at the systolic for aneurysm ipsilateral (left column) and contralateral (right column) to the stenosis at both the proximal and distal part of the neck.

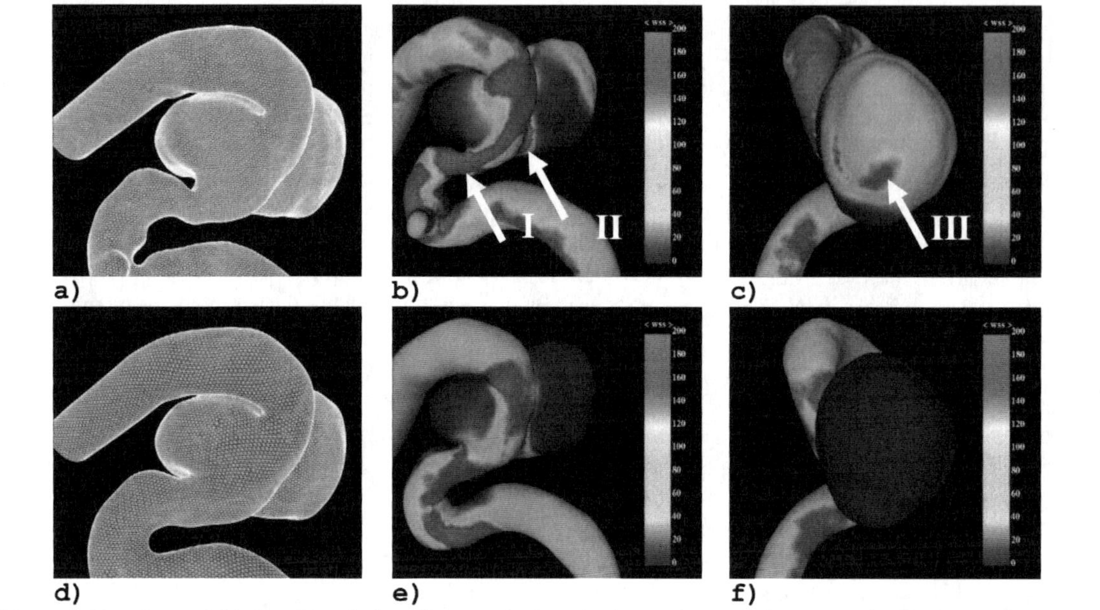

Figure 6. Vascular model before (a) and after (d) removal of proximal stenosis. Wall shear stress distributions before (b and c) and after (e and f) virtual intervention.

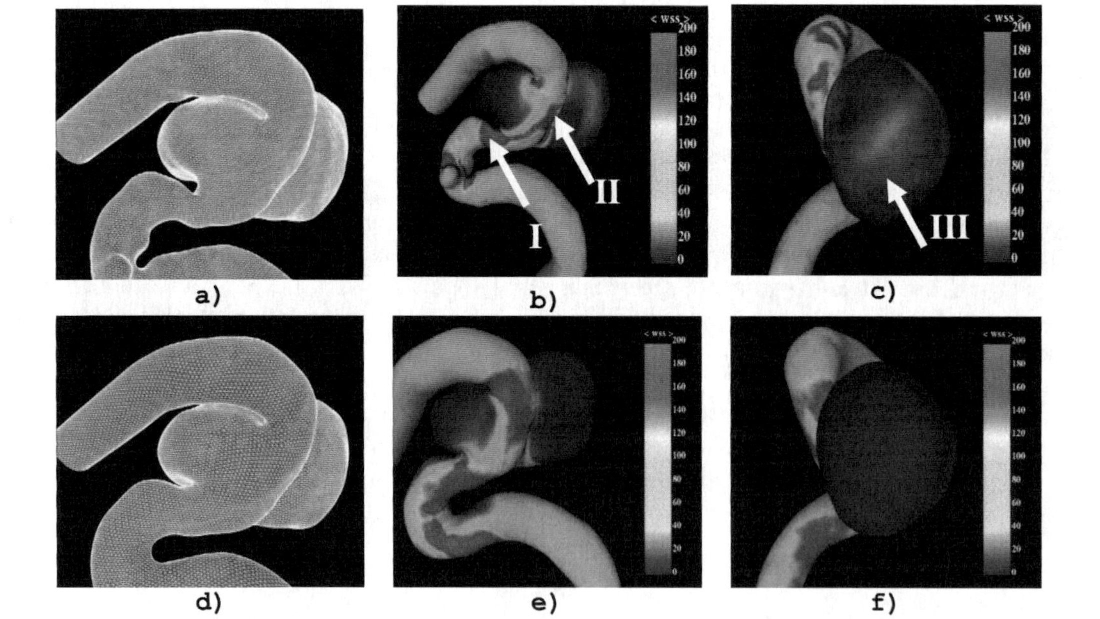

Figure 7. Vascular model before (a) and after (d) removal of proximal stenosis, considering lower flow rate before the intervention due to the stenotic plaque. Wall shear stress distributions before (b and c) and after (e and f) virtual intervention.

For severe stenoses, which are beyond the scope of this work, transition to turbulence begins to play an important role. Varghese et al. investigated in detail the hemodynamic characteristics through a 75-degree stenosis using an idealized model of an eccentric stenosis using direct numerical simulations [20]. However, those models did not include distal aneurysms. In a previous work, Varghese et al. compared the prediction of post-stenotic velocity field and wall shear stress as well as shear stress peak at the throat of the stenosis in a pulsatile regime through an idealized stenotic vessel (75% axisymmetric stenosis) using four different turbulent models implemented in a popular commercial CFD code, with respect to two previous experiments that described stenotic flows [21].

They found discrepancies between the different turbulent models. The low Reynolds number *k-ω* turbulent model was closer to the experimental results. All of them predicted low wall shear stress values distal to the stenosis as a consequence of high jet concentration, which is in agreement with our results.

Patel et al. reviewed several turbulence models for low Reynolds number flows [22]. They found that many of the models lacked generality and were restricted by the choice of empirical constants, being the Chien model the one that performed better compared to the others. All *k-ε* models use wall functions to obtain near-wall results. Since accurate estimation of wall shear stress in one of the main goals in the problem under study, such models may not be the wisest choice. From that point of view, *k-ω* models may be more suitable for stenotic flows. However, the transitional nature of this kind of flows suggest the use of direct numerical simulations to resolve the flow, although that will be a complex undertaking in realistic diseased 3D vessels under pulsatile flow conditions [23].

Additionally, we selected a case from our data base corresponding to a patient with a cerebral aneurysm right distal to an artery stenosis. The diseased artery exhibited a two-lobe aneurysm, impacting the main jet into the dome of one of them. The aneurysm formed right downstream of the stenosis and the narrowed jet produced a small impaction zone with high wall shear stress values. The stenosis was virtually removed by means of the use of Laplacian filters and the intraaneurysmal hemodynamics was compared before and after the intervention. Wall shear stress at that dome reduced about one order of magnitude after the plaque was removed.

The methodology has some limitations. The study only included low and mild stenoses. Severe stenosis would require higher resolution models to correctly depict transition to turbulence and also to consider flow correction after the intervention. Although the removal of a high degree stenotic plaque may significantly affect the intraaneurysmal hemodynamics, the post-intervention flow may return to its higher normal value, resulting in higher forces exerted on the aneurismal wall, and increasing the risk of rupture. Vascular models with aneurysms distal to a severe stenosis will be studied in detail. Another limitation is that hemodynamic characteristics observed in ideal cases may not necessarily be representative of real cases. For that purpose, a patient-specific case was also studied. Although the correct vascular geometry reconstruction results in realistic blood flow simulations [24], the use of non-patient-specific flow rate waveforms imposed at the inlet of our models may incorporate some errors. However, this limitation is suffered by most computational studies since flow rate measurements are not routinely acquired in clinical evaluations. This limitation is partially overcome by wall shear stress normalization at inflow segments, which allows comparison between different models. Other model assumptions had limited impact on intra-aneurysmal hemodynamic characterization, as it was shown in a previous work [4].

Conclusions

The present work shows the difference between the intra-aneurysmal hemodynamics before and after the removal of a proximal stenotic plaque. Both idealized and patient-specific models showed significant differences that may impact on the risk of rupture.

Acknowledgements

Marcelo Castro wants to acknowledge CONICET (Consejo Nacional de Investigaciones Científicas y Técnicas) for financial support (PIP CONICET 2011-2013 Nr. 287).

References

[1] Espinosa G, Dzieciuchowicz L, Grochowicz L. Endovascular treatment of carotid stenosis associated with incidental intracranial aneurysm. Ann Vasc Surg 2009:23:688.e1-688.e5

[2] Kappelle LJ, Eliasziw M, Fox AJ, et al. Small unruptured intracranial aneurysm and management of symptatic carotid artery stenosis: North American symptomatic carotid endarterectomy trial group. Neurology 2000:55:307–309

[3] Pappada G, Fiori L, Marina R, et al. Incidence of asymptomatic of asymptomatic berry aneurysms among patients undergoing carotid artery endarterectomy. J Neurosurg Sci 1997:41:257-262
[4] Rouleau, PA, Huston J, Gilbertson J, Brown RD, Meyer FB, Bower TC. Carotid Artery Tandem Lesions: Frequency of Angiographic Detection and Consequences for Endarterectomy. Am J Neuroradiol 2009:20:621-625
[5] Adams HP. Carotid stenosis and coexisting ipsilateral intracranial aneurysm: a problem in management. Arch Neurol 1977:34:515-516
[6] Navaneethan SD, Kanna VS, Osowo A, Shrivastava R, Singh S. Concomitant intracranial aneurysm and carotid artery stenosis: A therapeutic dilemma. Sothern Med J 2006:99(7): 757-758
[7] Cebral JR, Löhner R, Choyke PL, et al. Merging of intersecting triangulations for finite element modelling. J Biomech 2001:34:815–819
[8] Castro MA, Putman CM, Cebral JR. Hemodynamic Patterns of Anterior Communicating Artery Aneurysms: A Possible Association with Rupture. Am J Neuroradiol 2009:30(2):297-302
[9] Cebral JR, Castro MA, Appanaboyina S, Putman C, Millán D, Frangi A. Efficient Pipeline for Image-Based Patient-Specific Analysis of Cerebral Aneurysms Hemodynamics: Technique and Sensitivity. IEEE - Trans Med Imag - Special Issue on Vascular Imaging 2005:24(4): 457-467
[10] Yim PJ, Vasbinder B, Ho VH, et al. A deformable isosurface and vascular applications. Proc SPIE 2002:46841390 –1397
[11] Castro MA, Putman CM, Cebral JR. Patient-specific computational modeling of cerebral aneurysms with multiple avenues of flow from 3D rotational angiography images. Acad Radiol 2006:13:811–821
[12] Löhner R. Automatic unstructured grid generators. Finite Elements Analysis Design 1997:25: 111–134
[13] Löhner R. Extensions and improvements of the advancing front grid generation technique. Comp Meth Appl Mech Eng 1996:5:119–132
[14] Löhner R. Regridding surface triangulations. J Comp Phys 1996:126:1–10
[15] Mazumdar JN. Biofluid Mechanics. World Scientific 1992, Singapore, 41-104
[16] Cebral JR. Castro MA, Putman CM. Alperin, N., Flow-area relationship in internal carotid and vertebral arteries. Physiol Meas 2008:29(10):585-594
[17] Sherman TF. On connecting large vessels to small. The meaning of Murray's law. J Gen Physiol 1981:78:431–453
[18] Womersley JR. Method for the calculation of velocity, rate of flow and viscous drag in arteries when the pressure gradient is known. J Physiol 1955:127:553–563
[19] Taylor CA. Hughes TJR. Zarins CK. Finite element modeling of blood flow in arteries," Comp Meth Appl Mech Engin 1998:158:155–196
[20] Varghese SS, Frankel SH. Fischer PF. Direct numerical simulation of stenotic flows. J Fluid Mech 2007:582:253-280
[21] Varghese SS, Frankel SH. Numerical modelling of pulsatile turbulent flow in stenotic vessels. J Biomech Eng 2003:125:445-460
[22] Patel VC, Rodi W, Scheuerer G. Turbulence models for near-wall and low Reynolds number flows : A review. AIAA J 1984:22:1308-1319
[23] Stroud JS, Berger SA, Saloner D. Numerical analysis of flow through a severely stenotic carotid artery bifurcation. J Biomech Eng 2002:124:9-20
[24] Cebral JR, Pergolizzi RS, Putman CM. Computational fluid dynamics modeling of intracranial aneurysms: Quantitative comparison with cerebral angiography. Acad Radiol 2007:14:804-813

9th Argentinean Bioengineering Society Congress (SABI 2013) IOP Publishing
Journal of Physics: Conference Series **477** (2013) 012004 doi:10.1088/1742-6596/477/1/012004

Wall motion and hemodynamics in intracranial aneurysms

Marcelo A. Castro[1]
CONICET. Universidad Tecnológica Nacional, Facultad Regional Buenos Aires
Grupo de Investigación y Desarrollo en Bioingeniería
Medrano 951, Ciudad Autónoma de Buenos Aires, CP 1179, Argentina

E-mail: marcelo.a.castro@gmail.com

María C. Ahumada Olivares
Universidad Favaloro, Facultad de Ingeniería, Ciencias Exactas y Naturales
Solís 453, Ciudad Autónoma de Buenos Aires, CP 1078, Argentina

Christopher M. Putman
Inova Fairfax Hospital, Fairfax Radiological Consultants
3300 Gallows Road, Falls Church, VA 22042, USA

Juan R. Cebral
George Mason University. College of Sciences
4400 University Drive, Fairfax, VA, 22030, USA

Abstract. The optimal management of unruptured aneurysms is controversial and current decision making derived from the International Study of Unruptured Intracranial Aneurysms is based on aneurysm size and location. Wall shear stress triggers biomechanical responses of endothelial cells, which are associated to cerebral aneurysm initiation, progress and bleeding. Early identification of potential risk factors may contribute to decide the treatment and improve patient care. Previous studies have shown associations between high aneurysm wall shear stress values and both elevated risk of rupture and regions of aneurysm growing. Based on the assumption that damaged regions of the endothelium have different mechanical properties, regions with differentiated wall displacement amplitudes are expected. A previous approach based on the analysis of bidimensional dynamic tomographic angiography had been designed to investigate those correlations, but its main limitation was that wall motion was measured in a selected plane. In this work a high time and spatial resolution four-dimensional computerized tomographic angiography image of an anterior communicating artery aneurysm was acquired and analyzed to identify and characterize wall motion and intra aneurysmal hemodynamics. All three-dimensional images were filtered and segmented and wall displacement was estimated within the aneurysm sac and compared to wall shear stress distributions from patient-specific unsteady finite element blood flow simulations. Regions were high wall motion was detected are in close agreement with regions were high wall shear stress values were obtained from the numerical blood flow simulations.

[1] To whom any correspondence should be addressed.

Introduction

Stroke is the leading cause of long-term disability and the third cause of death in the Western World. Subarachnoid hemorrhage is one of the most severe types of stroke, which usually occurs when an intracranial aneurysm ruptures [1]. This kind of aneurysms tends to initiate at or near arterial bifurcation, mostly in the circle of Willis. The optimal management of unruptured aneurysms is controversial and current decision making is mainly based on aneurysm size and location, as derived from the International Study of Unruptured Intracranial Aneurysms (ISUIA) [2]. However, it is widely accepted that hemodynamics, particularly the wall shear stress (WSS), plays an important role on the development, growth and rupture of cerebral aneurysms. Previous studies have shown that ruptured aneurysms tended to have small impaction zones and complex or unstable flow patterns [3]. An association between concentrated jets, high WSS and high rupture rate was found in two previous studies where two cohorts of patients with cerebrovascular networks harboring aneurysms were analyzed using an image-based patient-specific computational fluid dynamics (CFD) methodology in the anterior communicating artery [4] and a set of terminal aneurysms [5]. Other authors analyzed twenty middle cerebral artery aneurysms and also found high WSS values in the group of ruptured aneurysms accompanied with low WSS in their domes, which would suggest that low WSS values may be responsible for aneurysm rupture [6]. However, in that study CFD models were truncated close to the aneurysm neck resulting in a simplified flow lacking of secondary flows. The effect of parent artery on intra aneurysmal hemodynamics was previously studied and higher WSS in the aneurysm domes was found to be systematically related to those secondary flows [7,8]. Another study showed that more than 80% of bleb formation occurs in regions of high WSS [9]. Recently, independent works showed that aneurysms tend to initiate in regions of moderate and elevated WSS [10] and regions of high WSS or high WSS spatial gradient [11]. Those findings have encouraged researchers to explore possible connections between hemodynamics, wall motion, wall weakness and aneurysm rupture. Many of the quantitative results of pulsation measurement reported in the literature correspond to experiments with phantoms, simulated images or experimental models [12-15]. A Methodology to estimate wall motion from X-ray dynamic imaging and impose the time-dependent deformation on the vascular CFD models reconstructed from 3D angiographic images was presented [16]. The blood flow characterization obtained from numerical integration of the Navier-Stokes equations in a rigid wall model did not significantly differ from that in compliant models where the deformation field was extrapolated from the bidimensional measurements [17,18]. The purpose of this work is to present a methodology to both estimate regions of high wall motion and reconstruct CFD vascular models from 4D computerized tomographic angiography (CTA) data sets.

Materials and methods

1.1. Images and modeling

A patient with a cerebral aneurysm in the anterior communicating artery was selected from our data base. A four dimensional computerized tomographic angiography image containing nineteen time points along the cardiac cycle was acquired using a Philips Integris System (Philips Medical Systems, Best, The Netherlands). Each three dimensional volume consisted of 512 x 512 x 205 pixels with a spatial resolution of 0.3125 x 0.3125 x 0.6125 mm covering a field of view of 16.0 x 16.0 x 12.81 cm. The imaging protocol was approved by the institutional review board and informed consent was obtained from the subjects. The data was exported into a PC for mathematic vascular modeling using a previously presented methodology [3,4,19]. Figure 1 shows three axial slices for the first time point. Images were cropped (Nx x Ny x Nz) and averaged in order to reduce the noise and the computational cost. In lower regions masks were applied to all images in order to differentiate internal carotid arteries from extravascular structures and properly reconstruct parent arteries, which is needed for a realistic blood flow simulation [7,8]. The vascular model was reconstructed from the reduced noise image obtained after averaging the nineteen three dimensional volumes. A high-quality volumetric finite element grid composed of tetrahedral elements with an advancing front technique was generated

Journal of Physics: Conference Series **477** (2013) 012004 doi:10.1088/1742-6596/477/1/012004

[20-22]. Element size was adjusted in order to approximately maintain the same number of elements in both large and small arteries (see Figure 2).

1.2. Numerical simulations

Finite element blood flow numerical simulations were performed. Blood was modelled as an incompressible fluid with density 1.0 g/cm^3 and viscosity 0.04 Poise. The governing equations were the unsteady Navier-Stokes equations in 3D [23]. Vessel walls were assumed rigid, and no slip boundary conditions were applied at the walls. Pulsatile flow conditions derived from phase-contrast magnetic resonance measurement in a healthy subject were imposed at the inlet of the model. Flow waveforms were scaled with the inlet area to achieve a mean WSS of 15 dyne/cm^2 at the inflows of the model according to a typical mean WSS value, and Murray´s and Poiseuille´s laws [24,25]. Fully developed pulsatile velocity profiles were prescribed with use of the Womersley solution [26,27]. Assuming that all distal vascular beds have similar total resistance to flow, traction-free boundary conditions with the same pressure level were applied at outlet boundaries. Both Newtonian and non-Newtonian rheologies were considered. For the non-Newtonian case, the Casson model was incorporated into the simulations by considering a velocity dependent apparent viscosity. The Navier-Stokes equations were numerically integrated by using a fully implicit finite-element formulation [28]. Our methodology is based on a projection scheme which arises from the hyperbolic character of the advection operator and the elliptic character of the pressure-Poisson equation. A fully implicit finite element formulation that yields to stable solutions for arbitrary time steps was utilized. The discretized momentum equation is solved using a generalized minimal residual (GMRES) algorithm, while the pressure equation, which is obtained by taking the divergence of the momentum and considering the incompressibility constraint, is solved using an incomplete lower-upper (ILU) preconditioned conjugate gradient solver. The algorithm is iterated until convergence is achieved at each time step. Two cardiac cycles using 100 time-steps per cycle were computed, and all of the results reported correspond to the second cardiac cycle.

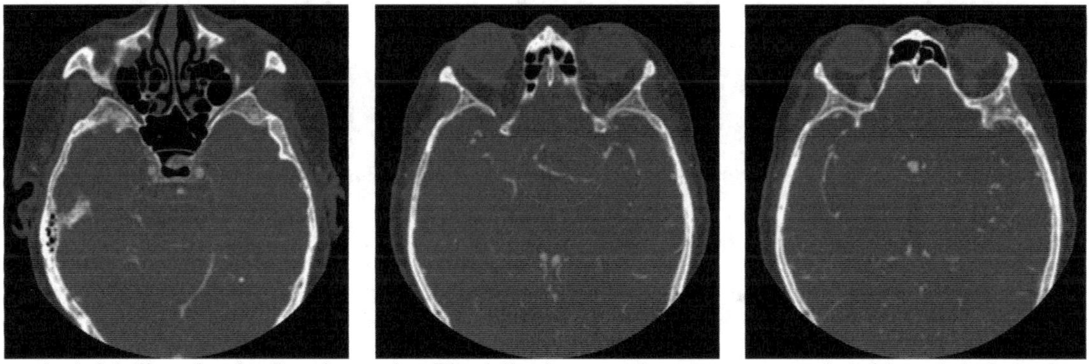

Figure 1: Axial slices for the first time point. The following vascular structures are circled in red. Left: both internal carotid arteries along with the basilar artery. Middle: both A1 segments of the anterior cerebral arteries, the anterior communicating artery and the lower portion of the aneurysm. Right: both A2 segments of the anterior cerebral arteries along with the upper portion of the aneurysm. See Figure 2 for A1 and A2 segments.

1.3. Wall motion and wall shear stress

In order to detect those regions where wall motion is not negligible the time evolution of segmented clusters at every slice within the volume of interest was analyzed. For each of the N=19 three dimensional images (Nx x Ny x Nz) the same selected number of NS slices covering the whole aneurysm was chosen for the analysis. A set of NS images Nx x Ny x N was created to analyze the time evolution of pixel intensity in all NS slices. Four classes were identified, two of which corresponded to the background (*C1* and *C2*), and the other two are associated to the border of the vascular domain (*C3*) and the vasculature (*C4*). A four-class fuzzy c-means algorithm was utilized to create the four images corresponding to the membership probability associated to each of the *NS* images, where the pixel intensity *I* ($0<I<1$) represents the probability of that pixel to belong to the corresponding class. Particularly, for the image associated to *C4*, the pixel intensity represents the probability of that pixel to belong to the vasculature. It is expected that those probabilities should significantly change for pixels close to a moving boundary. Consequently, the maximum difference of the probability along the *N* time slices was computed for each of the *Nx*Ny* pixels for each of the *NS* images. This information was used to reconstruct a *Nx* x *Ny* x *NS* image where the pixel intensity represents the maximum difference in time of the probability to belong to class *C4*. Since *C1* and *C2* clusters correspond to the lowest intensities and *C4* to the highest ones, pixels that change from vasculature to background due to wall motion are expected to have significant different probability to belong to class *C4*. That image was segmented using different thresholds. Regions that are likely to change from vasculature to background were isolated and compared to shear stress distribution over the wall.

Maps of WSS magnitude were created to visualize the distribution of shear forces on the aneurysm wall at the systolic peak where the maximum values are expected. Regions of elevated, mild and low WSS were investigated. Additionally, WSS computed using Newtonian and Casson rheologies were compared. Although in this preliminary study the model walls were rigid, it is expected that flow pattern should remain roughly unchanged for moderate wall displacements [16,17]. This hypothesis will be further tested in large aneurysms with high deformations using dynamic tomographic angiography images. Streamlines were created and the velocity field was imaged in order to investigate the relation between the location of the impaction zone, the maximum wall motion and the elevated WSS values and the flow characterization.

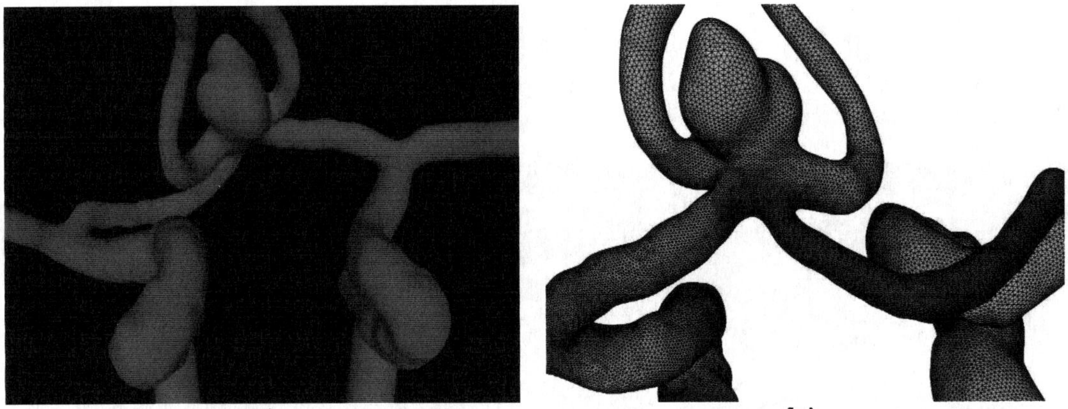

a) **b)**

Figure 2: a) Vascular model containing both internal carotid arteries, anterior cerebral arteries and middle cerebral arteries (anterior view); b) Grid of triangles (posterior view). A1 segments are proximal to the AComA, while A2 segments are distal to the AComA.

Results

This first paragraph explains the results of the clustering algorithm used to detect regions of high wall displacements. Figure 3a corresponds to a selected axial slice and a given time were the brightest spots correspond to both A2 segments of the anterior cerebral arteries (upper left corner) and the upper region of the anterior communicating artery aneurysm (brightest and largest spot). Visual inspection of that image revealed a background (extravascular structures) with a large range of pixel intensities, while the highest intensities corresponded to the vasculature. A transition takes place in the boundary between the vasculature and the background. Figures 3b through 3e correspond to the probability maps for classes *C1* through *C4*, respectively. Darker regions correspond to pixels that unlikely belong to a particular class. Figure 4a and 4b correspond to the same selected slice but at different times. Slightly changes in the aneurysm morphology can be observed. Figure 4c shows the maximum difference in the probability that a given pixel belongs to class *C4* at that slice. Pixels with higher intensities have significant different probability of belonging to class *C4* along the cardiac cycle. Those pixels are mainly located at the aneurysm wall. Additionally, non-zero values are observed in other regions, which is due to the typical noise observed in these images. It is expected that higher threshold should discriminate between changes produced by noise and those that arise from the wall displacement.

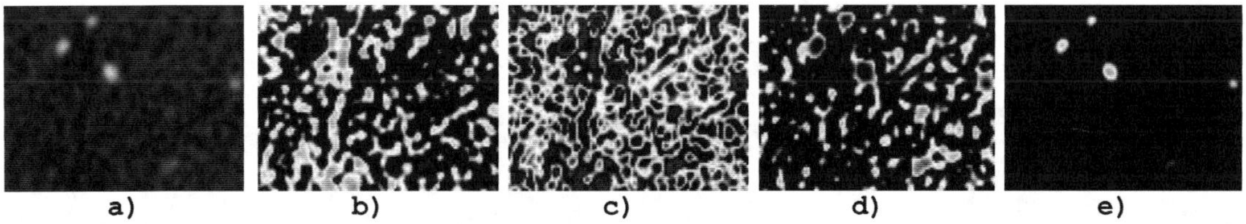

Figure 3: a) Selected axial slice at a given time; b) Membership probability for class *C1*; c) Membership probability for class *C2*; d) Membership probability for class *C3*; e) Membership probability map for class *C4*.

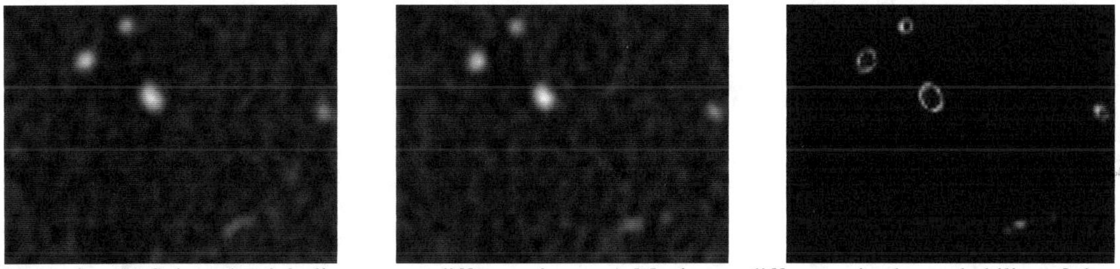

Figure 4: a,b) Selected axial slice at two different times; c) Maximum difference in the probability of those pixels to belong to class *C4* (Vasculature).

The image that represents the maximum change along the cardiac cycle in the probability of a given pixel to belong to class *C4* (Figure 4c) can be segmented in order to identify those pixels that experiment the highest changes. Changes due to noise are not extremely high and are observed in small disconnected regions. On the other hand, changes due to wall motion are associated to higher values and are observed in larger areas. Therefore, an intensity-based segmentation for different thresholds (T1=0.325, T2=0.400) was performed in order to investigate the location where the highest displacements take place. The segmented pixels are shown along with the WSS distribution at the systolic peak for the Newtonian rheology (Figure 5a) and the Casson rheology (Figure 5b). As it was expected, the lower the threshold, the larger the segmented region. The grey regions in Figure 5c and Figure 5d correspond to pixels that experiment a change in the probability of belonging to class *C4* higher than 0.325 and 0.400, respectively, displayed over the Newtonian WSS distribution. For example, in Figure 5d, if any of those grey pixels belongs to class *C4* with a probability of 0.80 at a given time, there exists another time when that probability drops below 0.40. Figure 6 compares WSS distribution, the velocity field and the region where the largest wall displacement is more likely to occur, for the Newtonian case. That region is located close to WSS peak slightly displaced in the flow direction in region of elevated WSS (see Figures 6b).

Streamlines were computed in order to study whether or not the aneurysm behaves as a bilateral aneurysm (e.g., it receives blood from two differences sources). Blue streamlines are originated in the right internal carotid artery and red ones do in the left carotid artery (see Figure 6d). Complex velocity field observed in Figure 6d is due to the collision of both jets. However, intraaneurysmal flow pattern is mostly dominated by the left inflow while high WSS values occur where both jets collide.

Figure 7 compares WSS distributions for the Newtonian and Casson rheologies. Figures 7a and 7c show the Newtonian WSS distribution at the systolic peak from the front and back view, respectively. Analogously, Figures 7b and 7d show the difference between both rheologies. No significant difference in the absolute value of the WSS is observed in black regions (higher velocities). Magenta represents regions where the Casson WSS is as much as twice as the Newtonian WSS. It is observed that Cason WSS is larger in regions of low WSS, which are located far from the impaction zone and where the wall exhibit larger amplitudes of motion. On the other hand, blue represents regions where the Newtonian WSS is as much as twice the Casson WSS. The region where the highest WSS values are observed at the systolic peak has an average of 68.6 dyn/cm^2 and 69.2 dyn/cm^2 for Casson and Newtonian rheologies, respectively. Therefore, Newtonian WSS is 0.87% higher than Casson WSS. On the other hand, in the region where the lowest WSS values are observed, the average WSS values are 1.22 dyn/cm^2 and 0.51 dyn/cm^2, for Casson and Newtonian rheologies, respectively. Therefore, Newtonian WSS is 42.8% lower than Casson WSS, in agreement with the theory [18].

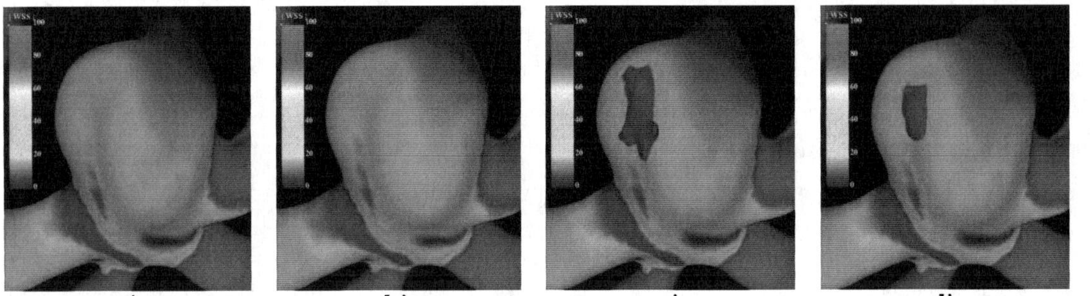

a) b) c) d)

Figure 5: WSS distribution at the systolic peak for Newtonian rheology (a) and Casson rheology (b). Newtonian WSS distribution at the systolic peak along with the segmented region where higher wall displacements are expected for T1=0.325 (c) and T2=0.400 (d).

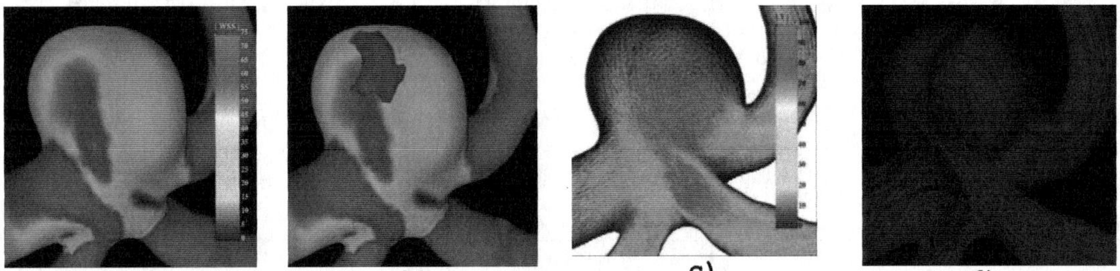

a) b) c) d)

Figure 6: a) WSS distribution at the systolic peak; b) WSS distribution at the systolic peak along with the segmented region where higher wall displacements are expected (T1=0.325); c) Velocity field; d) Streamlines coloured according to the source of inflow. Blue streamlines are originated at the right internal carotid artery, and red to the left internal carotid artery.

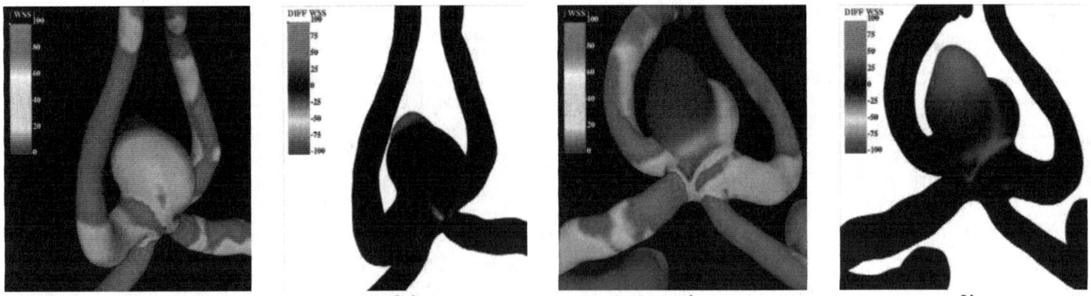

a) b) c) d)

Figure 7: a) Newtonian WSS distribution at the systolic peak (front view); b) Difference between WSS distributions at the systolic peak (front view); c) Newtonian WSS distribution at the systolic peak (back view); b) Difference between WSS distributions at the systolic peak (back view). In Figures (b) and (d), magenta represents regions where the Casson WSS is as much as twice as the Newtonian WSS, while blue represents regions where the Newtonian WSS is as much as twice the Casson WSS. No significant difference in the absolute value of the WSS is observed in black regions.

Discussion

The purpose of this work was to design and test a methodology to estimate both the aneurysm wall shear stress, and aneurysm wall motion from 4D computerized tomographic angiography images. A test case was randomly selected from our data base. The patient had a large aneurysm with multiple blebs in the anterior communicating artery. The high temporal and spatial resolution set of images consisted of nineteen three dimensional images along the cardiac cycle. Images were averaged to reduce the impact of noise on the reconstruction of the vascular model used to generate the finite element unstructured volumetric grid for the blood flow simulation and wall shear stress estimation. In order to estimate the localization of regions that exhibit an important wall displacement, the time evolution of individual slices across the aneurysm was analyzed and a fuzzy c-means clustering algorithm was applied to simultaneously segment all time points in four different classes for each slice. The probability maps for each class were recorded. Pixels within the vascular domain but close to a moving boundary exhibited a change in that probability. That difference was segmented using different thresholds. The regions exhibiting important displacements were overlapped with the WSS distribution. Maximum WSS values at the systolic peak occurred close to the region where important displacements took place. Particularly, this anterior communicating artery aneurysm behaved as a bilateral aneurysm, which means that received blood from both A1 segments of the anterior cerebral arteries [4], and high wall motion occurred close to where both jets collided. This strategy estimates the wall motion from a 3D analysis and overcomes some of the limitations from a previous methodology where the motion was measured in a plane and extrapolated to the aneurysm [16]. WSS distributions remain unchanged when different rheologies are used. However, absolute values in specific regions are observed. Vascular regions under high velocities have high WSS values associated in the wall therefore, Casson effect is almost negligible. On the other hand, in regions under extremely low WSS, non-Newtonian effects are more important. However, that does not occur in regions where the wall exhibit large amplitudes of motion.

The methodology has some limitations. First, the computational fluid dynamics simulation was performed under the assumption of rigid walls. Although this assumption is not accurate, it was observed in a previous work that moderate wall displacements along the cardiac cycle do not significantly change the WSS distribution over the wall of the aneurysm sac [17]. However, in future works wall motion will be also included in the CFD blood flow simulations. Second, bilateral anterior communicating artery aneurysms have flow patterns that depend on inflow conditions. We imposed the same waveform at both internal carotid arteries, but scaled according to their cross-sectional areas based on a typical wall shear stress value, and both Murray′s and Poiseuille′s laws [23]. However, it was previously observed that differences in either those mean flow rates at the inflow segments or waveform phases may relocate the region where the maximum wall shear stress appeared. In this particular case, a sensitivity study should be carried out to determine whether or not the location of that maximum may be affected. Finally, in order to estimate the probability to belong to a given class, images were not pre-processed. Consequently, pixels in the background that changed from one class to another due to the noise were also detected as possible candidates. However, those regions were isolated and the change in the probability was not as large as that observed in boundary pixels. It was observed that higher thresholds kept the large regions where displacements occurred due to wall motion, and discarded those small regions due to the noise.

Conclusions

The present work shows that four dimensional computerized tomographic angiography images can be used to estimate the location of regions with both important wall motion and high wall shear stress by means of a combination of image segmentation algorithms and image-based computational fluid dynamics blood flow simulations. The results suggest that aneurysm wall tends to exhibit higher displacements along the cardiac cycle in regions with elevated wall shear stress and complex flow pattern.

Acknowledgements
Marcelo Castro wants to acknowledge CONICET (Consejo Nacional de Investigaciones Científicas y Técnicas), MINCyT (Ministerio de Ciencia y Tecnología – Project PICT Bicentenario # 984), UTN (Universidad Tecnológica Nacional, Project PID # 1579), and Fundación Florencio Fiorini, Buenos Aires, Argentina, for financial support.

References

[1] Hwang S, Kim K, Nam T. "A pseudoaneurysm appeared after rebleeding," Journal of Korean Neurosurgical Society, 41, 134-136 (2007)
[2] Wiebers DO, et al. "International study of unruptured intracranial aneurysms investigators. Unruptured intra-cranial aneurysms: natural history, clinical outcome, and risks of surgical and endovascular treatment," Lancet, 362, 103–110 (2003)
[3] Cebral JR, Castro MA, Burgess JE, Pergolizi RS, Sheridan MJ, Putman CM. "Characterization of cerebral aneurysmfor assessing risk of rupture using patient-specific computational hemodynamics models," Am J Neuroradiol, 26, 2550 –2559 (2005)
[4] Castro MA, Putman CM, Cebral JR. "Hemodynamic Patterns of Anterior Communicating Artery Aneurysms: A Possible Association with Rupture," Am J Neuroradiol, 30(2), 297-302 (2009)
[5] Castro MA, Putman CM, Radaelli A, Frangi AF, Cebral JR. "Hemodynamics and rupture of terminal cerebral aneurysms," Acad Radiol, 16(19), 1201-1207 (2009)
[6] Shojima M, Oshima M, Takagi K, Torii R, Hayakawa M, Katada K, Morita A, Kirino T. "Magnitude and role of wall shear stress on cerebral aneurysm: computational fluid dynamic study of 20 middle cerebral artery aneurysms," Stroke, 35, 2500-2505 (2004)
[7] Castro MA, Putman CM, Cebral JR. "Computational fluid dynamics modeling of intracranial aneurysms: effects of parent artery segmentation on intra-aneurysmal hemodynamics," Am J Neuroradiol, 27, 1703-1709 (2006)
[8] Castro MA, Putman CM, Cebral JR. "Effects of Parent Vessel Geometry on Intraaneurysmal Flow Pattern," Proc. SPIE Medical Imaging 2006: Physics of Medical Imaging Image Reconstruction, 6143, 123-131 (2006)
[9] Cebral JR, Sheridan MJ, Putman CM. "Hemodynamics and Bleb Formation in Intracranial Aneurysms," Am J Neuroradiol, 31, 304-310 (2010)
[10] Castro MA, Putman CM, Cebral JR. "Computational analysis of anterior communicating artery aneurysm shear stress before and after aneurysm formation," Journal of Physics C.S., 32, 1-7 (2011)
[11] Kulcsár Z, Ugron A, Marosfo M, Berentei Z, Paal G, Szikora I. "Hemodynamics of Cerebral Aneurysm Initiation: The Role of Wall Shear Stress and Spatial Wall Shear Stress Gradient," Am J Neuroradiol, 32(3), 587-594 (2011)
[12] Yaghmai V, Rohany M, Shaibani A, Huber M, Soud H, Russell EJ, Walker MT. "Pulsatility imaging of saccular aneurysm model by 64-slice CT with dynamic multiscan technique," J Vasc Intervent Radiol, 18, 785–788 (2007)
[13] Boecher-Schwarz HG, Ringel K, Kopacz L, Heimann A, Kempski O. "Ex vivo study of the physical effect of coils on pressure and flow dynamics in experimental aneurysms," Am J Neuroradiol, 21, 1532–1536 (2000)
[14] Ueno J, Matsuo T, Sugiyama K, Okeda R. " Mechanism underlying the prevention of aneurismal rupture by coil embolization," J Med Dental Sci., 49, 135–141 (2002)
[15] Zhang C, Villa Uriol MC, De Craene M, Pozo JM, Frangi AF. "Time-resolved 3D rotational angio-graphy reconstruction: Towards cerebral aneurysm pulsatile analysis," Int J Comput Assist Radiol Surg, 3, S44–46 (2008)
[16] Oubel E, Cebral JR, De Craene M, Blanc R, Blasco R, Macho J, Putman CM, Frangi AF. "Wall motion setimation in Intracranial aneurysms," Physiol Meas, 31, 1119-1135 (2010)

[17] Dempere-Marco L, Oubel E, Castro MA, Putman CM, Frangi AF, Cebral JR. "Estimation of wall motion in intracranial aneurysms and its effects on hemodynamic patterns," Lecture Notes in Computer Science, 4191, 438-445 (2006)

[18] Castro MA, Putman CM, Cebral JR. "Computational hemodynamics of cerebral aneurysms: Assessing the risk of rupture," VDM Verlag (2008)

[19] Yim P, Vasbinder GB, Ho VB, Choyke PL. "Isosurfaces as deformable models for magnetic resonance angiography," IEEE – Trans Med Imag, 22(7), 875-881 (2003)

[20] Löhner R. "Extensions and improvements of the advancing front grid generation technique," Comp Meth App Mech Eng, 5, 119–132 (1996).

[21] Löhner R. "Regridding surface triangulations," J Comput Phys, 126, 1–10 (1996)

[22] Löhner R. "Automatic unstructured grid generators," Finite Elements Analysis Design, 25, 111–134 (1997)

[23] Mazumdar JN. "Biofluid Mechanics," World Scientific, Singapore (1992)

[24] Cebral JR, Castro MA, Putman CM, Alperin N. "Flow-area relationship in internal carotid and vertebral arteries," Physiol Meas, 29(10), 585–594 (2008)

[25] Sherman TF, "On connecting large vessels to small. The meaning of Murray's law," J Gen Physiol, 78, 431–453 (1981)

[26] Womersley JR, "Method for the calculation of velocity, rate of flow and viscous drag in arteries when the pressure gradient is known," J Physiol, 127, 553–563 (1955)

[27] Taylor CA, Hughes TJR, Zarins CK. "Finite element modeling of blood flow in arteries," Comp Meth Appl Mech Engin, 158, 155–196 (1998)

[28] Cebral JR, Castro MA, Appanaboyina S, Putman C, Millán D, Frangi A. "Efficient Pipeline for Image-Based Patient-Specific Analysis of Cerebral Aneurysms Hemodynamics: Technique and Sensitivity," IEEE - Trans Med Imag - Special Issue on Vascular Imaging, 24(4), 457-467 (2005)

In-vitro hemodynamic measurements and in-silico simulation of a physical model of arterial bifurcation

D Suárez-Bagnasco[1], G Balay[1], L Cymberknop[3], R L Armentano[2], C Negreira[1].

[1] Laboratorio de Acústica Ultrasonora, Instituto de Física, Facultad de Ciencias (FCIEN), UDELAR, Montevideo, Uruguay.
[2] Departamento de Fisiología, Facultad de Medicina (FMED), UDELAR, Montevideo, Uruguay.
[3] Departamento de Electrónica, Facultad Regional Buenos Aires, Universidad Tecnológica Nacional (UTN), Buenos Aires, Argentina.

E-mail: dsuarez@fisica.edu.uy

Abstract. The quantification of fluid-structure interactions in arterial walls (from a biomechanical standpoint) requires a complete characterization of blood flow, shear stress in the interface between blood and endothelium, wall elasticity and wall stresses distribution. These interactions play an important role in pathogenic mechanisms of cardiovascular diseases, such as atherosclerosis and arteriosclerosis. A specific hemodynamic work bench simulator is used for an in-vitro characterization of the biomechanics and hemodynamics of a custom constructed physical model of arterial bifurcation, under near-physiologic pulsated flow conditions. The model was constructed using fresh porcine aorta. Some experimental in-vitro results are shown. In-vitro results are compared with in-silico results obtained from a simple CFD model of the abovementioned bifurcation. Experimental data allow construction of meshes and parameter estimation of the biomechanical properties of the arterial wall, as well as boundary conditions, all suitable to be employed in CFD and FSI numerical simulation. In-silico simulations enables the estimation of some parameters that cannot be easily obtained by means of in-vitro experimental measurements (for example wall shear stress).

1. Introduction

Arterial behavior in-vivo, as well as the biomechanical properties of the arterial wall, are influenced by the interaction between blood flow and the arterial wall endothelium, arterial wall elasticity and the stress-strain distribution inside the arterial wall and surrounding tissues, amongst other factors (e.g. viscosity).

To quantify these interactions both from biomechanical and hemodynamic standpoints, a complete characterization and modeling of the dynamic of arterial wall, blood flow, shear wall, wall elasticity and wall strain-stresses are needed. Characterization can be done in-vivo and in-vitro. In-vivo by means of invasive and non-invasive (tonometry, pulsed Doppler, pulse wave velocity, elastographic) methods. In invasive methods, suitable instrumented animal specimens are used to perform the measurements on a given arterial segment. In-vitro, the characterization can be done by means of a work bench simulators in which arterial segments are intercalated in closed fluid circuits that emulates the main features of systemic arterial circulation [1] [2] [3] [4]. A sound in-vitro characterization of arterial segments (including pressure and flow patterns) may furnish valuable information to be applied in-vivo [5]. A specific experimental set-up is needed in order to be able to work with arterial bifurcations under near physiological pulsated flow conditions. Measurements obtained from cryopreserved [6] and fresh arterial bifurcation samples subjected to different hemodynamic regimes allows the characterization of the biomechanical properties of arterial segments. These are obtained by parameter estimation and can be employed in a digital simulation model of the interaction between structure and fluid. Velocity profiles obtained in the work-bench can be compared with the ones obtained by computer fluid dynamics (CFD) simulations with moving wall.

The main purposes of the present paper are:
-To describe the in-vitro hemodynamic work bench simulator used to work with bifurcations.
-To show some experimental results for a physical model of arterial bifurcation surgically implemented from a sample of fresh porcine aorta.
-To compare some in-vitro with the in-silico results obtained from a simple CFD model of the abovementioned bifurcation.
-To show some results obtained in-silico that cannot be obtained easily by means of experimental in-vitro measurements.

2. Hemodynamic work bench simulator

The closed circuit of the hemodynamic work bench simulator [1] is composed by a pump, the tubing between the pump exit and the entrance to the arterial bifurcation, the arterial bifurcation being studied, tubing between each branch of the bifurcation exit and the entrance to a fluid reservoir, adjustable constrictions on each branch, a fluid reservoir with mean pressure adjustment, and a returning tubing from the reservoir to the pump (Figure 1).

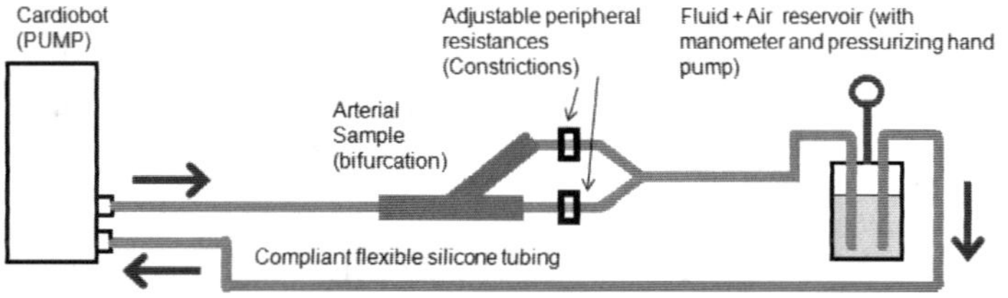

Figure 1. Schematic diagram. Compliant silicone tubing is used for the circuit.

An artificial heart named Cardiobot is used as the pulsated pump, with frequency and ejection volume adjusted at demand. Cardiobot is an electronically controlled programmable pump, specially developed [2] [7] in order to generate different flow and pressure patterns. For each branch, there is an independent restriction located far away from the corresponding branch output. Each restriction introduces localized mechanical impedance mismatch that produces reflected waves (Figure 2).

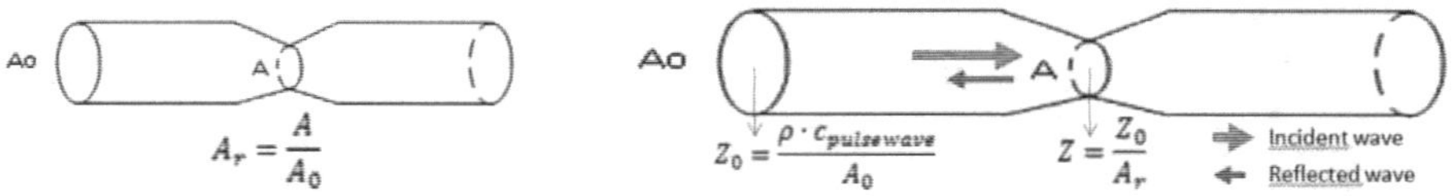

Figure 2. Simplified model for mechanical impedance change at constriction. A_0 : vessel cross sectional area before constriction. A: vessel cross sectional at constriction.

Each restriction introduces a non linear loss of charge (Δp) that can be approximated by:

$$\Delta p = C_r \cdot \rho \cdot \frac{Q^2}{2 \cdot A_0^2} \tag{1}$$

Where

$$C_r = \frac{\zeta}{A_r^2} \tag{2}$$

and ζ is a loss of charge coefficient.

The combination of upstream and downstream pressure pulses produce global pressure patterns in the arterial sample. Measurements are done after a steady regime of pulsated flow is attained in the circuit. The already established dynamic regime and a time reference signal generated by the Cardiobot (trigger at the beginning of each systole), allows relocation and exchange of sensors during measurements (for example, A-Scan and Pulsed Doppler measurements can be done in the same region alternating probe placement). Samples are mounted on an arterial bifurcation fixing system (couplers and adjustable arms) and submerged in a physiological solution pool. This pool is rounded by an external water pool, where temperature control can be done (Figure 3).

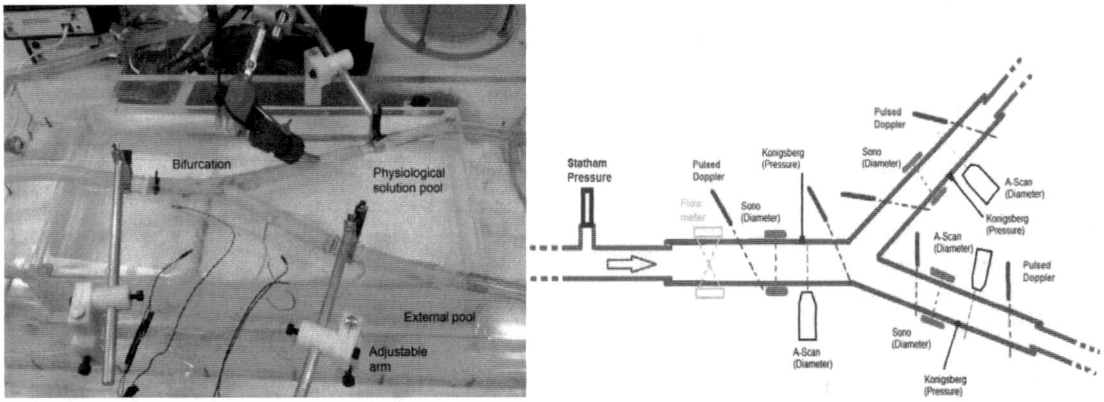

Figure 3. Left: Arterial bifurcation in the physiological solution pool. Right: schematic diagram of sensor placement.

The circuit was filled with physiological solution, and temperature control was done, with a set-up point of 37°C. The reservoir is a glass flask containing the fluid and air. It's mean pressure can be adjusted at constant values by means of a hand pump and a differential pressure sensor. With no flow and no transmural pressure in the circuit, A-Scan ultrasonic measurements are done over the bifurcations with a 15MHz Panametrics V313 transducer probe fixed on a mechanical graduated positioning system (Figure 4). A custom algorithm is used to obtain internal and external diameters and wall thickness [1].

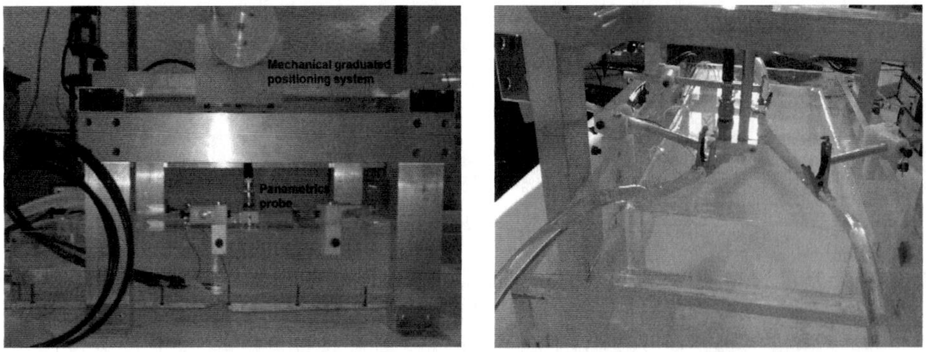

Figure 4. Mechanical graduated positioning system (in x, y and z axes) and Panametrics V313 probe during morphometric measurements.

After that, several hemodynamic conditions are simulated and measured, adjusting the flow pattern of the Cardiobot, reservoir pressure and the variable restrictions. External diameter measurements are taken at suitable points in each segment of the arterial bifurcation using a gold-standard System 6 Mainframe equipped with sonomicrometer ultrasonic modules [8] and ultra resolution techniques [5]. Pressure waveforms are obtained using Königsberg sensors (with high frequency response) located near diameter measurement points. Sensors are sutured so that sensing surface points to the lumen in order to obtain the hydrostatic component of the pressure. Velocity profiles at different locations are obtained using a multi-gate Doppler system (DOP, Signal Processing SA). Micro-spheres powder (scatterers) are used when the hemodynamic work bench simulator circuit is filled with a physiological solution or water-glycerine mixture fluid. Cross sectional flow is measured using suitable diameter perivascular transit time flowmeters (Triton flowmeters). A Statham pressure sensor (located in the tubing near the common segment) is used for monitoring and initial adjustments procedures. Signals are acquired at 3kHz using 12-bit data acquisition modules (NI USB-6009 and LabJack U3-HV) operated from a specially developed Matlab application. Two Tektronix TDS2024B, 100MHz, 4 channel digital oscilloscopes are used for initial adjustments and for monitoring purposes. Pressure and diameter samples are processed in a computer running a specific developed algorithm in order to obtain biomechanical properties of the wall by means of model parameter adjustment [9].

The hemodynamic work bench can work with arterial samples, phantoms or other man-made tubes or bifurcations, under static and dynamic conditions. Other experimental setups are designed and operated exclusively to characterize mechanical properties of the arterial or venous walls, but can't be used to study wave propagation and fluid-structure interactions. Bertram et al. [8] developed an experimental setup for arterial or tube segments that allows pulse propagation measurements, under near-physiological pressure pulse and flow patterns. Their setup employs rigid wall tubing. Two closed fluid compliance reservoirs (with a pressurized air chamber inside) are located before and after the sample. The first one attenuates pressure transients originated in the closure of the valves of the piston pump. The second one is intended to allow the modification of mechanical impedance. A Darcy type hydraulic restriction located after the second chamber introduces a linear loss of charge. However it can't be adjusted during operation. Our setup allows the adjustment of the two restrictions independently and during operation.

For the reported experiments, a physical model of arterial bifurcation was used, surgically implemented from a sample of fresh porcine aorta (Figure 5).

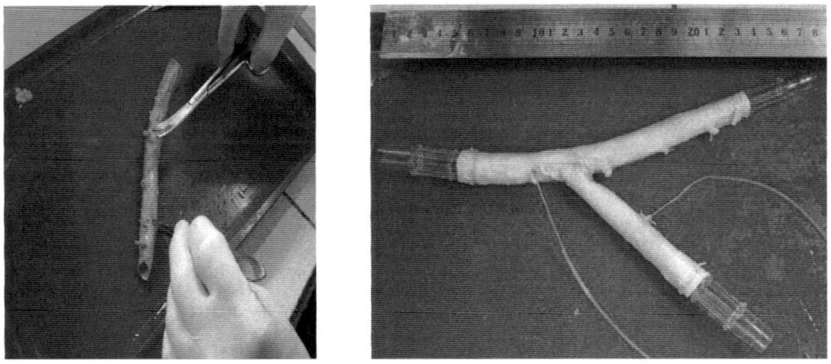

Figure 5. Left: Sample of fresh porcine aorta. Right: Surgically constructed physical model of arterial bifurcation. The common segment and the two branches are in the same plane, with a bifurcation angle of 60 degrees between branches. Tubing couplers and two Konigsberg sensors are shown.

3. Data obtained in-vitro: some experimental results

Static and dynamic measurements of diameters, pressures, flows and velocity profiles were done in several locations in the common segment and branches of the bifurcations. Figure 6 shows the simultaneous acquisition of diameter, pressure and cross sectional flow for a location in the common segment.

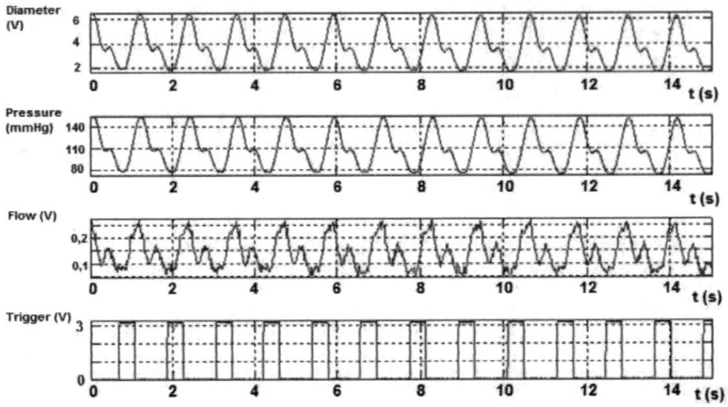

Figure 6. Diameter and flow raw data (in volts) and pressure waveform for common segment. Time reference signal generated by Cardiobot at each systole can be seen at the bottom.

For this experiment, a physiological solution was used in the circuit, with diluted suspension of microspheres (doppler scatterers). Figure 7 shows velocity profiles (without filtering) obtained by pulsed Doppler, in the same place. Vessel pulsation can be seen through the variation of the profiles width over the x axis (distance in mm).

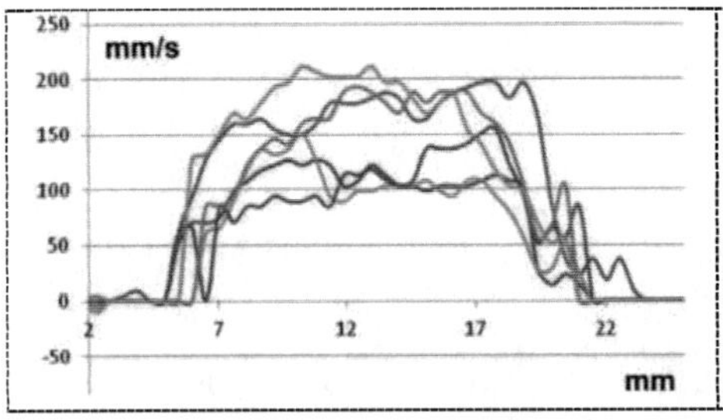

Figure 7. Raw velocity profiles obtained for the common segment at 40mm from the entrance. Vessel pulsation can be seen through the variation of the profiles width over the x axis (distance in mm). The segment presented a radius of 7,5 mm at zero transmural pressure.

4. Data obtained in-silico: some digital CFD simulation results

A simplified CFD model was constructed as a first approximation in order to compare in-vitro with in-silico results. The spatial domain used for finite-element digital simulation [10] was determined from geometrical measurements done to the surgically constructed physical model (Figure 8).

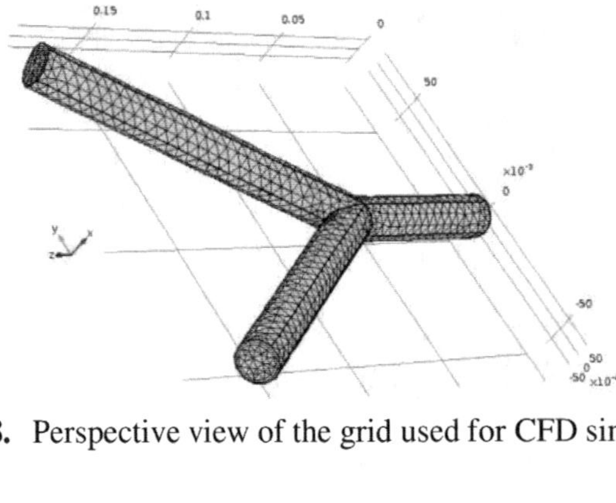

Figure 8. Perspective view of the grid used for CFD simulations.

A mean radius of 7,5mm was used for the common segment and each branch and a rigid wall condition was imposed (non elastic wall). Lengths were taken from the physical model as well as the bifurcation angle (implemented to be 60 degrees). Time dependent Navier-Stokes equations were solved with a suitable viscosity coefficient, using a commercial software. Fixed and impervious wall conditions were imposed. In the open ends (commons segment entrance and branch exits) pressure boundary conditions were applied, taken from in-vitro experimental measurements [11]. Figure 9 shows the pulsated pressure measured in-vitro at the entrance of the common segment.

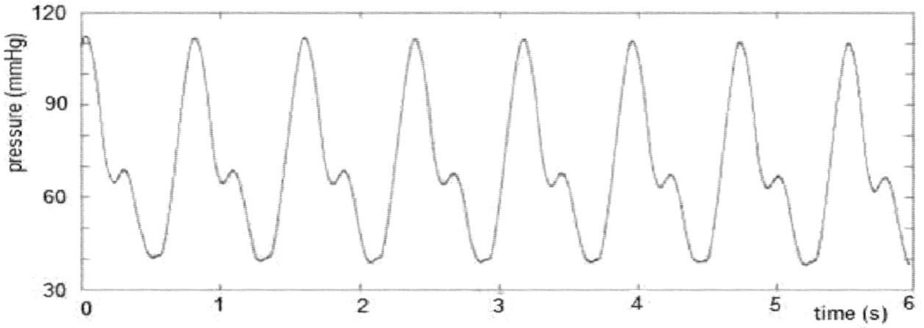

Figure 9. Input pressure waveform for the common segment entrance, with a pulsation frequency of 60 beats per minute.

Figure 10 shows the velocity profiles obtained at different times in the common segment at 40mm from the entrance.

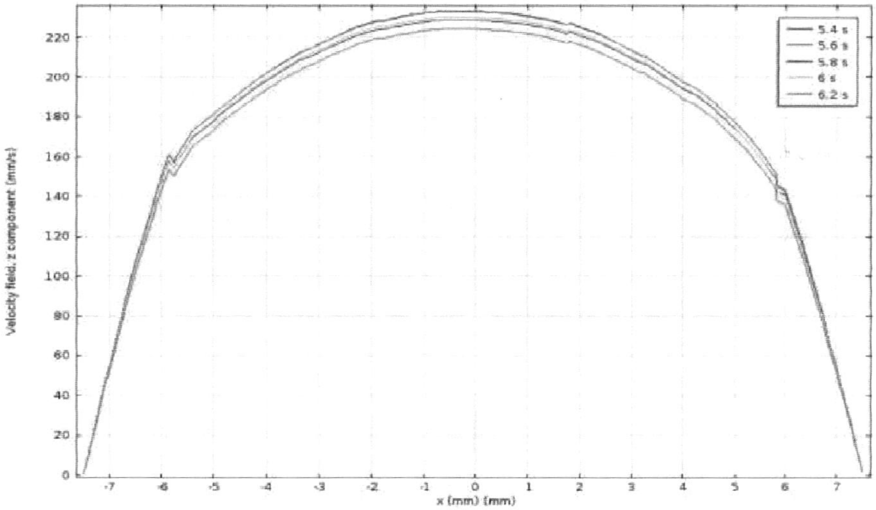

Figure 10. In-silico velocity profiles for the common segment. Velocities are expresed in milimeters per second (mm/s).

Figure 11 shows the velocity profiles obtained at the same instant at different cross sections. Flow inversion near the outer walls of the bifurcation branches can be seen.

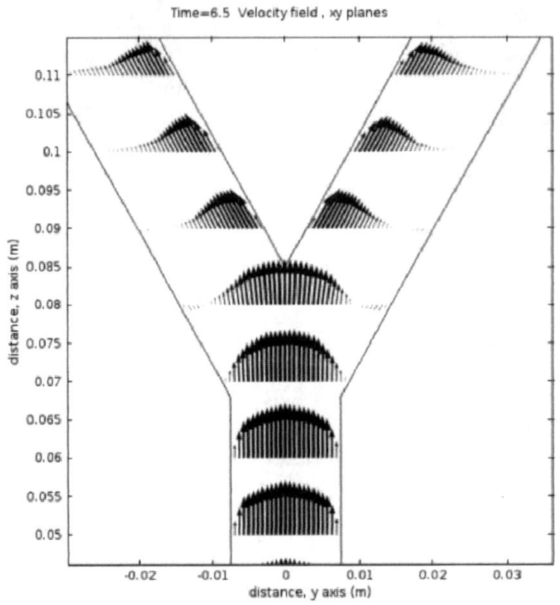

Figure 11. Velocity profiles near the bifurcation for a given time.

Figure 12 shows the relative magnitudes of the velocities for the whole bifurcation in the plane x = 0 at the same time instant of Figure10. Flow inversion can't be seen because this figure represents absolute values.

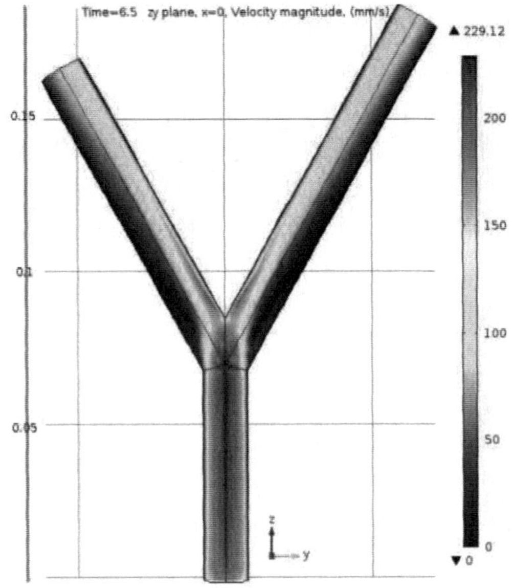

Figure 12. Relative magnitude of the velocity of fluid (mm/s) for a given time, in the plane x=0.

Figure 13 shows shear rate magnitude for the same time instant. The inner walls of bifurcation branches show maximum values. Taking into account fluid viscosity, shear stress can be calculated at the wall of the vessel.

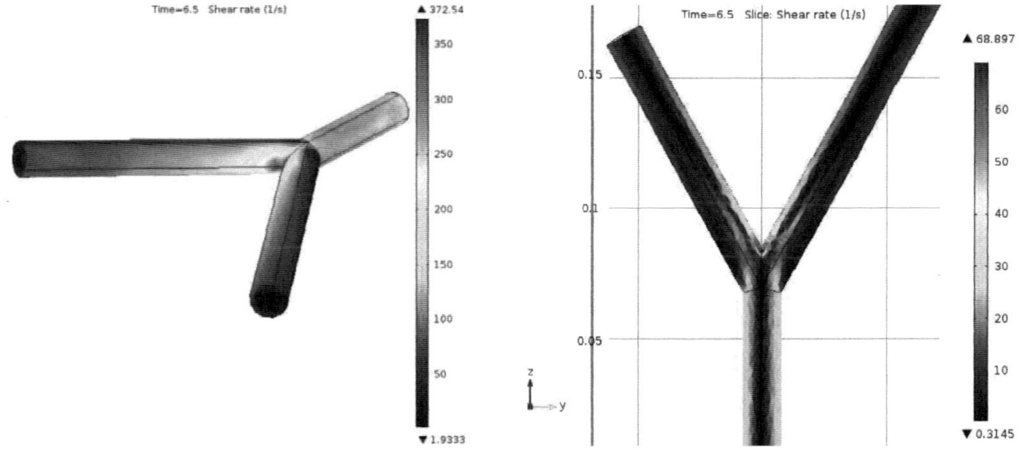

Figure 13. Left: Shear rate magnitudes on the walls of the vessel, for a given time. Right: Shear rate magnitudes for a given time, in the plane x=0.

5. Discussion

Near-physiological flow and pressure patterns can be obtained for different kind of arterial bifurcations by means of adjusting compliant tube lengths and hydraulic resistances as well as the Cardiobot pulsation patterns. This kind of flow and pressure patterns are not strictly necessary if the intention is a biomechanical characterization of the arterial wall. However, realistic flow and pressure patterns are necessary to study fluid-structure interactions in scenarios as near as possible to in-vivo ones.

Working in a steady pulsated regime with a time reference signal allowed us to take different physical measurements in the same point of the arterial sample locating there, at different time instants, different sensors and relating the signals recorded to the same phase of the periodic regime. This may help to overcome some experimental difficulties associated with the size of and possible interference between different sensors. The duration of the whole experiment must be short enough so that the properties of the bifurcation don't change significantly.

Although it is not completely equivalent to a real bifurcation, physical models like the one presented here, surgically constructed from a fresh artery, enables us to change at will the bifurcation angle, the segment lengths and the tapering.

The dimensions facilitated the execution of the in-vitro measurements, which can be later compared with digital simulation results. Inner radius of common segment, input coupler and the straight tubing are similar enough to minimize entrance lengths effects.

In-vitro experimental measurements afforded realistic waveforms of pressure that were used as boundary conditions at the entrance and exits of the bifurcation CFD model.

Velocity profiles registered at the common segment during in-vitro measurements, shows a time variation in their widths due to the radius variation of the arterial wall, as consequence of pulsed pressure and elastic compliance. Centerline velocities obtained in-silico (210 to 270 mm/s) for the common segment is greater than the one measured in-vitro (90 to 211 mm/s). This difference must be assessed taking into account experimental and numerical computation errors, pulsed doppler measurement limitations in resolution, wall radius variation, energy dissipation in the arterial tissues, and the rigid wall condition of the simplified CFD model.

The presented hemodynamic workbench simulator will enable us to work with real arterial bifurcations and to use the experimental data as inputs in realistic digital simulation of fluid-structure (FSI) interactions. This kind of digital simulation requires a knowledge of the biomechanical wall properties. These can be obtained by adjusting the parameters of a suitable viscoelastic model of the arterial wall to pressure-diameter data [8] [12] obtained in-vitro using the work bench simulator and real arterial samples.

Properly fine-tuned in-silico CFD+FSI digital simulation based on in-vitro experimental data allows the determination of physical magnitudes and hemodynamic indexes (like oscillatory shear index (OSI), temporal wall shear stress gradient (TWSSG), spatial wall shear stress gradient (SWSSG), etc. [8] [13]) that can't be obtained easily in experimental in-vitro measurements.

6. References

[1] Suárez-Bagnasco D, Armentano RL, Balay G, Cymberknop L, Brum J, Bia D, Negreira C, 2013 Measurement system for an in-vitro characterization of the biomechanics and hemodynamics of arterial bifurcations, *J. Phys: Conf. Ser.* **421** 012018.

[2] Brum J, Bia D, Benech N, Balay G, Armentano RL, Negreira C 2010 Setup of a cardiovascular simulator: application to the evaluation of the dynamical behavior of atheroma plaques in human arteries, *Phys. Proc.*, **3** 1095-1101.

[3] Balay G, Brum J, Bia D, Armentano RL, Negreira C 2010, Improvement of artery radii determination with single ultra sound channel hardware & in vitro artificial heart system, *Conf. Proc. IEEE Eng. Med. Biol. Soc.*, pp. 2521-4.

[4] Bertram CD, Pythoud F, Stergiopulos N, Meister JJ 1999 Pulse wave attenuation measurement by linear and nonlinear methods in nonlinearly elastic tubes, *Medical Engineering and Physics*, vol **21**, pp 155–166.

[5] Nichols W, O'Rourke M, Vlachopoulos Ch 2011 *McDonald's blood flow in arteries: theoretical, experimental and clinical principles*, (London: Hodder-Arnold).

[6] Armentano RL, Bia D, Cabrera Fischer E et al. 2006, An in-vitro study of cryopreserved and fresh human arteries: a comparison with ePTFE prostheses and human arteries studied non-invasively in vivo, *Cryobiology*, 52(1), pp17-26.

[7] Balay G 2012 Elasticity in arterial tissues, design of an in-vitro artificial heart and a new ultrasonic method of arterial elesticity assessment, *M.Sc. thesis (Physics), LAU-FCIEN-UdelaR.*

[8] Chandran K, Rittgers S, Yoganathan A 2012 *Biofluid mechanics: the human circulation,* 2nd ed, (Boca Raton: CRC Press).

[9] Gamero LG, Armentano RL, Barra JG, Simon A, Levenson J 2001 Identification of arterial wall dynamics in conscious dogs, *Exp Physiol*, vol **86** pp 519-528.

[10] Press W, Teukolsky S, Vetterling W, Flannery B 2007 Numerical *recipes: the art of scientific computing*, 3rd Ed., (NY: Cambridge Univ. Press).

[11] Anderson J, 1995 *Computational fluid dynamics: the basics with applications*, (NY: McGraw-Hill).

[12] Fung YC 1981 *Biomechanics: mechanical properties of living tissue*, (NY: Springer-Verlag).

[13] Hsiai T, Blackman B, Jo H 2010 *Hemodynamics and mechanobiology of endotelium*, ed T Hsiai, B Blackman, H Jo, (Singapore: World Scientific).

Acknowledgements

The first author acknowledges ANII for grant BE_POS_2010_1_2588 and PEDECIBA for research support.

FPGA-BASED PHASE AND MODULE IMPEDANCE METER

M A Gómez López[1,2], F A Migliorino[1], C B Goy[1,2,3], J B Acuña[1], M C Herrera[2,3]

[1]Departamento de Electricidad, Electrónica y Computación, Facultad de Ciencias Exactas y Tecnología, Universidad Nacional de Tucumán, Av. Independencia 1900, Tucumán, Argentina
[2]Laboratorio de Investigaciones Cardiovasculares Multidisciplinarias (LICaM), Departamento de Bioingeniería, Facultad de Ciencias Exactas y Tecnología, Universidad Nacional de Tucumán, Av. Independencia 1900, Tucumán, Argentina
[3] Instituto Superior de Investigaciones Biológicas (INSIBIO), Consejo Nacional de Investigaciones Científicas y Técnicas (CONICET), Chacabuco 461, Tucumán, Argentina.

E-mail: mherrera@herrera.unt.edu.ar

Abstract. A conductance catheter system measures the left ventricle volume by measuring the blood conductance in the cavity. It is known that there are some limitations and errors related to these assumptions. New trends claim that a nonlinear conductance-volume relationship is observed and volume measurement not only has a resistive behavior but also has a capacitive component. It is exclusively due to the myocardial muscle; even further, myocardial impedance varies in both phase and magnitude between end systole and end diastole decreasing the accuracy of volume measurement. The incorporation of an appropriate technology catheter system based on FPGA is proposed allowing measure impedances instead of conductances. Thus, the system can measure impedance module and phase in order to accurately estimate the left ventricle blood volume. The system is digitally designed and evaluated with known pattern impedances with electrical characteristics similar to those presented in the left ventricle (module range of 6.8Ω to 47Ω and phases from 2° to 90°).

1. Introduction

The development of the conductance catheter method has enabled the evaluation of left ventricular function in the human heart in vivo [1-3]. This method allows to measure real time ventricular conductance which becomes, using a conversion equation, in the left ventricular volume; its process includes a series of approximations. The first linear classic conductance-to-volume conversion equation was proposed by Baan et al. in 1989 [1]; this widely used linear equation assumes that the catheter-generated electric field is homogeneous inside the ventricle. Subsequently, a second conversion equation has been presented by Wei *et al* [4] in which a nonlinear conductance-volume relationship is presented in murine hearts using a single catheter section. This research group showed that if the measured chamber is not a symmetric cylinder and/or the catheter is not placed exactly along the central longitudinal axis of the chamber, linear classic equation restricts the accuracy of the volume estimation [5-6]. Furthermore, the accuracy of these equations is also determined by the contribution of the myocardium [7-8], known as the "parallel conductance Gp". Wei's research demonstrated that Gp varies in both phase and magnitude between end systole and end diastole. Using this concept, they proposed that both the capacitive and the resistive properties of the myocardium were substantial and neither should be ignored [9]. Hence, they proposed to measure the left ventricle admittance (magnitude and phase) instead the conductance. However, all these investigations were conducted in mouse hearts using a single section catheter. There is a lack of Gp studies in large or dilated hearts. To elucidate these aspects, it is necessary to build a system able to measure admittance (magnitude and phase) using a catheter for healthy and/or dilated human hearts. Considering that our research group has recently developed a based-FPGA (Field Programmable Gate Arrays) digital left ventricular volume system [10], this paper proposes to incorporate the phase and magnitude impedance measurement for each of five sections conductance catheter and to evaluate the performance of the system using known resistances and capacitances.

2. Theoretical background

2.1. Ventricular volume measurement

It has been stated that ventricular impedance/admittance is not purely conductive; instead it is in part capacitive. Since blood can be modeled with a resistance (R), the capacitance is a consequence of myocardium, which can be modeled as a parallel R-C impedance. Myocardium contribution can be known by measuring both: ventricular admittance phase (Φ_i [degree]) and module ($|Ymeas_i|$ [S]) [6] The phase values that result from myocardium contribution goes from 5° to 60° approximately, and myocardium admittance (Y_{mi} [Ω^{-1}]) can be calculated as follow:

$$Cm_i = \frac{|Ymeas_i| * \sin \emptyset_i}{2\pi f} \tag{1}$$

$$gm_i = \frac{Cm_i \times \sigma_m}{\varepsilon_m} \tag{2}$$

$$Ym_i = gm_i + j2\pi f Cm_i \tag{3}$$

where *Cmi* [F], σ_m, ε_m, gm_i [S] represent the myocardium capacitance, conductivity, permittivity and conductance respectively and f [Hz] is the frequency of the signal to be introduced into the body. The index i=1,…,5 belongs to each catheter section.

If myocardium admittance is determined, blood conductance (gb_i [S]) can be obtained using equation (4).

$$gb_i = |Ymeas_i| * cos\emptyset_i - gm_i \tag{4}$$

where gm_i [S] is obtained for each catheter section, ventricular volume can be estimated using any of the two equations that follow: a) Baan traditional equation [1] or b) Wei equation [4].

The first equation assumes that the electric field generated by the catheter system is homogeneous, and ventricular volume can be calculated using equation (5)..

$$Vol_i = \frac{1}{\alpha} * \rho * L^2 * gb_i \tag{5}$$

In this equation Vol_i [V] represents volume instantaneous signal, α is an empirical factor used for calibration, ρ [Ω*m] is the blood resistivity and L [m] is the distance between electrodes.

The second equation solves the problem that emerges from considering a homogeneous electric field inside the ventricle, and ventricular volume can be calculated using equation (6), where Υ is an empirical factor used for calibration.

$$Vol_i = \frac{\gamma}{\gamma - gb_i} \rho L^2 gb_i \tag{6}$$

The equations (1) to (6) are solved using a commercial software (LabView 2009, National Instruments Inc.), which receives ventricular admittance phase and module measured for each catheter section.

2.2. Signal characteristics

By means of a multi-electrode catheter, a current is injected into the left ventricle cavity, and the voltage between each pair of electrodes is measured (figure 1).

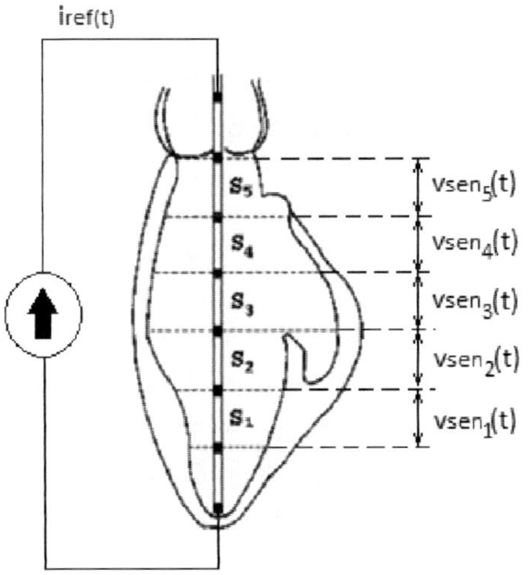

Figure 1. The reference current (*Iref(t)*) is injected between outer electrodes, and a voltage signal is measured between each pair of electrodes ($Vsen_i(t)$).

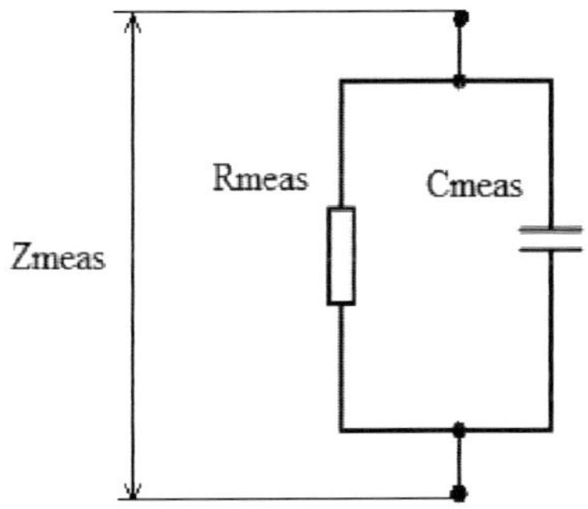

Figure 2. Circuital model of the impedance measured at each catheter section.

The injected current *Iref(t)* can be expressed by means of equation (7), where |*Iref*| is its module and *fref* [Hz] is the working frequency.

$$Iref(t) = |Iref|\cos(2\pi\, fref\ t) \tag{7}$$

Ventricular impedance (Z_{meas}), presented by equation (8), can be modeled as a parallel R-C circuit (figure 2) where the resistive part (R_{meas}) represents the sum of blood and myocardium resistances while the capacitive part (C_{meas}) model the cardiac muscle capacitance. Its reciprocal is the ventricular admittance (Y_{meas}) -equation (9)-. Ventricular impedance module and phase can be calculated using equations (10) and (11), respectively.

$$Z_{meas} = R_{\mathrm{meas}} // \left(\frac{1}{j2\pi f_{ref} C_{meas}}\right) = \frac{R_{\mathrm{meas}}}{j2\pi f_{ref} R_{meas} C_{meas}+1} \tag{8}$$

$$Y_{\mathrm{meas}} = \frac{1}{Z_{\mathrm{meas}}} = \frac{1}{|Z_{\mathrm{meas}}|} \angle -\emptyset_{meas} \tag{9}$$

$$|Z_{\mathrm{meas}}| = \frac{R_{meas}}{\sqrt[2]{1+(2\pi f_{ref} R_{\mathrm{meas}} C_{\mathrm{meas}})^2}} \tag{10}$$

$$\emptyset_{\mathrm{meas}} = -\tan^{-1}(2\pi f_{ref} R_{\mathrm{meas}} C_{meas}) \tag{11}$$

If R_{meas} and C_{meas} are varying during cardiac cycle; hence, Z_{meas}, $|Z_{meas}|$ and Φ_{meas} are also variable signals.

The current *Iref* flowing through the impedance Z_{meas} produce the voltage V_{sen} represented as a phasor and as function of time in equations (12) and (13), respectively.

$$V_{sen} = I_{ref} Z_{meas} = |I_{ref}||Z_{vent}| \angle \emptyset_{meas} \tag{12}$$

Journal of Physics: Conference Series **477** (2013) 012006 doi:10.1088/1742-6596/477/1/012006

$$v_{sen}(t) = |I_{ref}||Z_{\text{meas}}|\cos(2\pi f_{ref}t + \emptyset_{meas}) \tag{13}$$

$|Z_{meas}|$ can be modeled as a periodic signal with a fundamental frequency f_m equal to cardiac frequency, plus a constant value that is never zero. So, $V_{sen}(t)$ is an amplitude modulated signal where *Iref* is the carrier and $|Z_{meas}|$ is the modulating one.

3. System description

3.1. General scheme

The equipment acquires and processes signals from the five sections of the catheter and delivers voltage values that are proportional to the impedance modulus and phases (of each catheter section). Finally, it transmits the data to a PC using a LabView interface where admittances and volumes are calculated (figure 3).

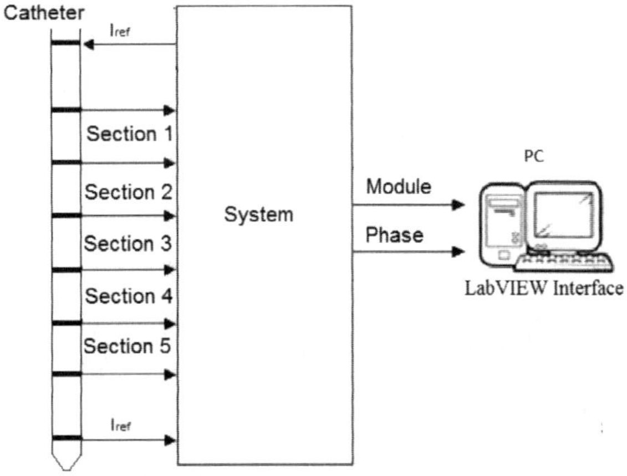

Figure 3. Connection diagrams (multielectrode catheter, FPGA-based system and PC). Iref, current flowing between outer electrodes. Each section voltage is processed and conditioned for transmission to the PC.

3.2. FPGA-based system

Figure 4 shows the schematic of the phase and module meter. It is composed of two parts, a digital one implemented in a FPGA and an analog one which is in contact with the biological element.

FPGA generates the digital reference current Iref(t) using a 12 bits DAC (Digital-to-Analog Converter) and a signal conditioner. Iref(t) is injected into the left ventricular between outer catheter electrodes. The signals at each catheter section are amplitude-modulated (AM) voltage signals, where the carrier is Iref(t) and the modulating signal is the biological signal. The AM signals, prior isolation, are digitized using an ADC (Analog-to-Digital Converter) and then they enter into FPGA for processing. The digital stage is a logical design configured in an Altera FPGA (FLEX 10K 70RC240; Altera Corp.) The logic equations are implemented with Quartus II version 9.0 development system (Web Edition). The design is carried out by combination of schematic and VHDL hardware description language forms. The logical design does the following tasks: 1) generate a 12-bits digital carrier signal using a ROM; 2) perform phase detection by measuring the time difference between the zero-crossing points between Iref(t) and Vsen(t); 3) calculate the phase; 4) demodulate Vsen(t) in order to detect the impedance module, and finally, 5) control the 12-bits ADC and the Analog Multiplexer.

Furthermore, analog stage is composed by: 1) a 12-bits DA R-2R converter (12-bits DAC); 2) a signal conditioner that consists of a 4th order band-pass filter with Sallen-Key structure, a multi-electrode catheter and an integrated 12-bits AD converter (12 bits ADC).

The Instrumentation amplifiers and Analog Multiplexer blocks are part of the previous system developed by Gómez López [2].

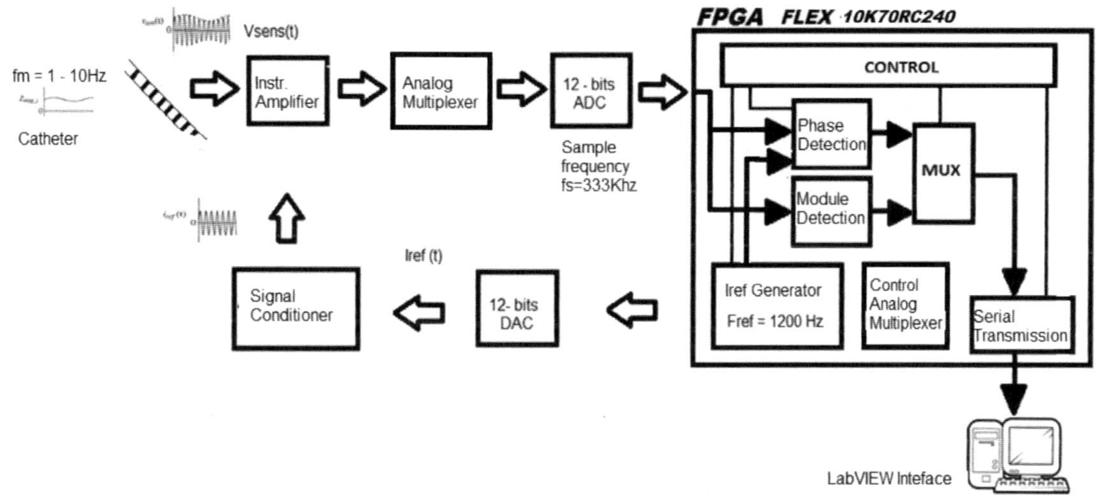

Figure 4. Schematic diagram for phase and module impedance measurement.

3.3. Analog stage

3.3.1. Precision 12 bits digital to analog converter (12 bits-DAC)
The 12-bit DAC is a resistor ladder network that uses only two resistance values (R and 2R). For its implementation three CD4053BC analog multiplexers (Fairchild) and several electronic switches are used. One input is connected to the reference voltage (5V) and the other is grounded (0V). In order to adapt the voltage levels between the FPGA (0V for a '0' logic level and 3.3V for a '1' logic level) and the multiplexers, three CD4071BC gates are used (National Instr.).

12-bit DAC has a 2mV resolution and a 11.8mV offset error. The ladder resistor network consists of several precision resistors of 2KΩ and 1KΩ with a 1% tolerance.

3.3.2. Signal Conditioner
The DAC output is a constant amplitude and 1200Hz frequency signal. It has high frequency harmonics introduced by the switching of the multiplexers and other components that make up the DA converter. So, this signal is entered to a band-pass filter with $fc_{LOW\text{-}PASS}$=3600Hz and $fc_{HIGH\text{-}PASS}$ = 400Hz.

For the low and high-pass filters design, a 2nd Order Butterworth filter with Sallen-Key structure is chosen, which results in a 4th order band-pass filter. Each filter is implemented using two LF353 operational amplifiers (Fairchild) and a RC resistor network.

3.3.3. Analog to digital converter (ADC)
The analog to digital converter is implemented using a 12 bits ADS7800 (Burr-Brown) integrated circuit and a 3us sampling time. It is a successive approximation AD converter with a sample and hold circuit and two ± 5V or ± 10V analog inputs. The RC (Read / Convert) active-low input signal initiates the conversion. While the conversion is performed, the input lines are disabled, the output data becomes high impedance and BUSY output turns to low. When data is available BUSY output turns to high level which indicates to the control logic that a binary word can be read.

The frequency of the analog signal that enters the converter is fref = 1.2KHz and the sampling rate is 333kHz, therefore the minimum phase that can be measured is:

$$\emptyset_{MIN} = \frac{T_S}{Tref} \times 360° = \frac{3\mu s}{834\mu s} \times 360° \tag{14}$$

$$\emptyset_{MIN} = \frac{3\mu s}{834\mu s} \times 360 \approx 2° \tag{15}$$

3.4. Digital stage

Figure 5 depicts the logic blocks of the digital stage that are implemented on the EPF10K70RC240 device (Altera). The 25.175MHz clock signal (Clock) is provided by a crystal oscillator from an educational board (UP2 Education Board, Altera). Iref [11 ... 0] is a 12-bits output bus that reconstructs a 1.2KHz sinusoidal signal using a 12-bit DA converter and provides the current injected into the catheter. VAM [11 ... 0] is a 12-bit bus coming from the output of the ADS7800 analog to digital converter. VAM represents the sensed and digitized AM signal at each catheter section. BUSY is an input digital signal that comes from the 12-bits ADC and indicates to the control logic that conversion data are ready to be read. RC is the output signal that comes from digital interface and indicates the start of the 12-bits ADC conversion. This signal is synchronized with the clock signal. Cmux [2 ... 0] is a 3-bits bus and it is connected to the CD4097 Analog Multiplexer (Texas Instruments) control inputs. This control inputs allows multiplexing the five channels of the catheter. TX is an output signal needed for RS232 serial transmission.

The ROM MEMORY generates the 12 bits current (carrier). The 333 kHz FREQUENCY DIVIDER generates a 333 KHz synchronic signal using a 25.125 MHz external clock. ADC CONTROL indicates the start of conversion. ANALOG MUX CONTROL commands the control input of a CD4097 analog multiplexer. INPUT stores VAM input signal temporarily, until new data are ready, besides this block works in synchronization with the ADC CONTROL block. PHASE detects the zero-crossing of the reference signal (Iref(t)) and VAM voltage signal, and then this block calculates the phase difference among them. MODULE demodulates VAM by means of positive peak detection. MULTIPLEXING AND TRANSMISSION condition the phase and module signals of the impedances presented at each catheter section for serial transmission to the PC.

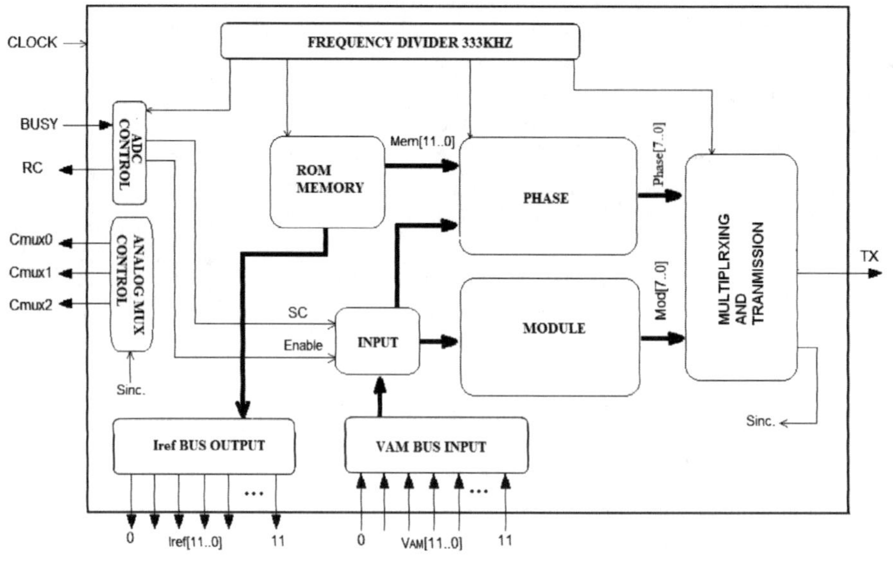

Figure 5. FPGA digital stage.

3.4.1. Phase detection

Phase detection is based in the method of zero-crossing detection. Having determined the zero crossings of the Iref [11 ... 0] reference signal and VAM [11 ... 0] sensed signal, this block is responsible for calculating the phase between the two signals. It consists of a JK flip-flop and a counter, when it detects a zero in Iref is activated the set input of the flip-flop and when it detects a zero in VAM is activated reset input of the flip-flop. The Iref [11 ... 0] signal is taken as a reference to measure the phase. The output of the flip-flop is the enabler of an 8-bits counter with count up. The account obtained while is enabled is proportional to the time between the zero crossings of the two signals.

3.4.2. Detection Module

Module detection consists in demodulating VAM [11 ... 0] by means of detecting the peaks of the modulated Iref[11 ... 0] signal. An algorithm that compares three VAM consecutive points and stores the larger one is implemented. This comparison is carried out using two magnitude comparators (Library Parameterized Module Quartus) that compare the magnitude of a point with the magnitude of the other two ones. A moving average filter is used to prevent false peaks detections due to noise or interference from other systems.

3.4.3. Multiplexer and transmission

Once the voltage values proportional to module and phase of each of the five catheter sections are determined, they are multiplexed and serially transmitted to the PC. A RS232 serial transmission is used for communication between the FPGA device and PC. This transmission allows asynchronous communication between the two devices, but a previous RS232 to USB conversion is needed. The duration of each bit transmission is determined by the speed at which the data transfer is performed. The digital multiplexer and transmitter were developed in VHDL language and they work synchronously with a 9600Hz frequency.

Transmission sequence is the next one: phase and impedance module of the first section, phase and impedance module of the second section, and this sequence continues until the module and phase values of the five impedances presented at each catheter sections are transmitted. This cycle is continuously repeated.

4. Phase and module impedance meter evaluation

System evaluation is implemented in three parts. First, the phase difference between I_{ref} and VAM is measured. In order to do this a phase shifter circuit is used, phase values are obtained and visualized in the PC by means of the software implemented in LabView and results are compared with others obtained using an digital oscilloscope. Second, the ability of the system to demodulate VAN and to detect signals peaks is evaluated. Besides, a frequency sweep of the modulating signal is carried out in order to assess the system bandwidth by measuring demodulated signal amplitude and frequency. Finally, VAM phase and module obtained, using the system, for known pattern impedances (parallel R-C circuits) are compared with the real values.

4.1. Phase measurement

The measurement is performed using a phase shifter circuit, that consist of a TL081 operational amplifier (Texas Instruments) and a R-C network, in order to obtained phase differences that goes from 2° to 150°. Phase values are obtained and visualized in the PC by means of the software implemented in LabView and they are compared with others obtained by means of observing Lissajous figures on a two channels digital oscilloscope. One channel is connected to the phase shifter input and to Iref(t), while the other is connected to phase shifter output Iref(t+phase) an to Vsen(t). Results are presented in figure 6.

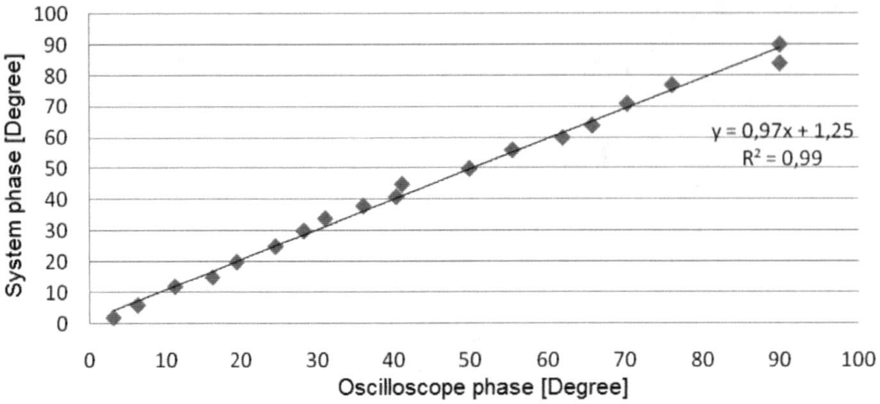

Figure 6. System vs Oscilloscope phase measurement.

The correlation value of 0.99 implies that the system has a linear response regarding to phase measurements made with the oscilloscope.

4.2. Module measurements

Measurement modules are implemented by means of an AM modulator and a phase shifter. Module measurements are carried out using an amplitude modulated signal which is generated with characteristics (amplitude and frequency) similar to those of biological signals. The AM signal is then processed and visualized on the PC. Besides, impedance frequency and module are measured with the system and with the oscilloscope are compared.

The AM modulator consists of an analog multiplier MC1495 (Motorola Semiconductor). Iref(t) is phase shifted and enters the modulator as a carrier signal. A signal generator GWINSTEK GFG – 8219 produces a signal which is similar to a biological one, this signal is used as modulating signal. A frequency sweep of the modulating signal is carried out, and the maximum and minimum frequency at which impedance module can be determined without information loss is determined. The relative error of the obtained module value with regard to the oscilloscope result is less than 6% when frequency goes from 1 to 90Hz; beyond this frequency, the relative error rises to values higher than 13%.

With respect to modulus measurement, minimum and maximum peaks that are measured by the system are similar to those which that are measured by oscilloscope. This situation does not change with frequency variations.

4.3. Bench test using known pattern impedances

Identical pattern impedances are connected to each one of the catheter sections and their phase and module are measured. This pattern impedances are adopted considering that biological conductance goes from $0.022\Omega^{-1}$ to $0.150\Omega^{-1}$, and that the phase shifts that result from myocardium capacity goes from 5° to 60°. Table 1 and table 2 show voltage values that represent pattern impedances phase and module measured at each catheter section. Measurements are performed using a 95mA current and the results are displayed on the PC using the software implemented in LabView.

The 0.99 and 0.91 correlations allow us to affirm that modules and phases are directly proportional to impedance modulus and phase respectively in each section of the catheter.

Table 1. Modules measurement for each one of the five sections

Pattern impedance Zp	Module [Ω]	Section 1 [V]	Section 2 [V]	Section 3 [V]	Section 4 [V]	Section 5 [V]
Zp1	18	1,86	1,86	1,86	1,86	1,86
Zp2	18	1,72	1,72	1,72	1,72	1,72
Zp3	47	4,90	4,90	4,90	4,90	4,90
Zp4	6,8	0,88	0,88	0,88	0,88	0,88
Zp5	33	2,43	2,43	2,43	2,43	2,43

Table 2. Phase measurement for each one of the five sections

Pattern impedance Zp	Phase [degree]	Section 1 [V]	Section 2 [V]	Section 3 [V]	Section 4 [V]	Section 5 [V]
Zp1	-16	-16	-15	-15	-16	-15
Zp2	-30	-31	-31	-31	-32	-32
Zp3	-10	-13	-13	-13	-13	-12
Zp4	-6	-7	-7	-7	-8	-7
Zp5	-49	-48	-48	-48	-49	-48

5. Discussion and conclusions

A FPGA based system that can measure impedance phase and module of each section of a conductance catheter system has been built. The functionality of the digital and analog stage has been evaluated with laboratory tests using 6.8Ω to 47Ω input impedances at each section. These impedances have 6º to 49º phases, although the system detects phases from 2º to 90º. The high correlation values - of 0.97 and 0.91- for phase measurements demonstrates the system linear response. It is concluded that the system is suitable for acquisition and sensing of impedance signals and for processing and calculating the module and phase of them.

For small angles (2° to 5°), the percentage error is greater because the measurement is near of the minimum value that can be measured. This behavior can be improved by increasing the number of samples needed to digitize the sensed input voltage. For low impedance modules, the percentage error is greater because the signal/noise ratio is increased. This behavior can be improved through high quality input digital filters.

A complete system evaluation requires impedance measurements (module and phase) from real biological impedances.

6. References

[1] Baan J, Van Der Velde E T, Steendijk P and Koops J 1989 Calibration and application of the conductance catheter for ventricular volume measurement. Automédica.11 357-65

[2] Schreuder JJ, Biervliet JD, van der Velde ET, ten Have K, van Dijk AD, Meyne N, Baan J. Systolic and diastolic pressure-volume relationships during cardiac surgery. J Cardiothor Vascui Anesth 1991;5:539-545.

[3] Senzaki H, Chen C, Kass DA. Single-beat estimation of end-systolic pressure volume relation in humans. Circulation 1996;94:2497-2506.

[4] Wei C L, Valvano J W, Feldman M D and Pearce J A 2005 Nonlinear conductance-volume relationship for murine conductance catheter measurement system. IEEE Trans. On Biomedical Engineering.52 10 1654-61

[5] Wei C L, Valvano J W, Feldman M D, Nahrendorf M, Pearce JA 2003 3D Finite Element complex Domain Numerical Models of Electric Fields in Blood and Myocardium. Annual Int. Conf. of the IEEE Engineering in Medicine and Biology – Proc.1 62-5

[6] Wei C L and Shih M H 2009 Calibration capacity of the conductance-to-volume conversion equations for the mouse conductance catheter measurement system. IEEE Trans. On Biomedical Engineering.566 1627-34.

[7] Wei C L and Wu P Y 2008 Investigation of mouse conductance catheter position deviation effects on volume measurements by finite element models.Proc. of the 30th Annual Int. Conf. of the IEEE Engineering in Medicine and Biology Society, EMBS'08 - "Personalized Healthcare through Technology" 2008, Article number 4649427 1399-1402.

[8] Wei C L, Kan C D, Wang J N, Wang Y W, Chen C H and Tsai ML 2011 Does Conductance Catheter Measurement System Give Consistent and Reliable Pressure–Volume Relations in Rats? IEEE Trans. on Biomedical Engineering 58 6

[9] Wei C L, Valvano J W, Feldman M D, Nahrendorf M, Peshock R, Pearce J A 2007 Volume catheter parallel conductance varies between end-systole and end-diastole. IEEE Trans. On Biomedical Engineering 54 8 1480-1489

[10] Gómez López M A 2008 Equipo para mediciones en el sistema cardiovascular: Tecnología FPGA aplicada a la determinación de volúmenes cardíacos Tesis Maestría (Tucumán: Universidad Nacional de Tucumán)

Acknowledgments

This work was supported in part by grants from the UNT Research Council (CIUNT), 26/E422-1 Research Program and INSIBIO (CONICET).

Noninvasive home telemonitoring for patients with heart failure: A review of medical consensuses

L M Yanicelli[1], C B Goy[1,2,3], M A Gómez López[2] and M C Herrera[1,3]

[1] Departamento de Bioingeniería, Facultad de Ciencias Exactas y Tecnología, Universidad Nacional de Tucumán, Av. Independencia 1900, Tucumán, Argentina.
[2] Departamento de Electricidad, Electrónica y Computación, Facultad de Ciencias Exactas y Tecnología, Universidad Nacional de Tucumán, Av. Independencia 1900, Tucumán, Argentina.
[3] Instituto Superior de Investigaciones Biológicas (INSIBIO), Consejo Nacional de Investigaciones Científicas y Técnicas (CONICET), Chacabuco 461, Tucumán, Argentina.

E-mail: luciayanicelli@gmail.com

Abstract. Heart Failure (HF) is the leading cause of death in Western World, producing the highest number of hospitalizations in patients over 65 years. These continuous hospitalizations generate high costs to the health system. Studies point out that multidisciplinary post-discharge treatment programs, i.e. telemonitoring or structured telephone support, can avoid a large portion of readmissions related to HF, with benefit also in mortality, length of hospital stay and quality of life. Considering the impact of telemonitoring in HF, this paper aims to identifying the different world opinions and recommendations of the participants in the management, treatment and monitoring of HF. This purpose is achieved by studying the different HF consensuses. From this study it was possible to detect a number of aspects (parameters to be monitored, health education) that should be considered when a telemonitoring system for HF patients is designed.

1. Introduction

Cardiovascular diseases (CVD) are the leading cause of death in Western World. They provoke 40% of all deceases [1]. Meanwhile, in Argentina 33% of deaths from known causes are caused by CVD, while Heart Failure (HF) is the first responsible entity, this situation has remained stable over the last 20 years. HF caused 33422 deaths in 2011 in Argentina, which represents about 35% of the deaths from diseases of the circulatory system [2, 3].

All studies indicate a substantial economic impact of this entity in the health systems of most countries. The growing concern about various aspects of this clinical syndrome is reflected in part by the numerous publications mainly from the United States and Europe. Congestive Heart Failure (CHF) is the most common medical reason for inpatient hospitalization in adults over age 65 and reducing the rate of hospital admission in this patient population has become a national priority in the U. S. [4]. This is because the high costs generated by continuous hospitalizations to the health system; projections show that by 2030, the total cost of HF will increase almost 120% to $70 billion from the 2013 estimated total cost of $32 billion [5]. In Argentina high costs are generated because of hospitalizations, but especially by readmissions of patients for worsening HF, which constitutes 70% of all admissions for HF [6].

Common reasons for HF rehospitalizations include delays in symptom recognition, medication and dietary noncompliance, and lack of knowledge and skills for competent self management [7]. Studies point out that multidisciplinary post-discharge treatment programs, i.e. telemonitoring or structured telephone support, can avoid a large portion of readmissions related to HF, with benefit also in mortality, length of hospital stay and quality of life [8]. In addition, for many patients, frequent clinic visits may be impeded by different barriers such as transport, cost, or other diseases. That is why, in these situations, the telemonitoring systems are attractive options for closely monitoring the first signs of patients decompensation. A 2013 review showed a significant CHF-related re-admission benefit from telemonitoring and speculated this intervention may be of particular benefit for patients with physical, geographic, or socio-economic barriers to specialty care. At the same time this work indicated that recognizing the potential benefits to be realized through the expanded use of telemedicine, health insurers are developing new business models incorporating telemonitoring as a component of comprehensive chronic disease management programs [4].

Because of the impact of telemonitoring in HF, this paper aims to identifying the different world opinions and recommendations of the participants in the management, treatment and monitoring of HF. From this study a number of useful guidelines to be considered when designing a telemonitoring system emerge.

2. Theoretical background

To understand and evaluate worldwide existing opinions a series of instruments called consensuses are used. A consensus is an agreement that is reached around a theme by the consent between the members of a group or between groups. The medical consensuses are useful since they have the potential to promote the appropriate use of procedures and generally to improve efficiency in the provision of healthcare.

Most HF consensus recommendations provided by the experts were classified according to two criteria which are detailed in tables 1 and 2. A first classification is an evaluation of the degree of agreement between panelists (table 1), while the second one depends on the level of evidence (table 2).

Table 1. The class of recommendation.

Class I	Evidence or general agreement that a given procedure or treatment is beneficial, useful and effective.
Class II	Conflicting evidence or a divergence of opinion about the usefulness or efficacy of the procedure or treatment.
Class IIa	Weight of evidence is in favour of usefulness or efficacy
Class IIb	Usefulness or efficacy is less well established by evidence or opinion.
Class III	Evidence or general agreement that the procedure or treatment is not useful or effective and in some cases may be harmful.

Table 2. The level of evidence.

Level of Evidence A	Data derived from multiple randomized clinical trials or meta-analyses.
Level of Evidence B	Data derived from a single randomized clinical trial or nonrandomized studies.
Level of Evidence C	Consensus of opinion of experts and/or small studies.

3. Methods
In early 2013 HF consensuses have been searched in the following databases: Annual Reviews, EBSCO, Nature.com, Ovid, Sage Journals, Science Direct, Springer Link, Wiley-Blackwell and in a general purpose search engine. Keywords used in the search were "heart failure consensus" in Spanish and in English. This search produced a series of works from major national and international societies and scientific organizations that have been published in different journals. The first selection criterion was to study only the consensuses published from the year 2000, because the previous works did not mention the telemonitoring. Another adopted criterion consisted in to exclude papers which had updates. From the analysis of the titles and abstracts of papers, several works that studied the HF from an exclusively clinical perspective (such as the incidence of certain drugs in patients or recommendations for cardiac resynchronization therapy, among others) were excluded.

After the selection and exclusion process, eleven consensuses that meet the requirements listed above were identified. These consensuses selected were classified into two groups according to the type of recommendation they provide: consensuses of general recommendations for telemonitoring and specific consensuses. Four specific consensuses were found. They deal with issues that should be considered in the telemonitoring systems design, so they will be studied with main emphasis in those topics. Between the general consensuses there is one from the Heart Failure Society of America (HFSA) from 2008 [9], three from the Canadian Cardiovascular Society (CCS) [10-12], one that studies the treatment of HF in African Americans written by Franciosa J A et al. [13] and two that belong to South American societies of cardiology, one from the Argentine Society of Cardiology (SAC) [14] and one from the Uruguayan Society of Cardiology (SUC) [15,16]. The four remaining papers are specific consensuses that were studied only in some subjects such as health education [17-19], functional capacity [20] and patient's symptoms [18].

After defining the consensuses to study, their medical opinions and recommendations were analyzed. First, the opinions of the participants about the benefits of telemonitoring in HF patients were studied. Then, the parameters that should be monitored were studied by comparing the recommendations given by different consensuses. Finally, educational aspects that need to be taken into account in order to improve patient self-management were studied.

4. Findings
The findings were divided into three sub-sections.

4.1. Telemonitoring
The seven consensuses of general recommendations indicate that in order to improve patients quality of life and to detect patients decompensation, the telemonitoring and monitoring of certain parameters (blood pressure, body weight, symptoms, etc.) are useful. This is a consequence of that the HF signs and symptoms gradually appear along the disease and they need to be detected in time for proper medical attention, like medication readjustment or other actions. The aim pursued by these actions is to prevent possible patients decompensation that may result in hospital readmissions.

Most consensuses establish the need for a health monitoring of HF patients after discharge. The time suggested for first check-up varies according to the consensus: 72 hours in severe cases [9], first week [9, 16], second week [12] or fourth week [10], although this last consensus clarified that it is more beneficial sooner if feasible. The reason provided for check-ups after discharge, is the high probability in the emergence of various symptoms and discomforts that may lead to patient decompensation. While this corresponds to a first check-up, all consensuses recommend a frequent monitoring of patient throughout the disease, and they mention again that the management programs of HF (telemonitoring) is a means to achieve this aim. Some of the benefits mentioned by the consensuses about telemonitoring are detailed in table 3, but they also indicate a decrease in mortality [9, 13], reduced health care costs [14] and that the telemonitoring is a solution for those patients that are often unable to access the clinic [9, 13].

Table 3. Benefits of telemonitoring in HF patients.

	HFSA 2008 [9]	CCS 2006 [10]	CCS 2007 [11]	CCS 2012 [12]	Franciosa J A et al. 2010 [13]	SAC 2010 [14]	SUC 2000 [15,16]
Quality of life and clinical stability	X	X			X	X	X
Early recognition of symptoms	X	X	X	X	X	X	X
Hospital readmissions reduction	X	X		X	X	X	X
Adherence[a]	X	X	X			X	X
Self-care	X					X	X
Assessment treatment progress		X		X		X	X
Patient education	X	X	X	X	X	X	X
Communication with the physician	X	X			X		

[a] The medication is taken correctly, there are dietary compliance, etc.

4.2. Parameters to be monitored

This section details the parameters that the consensuses suggest to monitor in the following of HF patients.

4.2.1. Blood pressure (BP). All consensuses put great emphasis on the BP control. On one hand, this is because one of the symptoms which often occur along the HF is orthostatic hypotension, which is a cause of syncope or fainting. While on the other hand, the consensuses say that knowledge of the BP variation during the different days, allows the doctor to make a reasonable judgment about the medication dose. The table 4 shows the levels and variations in BP that are risk indicators according each consensus.

Table 4. The levels and variations in BP.

HFSA (2008) [9]	No information
CCS (2006) [10]	A fall of greater than 20 mmHg in systolic BP or greater than 10 mmHg in diastolic BP on standing (Orthostatic hypotension).
CCS (2007) [11]	BP should be controlled to less than 140/90 mmHg in most individuals and less than 130/80 mmHg in those with diabetes and/or kidney disease.
CCS (2012) [12]	No information
Franciosa J A et al. (2010) [13]	They consider African Americans with hypertension a higher-risk cohort of hypertensives and, like those with diabetes, BP should be lowered to a target level of <130/80 mm Hg.
SAC (2010) [14]	No information
SUC (2000) [15]	BP should be controlled to less than 140/90 mmHg in hypertensive patients. In diabetic patients, BP should be less than 130/85 mmHg (Class I, Level A).

The CCS consensuses from 2006 [10] and from 2012 [12] recommend measuring BP sitting and standing. This is because of the Orthostatic Hypotension can be detected from the knowledge of the BP variation in the different positions. This society indicates that supine BP should be measured after a patient has rested for 15 min. This consensus also indicates that standing BP should be measured within three to five minutes after the patient has been in this position, because BP may drop immediately after standing. Furthermore, only the CCS consensus from 2006 refers to the frequency of

BP measurement, indicating that repeated measurements at different times of the day are advisable because orthostatic hypotension is not consistently present in an individual. Orthostatic hypotension may be more likely to be observed in the morning.

4.2.2. Heart Rate (HR). Another parameter that all consensuses except one [13] suggest to monitor is the resting HR, because it may be increased although BP is low, normal or high. Furthermore, physicians indicate that HR variability is a marker of autonomic imbalance and it has been demonstrated that its alteration is also predictive of prognosis [14]. In addition, the CCS [12] recommends that HR should be closely followed when combinations of drugs affecting the renin-angiotensin-aldosterone system are used. This follow will allow the identification of the side effects the drugs can have and of the types of drugs should be given to the patient.
In regard to the normal Heart Rate range, first the SUC [15] mention that if the HR decreases to less than 50 beats/min, the suspension or at least the reduction of the dose of the drug that cause such effect must be considered. Furthermore, the CCS [11] indicates that one of the risk factors for increased mortality includes HR of more than 84 beats/min at rest.

4.2.3. Weight. All consensuses recommend weight monitoring because it is an indicator of fluid retention when it increases sharply in a few days. One of the consensuses indicates that morning weight should be daily monitored in HF patients with fluid retention or congestion that is not easily controlled with diuretics, or in patients with significant renal dysfunction (Class I, Level C) [10]. In addition, two consensuses establish a daily frequency for monitoring of weight [12, 13] and only one suggests a weekly frequency [15]. The rest of the consensuses no make reference to the frequency of measurement of this parameter.

Regarding to weight fluctuations, the SAC [14] considers that patients with a weight gain of 2 Kg or more in 3 days should be instructed to increase the dose of diuretics (Class I, Level C). While the SUC [15] indicates that if the weight gain is 2 Kg or more, the patient should consult a doctor promptly.

4.2.4. Electrocardiogram (ECG). In reference to the ECG the SAC recommended continuous ambulatory electrocardiographic monitoring (Holter) for the detection of paroxysmal tachycardia or the presence of intermittent bradyarrhythmias which cause or aggravate HF symptoms [14]. Another consensus recommends ECG monitoring as a necessary measure for the initial diagnosis of HF and to identify rhythm abnormalities (atrial fibrillation, flutter or bradycardia and ventricular tachycardias), acute coronary syndromes, RV, LV or atrial hypertrophy or strain, as well as myopericarditis. Moreover, if there are cardiac arrhythmias it should be evaluated by a 12-lead ECG and continuous ECG monitoring [11]. In no case, the rest of the consensuses mention continuous ECG monitoring.

4.2.5. Symptoms. Another aspect analyzed by the seven general consensuses and the specific consensus of Goodlin S J et al. [18] is the detection of symptoms (subjective elements perceived only by the patient), because they are an indicator of health status or patients quality of life. Some consensuses recommend using questionnaires to assess symptoms. These questionnaires have been validated and are effective especially in those patients suffering from HF. The HFSA [9] recommends using the following questionnaires: The MOS 36-item short-form health survey (SF-36), el SF-12, The EuroQol-5D, The Chronic Heart Failure Questionnaire, The Minnesota Living with Heart Failure Questionnaire and The Kansas City Cardiomyopathy Questionnaire. While, the SAC [14] suggests using the questionnaires that are listed below: The Chronic Heart Failure Questionnaire, The Yale Scale and The Minnesota Living with Heart Failure Questionnaire. The other consensuses not specify any particular questionnaire to assess symptoms.

4.2.6. Functional status. Another parameter that most of the consensuses suggest to monitor is patient functional status. While not all consensuses directly recommend perform the known functional classification the New York Heart Association (NYHA), absolutely all of them use it to provide their recommendations about exercise, medication, etc. This is because the functional classification has an important prognostic value and is used as the decisive criterion in the choice of certain therapeutic interventions, both medical and surgical. With the periodic evaluation of the patient functional class, doctors can know the patients progress and their response to treatment. Besides this classification, there are other tools used and/or recommended by the consensuses to assess functional status as shown in table 5.

Table 5. Tools used and/or recommended by the consensuses to assess functional status.

	NYHA	Ergometry	6MWT[a]	CPX[b]	IPAQ[c]	Duke Activity[d]	V'O2 Max[e]	Pedometers[f]
HFSA (2008) [9]	X		X	X	X	X		X
CCS (2006) [10]	X		X	X				
CCS (2007) [11]	X							
CCS (2012) [12]	X							
Franciosa J A et al. (2010) [13]	X	X						
SAC (2010) [14]	X	X	X				X	
SUC (2000) [15]	X	X	X	X				
Piepoli M F et al. [20]	X	X	X	X				

[a] Six-minute walk test

[b] Cardiopulmonary exercise testing

[c] The International Physical Activity Questionnaire

[d] Duke Activity Index scale

[e] Maximal oxygen consumption

[f] Physical activity detectors

4.3. Health education

All consensuses underline the importance of proper health education to patient, family and caregivers. This is necessary in order to achieve the correct self-care and thereby avoid diverse problems which are generated because of ignorance of caregivers on how to deal with common situations of this kind of illness.

Only the HFSA focuses specifically on health education [17]. This consensus defines health literacy as the degree to which individuals have the capacity to obtain, process, and understand basic health information and services needed to make appropriate health decisions. This health education includes the ability to: comprehend complex vocabulary, share personal information with health care providers, make decisions about healthy lifestyle habits, and participate in self care and chronic disease management while navigating a complex health care system. Furthermore, this consensus mentions a number of studies that conclude that ''low literacy is associated with several adverse health outcomes'', including increased risk of hospitalization and lack of knowledge about health services. In summary, the HFSA indicates that limited health literacy is associated with decreased knowledge of one's medical condition, poor medication recall, nonadherence to treatment plans, poor self-care behaviours, poorer physical and mental health, increased hospitalizations, and increased mortality. This society also state that instruct patients about t good health practices can help them to improve self-care and achieve positive health benefits.

SAC consensus [14] mentions that the literature suggests that only 20-60% of HF patients adhere to treatment, therefore the patients and their family should be informed about the disease, its symptoms and signs of decompensation and should be encouraged to self-care and early consultation. Moreover, this society emphasizes that health education is a continuous process and not of a single medical check-up (Class I, Level C).

Table 6 was made for summarizing the reasons why each of the consensuses recommend and continuously mention the need to educate patients, their families and caregivers.

Table 6. Benefits of health education to patients and their caregivers.

	Quality of life	Self-care	Decrease hospitalizations	Recognition of symptoms	Adherence	Medication recall
HFSA (2008) [9]			X		X	
CCS (2006) [10]		X		X		
CCS (2007) [11]		X		X		X
CCS (2012) [12]	X	X		X		
Franciosa J A et al. (2010) [13]			X			
SAC (2010) [14]		X		X	X	
SUC (2000) [16]	X	X	X	X		
HFSA (2010) [17]	X	X	X	X	X	X
Goodlin S J et al. (2004) [18]	X				X	X
Barisani J L et al. (2010) [19]			X		X	

The consensuses name different subjects on which the HF patients should be educated. Some recommendations are directed to the understanding of the disease and of the symptoms that could appear, whereas other recommendations of the consensuses are specific topics about to lead a healthy life according to possibilities, such as physical activity, smoking, alcohol, nutrition, among others.

5. Discussion

Knowing that HF is a disease that causes a progressive deterioration in the patients, and considering that the studied consensuses recommend to follow the patients after discharge, the need to continuously monitoring HF patients is established. Moreover, the study of the consensuses indicate that the telemonitoring can have many benefits including quality of life, medication recall, adherence to treatment, self-care behaviours, decrease in hospitalizations, decrease in mortality [9,13], reduced healthcare costs [14] and it can be a solution for those patients that are often unable to access the clinic [9,13]. For these reasons the telemonitoring proves to be a correct tool for continuous HF patients monitoring.

Taking into consideration consensuses recommendation about the care of HF patients, it can be deduced that an adequate telemonitoring system should be capable of allow early detection of symptoms and of physiological parameters variations that indicate a possible decompensation of the patients. These indicators should be properly communicated to the medical team for enabling them to act against dangerous situations and to avoid hospital readmissions. Besides, the decrease of hospitalizations would also decrease healthcare costs and with the appropriate preventive actions would reduce the mortality of HF patients. Moreover, it the telemonitoring system provides adequate health education for patients and their caregivers, it would contribute to improve their quality of life. This can be said because it has been shown that with the health literacy leads to improvement of patient self-care behaviour and to greater adherence [17].

Based on the similarities found from the study of consensuses, a serie of parameters of interest that should be monitored with the telemonitoring systems can be defined. These parameters are the BP, HR, body weight, symptoms and functional capacity. As far as the ECG is concerned, is recommended that it should be continuously monitored only in patients with arrhythmias so it is not included within the parameters of interest.

Finally it could be argued that telemonitoring system designers should take into account at least two aspects which arise from the recommendations of the consensuses. For one hand the system should be able to monitor the parameters of interest mentioned above. On the other hand, these telemonitoring systems should provide an adequate means of health education for patients and their caregivers. If the telemonitoring systems meet these characteristics, according to the studies mentioned in the consensuses, the patients, their family, their caregivers and even the health system will benefit.

References

[1] Villalba E, Peinado I, Arredondo M T, Meneu T, Sala P and Guillén S 2008 Self care system for heart failure out of hospital patients *2nd Int. Conf. on Pervasive Computing Technologies for Healthcare* (Tampere, January 30 - February 1 2008)

[2] Sosa Liprandi M I, Gonzalez M A and Sosa Liprandi A 1999 Perspectiva de la insuficiencia cardíaca en la Argentina *Medicina* **59** 787-92

[3] Marconi E H et al. 2012 Información Básica – Año 2011 *Estadísticas Vitales* **55** 1-135

[4] Smith A C 2013 Effect of telemonitoring on re-admission in patients with congestive heart failure *Medsurg Nursing* **22** 39-44

[5] Go A S et al. 2012 Heart disease and stroke statistics-2013 update: a report from the American Heart Association *Circulation* **127** 6-245

[6] Tacchi H H 2006 Insuficiencia cardíaca aguda descompensada *Insuficiencia Cardíaca* **1** 93-100

[7] Finkelstein J, Wood J, Cha E, Orlov A and Dennison C 2010 Feasibility of congestive heart failure telemanagement using a wii-based telecare platform *32nd Annual International Conference of the IEEE EMBS* (Buenos Aires, August 31 - September 4 2010) pp 2211-14

[8] Saponara S, Donati M, Bacchillone T, Fanucci L, Sanchez-Tato I, Carmona C and Barba P 2012 Remote monitoring of vital signs in patients with chronic heart failure - Sensor devices and data analysis perspective *Sensors Applications Symp. (SAS), 2012 IEEE* (Brescia, Italy, 7-9 February 2012) pp 1-6

[9] Hauptman P J et al. 2008 The heart failure clinic: a consensus statement of the Heart Failure Society of America *Journal of Cardiac Failure* **14** 801-15

[10] Arnold J M O et al. 2006 Canadian Cardiovascular Society consensus conference recommendations on heart failure 2006: diagnosis and management *Canadian Journal of Cardiology* **22** 23-45

[11] Arnold J M O et al. 2007 Canadian Cardiovascular Society consensus conference recommendations on heart failure update 2007: prevention, management during intercurrent illness or acute decompensation, and use of biomarkers *Canadian Journal of Cardiology* **23** 21-45

[12] McKelvie R et al. 2012 *The Canadian Cardiovascular Society Quality Indicators E-Catalogue for Heart Failure A CCS Consensus Document* ed The Canadian Cardiovascular Society (Ottawa: The Canadian Cardiovascular Society) pp 1-34

[13] Franciosa J A, Ferdinand K C and Yancy C W 2010 Treatment of heart failure in african americans: a Consensus Statement *Congestive Heart Failure* **16** 27–38

[14] Barisani J L et al. 2010 Consenso de diagnóstico y tratamiento de la insuficiencia cardíaca crónica *Revista argentina de cardiología* **78** 166-81

[15] Alagia F et al. 2004 Segundo Consenso Uruguayo de Cardiología (primera parte) *Suplemento de*

la revista Noticias 1-16

[16] Alagia F et al. 2005 Segundo Consenso Uruguayo de Cardiología (segunda parte) *Suplemento de la revista Noticias* 1-8

[17] Evangelista L S, Rasmusson K D, Laramee A S, Barr J, Ammon S E, Dunbar S, Ziesche S, Patterson J H and Yancy C W 2010 Health literacy and the patient with heart failured implications for patient care and research: a consensus statement of the Heart Failure Society of America *Journal of Cardiac Failure* **16** 9-16

[18] Goodlin S J et al. 2004 Consensus statement: palliative and supportive care in advanced heart failure *Journal of Cardiac Failure* **10** 200-9

[19] Barisani J L et al. 2010 Consenso de insuficiencia cardíaca aguda y avanzada *Revista Argentina de cardiología* **78** 264-81

[20] Piepoli M F et al. 2011 Exercise training in heart failure: from theory to practice a consensus document of the Heart Failure Association and the European Association for Cardiovascular Prevention and Rehabilitation *European Journal of Heart Failure* **13** 347–57

A wireless goniometry system

G Campiglio [1], J Mazzeo [1], and S Rodriguez[2]

[1]Instituto de Ingeniería Biomédica, Universidad de Buenos Aires, Argentina
[2]Engineering Student, Universidad de Buenos Aires, Argentina,

E-mail: gcampi@fi.uba.ar

Abstract. Gait laboratories comprise a diversity of devices of medium and high complexity with which it is possible to obtain records that quantify spatio-temporal patterns of human movement. Scientific knowledge achieved on the basis of such records in the normal and pathological gait is being applied to the diagnosis, treatment and monitoring of various diseases, as well as to the design of better methods for rehabilitation of motor disorders. In this paper we present a goniometry system designed as part of the gait laboratory of the Institute of Biomedical Engineering, Faculty of Engineering of the University of Buenos Aires. This system allows the measurement of angles in three dimensions between different body segments. Angles are sensed by means of electronic goniometers and presented real-time on a computer screen through a wireless communication channel.

1. Introduction

Human walking is a complex process involving many body segments moving harmoniously under the control of the central nervous system. Their study, multidisciplinary, is approached from the point of view of biomechanics based on records of the kinematics and dynamics of the different body segments set in motion. For some years we have had technology that allows such records in upper and lower limbs.

It is of central interest for the kinematic study to know the angular variations of the body joints together with the tilt twist and swing of each segment. Additionally, the study involves dynamic ground reaction forces as well as the moments and muscle forces actuating on the joints [1], [2]. This paper describes the design and implementation of a wireless system for recording real-time angle measurements obtained on the basis of electronic goniometers.

The immediate context of application of this design is the research project on human walking carried out by the Institute of Biomedical Engineering at Buenos Aires Faculty of Engineering.

2. Materials and Methods

The system provides simultaneous recording of two channels providing the ability to measure bending and twisting movements on a single joint or simultaneous recording of angles in a single plane into two different joints.

Its main components are: (a) sensors (b) portable electronic module (c) acquisition and display software in computer

2.1 Sensors

Traditional goniometers are useful for the static measurement of joint angles, for example they can be used to measure the maximum knee angle flexion achieved by a patient under treatment. Its structure is similar to rulers that must be aligned to the respective body segments trying to keep its center of rotation located on the joint that is evaluated.

In recent years electronic goniometers allow joint angles dynamic measurements in function of time while the subject performs an activity. The technologies mainly used in these devices are: potentiometric goniometers, flexible goniometers and others flexible developed with fiber-optic.

Electro-potentiometric goniometers are based on the resistive changes of a potentiometer that moves integrally with respective rod-like structures attached to the studied body segments. The most serious problem is its rigidity as it complicates fixation to the body which promotes measurement errors.

Flexible electrogoniometers are relatively small and easy to install. They offer simultaneous measurement of angles in two planes of motion, which allows measuring flexion/extension in the sagittal plane and abduction/adduction in the frontal plane.

Fiber optic goniometers are based on a curvature dependant magnitude of light transmission losses. In one of the extremes of a short optic fiber there is a light source and in the other sensors that detects a measurable change in the light intensity [3]. There is a commercial device, the Joint Shape Angle Sensor, which is manufactured by the Canadian company Measurand [4].

The goniometers that we used for this work belong to the flexible type and they are manufactured by the company Biometrics Ltd [5]. This goniometers, for signal conditioning purposes, may be analyzed as a Wheatstone-bridge type structure as suggested in the technical manual provided by the manufacturer. Using a power source of 5 volts, a 10 μV/degree sensitivity is obtained on each measuring plane.

2.2 Portable Module

Its operation is described by the functional block structure shown in FIG 1. Signals from the sensors are firstly processed in an analog stage for conditioning and amplification. Then it is digitized to be transferred to a PC. Sampling and transmission are controlled by a microprocessor.

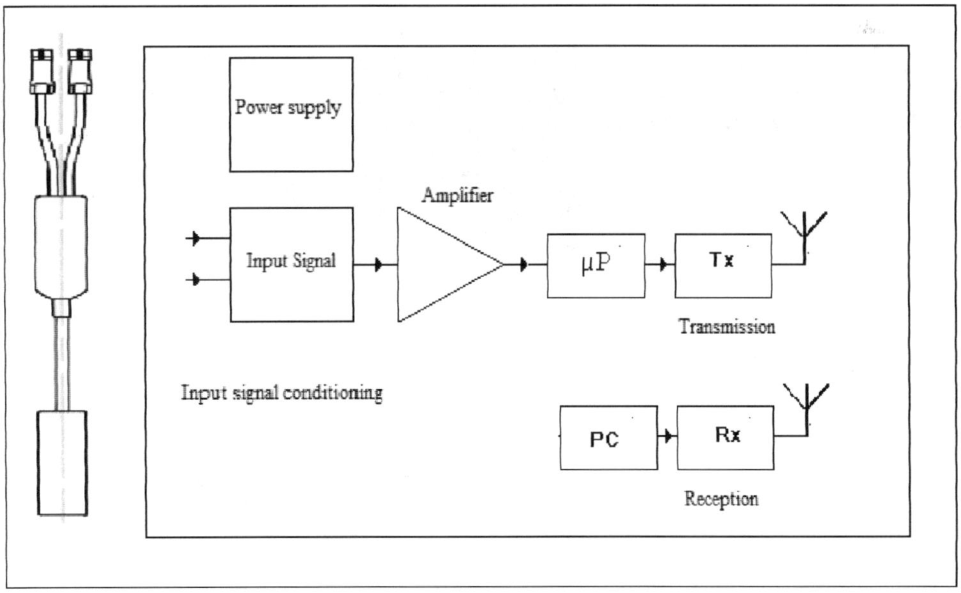

Figure 1. Block Diagram of the system.

The signals produced by the goniometer are electric potentials relatively small and require amplifiers to facilitate quantification .The primary purpose of the amplifier is increasing the level of the signal obtained from the sensors excluding any noise. The main noise source we observed is 50/60 Hz power line interferences in common-mode.

The analog stage is based on an INA instrumentation amplifier which features high common mode rejection and very low offset voltages which fit very well to the requirements of this project.

An Atmel ATmega328 microprocessor performs the functions of sampling digitizing and data formatting according to the transmission protocol designed for this application. This processor incorporates on a RISC architecture of 8 bits, 10-bit analog inputs and an USART (Universal Synchronous / Asynchronous Receiver Transmitter) interface.

MaxStream XBee modules are used for wireless connectivity both for transmission and for reception. They work on radio frequency in the 2.4 GHz band with 802.15.4 communication protocol. Their indoor range is 30 m and it extends up to approximately 100 m outdoors with a 2mW transmission power. We use in our application point to point configuration and AT commands. The portable module is powered by lithium ion batteries.

The radio transmitter is located inside the portable module while the receiver is wired to the computer via an USB port. Our ad-hoc communication protocol includes coding of samples for transmission. A sampling frequency of 200 Hz was chosen according to usual bandwiths in gait analysis.

By combining its wireless communication capabilities with laptop computers this system gives the possibility of obtaining outdoor records. Its ease of transportation also allows sharing it between different laboratories.

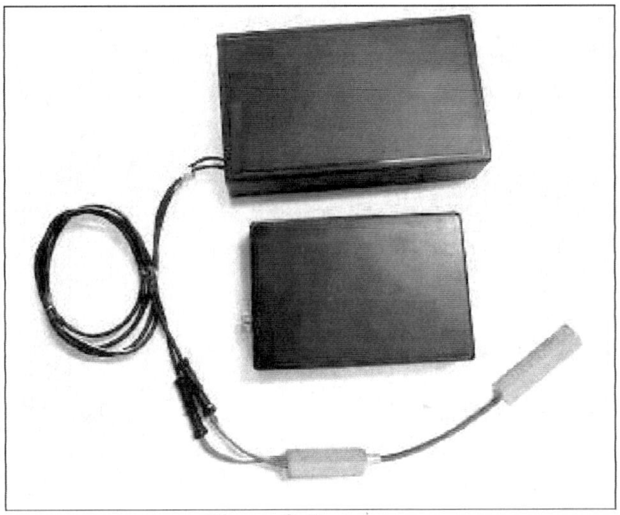

Figure 2. Prototype

Figure 2 shows the developed prototype that consists of the transmission and reception units and the goniometric sensor. The sensor has two output connectors, one measures flexion/extension and the other radial/ulnar deviation. According to manufacturer's specifications, crosstalk between both outputs is less than ± 5% over an angular range of ± 60°, accuracy is ±2° over a range of ±90° and repeatability is ±1° over a range of 90°. Our system comprises two channels that we currently use for measuring flexion/extension of right and left knees.

Journal of Physics: Conference Series **477** (2013) 012008 doi:10.1088/1742-6596/477/1/012008

2.3 Acquisition Software

Computer software, developed by this group, allows real-time monitoring of joint angles and optional storage in customized disk files. It includes data-base features for managing age, sex, pathology, date and time. This data is also embedded in every signal file.

Off-line analysis is accomplished by another piece of software that was also developed by this group. It offers several visualization tools to facilitate the study of signals.

3. Discussion

Figure 3 illustrates one of the measurements obtained by means of this system. It is a screen copy of our signal analysis application and corresponds to knee angles during treadmill walk of a healthy subject.

Flexion/extension angles were measured versus time in left and right leg by two attached flexible sensors. The treadmill was set to 2.5 km/h for comfortable walking speed.

In accordance to what other authors reported about other type of wearable sensors [6], we observed that sensor fixation is a critical factor to consider when using flexible goniometers. In our experiments we detect very high noise levels if a meticulous procedure is not carried out. Noise may be much higher than the signal itself in that case.

We observed that this type of sensors caused minimum motion discomfort.

4. Conclusions

The main motivation for developing the system presented in this paper was our interest in opening a gait laboratory. Our goal was to design a wireless goniometry system under the following premises: long term records, a minimum of two acquisition channels, simple use and easy outdoors transportation. This system was developed and tested on healthy individuals obtaining good behaviour and results that are consistent with the undertaken premises.

Among our future work we foresee a comparative test of accuracy against other recording methods and the development of complementary software featuring more elaborated tools for off-line analysis.

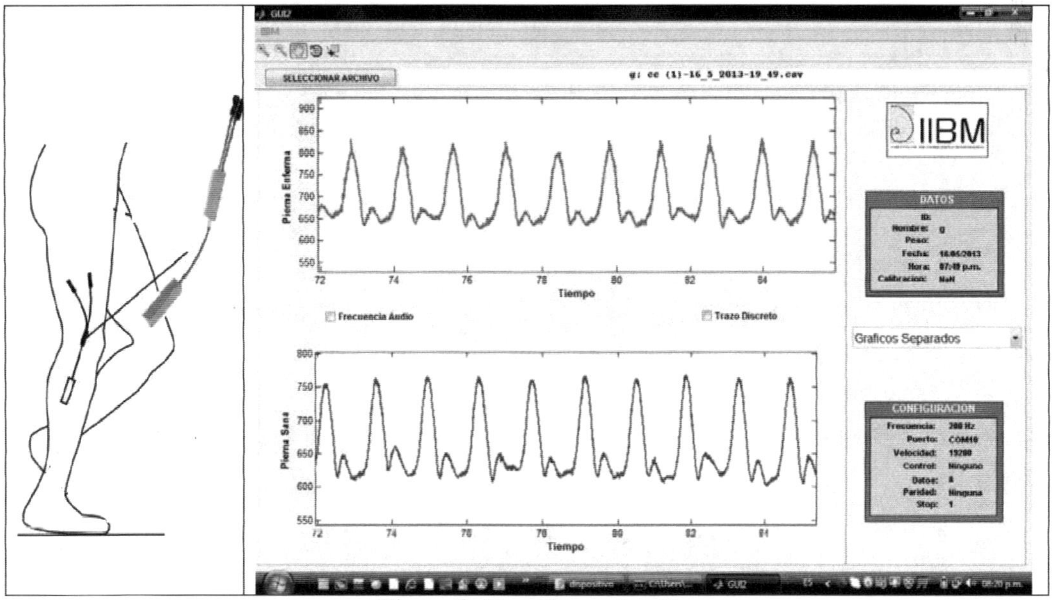

Figure 3. Knee flexoextención

References

[1] Benedetti M. G., Catani F., Leardini A., Pignotti E., Giannini S. 1998 Data management in gait analysis for clinical applications *Clinical Biomechanics* 13 204-215

[2] Prat J. M., Sánchez-Lacuesta, J. 2002 Biomecánica de la marcha humana normal y patológica. IMPIVA Generalitat *Valenciana, Instituto de Biomecánica de Valencia*

[3] Bell JA, Stigant M.2008 Validation of a fibre-optic goniometer system to investigate the relationship between sedentary work and low back pain. International Journal of Industrial Ergonomics 38 11-12 934-941.

[4] http://www.measurand.com/shapesensor.htm

[5] http://www.biometricsltd.com/gonio.htm

[6] Tik-Pui D., Chan Y, 2010 The Use of Wearable Inertial Motion Sensors in Human Lower Limb Biomechanics Studies: A Systematic Review Sensors 10 11556-11565 doi:10.3390/s 101211556

Diamond-Based Thin Film Bulk Acoustic Wave Resonator for Biomedical Applications

M Zalazar[1] and F Guarnieri[1,2]

[1] Facultad de Ingenieria, Bioingeniera, UNER, Ruta 11 km 10, Oro Verde (3100), Argentina.
[2] CIMEC, INTEC (UNL-CONICET), PTLC, El pozo, Santa Fe (3000), Argentina.

martin.zalazar@fulbrightmail.org

Abstract. Nowadays it is in constant growing the development of thin film bulk acoustic resonators. If the piezoelectric material is going to be implanted in the human body, an important requirement is the biocompatibility of the implant. In this regard, Aluminum Nitride (AlN) has emerged as an attractive alternative for use in biomedical MicroElectroMechanical Systems. Ultrananocrystalline Diamond (UNCD) is a promising material to be used in biomedical applications, due to its extraordinary mulifunctionality; it is exceptional for implantable medical devices requiring stringent biological performance. Since both UNCD and AlN films can be processed via photolithography processes used in microfabrication, the integration of UNCD and AlN films provides the bases for developing a new generation of biocompatible Bio-MEMS/NEMS. Research and development was conducted to produce implantable MEMS devices: Pt/piezoelectric AlN/Pt layer heterostructure was grown and patterned on the UNCD membrane with a Ti adhesion layer. By applying voltages between the top and bottom Pt electrodes layers the piezoelectric AlN layer is energized. The feasibility of the fabrication of biocompatible AlN/diamond-based FBAR structure has been demonstrated.

1. Introduction

Nowadays it is in constant growing the development of thin film bulk acoustic resonator (FBAR). This technology was born as a direct extension of the quartz crystal resonators. These are bulk acoustic wave-based (BAW) thickness mode wave resonators, allowing a reduction in the piezoelectric thickness of the resonator thus conducting to an increased resonance frequency. Therefore, it is reached a high sensibility in gravimetric applications in addition to their compatibility with the integrated circuit technology.

If the piezoelectric material is going to be implanted in the human body, an important requirement is the biocompatibility of the implant. It would be desirable to have a biocompatible piezoelectric material, thus avoiding the need for packaging that represents a significant portion of the cost to the device. In this regard, Aluminum Nitride (AlN) has emerged as an attractive alternative to other piezoelectric materials as have been shown by several authors [1][2]. AlN is a very attractive piezoelectric material for use in biomedical MicroElectroMechanical Systems (BioMEMS); it is biocompatible, exhibits high resistivity, high breakdown voltage, high acoustic velocity and it can be

[1] To whom any correspondence should be addressed.

grown by reactive sputtering technique at relatively low temperature thus being compatible with CMOS device technology [3][4].

It is also becoming important to be able to integrate piezoelectric films with materials used in medical devices. Ultrananocrystalline diamond (UNCD) in thin film form, is a multifunctional material, which is extremely bioinert and biocompatible [5][6][7][8]; in addition, UNCD is capable of being integrated with CMOS technologies. For example, UNCD has been developed as an hermetic bioinert/biocompatible encapsulating coating for a Si-microchip [9] to enable implantation inside the eye on the retina, as a critical component of an artificial retina, and as a biocompatible coating for glaucoma valves [10] and dental implants [11].

Since both UNCD and AlN films can be processed via photolithography and reactive ion etching processes used in fabrication of Micro/Nanoelectromechanical Systems (NEMS), it has been shown that the integration of UNCD and AlN films provides the bases for developing a new generation of biocompatible Bio-MEMS/NEMS [12].

Because diamond-based substrates exhibit the highest sound velocity among all materials [13] and AlN exhibits the highest phase velocity among all piezoelectric materials [14], the AlN/diamond heterostructure provides a very promising platform for fabrication of a wide range of devices from SAW to microfluidics. Zalazar et al. [12], reported the integration of films on UNCD layers exhibiting a piezoelectric coefficient of about 5.3 pm/V, one of the highest for AlN film demonstrated today.

This work is focused on the development of an AlN/UNCD-based thin film bulk acoustic wave resonator for biomedical applications. In this case, a FBAR piezoelectric biosensor was fabricated and characterized. The device has been previously modelated and simulated under different conditions by the author [15]; even though it was carried out the analysis of the behaviour of a FBAR piezoelectric mass sensor, it can be extended to new types of sensor devices.

2. Geometry design

The simplest configuration of the membrane FBAR involves the acoustic resonance cavity formed by the creation of an air cavity underneath the resonating membrane by etching completely a Si substrate. To acoustically isolate the resonance cavity, parts of the substrate are removed to create a freestanding membrane. The membrane is composed of a Pt/Ti/AlN/Pt/Ti/UNCD multilayer on a silicon substrate as shown in the scheme of Figure 1a. Pt layers serve as bottom and top electrode to apply voltage to excite the piezoelectric effect on the AlN layer for actuation of the FBAR device. Top electrode (Pt/Ti) is similar to the bottom electrode.

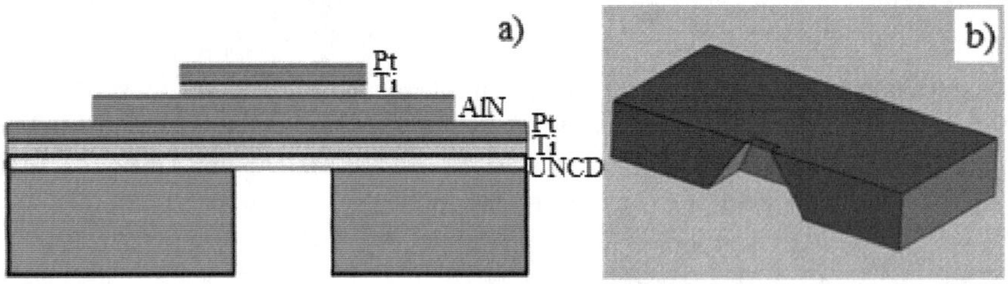

Figure 1. UNCD-based AlN FBAR. a) Scheme of the proposed heterostructure of Pt/Ti/AlN/Pt/Ti/UNCD/SiO2/Si and b) 3D longitudinal cut image of the freestanding FBAR heterostructure.

Devices composed of multilayer structures where the layers exhibit different thermal expansion coefficients often suffer from intrinsic residual stresses, which can be reduced using low temperature processes. In addition, being a low temperature process makes it promising for integrating with the IC technology.

An scheme of the FBAR device can be observed in Figure 1b. It can be seen a longitudinal cut of the freestanding heterostructure membrane composed by a Pt/Ti/AlN/Pt/Ti/UNCD multilayer on the silicon substrate.

3. Mask design and fabrication

The Layout Editor of the Tanner EDA software was used for the masks design. Layout is essentially a drawing process where the 2D drawings of the geometries will end up on the desired mask. Layout tools are essentially CAD (computer aided design) drawing tools, but include additional useful features.

Three different masks for the entire device were designed and fabricated:

- Mask 1: cavity on Si
- Mask 2: piezoelectric AlN
- Mask 3: top electrode

Figure 2 shows the mask for the cavity on the Si substrate. It can be observed the four different sizes used in this experimental step in order to compare their performances. Square membranes were designed for the FBARs.

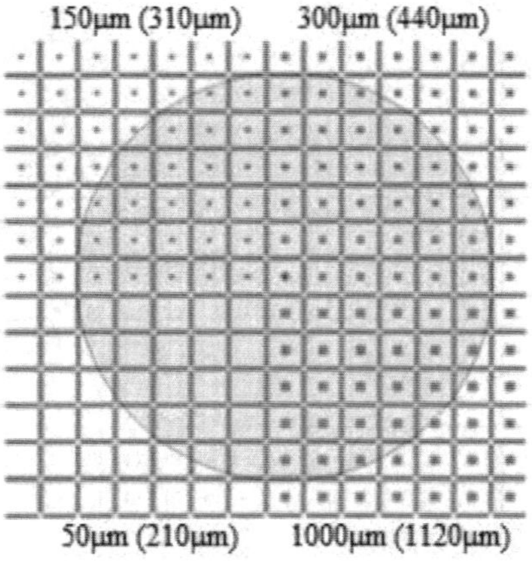

Figure 2. Mask layout for the Si cavity. It can be observed the four different sizes used in this experimental step, where the value in microns is the calculated length of the square while in brackets is the real length due to undercuts.

For the cavity design, the substrate material as well as the used etching method has to be considered. In the case of silicon, a common etching method is the wet anisotropic etching; potassium hydroxide (KOH) is an strong base capable of attack the Si substrate in an anisotropic way. N-type Si (100) wafers with surface polished to mirror finish as substrate were used, so the KOH will etch the Si preferably in the [100] plane. It produces the characteristic V anisotropic etch of the walls, forming an angle of 54.7° with respect to the main surface (Figure 3).

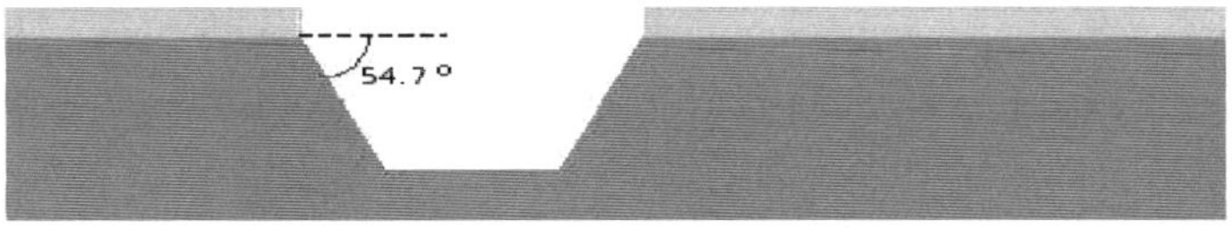

Figure 3. Si (100) showing characteristic V anisotropic etch of the walls forming an angle of 54.7° with respect to the main surface.

The used masks have a multilayer structure with the following layers: PR/Cr2O3/Cr/glass. The photoresist (PR) is the last layer and the Cr2O3 serves as antireflective film. The glass is highly flat and made of borosilicate glass. The fabrication of the masks was done by a Laser Pattern Generator (LW405). The pattern was generated over the photoresist on the chrome layer and the glass (square shape of 125 mm) during 24 hs. After that, the photoresist was developed (developer 351) and a wet etching of the chrome film was done.

4. FBAR fabrication

The freestanding membrane fabrication on the Si substrate involves several steps:

4.1 UNCD growth

A UNCD film on the front of the silicon wafer (N-type mirror polished Si (100)) was deposited. UNCD layers were grown using a MPCVD system (Lambda Technologies), powered by a 915 MHz magnetron generator with 2300 W power. It generates a plasma from a mixture of Ar (99%)/CH4 (1%) gases producing C2 dimers as the growth species at a pressure of 80mTorr. This layer was used as a substrate for the AlN and also serves as a stop layer for the anisotropic wet etching of the Si by using KOH.

4.2 Si3N4 mask

Silicon Nitride (Si3N4) was used as a mask for the wet etching of the Si using KOH. Commercial, polished, low stress Si3N4 residual (LSLPCVD, commercial grade, University wafer Inc) was used. Si3N4 pattern for the cavities was performed by photoresist spin coating (S1813), UV exposure for 5 seconds (mask aligner Karl Suss MA6), resist development (351 developer diluted in a 3: 1 ratio in DI water) and dry etching of the Si3N4 in a Tetrafluoromethane (CF4) plasma (Oxford CS-1701 RIE).

4.3 Pt/Ti deposition

The Pt layer was grown by magnetron sputter deposition on top of a Ti film deposited on the SiO2 surface as an adhesion layer. The Pt bottom electrode was grown by sputter deposition (AJA ATC Orion thin film deposition system). The Ti film provides an adhesion layer for growing highly c-axis (002) oriented AlN films. Films were deposited with a target RF power of 150W, pressure of 3 mTorr, Ar flow rate of 26sccm and base pressure of 1e-7 Torr. The thicknesses of the Platinum and Titanium were 150nm (720s deposition) and 10nm (310s deposition) respectively. In addition, this coating provides seeding capability and serves as a buffer layer, lowering the discrepancy in lattice parameter between substrate and film. It also helps to improve the adhesion of the AlN to the UNCD

4.4 AlN deposition

AlN films for MEMS devices are most often synthesized in the Wurtzite hexagonal crystallographic structure with c-axis (002) orientation perpendicular to the surface (diffraction peak at 36°), which is the orientation that yields the highest piezoelectric constant in addition to a high acoustic wave velocity [16].

AlN (002) oriented films were grown on the Pt layers at 500°C C at a pressure of 3mTorr, using reactive sputter deposition (AJA ATC Orion thin film deposition system), via sputtering material from an Al metallic target. Ar-plasma was used to produce Ar ions to sputter the Al material in a N2 atmosphere to produce the AlN films. Aluminum target with a diameter of 4" and a purity of 99.999% was used.

Several experiments were performed until reach to the crystallographic orientation (002) of the AlN with the diffraction peak in 36°. AlN layers with thickness in the 260-420 nm range were synthesized on silicon oxide (SiO2) and silicon nitride (Si3N4) were used as intermediate layer grown on Si.

4.5 AlN pattern

The pattern was transferred to the AlN film by spin coating of the photoresist (ma-N 415 negative photoresist), UV exposure for 30 seconds (mask aligner Karl Suss MA6), resist development (533) and dry etching using anisotropic Cl2-based reactive ion etching (Oxford CS-1701 RIE).

4.6 Electrode pattern

Patterning of Pt/Ti top electrode was done by applying lift off techniques. Firstly, a spin coating with S1813 on the frontside of the wafer, UV exposure and resist development were done. Finally a Pt/Ti sputter deposition and posterior removal of the photoresist was performed.

4.7 Cavity
The cavity etching was carried out on the backside by using KOH (30%). Hermetic protection to avoid frontside etching was used. Time etching was estimated in 9 hs at a temperature of 85°C using 250 rpm stirring.

The first cavity prototype can be seen in Figure 4, using only UNCD as membrane on the Si wafer substrate:

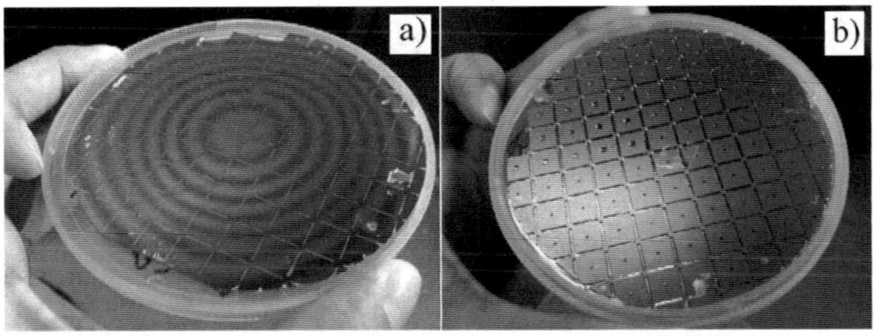

Figure 4. First cavity prototype. a) Frontside of the UNCD film and b) backside showing the different sizes of the cavities within dices.

5. Characterization

5.1 X-ray diffraction (XRD)
The XRD θ-2θ diffraction pattern taken from the XRD analysis of AlN/Pt/Ti/UNCD/SiO2/Si multilayer were scanned between 20° and 80°. Figure 2 shows an XRD θ-2θ scan of the AlN/Pt/Ti/UNCD/SiO2/Si layered film exhibiting high c-axis orientation for the AlN film. The presence of hexagonal AlN (002) diffraction peak at 36.05° can be noted, revealing the presence of mainly a (002) textured AlN film. Also, two Pt peaks, (111) at 40.05 and (002) at 46.5 are present. The weak peaks at 44.7 and at 65.1 correspond to Al (002) and Al (202), respectively. The strong narrow peak at 33.2 comes from the silicon wafer substrate and is the result of Si crystalline imperfections.

The X-ray wavelength was λ=1.540598A (Kα1, Long Fine focus, Cu Anode radiation). By using Bragg's law, it was calculated the c-lattice constant for the AlN layer. Its value was 4.9681 Å, which is in a good agreement with the value of the AlN found in the literature [17].

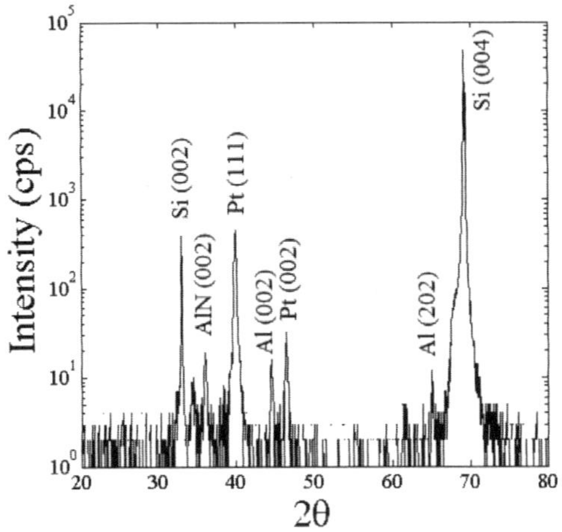

Figure 5. XRD spectrum of an AlN/Pt/Ti/UNCD/SiO2/Si heterostructure showing the characteristic peaks of the Pt layers and Si substrate, in addition to the critical AlN (002) peak that reveals the high orientation of AlN necessary to yield the high piezoelectric coefficient.

5.2 Optical Microscope
A fast analysis tool is the optical microscope. The obtained images help to evaluate the quality of the KOH etching. The morphology of the membranes and cavities are shown in Figure 6. The difference in the square membranes between frontside and backside of the Si wafer, was caused by the classical

pyramid shaped Si material; it obeys the preferential anisotropic etching direction (54.7°) of the (100) crystal plane of the Si wafer.

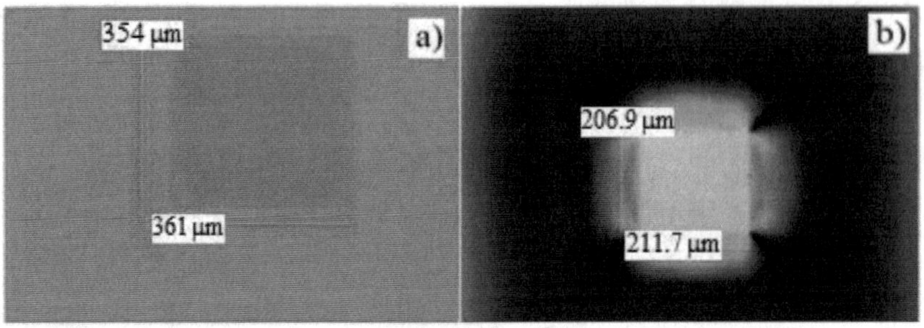

Figure 6. Square UNCD membranes. a) Membrane (mask 300 µm) seen from the frontside, and b) membrane (mask 150 µm) seen from the backside (cavity).

It can be seen that the size of each membrane is moderately higher than the expected due to the generated undercuts below the membranes. Undercuts increase the squares sizes in a range between 50-170 µm. The calculated etching rate for the Si using KOH was 1.16 µm/min

5.3 Scanning Electron Microscope

The scanning electron microscope (SEM) is a routinely used instrument for the examination of fine detail of a variety of samples. The instrument is, in simple terms, analogous to an optical microscope: the electron source (gun) is equivalent to the light source and the glass lens is replaced by electromagnetic lens. Information from the sample is collected and displayed on a viewing screen for visual interpretation.

In Figure 7a, a cross section of the AlN/Pt/Ti/UNCD membrane (cantilevered) is depicted. It exhibits a columnar microstructure. These columnar crystals are perpendicular to the substrate surface in agreement with the XRD results where the AlN film is highly textured. The piezoelectric film reached a thickness of 417nm after 5hs of growth thus arriving to a deposition rate of 83.4 nm/h (Figure 7b). This rate was affected by the low Ar flow used, reducing the amount of heavy ions impacting on the target. The UNCD layer used was 420 nm in thickness.

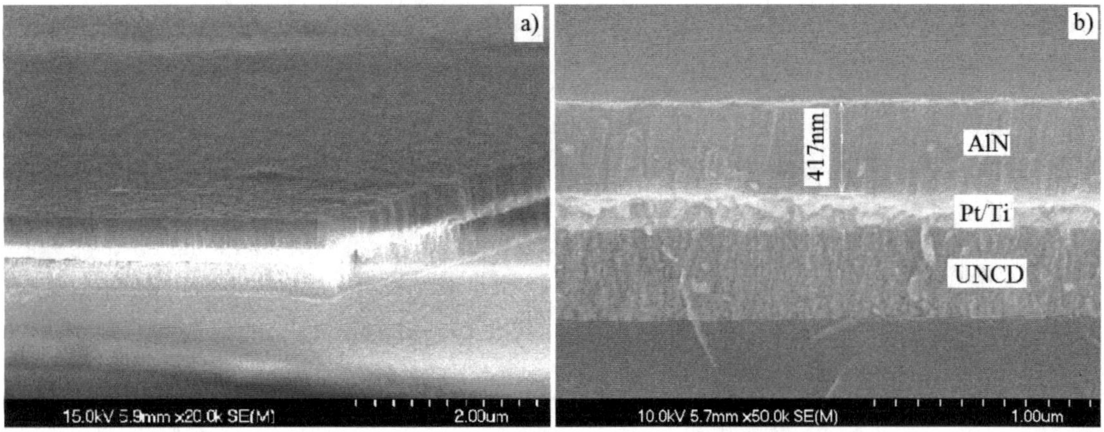

Figure 7. SEM cross section of the film for the AlN/Pt/Ti/UNCD membrane. a) Cantilevered membrane and b) measurements on the SEM image.

A clarified view of the generated Si cavity can be appreciated in the SEM image of the freestanding AlN/Pt/Ti/UNCD membrane on the Si substrate (Figure 8). It can be seen an in-focus image of the tilted Si walls (54.7°) where the big square have the dimensions of the mask and the smaller is the AlN/Pt/Ti/UNCD membrane. The white frame surrounding the big square is produced by the generated undercuts under the UNCD.

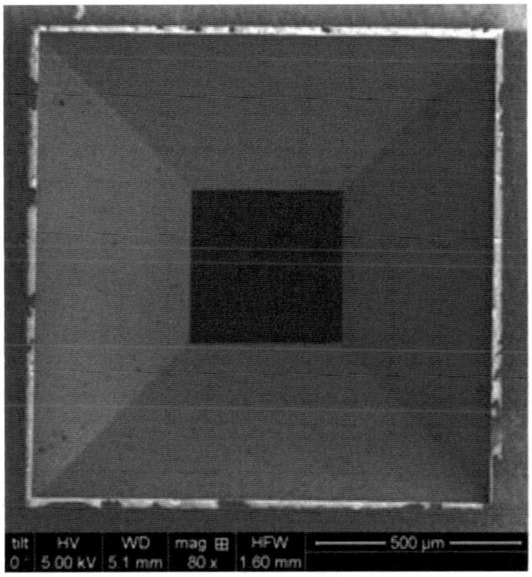

Figure 8. SEM image of the freestanding AlN/Pt/Ti/UNCD membrane on the Si substrate.

6. Conclusions

The feasibility of the fabrication of AlN/diamond-based FBAR structure has been demonstrated.

A direct method for the deposition of AlN on Pt/Ti/UNCD/SiO2/Si substrate avoiding polishing steps and thus reducing fabrication times has been demonstrated. It have been shown that thin AlN films with high (002) orientation can be produced

The columnar structure of the obtained AlN shows a highly textured film thus allowing the fabrication of FBAR resonators as well as piezoelectric actuators. The Pt film has proven to be a good buffer layer serving also as bottom and top electrodes.

Material compatibility with IC fabrication opens the way for monolithic integration of the traditionally incompatible IC and electro-acoustic technologies.

FBAR based on AlN/diamond materials is a very promising structure for biomedical applications. It is the opinion of the author that a huge field for biomedical devices based on AlN/UNCD is awaiting to be explored.

Acknowledgments

Authors thank the financial aid received by the Consejo Nacional de Investigaciones Científicas y Técnicas (CONICET), Argentina. Furthermore, authors thank the collaboration of Orlando Auciello and Pablo Gurman from the department of Materials Science and Engineering and Bioengineering, University of Texas at Dallas, USA.

References

[1] Chung K H, Liub G T, Duhb J G and Wang J H 2004 *Surface & Coatings Technology* pp 188-

189 745.
[2] Chen C C Lin C T Lee S Y Lin L H Huang C F and Ou K L 2007 *Applied Surface Science* 253, 5173.
[3] Chou C H Lin Y C Huang J HTai N H and Lin I 2006 *Diamond Relat. Mater.* 15, 404.
[4] Dubois M and Muralt P 1999 *Appl. Phys. Lett.* 74, 3032.
[5] Auciello O and Sumant A V 2010 *Diamond Relat. Mater.* 19, 699.
[6] Auciello O and Shi B 2010 *Biol. Med. Phys. Biomed.* 63.
[7] Bajaj P Akin D Gupta A Sherman D Shi B Auciello O and Bashir R 2007 *Biomed. Microdevices* 9, 787.
[8] Auciello O Birrell J Carlisle J A Gerbi J E Xiao X Peng B and Espinosa H D 2004 *J. Phys.: Condens. Matter* 16, 539.
[9] Xiao X Wang J Liu C Carlisle J A Mech B Greenberg R Guven D Freda R Humayun M S Weiland J and Auciello O 2006 *J. Biomed. Mater. Res. B Appl. Biomater.* 77B, 273.
[10] Zhou D D and Greenbaum E (Eds.) 2010, Implantable Neural Prostheses vols. 1 and 2 *Techniques and Engineering Approaches*, Springer (2010).
[11] Auciello O Gurman P Berra A Saravia M J and Zysler R Ch. 6, in *Diamond- Based Materials for Biomedical Applications*, ed Narayan R Woodhead Publishing Ltd, Cambridge, UK (in press, 2012).
[12] Zalazar M Gurman P Park J Kim D Hong S Stan L Divan R Czpalewski D and Auciello O 2013 *Appl. Phys. Lett.* 102, 104101.
[13] Assouar M B Elmazria O Kirsch P and Alnot P 2007 *J. Appl. Phys.* 101, 114507.
[14] Ishihara M Nakamura T Kokai F and Koga Y 2002 *Diamond Relat. Mater.* 11, 408.
[15] Zalazar M Guarnieri F 2010 *Mecánica Computacional*, vol XXIX, 6665-6684.
[16] Meinschen J Behme G Falk F Stafast H 1999 *Appl. Phys. A: Mater. Sci. Process.* 69, 673.
[17] Hirama K Taniyasu Y and Kasu M 2010 *Jpn. J. Appl. Phys.,* Part 1 49, 04DH01.

Use of sensors systems to monitor the mobility of elderly

M.R. Marques[1]; T.M. Raymundo[2]; C.S. Santana[3]

[1]Master student at Post Graduate Program Interunits Bioengineering EESC/FMRP/IQSC – University of Sao Paulo, Brazil
[2]Master of Science; PhD student at Post Graduate Program Interunits Bioengineering EESC/FMRP/IQSC – University of Sao Paulo, Brazil.
[3]Professor Dr. at Neuroscience and Behavioral Science Department – Ribeirao Preto Medical School, University of Sao Paulo and Post Graduate Program Interunits Bioengineering EESC/FMRP/IQSC – University of Sao Paulo, Brazil.

E-mail: mariana.ribeiro.marques@usp.br

Abstract. Even presenting risk of falls or mobility difficulties, many seniors prefer to live independently as long as possible. Thus, there is a need for solutions to enable independent living safely and to strengthen family support. Among them are the collection devices that assist in monitoring and quality of life of seniors. This paper aims to conduct a survey of devices developed to capture signal monitoring and mobility aid for the elderly. The devices were grouped into a main category (sensors) followed by secondary categories in relation to the type of use of the sensors (aid device with sensors, robotics and smart homes). Data were analyzed descriptively, listing the products with the types of use. The study raised a total of 22 products. It is observed that most of the feedback systems is to use sensor systems alone, without being used in sets with other technologies like assistive devices or robotics. Most of the technologies has been developed for detecting falls, and fewer tasks are transfer and mobility in the community. We emphasize the importance of unity of catchment systems signals with assistive technologies, in addition to monitoring the elderly provide direct aid in performing the task performed.

1. Introduction

The population aging bring consequences at a global level, which are a major challenge today. The WHO reports that in 2025 there will be 1,2 billion people over 60 years all over planet [1]. One of the effects of aging is the increase in health problems and disabilities, but there are many possibilities for prevention and improved functional capacity and physical limitations, cognitive and sensory resulting from this process [2], such as the use of technology, systems monitoring and optimization of daily activities.

Even with disabling health conditions, many seniors prefer to live independently as long as possible and thus there is a need to seek solutions to enable independent living by improving the security and family support [3].

Currently one of the areas targeted for investment in the quality of life of seniors is the gerontechnology. Some of the goals of the gerontechnology area are prevention or retardation of declines associated with age, care conditions and deficiencies installed and improvement in quality of life [4], for example, through assistive technologies to perform daily tasks, or catchment systems for monitoring. The devices to capture signals with sensors are easy to use tools that can identify the vulnerability, functional decline, predict mortality, as well as screening of conditions of life and health, can assist in the evaluations of the elderly for optimization of health services. Therefore, can enable the prioritization of care and improve the orientation on care [5]. Recently, the use of sensors is very present to monitor elderly people mainly in their homes. Understood as an early approach of disease, disability and risks based on the assumption that patterns of abnormal behavior can be captured by environmental sensors [6].

Technological innovations such as smart homes known as Smart Houses are promising solutions. These innovations should improve the quality of housing, care and life of residents, allowing seniors to live more independently and reduce the burden on health services [7].

With the development of technologies that aim to care and due to the high frequency of falls in the elderly, a very investigated area is the monitoring of falls by sensors. The evaluation of risk of falls with simple methods and objectives are necessary and, assessment schemes based on sensors, can differentiate the elderly who fell from those who did not have [8].

Mobility is represented in the International Classification of Functioning Disability and Health (ICF) by classifier d460 – Moving around in different locations, that can be subdivided into: d4600 – Moving inside the house; d4601 - Move inside buildings than own home; d4602 - Move out of your home and other buildings [9] (Figura 1).

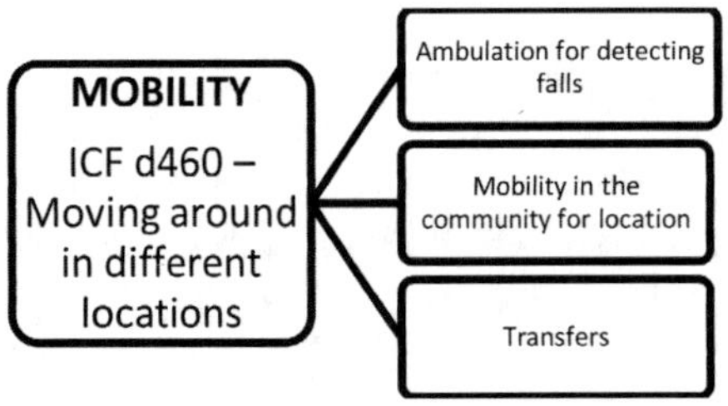

Figure 1. Classification system of the devices on the mobility activity monitored based on ICF [9]

This study was guided by the problem of elderly mobility and considering the need to investigate what has been researched and developed on this thematic. Insofar as the technological resources become important for improving quality of life and promote independent living for elderly, enabling monitoring, assistance and intervention in cases of emergency, this study aims to conduct a survey and description of devices with sensors for monitoring and assisting in the mobility of these subjects. Moreover, given the great diversity of use of the devices to capture signals, this study classified the products found according to their type of use (assistive devices with sensors, robotic systems).

2. Methods and materials

This is an exploratory and descriptive study.

For data collection, we performed a study of the mobility sensors through searching on websites of laboratories for developing technologies for the elderly, journalistic news sites, as well as in scientific papers in the area of health and bioengineering in the Pubmed and Scielo using the keywords: sensors, sensor networks, monitoring technologies; gerontotechnology; monitoring elderly and falls monitoring, in Portuguese and English. We included both devices available in the market such as those under development published from 2009 through 2013.

Were selected devices that monitored or assisted the mobility activity being excluded monitoring devices only routine, vital signs, among others. The data were classified according to their type of use as shown in Figure 2.

The main category, sensors systems, in accordance with figure 2 refers to the sensors themselves. Sensors are "devices that can detect and record the electromagnetic radiation in certain range of the electromagnetic spectrum, and generate informations that can be turned into a product that may be interpreted, either in picture form, in graphical or table". [10]

The subcategories of sensors systems were chosen in relation to the use of these devices together with other types of technology. Considering that not necessarily all sensors are used in conjunction with other technology, these, used alone, are classified in the item sensor systems. T hose who have use in other types of devices may be classified as:

1) Assistive devices with sensors: are capitation systems (sensors) associated with a support tool, such as a inteligente bengal beyond to informe the way it should be done and also assists in the gait. According to ISO 9999 [9] assistive products are "any product (including devices, equipment, tools, technology and software) especially produced or generally available, which is intended to prevent, compensate, monitor, mitigate or neutralize weaknesses, limitations in activity and restrictions in participation".

2) Robotic Systems: are machines with integrated technology systems and controlled as well as monitoring establish a direct relationship with the subject. A robot is a reprogrammable manioulator designed to move material parts, tools or special devices in variable programmed moviments for performing a variety of tasks [11].

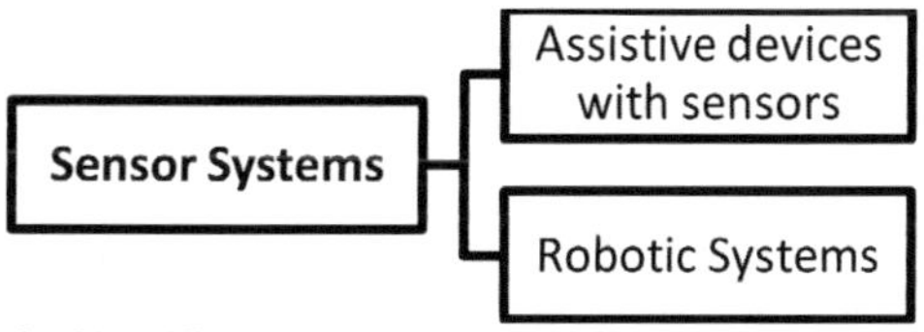

Figure 2. Classification according to the use of sensor systems

Data analysis was carried out through the description of the devices and their category was correlated with the type of use and the tasks involved.

3. Results

We surveyed a total of 22 sensors facing mobility, the table 1 shows the devices on the classification of the type of use and type of monitored mobility task. The devices collected in this study refers to technologies that monitor ambulation of subjects both to detect falls at home or away from home, as the mobility in the community that assist in the localization of the elderly and aid transfers (moving passage from sitting to standing and vice versa).

Table 1. Devices and classifications for the type of usage and activity monitored

SENSORS SYSTEMS	
ACTIVITY MONITORED	**DEVICES**
Ambulation for detecting falls	Sensor system adhesive skin [12]; Camera system of artificial intelligencel[13]; Pre-impact detection sensor [14]; HONEY- Ambient sensor [15]; Moviment sensor-QuietCare[16]; eNeighbor[17]; CARE system[18]; eCAALYX system[19]; HOPE system[20]; ROSETTA system[21]; SOFTCARE system[22]; Prevision system of fall [23]; Gait

	Measurement System [24]
Mobility in the community	Smartphone app Teo My Geo[25]; REMOTE system[26]; Tactile Belt for locomotion [27]
ASSISTIVE DEVICES WITH SENSORS	
MONITORADE ACTIVITY	**DISPOSITIVOS**
Ambulation for detecting falls	Cellphone BP[28]
Mobility in the community	Intelligent Bengal "The Aid"[29]; Intelligent wheelchair[30]; VA-PAMAID [31]
Transfers	Intelligent wheelchair[30]
ROBOTIC SYSTEMS	
ATIVIDADE MONITORADE ACTIVITY	**DISPOSITIVOS**
Ambulation for detecting falls	NAO – robô[32]
Mobility in the community	
Transfers	ASTRO – robot [33]

Regarding the type of use of sensors aimed at monitoring of mobility, was observed, according to graph 1, which approximately 75% are the sensors themselves, used alone, others used in conjunction with assistive devices and robotic systems appear in smaller amounts.

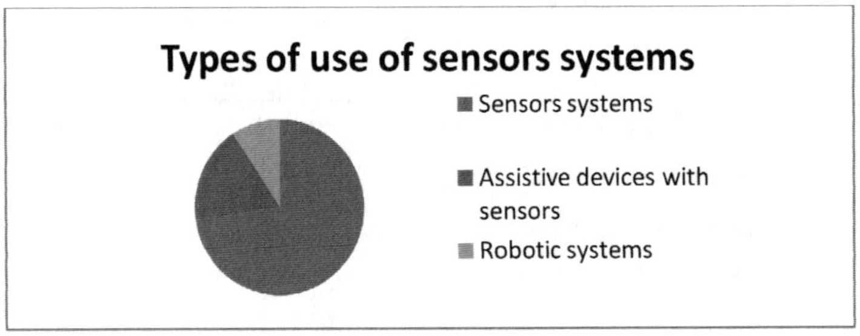

Graph 1. Classification according to the type of use of sensors

Regarding mobility activities monitored or helped by the catchment signals systems, most, about 75%, is for detect falls. Already mobility in the community to aid in the location and aid in transfers appears in smaller quantities, as shown in Graph 2.

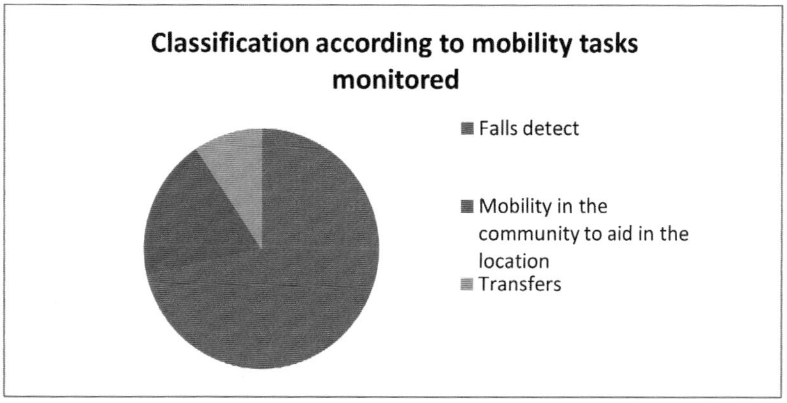

Graph 2. Classification according to mobility tasks monitored

Among the sensors systems it's possible to observe that only 3 of the 16 in the category, are for monitoring of mobility for location the person and the rest 87% are focused on the detection of falls. Most sensors to detect falls are used in systems with many interconnected sensors, as eNeighbor, CARE System, System eCAALYX, HOPE System, System ROSETTA, SoftCare System and Prediction System of falls. Some examples can be seen in figure 3.

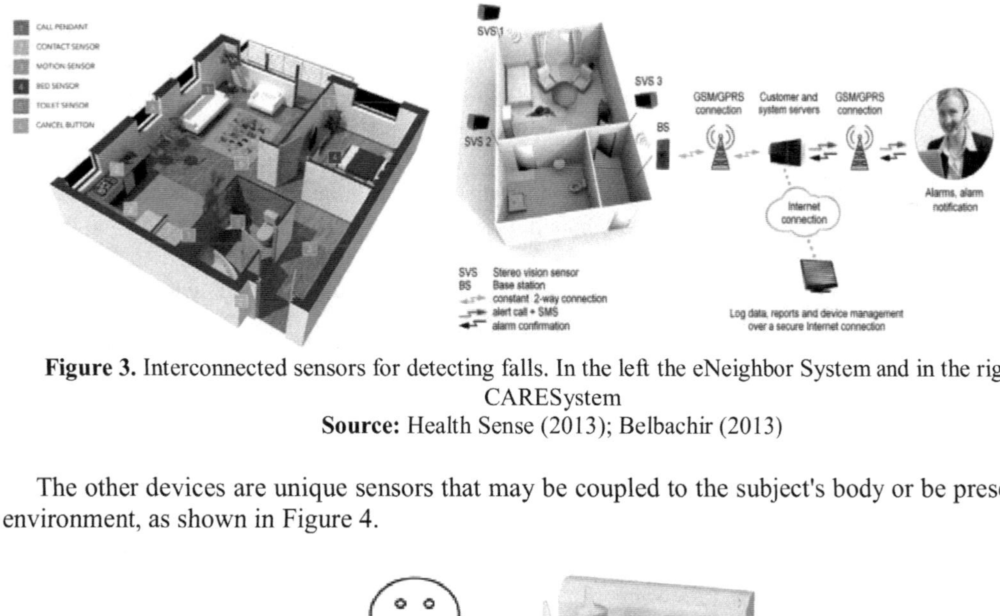

Figure 3. Interconnected sensors for detecting falls. In the left the eNeighbor System and in the right the CARESystem
Source: Health Sense (2013); Belbachir (2013)

The other devices are unique sensors that may be coupled to the subject's body or be present in the environment, as shown in Figure 4.

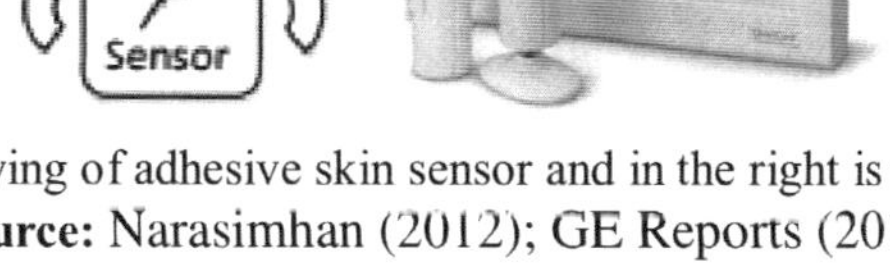

Figure 4. In the left the drawing of adhesive skin sensor and in the right is the motion sensor QuietCare
Source: Narasimhan (2012); GE Reports (2013)

Regarding sensor systems for localization during mobility, the program for Smartphone Teo My Geo (Figure 5) helps caregivers to locate the elderly when it leaves your home and also allows recording data of the health conditions of the elderly. Already the REMOTE system is a system that includes sensors and actuators that monitor the movement and location of the elderly and also allows you to monitor vital signs, posture, body temperature and environment. Finally the tactile belt with sensors (right on Figura5) uses technology of GoogleMaps to set the path and through vibrations indicates to the subject how much more he must still walk and the direction, allowing caregivers to monitor the elderly and reprogram the paths. It's suitable for people with visual impairments and cognitive.

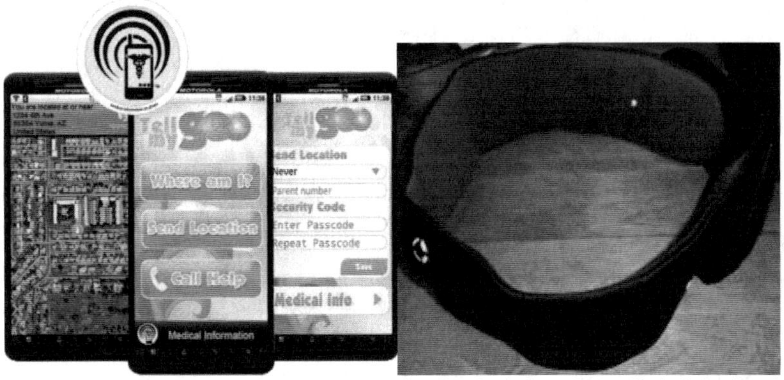

Figure 5. Program to Smartphone Teo My Geo in the left and Tactile belt in the right
Source: Moore (2013); Zelek (2012)

Among the devices with support for mobility with sensors found, has the cell phone BP used to detect falls, which detects sudden movements that may indicate a fall, and when this happens the phone automatically dials for different pre-programmed numbers to search for aid. This phone also allows you to organize schedules for drugs beyond communications activity. The other devices were found to intelligent Bengal "The aid", the intelligent wheelchair and walker robotic VA-PAMAID (Figure 6) that are geared for mobility in the community. The Bengal 'The aid " provides through a GPS system , and a wireless headset which way the elderly must take to reach its destination, and also monitor the vital signs . The intelligent wheelchair has an robotic system with detection module and information to aid in mobility, system to aid in transfers , adjustment of backrest and seat for different activities , automatic pressure relief , and a telehealth system integrated with Tablet serving as a channel communication to the distance between the subject and his family and caregivers . Already the walker with the robotic system , is a device designed to provide physical support and obstacle avoidance , as well as assistance in mobility for individuals with visual deficits. These last two devices were classified as assistive device, despite having robotic technology, because their primary objective is to assist in a specific task, the mobility.

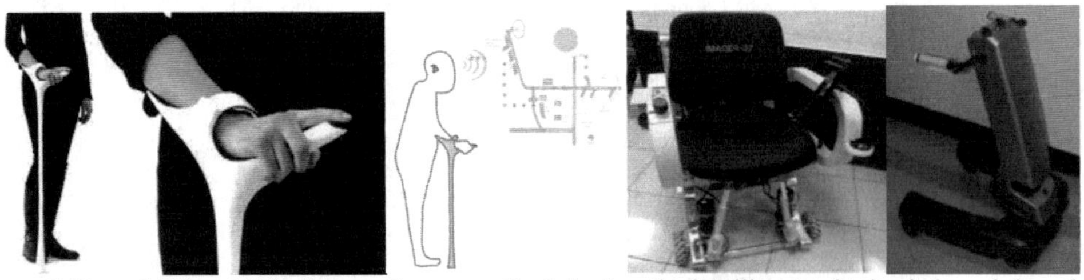

Figure 6. Bengal "The aid" in the two figures on the left, then the intelligent robotic chair and the robotic walker VA-PAMAID
Source: Designboom (2011); Hsu et al. (2012); Rentschler (2004)

Robotic systems appeared in smaller quantities, one of them, the robot NAO, is facing the detection of falls and the other, the robot ASTRO, for help in transfers (Figure 7).

The NAO robot has different sensors and devices, including cameras, microphones, distance sensors, voice synthesizer and speakers, and can capture sensor alarms at different locations in the house and getting to the location to verify one complication. The robot ASTRO has a lever technology that assists in transfers, plus a place to help to load objects during locomotion. It also allows communication with caregivers and family members remotely via a monitor that can also provide information on health care.

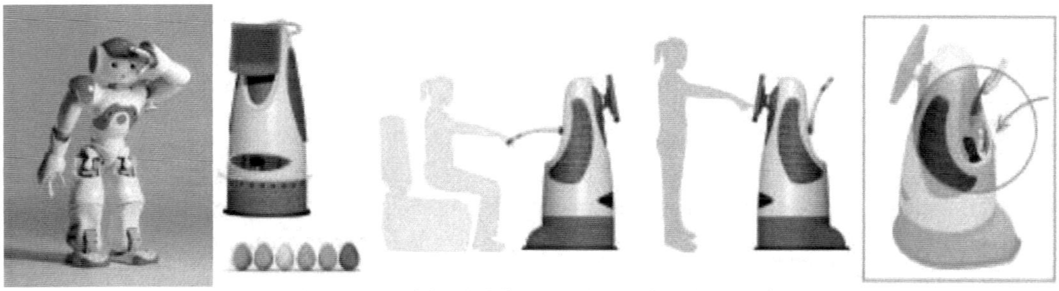

Figura 7. Robô NAO à esquerda e ASTRO à direita
Fonte: Bäck et al (2012); Cavallo (2013)

Thus, from this sample there is a variety of types of use of sensors and tasks related to mobility that they can help, such as detecting falls, the looking for help, assistance to locating paths during mobility in the community, assistance in transfers, and support to load objects during locomotion.

4. Discussion

Most devices were sensor systems used alone, which are used as a set of this devices. However, when used in conjunction with other technologies such as assistive devices and robotic systems appeared in smaller amounts (25% of the collected products).

The robotics itself already has into their devices the signal capture systems. On the other hand, assistive technologies (assistive devices) defined as equipment and services to optimize functional problems encountered by individuals with disabilities [34] and that aims at reducing the gap between the ability of the subject and the control their environment. [35], generally haven't sensors capturing signals in their development. The association between the sensors and the assistive technology has great potential to aid the elderly since they can promote the direct aid to tasks performance, facilitate monitoring, enable of risk perception and assistance in case of emergency.

Through the products collected, it was observed that most of the systems is directed to the detection of falls. This is in agreement with the high incidence of falls in this portion of the population. Aging is a dynamic and progressive process, which presents both morphological, biochemical and functional properties, which can generate functional losses, including postural instability, which occurs due to changes in sensory and motor systems, leading to a greater tendency to fall. [36]

Still, the task of transferring from sitting to standing and standing to sitting also creates risks for the elderly who may suffer a fall during movement due to decreased muscle strength or joint diseases. In the study of Ganança et al. [37] tasks that presented more risk for falls were: ambulation (53.1%), descending / climbing stairs (10.9%), postural transfer (9.4%) and bathing (6.3%). Thus, despite appearing less compared with ambulation, transfers are movements that deserve attention when considering the development of technologies for mobility of seniors.

In the same sense, is known about the prevalence of cognitive decline, especially memory deficits, arising from aging itself as well as chronic diseases such as Alzheimer's. Among the difficulties presented by the elderly with memory loss has the difficulty in locating yourself at home as well as in the community [38]. In this perspective, it is important to think of technology can, through signals catchment systems, provide the location of the elderly and help you reach your destination, as well as provide information to families and caregivers of the place where the senior been to prevent that he get lost. These technologies that enable the location and direction of the elderly were seen in some of the products found in this study.

Finally, it is noteworthy that most of the products described has multiple functions, not just monitoring or assistance in mobility tasks. It is observed that many devices also monitor vital signs, health conditions, carrying out tasks in the home environment, and promote the possibility of distant communication with family members and caregivers. This shows an improvement of technologies which aim to encompass the entire subject, and can meet different demands.

This study allowed a view of the current landscape of signals catchment technologies developed to aid the monitoring of the mobility tasks of elderly, and also support for both motor and cognitive tasks of spatial orientation. From this, it is possible facilitate future studies on these equipments, enabling the identification of devices still produced in less quantities, as well as the improvement of those already deployed.

5. Finals Considerations

This study enabled the survey and description of the use of signals collection devices for monitoring and assistance for mobility tasks to the elderly. Despite the few products collected, considering the short time involved in the collection (from 2009 to 2013), the findings contributed to the reflection of aspects that can be covered by the sensor devices, for example, their use in combination with technologies assistive devices or support that may offer a wider care to family and the elderly. It was possible to verify the search for solutions for the risk of falls in the elderly, that was the task more monitored by the systems. This demonstrates the concern of researchers and development teams of technologies to meet the great demand of this population. This study suggests the need to conduct a survey with more devices to check other areas of monitoring and assistance to elderly mobility by this tecnologies.

References

[1] Sousa L, Galante H e Figueiredo D 2003 Quality of life and well-being of the elderly: an exploratory study in the Portuguese population. *Rev. Saúde Pública* **37** 363-371.
[2] Sass G G, Alvarenga M R M, Oliveira M A C e Facenda O 2012 Information system for monitoring the health of elderly. *J. Health Inform.* **4** 209-215.
[3] Popescu M and Mahnot A 2012 Early Illness Recognition Using Multiple Instance Learning. *Methods Inf Med* **4**.
[4] Fozard J L 2012 The relevance of Gerontechnology for medical, biological and behavioral studies on aging *Gerontechnology* **11** 116-117.
[5] Saliba D, Elliot M, Rubesntein L Z, Solomon D H, Young R T, Kamberg C J, Roth C, MacLean C H, Shekelle P G, Sloss E M e Wenger N S 2001 The vulnerable elders survey: a tool for identifying vulnerable older people in the commnuty. *J Am Geriatr Soc.* **49** 1691-1699.
[6] Skubic M, Alexander G, Popescu M, Rantz M and Keller J A 2009 Smart Home Application to Eldercare: Current Status and Lessons Learned. *Tech and Health Care* **17** 183 – 201.
[7] Mohammadi M, Van de Koppel J, Van Beek M and Godfroij H 2012 Future smart (care) homes: An explorative research into the future trends regarding strategies for the smart living environment for older citizens. *Gerontechnology* **11** 186.

[8] Költzsch Y, Gövercin M, Spehr J, Gietzelt M and Marschollek M 2012 Identification of sensor-based parameters that predict falls of older people. *Gerontechnology* **11** 230-231.
[9] World Health Organization (WHO). ICF - International Classification of Functioning, Disability and Health. São Paulo, Edusp, 2008, 325p.
[10] Moreira M.A. Sensors Systems. National Institute for Space Research – INPE. 2001
[11] RIA – Robotics Institute of America.Carnegie Mellon The Robotics Institute. S.d. Disponível em: http://www.ri.cmu.edu/index.html.
[12] NARASIMHAN R 2012 Skin-Contact Sensor for Automatic Fall Detection *34th Annual International Conference of the IEEE EMBS* San Diego, California USA.
[13] Belshaw M, Taati B, Snoek J, and Mihailidis A 2011 Towards a Single Sensor Passive Solution for Automated Fall Detection *Conf Proc IEEE Eng Med Biol Soc.* 1773–1776.
[14] Zhao G, Mei Z, Liang D, Ivanov K, Guo Y, Wang Y and Wang L 2012 Exploration and Implementation of a Pre-Impact Fall Recognition Method Based on an Inertial Body Sensor Network *Sensors* **12** 15338-15355.
[15] Zhang Q, Ren L and Shi W 2013 HONEY: A Multimodality Fall Detection and Telecare System. *Telemedicine and e-health* **19** 415-429.
[16] GE reports. Home health monitoring growth eyed in QuietCare buy. 2009. Disponível em http://www.gereports.com/home-health-monitoring-growth-eyed-in-quietcare-buy/. Acessado em 15 de maio de 2013.
[17] Health sense 2011 eNeighbor® Remote Monitoring System Disponível em:<http://healthsense.com/index.php/products/remote-monitoring/eneighbor> Acesso em: 15 mai. 2013.
[18] Belbachir A.N. Ambient assisted living joint Programme - CARE. 2009. Disponível em http://www.aal-europe.eu/projects/case/. Acessado em 24 de maio de 2013.
[19]Hospedales M. Ambient assisted living joint Programme - eCAALYX. 2009. Disponível em http://www.aal-europe.eu/projects/ecaalyx/. Acessado em 24 de maio de 2013.
[20] Simon J.R. Ambient assisted living joint Programme - HOPE. 2009. Disponível em http://www.aal-europe.eu/projects/hope/ Acessado em 24 de maio de 2013.
[21]Karkowski I.P. Ambient assisted living joint Programme - ROSETTA. 2009. Disponível em http://www.aal-europe.eu/projects/rosetta/. Acessado em 24 de maio de 2013.
[22]Rodriguez A. Ambient assisted living joint Programme – SOFT CARE 2009. Disponível em http://www.aal-europe.eu/projects/softcare/. Acessado em 24 de maio de 2013.
[23] Y. KÖLTZSCH, M. GÖVERCIN, J. SPEHR, M. GIETZELT, M. MARSCHOLLEK. Identification of sensor-based parameters that predict falls of older people. Gerontechnology 2012;11(2):230-231.
[24] E. STONE, M. SKUBIC. Capturing in-home gait parameters using vision-based sensing. Gerontechnology 2012;11(2):260; doi:10.4017/gt.2012.11.02.648.00.
[25] Moore E A 2010 Android app Tell My Geo helps track loved ones Disponível em:<http://news.cnet.com/8301-27083_3-20025912-247.html> Acesso em: 15 mai. 2013.
[26] Maglaveras N. Ambient assisted living joint Programme – REMOTE. 2009. Disponível em http://www.aal-europe.eu/projects/remote/. Acessado em 24 de maio de 2013.
[27] J.S. ZELEK. Tactile personal navigation belt. Gerontechnology 2012;11(2):351; doi:10.4017/gt. 2012.11.02.195.00
[28]Campi M 2011 Company launches in the country cellphone for medical purposes Available in:<http://info.abril.com.br/noticias/tecnologia-pessoal/empresa-lanca-no-pais-celular-para-fins-medicos-22032011-32.shl> Access in: 15 may. 2013.
[29] Desingboom 2011 Available in:<http://www.designboom.com/technology/the-aid-by-egle-ugintaite-fujitsu-design-award-2011-grand-prize/> Access in: 15 may. 2013.
[30] P-E. HSU, C. GEISER, K-W. CHANG, J-M. LU, Y-L. HSU. Development of an intelligent robotic wheelchair. Gerontechnology 2012;11(2):340; doi:10.4017/gt.2012.11.02.250.00

[31] Rentschler, Andrew J (2004) Engineering and Clinical Evaluation of the VA-PAMAID Robotic Walker. Doctoral Dissertation, University of Pittsburgh. Disponível em http://d-scholarship.pitt.edu/6574/

[32] Bäck I, Kallio J, Perälä S and Mäkelä K 2012 An assistive humanoid robot for elderly care *Gerontechnology* **11** 357.

[33]Cavallo F, Aquilano M, Bonaccorsi M, Mannari I, Carrozza M C and Dario P 2011 *33rd Annual International Conference of the IEEE EMBS* Boston Massachusetts USA.

[34] Cook, A.M. & Hussey, S. M. (1995) Assistive Technologies: Principles and Practices. St. Louis, Missouri. Mosby - Year Book, Inc.

[35] McCreadie C, Tinker A. The acceptibility of assistive technology to older people. Ageing and Society. 2005;25:91–110.

[36] Guimarães LHCT, Galdino DCA, Martins, FLM, Vitorino DFM, Pereira KL, Carvalho EM. Comparison of the propensity for falls among elderly people who practice physical activity and sedentary elderly. Rev Neurociências. 2004

[37]Ganança F.F. et al. Circumstances and consequences of falls in elderly with chronic vestibular disorder. Revista Brasileira de Otorrinolaringologia. 2006;72(3):388-93.

[38]Okamoto I.H.; Bertolucci P.H.F. Memory loss in the elderly. Available in http://www.unifesp.br/dpsiq/polbr/ppm/atu5_01.htm. Access in 25 may. 2013.

9th Argentinean Bioengineering Society Congress (SABI 2013) IOP Publishing
Journal of Physics: Conference Series **477** (2013) 012011 doi:10.1088/1742-6596/477/1/012011

Development of Object Simulator for Radiation Field of Dental X-Rays

L F Silva[1], F C L Ferreira[1], F F Sousa[1], L X Cardoso[2], E D S Vasconcelos[3], L M Brasil[3]

[1]Federal University of Para (FUPA), Physics Department (PD), Maraba/PA, Brasil
[2]Faculty Sena Aires (FACESA), Engineering Department (ED), Valparaiso/GO, Brasil
[3]University of Brasília (UnB), Biomedical Engineering (EB), Brasília/DF, Brasil

E-mail: lilianf65@gmail.com, fernacarlaluan@gmail.com, ffs@ufpa.br, fiseandro@gmail.com, profafranio2011@gmail.com, lmbrasil@gmail.com

Abstract. In dentistry radiography is of fundamental importance to the dentist can make an accurate diagnosis. For this it is necessary to pay attention to the radiological protection of both the professional and the patient and control image quality for an accurate diagnosis. In this work, quality control tests were performed on X-ray machines in private dental intraoral in the municipality of Marabá, where they measured the diameters of the radiation field to see if these machines are in accordance with the recommendations, thus preventing the patient is exposed to a radiation field higher than necessary. We will study the results of each X-ray machine evaluated. For this we created a phantom to assess the size of the radiation field of X-ray dental, where we measure the radiation field of each device to see if they are in accordance with the recommendations of the ordinance No. 453/98 – MS.

1. Introduction

Many researchers from Brazil and around the world have published studies on radiological protection and the result of these studies, were created radiological protection standards and standardization body such as the International Atomic Energy Agency (IAEA), the TECDOC-796, published in 1995, which radiation dose is in radiodiagnosis and methods for its reduction [1].

Another document was published IAEA by the International Code of Practice for Dosimetry in Radiodiagnosis in 2007 [2]. Besides the IAEA has the International Commission on Radiological Protection (ICRP) founded in 1928, is considered the benchmark for many countries for the development of guidelines for radiological protection [3].

In Brazil the set of rules governing the operation of the radiology services is specified in ordinance No. 453 of the Ministry of Health - Brazil, dated 01/06/1998 - MH [4].

This ordinance was supplemented by resolution No. 64 of the National Health Surveillance Agency (NHSA) [5] and the document medical radiology: Safety and performance equipment [6].

These documents were established test parameters for execution quality beam radiology and proper criteria to be taken into account for assessing of the conformity equipment [7].

Services that work with radiology medical and dental care to comply with the specifications of ordinance No. 453/98 should do periodically for quality control analysis of radiological parameters. The evaluation of these parameters directly helps in control of radiology equipment.

The radiation dose received by the patient is linked to appropriate quality control and quality of a diagnostic x-ray beam. This dose should be reduced without loss to obtain the radiographic image.

Moreover, the major goal of radiation protection requires that: "The occupational exposures and public exposures arising from diagnostic radiology should be optimized to a value as low as practicable"[4].

1.1. Phantom

The phantom was created with the intention of using it in testing quality control of X-ray machines, of the training professionals and also to assess the overall quality of images for accurate diagnosis.

This testing tool has the function of reproducing characteristics of human tissues or organs in routine dosimetric in procedures radiology.

In phantom can also be introduced human tissues or organs. These phantoms, as well as help in the quality control of X-ray equipment, also assist in the training of professionals.

You could say that the phantoms are classified into: Phantom dosimetric Calibration Phantom and Phantom Image Anthropomorphic all tests used in quality control equipment, dosimetry and training professionals.

1.2. Radiation field

The radiation field of X-ray equipment should be limited (collimated) to the region of interest of the diagnosis. In apparatus for dental intraoral radiography is a diaphragm which limits the size of the radiation field to which the patient is exposed, the diaphragm is composed of a lead sheet with a central hole, trapped in the head tube.

Besides being limited by the diaphragm, the field size is also limited by a collimator, which is generally cylindrical; 20 cm in length approximately, consists of lead coupled to the head.

The first diaphragm is to limit the radiation beam being located near the exit of the X-ray tube. The collimator further limits the beam after its passage through the diaphragm.

In intraoral images the radiation field must have a maximum diameter of 6 cm at the output end of the collimator. The collimator must have a minimum length that varies from: 18 cm equipment to peak voltage less than or equal to 60 KVp, 20 inch voltage to between 60 and 70 KVp, 24 inch and if the voltage is greater than 70 KVp [4].

1.3. Radiographic film

The following certain standard radiographic films for its production, a pattern which "consists of a thin layer to the plastic base coated with a radiation sensitive emulsion. This emulsion comprises grains of silver bromide (AgBr) suspended in Gelatine "[8].

This emulsion has the function to absorb radiation during exposure to X-rays and produce a latent image, after developing, is transformed into radiographic images. As the grain silver bromide (AgBr) are more sensitive to radiation than visible light, then the dental X-ray films must be protected from light. Within the plastic base lead plate there is positioned one behind the film which has the function of reducing the dose to the patient.

The objective of this study is the evaluation of dental X-ray machines in eleven offices in the city of Maraba/PA.

It is noteworthy that the main objective of this work was to develop a phantom to assess the size of the radiation field of dental x-ray.

It is worth noting that the measures radiation field must be in accordance with ordinance No. 453/98/ MH, and the intraoral images, the size of the radiation field should not be more than 6 cm in diameter.

2. Materials and Methods

For the development of a phantom radiation field dental, we used raw materials and low cost dental use (Figure 1), which serve as support for the alignment of radiological films and to measure the diameter of the radiation field of dental x-ray.

For the cast of phantom test radiation field was used alginate to mold 4 movies together, forming a rectangle, to extract the mold of films together, put up the alginate films and over with the plastic and without protection. After the mold was applied to the self-curing acrylic, a 2:1 ratio of acrylic and liquid (Figure 1). The dimensions of the phantom is 9 x 11 x 1 cm^3 (external) and 8.2 x 6.2 x 0.5 cm^3 (internal).

In the tests we used the phantom and in it were inserted radiographic films. Subsequently the films were subjected to X-ray exposure of after exposure, the film was removed for revelation in a darkroom.

In the revelation of the films used to the Kodak brand and solution development time of the films in most offices were 1-2 minutes, and the office D which took more time for the revelation of the films taking 3 minutes to reveal them. But that does not harm or damage the film developed.

Figure 1. Phantom for the radiation field.

To measure the size of the radiation field of each X-ray equipment was analyzed according to the limit recommended by the ordinance No. 453/98 of the Ministry of Health/Brazil, acceptance limit of 6 cm and allowed ± 0.5 cm. For this, the four periapical films were aligned in phantom radiation field and directly exposed to the primary beam. We can see this in Figure 2.

Figure 2. Films lined up for evaluation of the radiation field

After development, the films were positioned and aligned on the phantom to measure the size of the diameter with a ruler of the field represented in Figure 3.

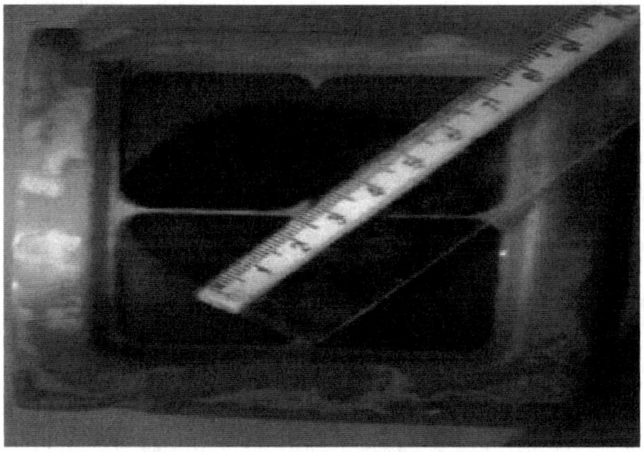

Figure 3. Measuring the diameter of the radiation field of the films revealed

3. Results and Discussion

According to tests conducted to assess the efficacy of phantom radiation field and in possession of the results obtained, it was observed that the phantom showed good results for the measurements of the radiation field.

After analysis of the phantom tests conducted in eleven X-ray equipment, it was observed that most of the equipment is within the standards recommended by NHSA.

Already equipment or X-ray equipment (E and M) whose sizes were measured radiation fields, respectively 7.0 cm and 7.3 cm, are releasing a radiation area larger than necessary to the image diagnosis which results, or are above the levels recommended by NHSA, thereby causing an increase in the absorbed dose in relation to patients and medical professionals.

Another aspect that needs attention is that some X-ray equipment also had a diameter of the radiation field smaller than 6.0 cm. The equipment (5.6 cm) and L (5.5 cm) below are recommended by NHSA. The results of the test radiation field using the phantom are detailed in Table 1.

Table 1 - Results of tests for measuring the diameter size of the radiation field.

Equipment	Radiation Field Measurement (cm)
A	5,6
B	6,0
C	6,0
D	6,0
E	7,0
F	5,6
G	6,0
H	7,3
I	6,0
J	6,0
L	5,5

It is noteworthy that the largest diameter field radiation found in X-ray equipment was the equipment H (7.3 cm), field diameter greater than recommended by the ordinance No. 453/98.

For information purposes, the revelation of the films was used a darkroom red and the eleven and portable X-ray equipment evaluated, four of them were to trigger analog equipment (devices D, E, F and I).

6. Conclusion

From the field test conducted in radiation X-ray equipment Maraba/PA, it can be concluded that the majority of dental X-ray equipment is within the standards of ordinance No. 453/98.

Thus, it can be seen that the percentage corresponding to 81.8% within the recommended standard of quality.

The phantom radiation field efficacy tests showed and proved how important helper for the correct positioning of film radiographic film used in dental radiography.

Thus presents itself as a safe method for testing and field measurements of radiation for dental X-ray machines.

This phantom, unlike the constituent material of the phantoms used in these tests may be made with good quality raw material and low financial cost, may be a good alternative for the quality control of these devices.

From these studies we will begin the program of quality control in Maraba/PA following the rules of NHSA. This is a very important step for the correct application of the conditions related to radiological protection both for professionals working in the field of radiology, as the population of Maraba/PA and the northern region of Brazil.

Here, we will expand our field of study with respect to diagnostic imaging which will best prepare, retrain and guide professionals on standards NHSA for quality control of equipment and images and also on the implementation of standards and legislation on radiological protection in force in Brazil.

Acknowledgment

The authors thank PROPESP/UFPA for financial support and the dental Marabá/PA for technical support.

References

[1] IAEA: TECDOC – 786 1995. Radiation dose in diagnostic radiology and methods for doses reduction. International Atomic Energy Agency. Viena.

[2] IAEA: Technical Report Series – TRS N° 457 2007. Dosimetry in diagnostic radiology: an international code of practice. International Atomic Energy Agency. Viena.

[3] ICRU REPORT 48 1992. Phantoms and Computational Models in Therapy, Diagnosis and Protection.

[4] BRASIL, MINISTÉRIO DA SAÚDE 1998. Diretrizes de proteção médico e odontológico. Portaria N° 453. 1 de junho de 1998.MS/SVS.

[5] BRASIL: MINISTÉRIO DA SAÚDE. AGÊNCIA NACIONAL DE VIGILÂNCIA SANITÁRIA 2003. Uma orientação técnica sobre guia de procedimentos para segurança e qualidade de imagem em radiodiagnóstico médico, Resolução - RE, No. 64 de 04 de abril.

[6] BRASIL: MINISTÉRIO DA SAÚDE. AGÊNCIA NACIONAL DE VIGILÂNCIA SANITÁRIA 2005. Radiodiagnóstico Médico: Segurança e Desempenho de equipamentos.

[7] BELINATO, W. 2010. Avaliação de Parâmetros Físicos em Radiologia Odontológica de Consultórios Públicos de Sergipe, Dissertação de Mestrado, Universidade Federal de Sergipe.

[8] DEZOTTI, M. S. G. 2003 Avaliação de Filmes Radiográficos Periapicais em Diferentes Condições de Processamento pelos Métodos Sensiométrico, Digital e Morfométrico. Tese de Doutorado, Universidade de São Paulo.

Journal of Physics: Conference Series **477** (2013) 012011 doi:10.1088/1742-6596/477/1/012011

Contatos:
Edgard Rogério de Siqueira Vasconcelos
Address: Q. 43, lot 78, house 68
District: East Sector
City: Gama-Brasilia
State: Distrito Federal
Country: Brazil.
E-mail: profafranio2011@gmail.com.
Telephones: 0xx55-61-3484-8899, 0xx55-61-8529-7603

Design, implementation and characterization of a BioMEMS testing device.

F Schaumburg[1], M C Perez[1,2], Zalazar M A[1], F A Guarnieri[1,2]
[1] Laboratorio de BioMEMS, Facultad de Ingeniería, Universidad Nacional de Entre Ríos, Ruta 11 Km 10,Oro Verde (3100), Entre Ríos, Argentina.
[2]Centro Internacional de Métodos Computacionales en Ingeniería (CIMEC), (INTEC-CONICET-UNL), Colectora Ruta Nacional 168 Km 472, Santa Fe, Argentina.

E-mail: schaumburg.f@gmail.com, mcperez78@gmail.com, martin.zalazar@fulbrightmail.org, aguarni@santafe-conicet.gov.ar

Abstract. Several issues have been reported regarding aqueous shunts used in the treatment of glaucoma. To attend to these problems an active implantable valve, based in MEMS (mirco electromecanic structure) processes, is proposed. This BioMEMS would be fed wirelessly and would be composed of an integrated circuit, an electrochemical actuator and microfuidics. In order to assess functionality and to optimize these blocks, it is desirable to test them prior to the development of the implant. In this paper a circuit for testing the energy source, the RF link, and the actuator of the active valve was developed and characterized.

1. Introduction

About 1% of world's population suffers from glaucoma, a disease of uncertain etiology. For that reason no cure was still developed and the available treatment options act over the main symptom: the elevated intra-ocular pressure. Those who suffer from this condition can choose among eye-drops, pills, and valves. Eye-drops are suitable during the first stages of glaucoma, but as the disease follows its course other measures are needed. Implantable valves are the final option, since they require surgery, are expensive, are molest because of their size and their conduits get obstructed frequently. Finally, as the disease follows its natural course, the amount of liquid to be drained increases. [1][2][3][8][9][10].

Thus, there is a need of valves allowing large, variable actuator displacement and, as a consequence, permitting big molecules to pass through the valve without risk of obstruction, and providing variable hydraulic impedance. Large displacements in the actuator involve the use of big amounts of energy that can only be provided by active systems. Active systems require an energy source and extra blocks, which can be translated to bigger size and complexity for the valve. This might explain why available glaucoma valves remain passive.

To serve these problems an active valve is proposed using MEMS technology. The nano and microscale size of the components will allow the addition of the extra blocks needed for the active system, which will deliver the energy that the actuator needs. An external power source will provide the energy via a 13.56MHz inductive coupling. A DBS-PPY electrochemical diaphragm fed by a potentiostat, will be used as an actuator. An application specific integrated circuit (IC), will condition the signal induced in the internal coil to power the potensiostat, and provide for the reference voltage. This reference will vary according to the required actuation level, and will be proportional to the duty cycle of the 13.56MHz amplitude modulated signal.

Other implants have already used the idea of including an IC and a telemetric link. Well known examples are pacemakers. In addition, other implantable devices have already implemented the usage of wireless power supplies. Such is the case of cochlear implants, devices for health information and patient identification, and retinal implants. However, there are no reports about a wirelessly powered active ocular implant for the treatment of glaucoma. [4] [5] [6] [7].

In this paper a testing circuit, called discrete transponder (TPD) is proposed, implemented and characterized. The TPD emulates the behavior of the IC, thus allowing the improvement of its design and, above all things, the characterization of the actuator and the RF link. Finally, it will provide the first information about the overall performance of the system.

2. Generation of the RF stimulation signal

The RF stimulation signal should have the characteristics listed in Table 1

Table 1. RF stimulation signal desired parameters.

Parameter	Value	Unit
Carrier frequency	13.56	MHz
Modulation frequency	485	KHz
Modulation	ASK	
Modulation Index (MI)	>30	%
Power	400	mW
Working cycle resolution	3.57	%

In order to generate the RF stimulation signal, an evaluation board provided by Texas Instruments, more precisely a TRF7960AEVM, was modified. The firmware of the module was rewritten in order to meet the features named in Table 1.

3. Design and implementation

3.1. Requirements for the TPD

The TPD should provide a means for testing PPY-DBS electrochemical actuators. So as to be able to fix the excitation voltage, a potentiostat should be used. In order to provide full electrical characterization of the actuator, a means for measuring the current delivered to the electrochemical cell´s working electrode should be included. Furthermore, the TPD should receive wirelessly information about the level of actuation. In order to avoid a high power RF source, the TPD must be fed by a source other than the RF power.

The supply block of the TPD should receive at least ±6V and provide the rest of the circuit with ±5V regulated. In order to stimulate the potentiostat, the TPD should generate a reference voltage proportional to the RF signal working cycle. This reference should be in the ±1V range, as higher voltages may damage the actuator.

3.2. TPD Blocks
To meet the requirements previously presented, the following blocks were considered for the TPD circuit.

3.2.1. RF acquisition. Comprising a tank circuit tuned to 13.56 MHz, and adapted to the impedance of the rectification block.

3.2.2. RF rectification, filtering and offset removal. The purpose of these blocks is to filter the carrier and to obtain the modulated signal. To achieve this, a diode bridge and a first order lowpass filter with cutoff at 1 MHz, was used.

3.2.3. Range correction. The amplitude of the induced signal depends on variables such as the distance and coil alignment. In order to fix the range to a known value, a transistor working in its cutoff and saturation regions was used.

3.2.4. Reference voltage generation. So as to obtain a voltage propotional to the RF signal working ratio (V_{ref}), a second order lowpass filter with cutoff at 66.6 Hz, was used. An offset correction stage was also included in this block.

3.2.5. Potentiostat. A potentiostat maintains the voltage between the working (WE) and the reference electrodes (RE) of an electrochemical cell, to a fixed reference value. This reference is provided by the previous block.

3.2.6. Working electrode current measurement. In order to measure this current, a transimpedance amplifier connected the WE was used.

3.2.7. Power supply. This block, generates ±5V required by the active blocks.

A block diagram of the circuit is shown in figure 1, where all the blocks were identified.

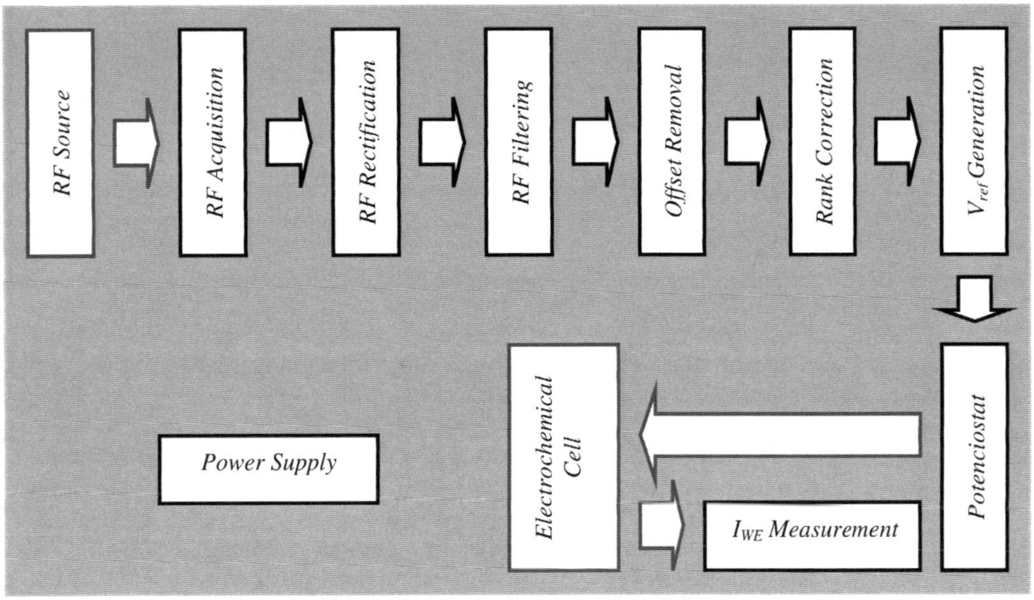

Figure 1. Block diagram for the TPD.

4. Characterization

To evaluate the behavior of the circuit when exposed to a RF stimulation signal, several test were performed. All the informed voltages are measured with respect to ground.

4.1. RF source characterization

The TRF7960AEVM module (the RF source) was characterized as regards to carrier and modulation frequency, IM and WC using an Instek GDS-806C oscilloscope. The setup used, is shown in figure 2. All different possible WC were set, and on and off times were measured. Figure 3 shows the waveform obtained for a WC = 75 %. Table 2 shows the measured values of the RF signal, and table 3 shows the calculated WC values.

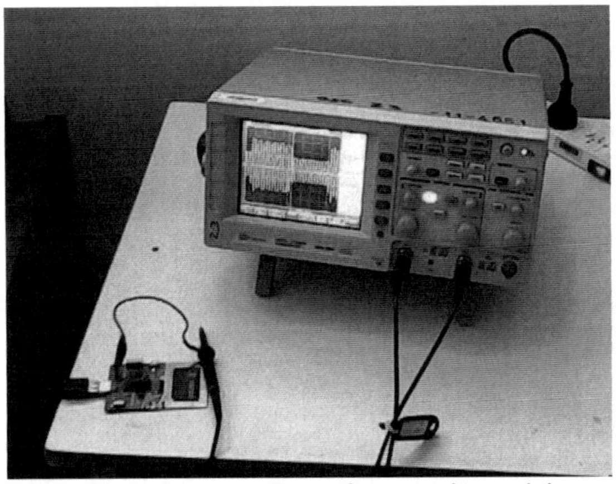

Figure 2. Laboratory Setup for experience 4.1.

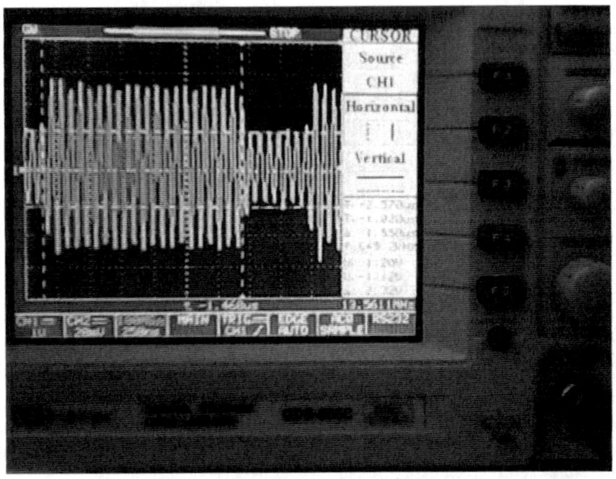

Figure 3. RF excitation signal with WC = 75%.

Table 2. RF excitation signal measured parameters.

Parameter	Value	Unit
Carrier frequency	13.5611	MHz
Modulation frequency	482.85	KHz
Modulation Index (MI)	45	%

Table 3. RF excitation signal measured parameters.

Theoretical WC (%)	T_{on} (ns)	T_{off} (ns)	T_{total} (ns)	Measured WC (%)	Error (%)
0	0	2065	2065	0	0
35.71	720	1360	2080	34.6153846	1.09461538
39.28	810	1250	2060	39.3203883	0.04038835
42.86	890	1190	2080	42.7884615	0.07153846
46.42	1031	1110	2141	48.1550677	1.73506773
50	1040	1030	2070	50.2415459	0.24154589
53.57	1110	960	2070	53.6231884	0.05318841
57.14	1190	880	2070	57.4879227	0.34792271

4.2. Induced amplitude vs. distance

Amplitude of the induced voltage was obtained for several distances between the RF source and the TPD. For this purpose, the TRF7960AEVM connected to a PC, an Atten ADS1102CAL oscilloscope, two graving arms, and a ±12V power source, were used. This setup can be seen in figure 4. WC of the RF excitation signal was set to 50%.

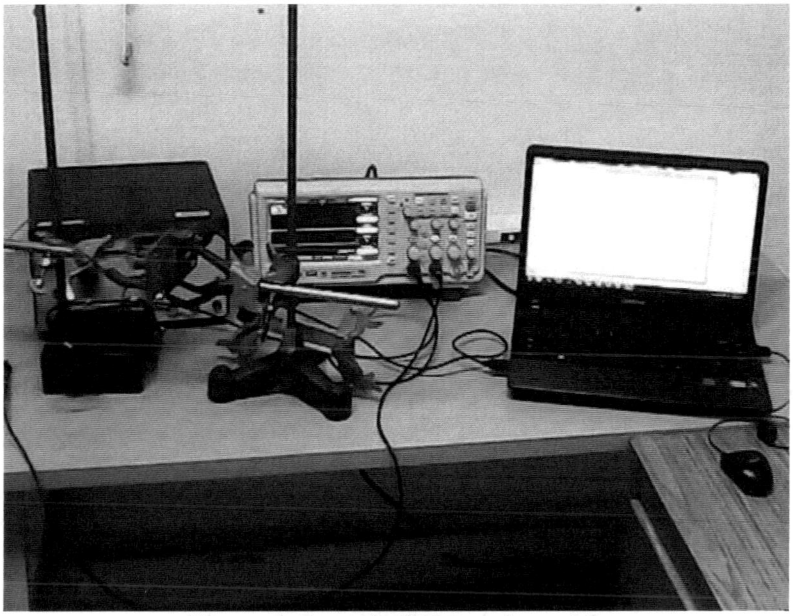

Figure 4. Setup for RF experience 4.2.

The measure was not performed in the tank circuit, because the oscilloscope probe capacitance would have tuned the circuit to another frequency. In return, the voltage in the RF Filtering block was measured. High and low levels of the square wave obtained in this block (the modulation signal) are plotted in figure 5 for different distances.

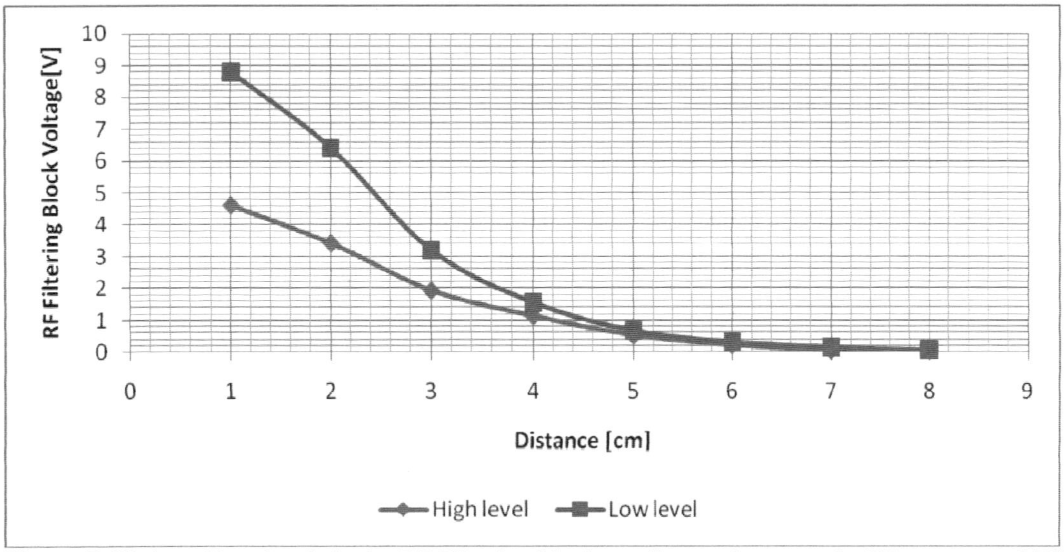

Figure 5. High and low levels in the RF Filtering block vs. distance between the RF source, and TPD.

4.3. Induced amplitude vs. coupling angle

The induced amplitude was obtained for several coupling angles. These, were measured between the projections of the coils planes. Setup used was similar to the one used in 4.2. The distance between the center of the coils of RF source and TPD, was equal to 4 cm. Results are shown in fig. 6.

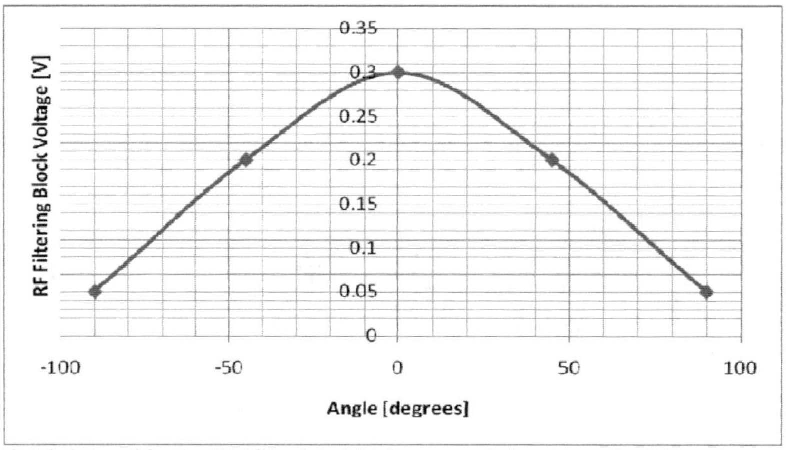

Figure 6. High level in the RF filtering block vs. angle between the RF source and TPD.

4.4. Range correction vs. distance

The voltage amplitude in the range correction block was obtained for several distances between the RF source and the TPD. Setup used was identical to the one in 4.2. Results are shown in fig. 7.

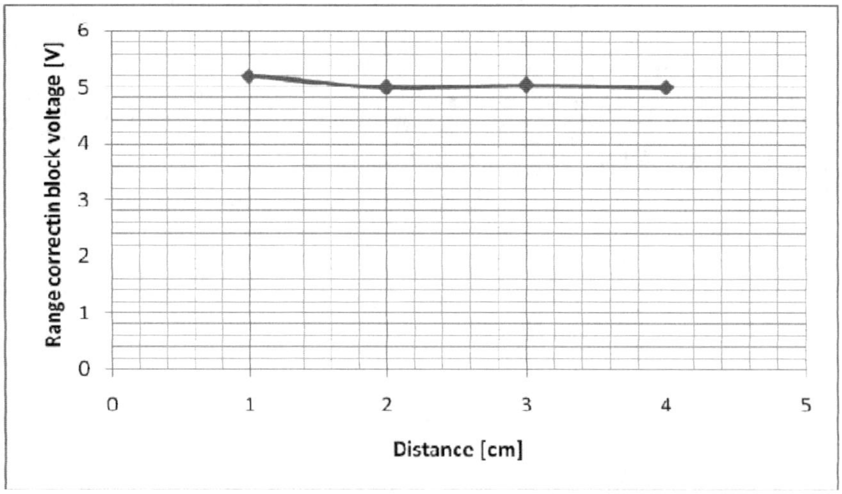

Figure 7. High level in the range correction block vs. distance between the RF source and TPD.

It has to be noted that the waveform obtained at this instance was not perfectly square, and changed with distance.

4.5. TPD WC vs. RF source WC

Using a setup similar to 4.2, but keeping the distance between the RF source and the TPD fixed to 2 cm, WC in the Range correction block of the TPD was measured for each possible WC of the RF source. In figure 8, results can be seen.

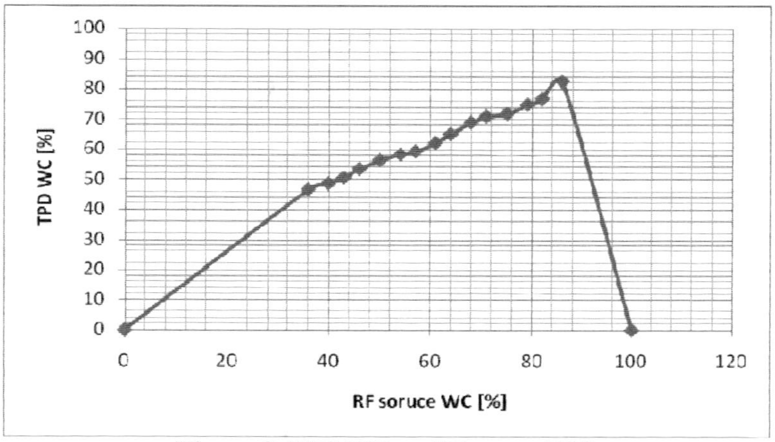

Figure 8. TPD WC vs. RF source WC.

4.6. V_{ref} vs. distance

V_{ref} was measured for several distances between the RF source and the TPD. Setup used was identical to the one in 4.2. Results are shown in figure 9.

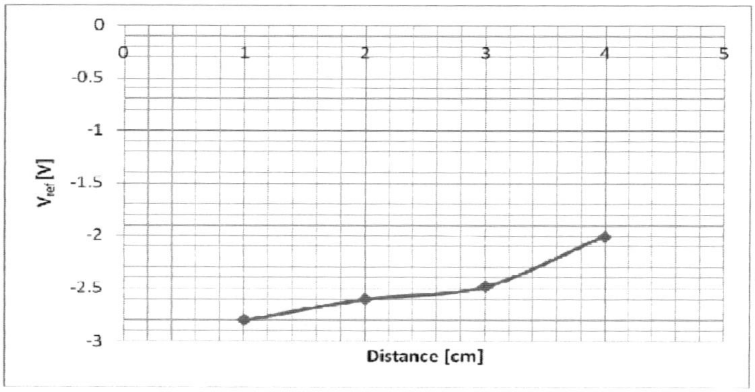

Figure 9. V_{ref} level vs. distance between the RF source and TPD.

4.7. V_{ref} vs. distance

V_{ref} was measured for all the possible values of WC. Setup used was identical to the one in 4.5 and results are shown in figure 10.

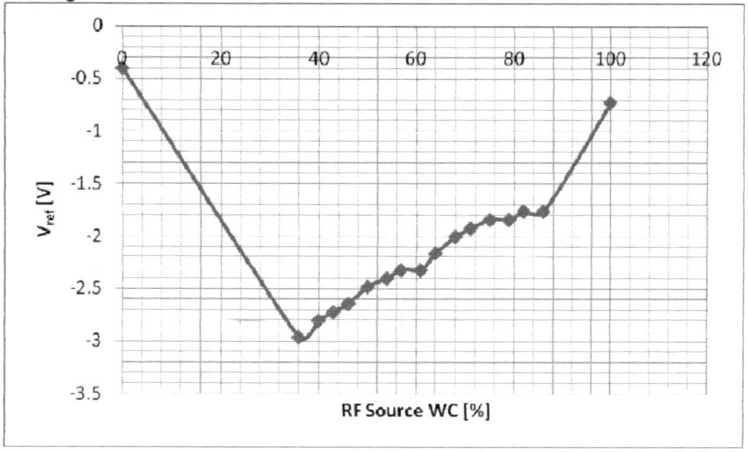

Figure 10. V_{ref} vs. RF source WC.

5. Discussion

The RF source modulation index, carrier and modulation frequencies took design values given in table 1. Perceptual error between software set working cycle, and measured working cycle was minor to 2% in all cases.

Test 4.2, showed typical $1/r^2$ decay in induced voltage, as distance increased. Test 4.3 showed that maximum coupling occurred when coils were parallel, and no coupling at all was found when angle between planes was 90°. This is logical since in the center of the RF source coil, magnetic flux is normal to it. This means that no flux will pass through the TPD coil when the angle is 90° and no voltage will be induced according to Faraday's Law. In the other hand, the contrary situation (parallel coils) will cause maximum induction.

Range correction block, was characterized in test 4.4. Although the waveform high level value was the expected, the shape was not perfectly square and changed with distance. This phenomenon had repercussions in test 4.6, where V_{ref} varied with distance even though it was expected no to.

Figures 8 and 10, show that V_{ref} and TPD WC change lineally with RF source WC except when WC = 0 and 100%. The reason is that a pure 13.56 MHz excitation signal, will not pass the RF Filtering block. If these values for WC are removed, linearity becomes evident (see fig. 11 and 12). In fact, R^2 parameters remain close to the unity in both cases.

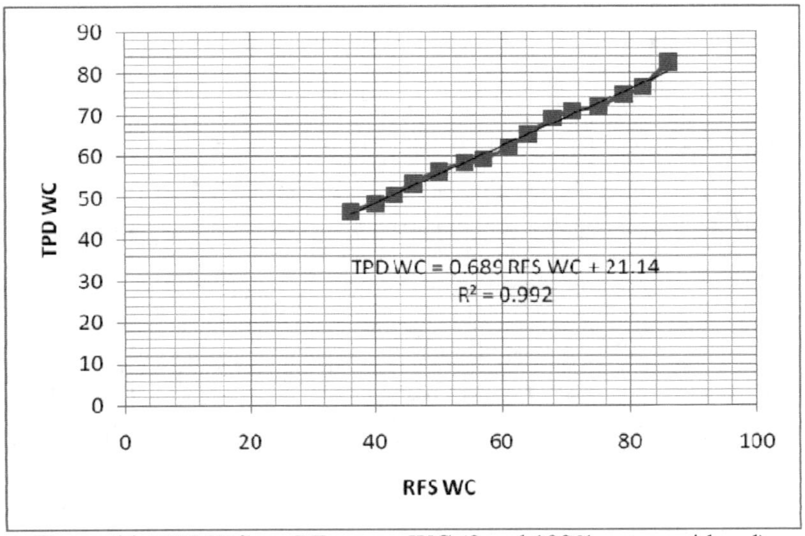

Figure 11. TPD WC vs. RF source WC (0 and 100% not considered).

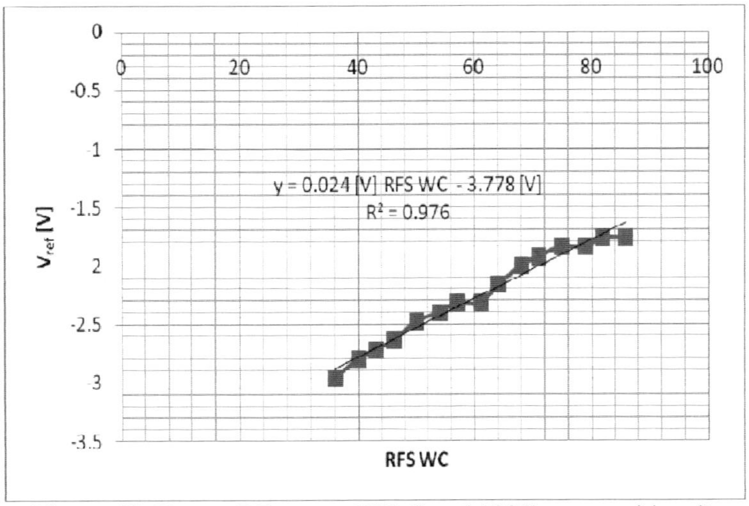

Figure 12. V_{ref} vs. RF source WC (0 and 100% not considered).

6. Conclusions

TRF7960AEVM behaved as expected regarding to MI, carrier frequency, modulation frequency and working cycle. The output power remains to be characterized.

Tests 4.2 to 4.7 were designed to test the RF link. TPD responses were the expected with two exceptions. First, a distance variation between the RF source and the TPD produced a V_{ref} variation. This undesired phenomenon may be solved by causing the transistor in the range correction block to enter saturation more easily, and thus obtaining square waveforms. A Darlington configuration should be considered in a future version. Second, working cycles of 0% and 100% will produce non linearities in V_{ref} vs. WC plot. This can be easily solved by not using those values for WC. Leaving those issues aside, the RF coupling performed properly. In order to achieve the desired potentiostat output, the offset correction stage should add 1.89 V to V_{ref}, so that plot at figure 12 is centered at 0 V.

Amendment of these problems, potentiostat characterization, and actuation of PPY-DBS electrochemical cells wirelessly are left for future research.

References

[1] Quigley H A 1996 Number of people with glaucoma worldwide *Br J Ophthalmol* **80(5)** 389-393.

[2] Alañón Fernández F J, Fernández Pérez J, Ferreiro López S 2003 Oftalmología En Atención Primaria *Formación Alcalá.*

[3] Huang M C, Netland P A, Coleman A L, Siegner S W, Moster M R and Hill R A 1999 Intermediate-term clinical experience with the Ahmed Glaucoma Valve implant *A J Ophthalmol* **127(1)** 27-33.

[4] Ashar B S and Ferriter A 2007 Radiofrequency Identification Technology in Health CareBenefits and Potential Risks *JAMA* **298(19)** 2305-2307

[5] Michael K and Masters A 2004 Applications of human transponder implants in mobile commerce *Proceedings of the 8th World Multiconference on Systemics, Cybernetics and Informatics* **5** 505-512

[6] Roessler G, Laube T, Brockmann C, Kirschkamp T, Mazinani B, Goertz M, Koch C, Krisch I, Sellhaus B, Trieu HK, Weis J, Bornfeld N, Röthgen H, Messner A, Mokwa W and Walter P. 2009 Implantation and Explantation of a Wireless Epiretinal Retina Implant Device: Observations during the EPIRET3 Prospective Clinical Trial *Invest Ophthalmol Vis Sci. 50(6)* 3003-8

[7] Moctezuma A, Tu J 2011 An Overview of Cochlear Implant System

[8] Beck A D, Freedman S, Kammer J and Jin J 2003 Aqueous shunt devices compared with trabeculectomy with Mitomycin-C for children in the first two years of life *American Journal of Ophthalmology*, **136**(6) 994–1000.
[9] Hau S and Barton K 2009 Corneal complications of glaucoma surgery *Current Opinion in Ophthalmology* **20(2)** 131- 136
[10] Minckler D S, Francis B A, Hodapp E A, Jampel H D, Lin S C, Samples J R, Smith S D and Singh K 2008 Aqueous Shunts in Glaucoma: A Report by the American Academy of Ophthalmology *Ophthalmology* **115(6)** 1089–1098

Dynamic adaptation software for virtual interconsultation

P.F.Solarz[1], **V.I.Rotger**[2], **A.A.Salas**[1]
[1]Instituto de Bioelectrónica, FM, U.N.T.
[2]Gabinete de Tecnologías Médicas, Dpto. De Bioingeniería, FACET, U.N.T.

E-mail: pablosolarz@gmail.com, vrotger@herrera.unt.edu.ar

Abstract: One of the most important problems in the interconsultation system originated in the rural sphere is to customize the behavior for each specialty, in order to achieve an efficient exchange among the health system actors. In previous experiences with application prototypes, the specialists participants in the requirements engineering process elaborate their specifications as a guide for rural health agent, resulting in so many interfaces as specialities they were consulted. This work present an application web design based on multilayer architecture that allows dynamic adaptation of behavior and user interfaces according to context conditions (such as specialty, interconsultation intended use, consultation, etc). The design also contains a system of electronic health record that essentially serves as a single repository of data from patient encounters and the consultations. The context conditions and the interconsulted speciality are described by metadata which determines the modules that are instantiated, composition of user interfaces and the parts of the patient medical record involved.

Keywords: Telemedicine, eHealth, Health Informatics, Software Architecture

Introduction

The medical information management is complex for several reasons, the variety of the languages used, the variety of submission format, the criticality of the information handled and the multiplicity of scenarios in which it operates.
In the Argentina's northwestern region and much of the country is scattered in offices, clinics and hospitals, and is present in various forms, from traditional one in paper, to different degrees of computerization, but without the adoption of standards that allow interoperability that makes it very difficult to reach.
On the other hand, as well as from the medical point of view it is desirable to have, under ethical and legal conditions that are required, as much information as possible to describe the situation of a patient. The representation of this has to be effective, providing what is needed in each case quickly and under an intuitive format to each specialty. These requirements, in the context of scenarios such as outpatient clinics, intensive care units, interconsultations as specialists, are just a few examples of different

[1] To whom any correspondence should be addressed.
[2] To whom any correspondence should be addressed.

interaction needs of patient information. For these reasons standards like HL7 CDA [1], OpenEHR [2], define specifications at an abstract level.

A very particular scenario is the medical interconsultations where the information exchanged among health workers is focused on a particular specialty. More specifically, in the interconsultations made at the health rural centers [3] is particularly important to make decisions about attending on site or derive, and whose resolution may be more or less urgent according to the patient care capabilities of the place, specialty, etc. Each specialty has particular aspects that are relevant information to make appropriate decisions. At the same time it depends on the capacity in terms of human resources, care's complexity level, transfer possibilities, etc.

In this paper we present a web application design based on multi-layer architecture that allows behavior dynamic adaptation and user interfaces according to context conditions (such as specialty, interconsultation intended use, consultation, etc..), which are described by metadata which determine the modules that are instantiated, the composition of user interfaces and parts of the medical records involved. An electronic health record was developed which serves, in interconsultation context, as repository of data from patient encounters and the consultations.

Rural interconsultation's scenarios

The dual phenomenon of isolation by the medical staff and the lack of accessibility for patients has been addressed in previous work which showed prototypes intended interconsultations multimedia web in general and specialized in dermatology [4][5]. The typical scenario posed for these prototypes is shown in Figure 1, where the actors are, on the one hand: a patient and a Rural Hospital's health worker, which require support service system providing information about the patient's condition. On the other hand, specialists or specialized institutions provide their points of view as interconsulters.

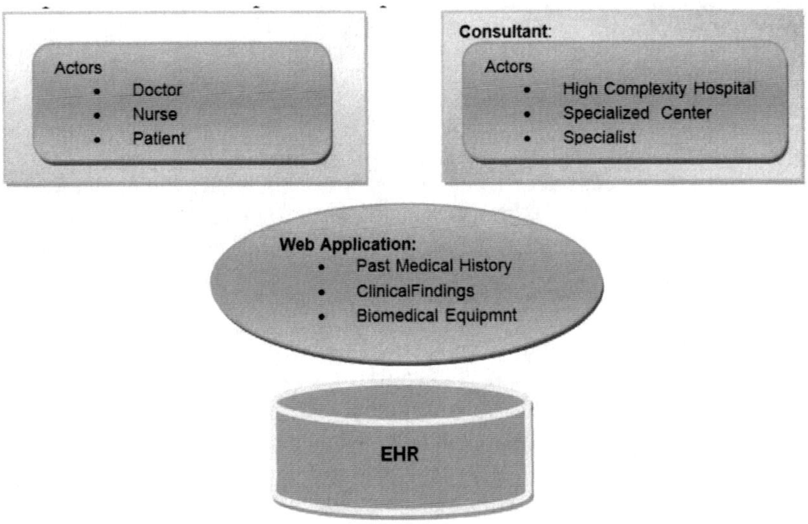

Figure 1: Medical interconsultation scenario

Regarding the information entered, one part comes from the data loaded in rural hospitals on web forms present in the application, designed by referents of each specialty. Another part comes from the output of diagnostic equipment, with different characteristics, which are integrated into the patient information (personal data, clinicians, etc.) that sets the basis of exchange between the actors. The web application manages information between the actors and the repository of electronic health record (EHR).

In previous experiences with the development of prototypes and then working with other specialties, was found that, under the scenario of rural interconsultation, they develop their specifications in a form that serve as a guide to the information required for health workers that establish the consultation. An example of this is the set of user interfaces developed for teledermatology [6], with attributes such as: Reason for visit, number of relatives with the problem, number of deaths with the problem, previous treatments, date of previous treatments, medications during treatment, photographs of injuries, etc, that are not necessarily specific to dermatology, and that can be shared by a group of specialties. Cardiology however, resulted in a more specific set: heart sounds, femoral pulses, palpation of the liver, plus some attributes are also shared as part of physical examination: heart rate, respiratory rate, blood pressure, etc. Specialized Interconsultations from pediatrics usually share perinatal data.

In short, from different specialties are generated some common and some specific attributes. In addition, many of the attributes are compounds: blood pressure (minimum, maximum, patient status, instrument, etc), Heart sounds (1st, 2nd, etc.), and such as, indivisible units.

These different points of view are given in the context of the goal of an unified medical record that can be raised in the current application or be part of a more general system of electronic health record, so regardless of specialty, they are part of a whole. The developed web application design is mainly oriented to rural teleconsultations that have two main parts. The first one: a contextually customized medical record ad hoc and the second one: interconsultations system that works synchronously or asynchronously, in which health workers hold conferences. Both the medical record and the interconsultation application are multimedia. The application, which is in use as a prototype, is developing parallel issues such as security and confidentiality of information. In this paper we address the first part architecture, the second one has been developed in previous work [5][6].

To complete the scenario of discussion is important to consider the available technology and its evolution. In rural areas it is important to have flexible adaptation systems for different interfaces such as PCs, tablets or notebooks without impacting the usability of it, allowing the use of the widest possible range given the limited choice.

Application Architecture

The web application is developed based on a multilayer architecture, showed in figure 2, with persisted data in MySQL relational database and XML (eXtended Markup Language) metadata files [7]. A DAO (Data Access Object) is responsible for the management and the access to both types of data using specific objects. The control layer manages the events that occur at the interface and connect them to the layer containing the business rules of the application.

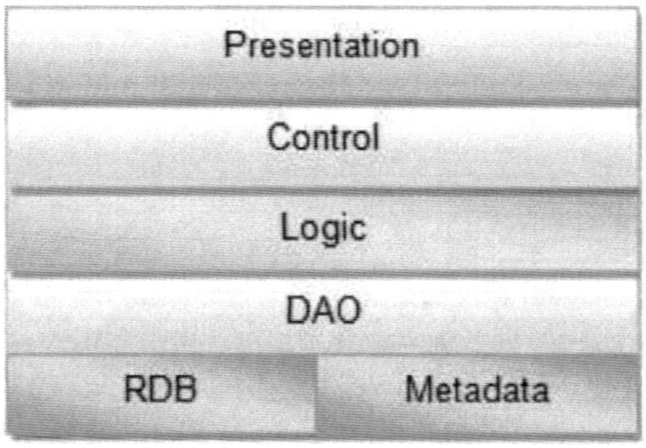

Figure 2: Software Architecture Layers

On the basis of the architecture, we found the data that is structured into a relational database implemented in MySQL. All information generated by the application is in this part. But this layer is also part of the data structure corresponding to each of the specialties in metadata XML files, which contains information about which of the structure elements should be instantiated.

Other metadata file contains information for the presentation's construction of the user interface in strict correspondence with the above, that is to say, according to each speciality profile. In both cases the metadata is organized into two layers: one composed of XML files that describe information units on clinical observations, called Minimal Unit Descriptors (MUD) and another with the XML files comprising the above according to each specialty, called Integration Descriptors (ID). Figure 3 shows a diagram of metadata layer (the same is repeated for the user interfaces), each integrator generally uses more than a minimum unit.

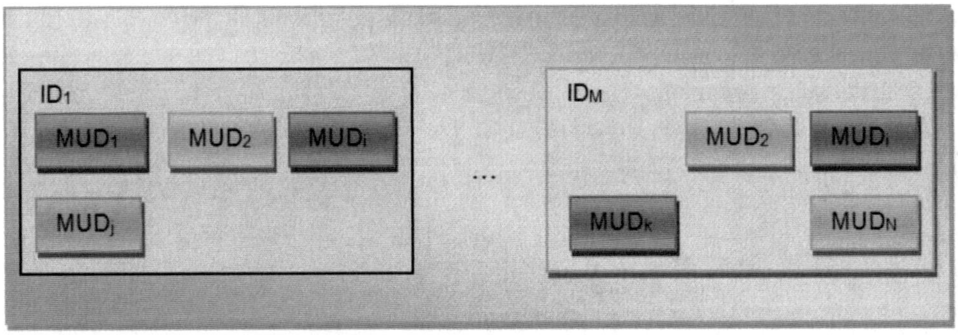

Figure 3: Metadata structure. MUD: Minimal Structure Descriptor. ID: Integration Descriptor

The data management layer, developed in Java, contains a DAO (Data Access Object) that consists of a set of classes that map the tables in a relational database with EclipseLink framework support and another set that solves data management built ad-hoc. This layer is also responsible, with the JAXB (Java Architecture for XML Binding) framework's help, for processing the XML files metadata that characterize the particularities of each specialty and each point of view that the application evolves.

The business layer, also developed in Java, contains the application's logic, by combining and processing the information. This can be deployed in different points of view: interconsultation problem oriented, patient-oriented medical records, statistics, etc. In this case the first was developed. The business layer manages the dynamic integration of information, which is the way in which the application is dynamically specialized, depending on one or more aspects of the context and it enables to guide the rural health worker in relation to the particular information to develop in the interview.

The control layer is composed mainly of JavaBeans that process the presentation layer's events, based on Java Server Faces framework.

Application operation

Figure 4 shows some examples of data types that describe clinical observations. In the example some of the data is shared between the profiles of medical history of diabetes and cardiology, and others are specific. In both cases the observations are integrated in the profiles described in the metadata.

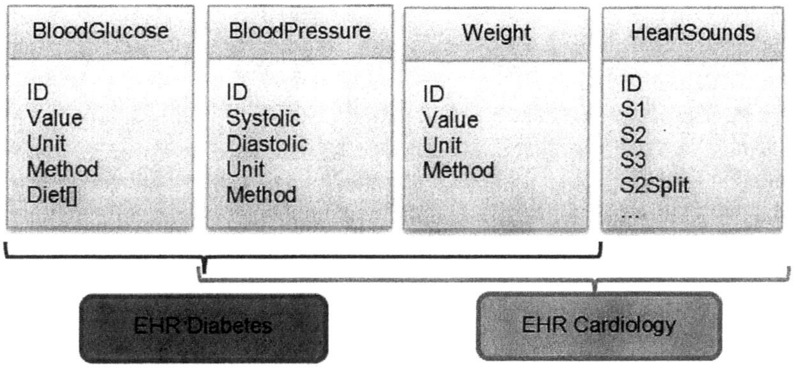

Figure 4: Class reuses by specialized EHRs views

Classes are minimum units (MU) that are at the data model's basis and they are reused at a higher level. They represent measures (blood pressure, etc.), specific parts of the medical history (personal background disease-specific families, diagnostics, etc.), And they have one MUD in the metadata. The composition of these units, to be part of a specific medical record, is done at runtime, with background information that allows to instantiate the integration descriptor.

Figure 5 shows XML descriptors, which are part of the metadata. In part A, the descriptors refer to classes that represent blood pressure and Apgar. In part B, is shown an integration's descriptor portion for pediatric cardiology.

Figure 6 shows the assembly in operation. On the left it is a diagram of the integration descriptor's relationship (ID) and descriptors of the minimum units (MUD). On the right is shown a similar descriptors' relationship, obtained from an engine of existing interface generation in the business layer, generating the components of the presentation file representing patient data.

Journal of Physics: Conference Series **477** (2013) 012013 doi:10.1088/1742-6596/477/1/012013

MUD Example

```
<table name="BloodPressure">
        <ehrAttribute="ID" label="Identifier" type="Integer" cardinality=1 / ehrAttribute >
        <ehrAttribute="Systolic" label="Systolic Pressure" type="Float" cardinality=1 / ehrAttribute >
        <ehrAttribute="Diastolic" label="Diastolic Pressure" type="Float" cardinality=1 / ehrAttribute >
        <ehrAttribute="Unit" label="Unit" type="String" cardinality=1 / ehrAttribute >
</table>
```

ID Example

```
<ehr name="Diabetes">
        <table name="BloodPressure"...    /table>
        <table name="BloodGlucose"         /table>

</ehr>
```

Figure 5a: Minimal Unit Descriptor MUD example, **5b:** Integrator Descriptor example

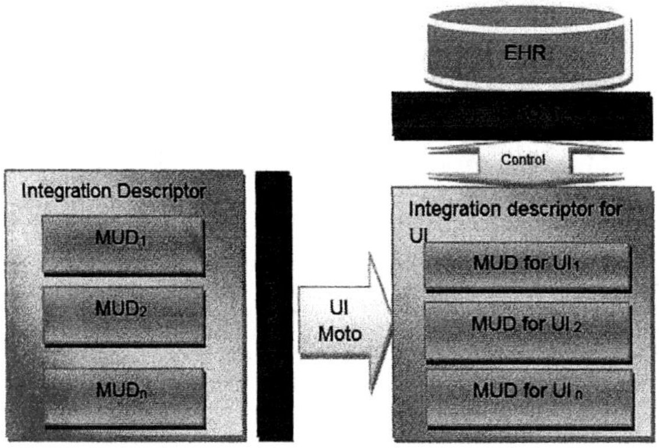

Figure 6: Architecture running. MUD: Minimal Unit Descriptor. UI: User Interface. DAO: Data Access Object. HER: Electronic Health Record.

Journal of Physics: Conference Series **477** (2013) 012013 doi:10.1088/1742-6596/477/1/012013

This architecture can be adapted to the requirements of each specialty, specified by its integration descriptor, adding or removing minimum unit descriptors.

Besides adaptability achieved by integration descriptors and minimum units descriptors, the control layer instantiate different prior descriptors' points of view such as: a medical record according to patient encounter events with some element of the health system (medical consultation, emergency, etc.), a medical record focused on the problem, statistics on some specialty, illness symptomatology etc. In this case we have developed the first aspect, exemplifying it in Figure 7 with a user interface where the integration descriptor contains general components for any specialty, as the event tree on the left that displays queries, medical records' general parts concerning each health system-patient encounter. Also the central and the left panels, which contain specific information, are a part of this descriptor. The content of these panels is shown with a dermatology example, which also has multimedia information.

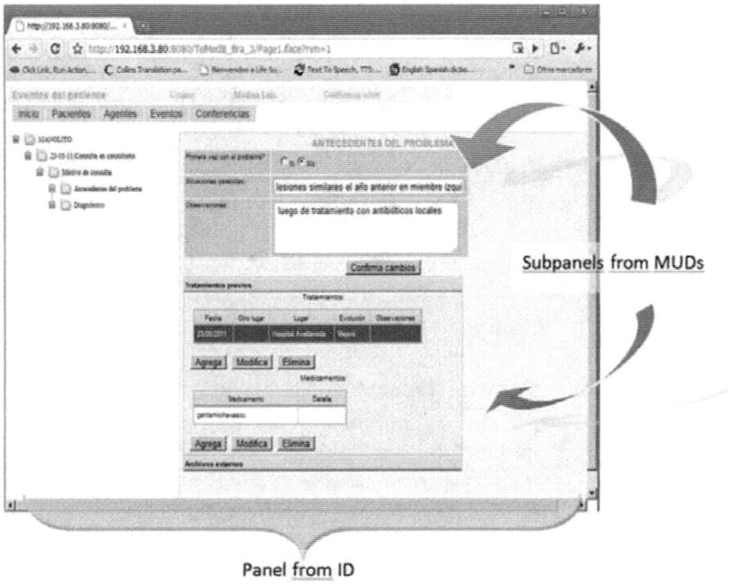

Figure 7: Dermato Panels

Conclusions

The architecture of a multilayer web application was presented, whose development was based on requirements derived from the rural telemedicine experience, particularly of interconsultation. The developed architecture is characterized by its flexibility, in the sense that its layers there are modules that are dynamically instantiated to achieve a particular behavior.

The adaptability model follows two ideas: the first one, it must behave according to the pre-established by specialists fulfilling the interconsultor role, the second one, still in development, refers to that the same application must be used from different points of view. The application is developed on a unique electronic health record system, regardless of whether the profiles adjust to the specialty that is consulted [8].

One of the most important challenges is to reconcile the common aspects to more than one specialty avoiding redundancy of information. This issue is being addressed through the creation of an information model from the smallest units, which refer specifically to clinical information elements, closed to openEHR archetype idea[2], but without abstracts reference models such as HL7 RIM [9] or the reference model openEHR [10]. From the point of view of future work orientations, immediate ones refer to the experiences development with various converged specialties, testing the application's adaptability and flexibility. Another aspect has to do with the system interoperability. While it was not the medical record development the initial target, the lack of electronic medical record reference in the healthcare system, led us to the development of this, with an evolutionary approach, guided by the needs of our application.

References

[1] Dolin RH, Alschuler L, Boyer S, Beebe C, Beilen FM, Biron PV, Shabo A. HL7 Clinical document architecture, release 2. J Am Med Inform Assoc. 2006;13(1):30–39. doi: 10.1197/jamia.M1888.

[2] Beale T. Archetypes, constraint-based domain models for future-proof information systems.Seattle, Washington, USA: Northeastern University, Boston; 2002. pp. 16–32.

[3] Sergio Litewka. Telemedicina: Un desafío para América Latina. Acta Bioethica. 2005; 11(2)

[4] Solarz P, Rotger V, Medina Ruiz L, Salas A. Aplicación Web para Telemedicina Rural . Anales del CAIS 2012 Congreso Argentino de Informática y Salud ISSN: 1853-1881. La Plata Buenos Aires. Septiembre de 2012. http://www.41jaiio.org.ar/sites/default/files/1_CAIS_2012.pdf.

[5] Pablo F. Solarz, Viviana I Rotger, Luis Medina Ruiz "Prototipo web para telemedicina"
Proceedings XVII Congreso Argentino de Bioingeniería y VI Jornadas de Ingeniería
Clínica, SABI 2009. Vol 1, Núm 1, pp 1-2, ISSN 1852-6292. 2009

[6] V I Rotger, P F Solarz, L M Ruiz, A Salas, M C García Mena and J M Olivera. Teledermatology: an experience in Tucumán. *J. Phys.: Conf. Ser.* 332 012053 doi:10.1088/1742-6596/332/1/012053. 2011.

[7] F, Martin. Patterns of Enterprise Application Architecture.. ISBN: 0-321-12742-0. 2002. Addison Wesley

[8] Blobel B. EHR architectures - comparison and trends. Stud Health Technol Inform. 2008;134:59-73.

[9] HL7 – Ref. Information Model. V3.
www.hl7.org/implement/standards/product_brief.cfm?product_id=77

[10] OpenEHR. Reference Model. Version 1.0.2.
http://www.openehr.org/wiki/display/spec/openEHR+1.0.2+UML+resources

Acknowledgments This work was financial supported in part by grants from the UNT Research Council (CIUNT), 26/E480/3

Biomedical equipment management software for small and medium-sized medical facilities.

M D Freyre[1,2], A E Gabosi[1,3] and D M Rodríguez[1]

[1]School of Exact, Physical and Natural Sciences (Facultad de Ciencias Exactas, Físicas y Naturales), National University of Córdoba (Universidad Nacional de Córdoba), Argentina – Biomedical Engineering.

[2]Maintenance department of Hospital Privado Centro Médico Córdoba, Argentina.

[3]Bioengineering department of Hospital de Niños de la Santicima Trinidad Córdoba, Argenina.

E-mail: matiasfreyre@magussingenieria.com.ar

Abstract It has been found that the medical facilities, up to and including those at level seven, under Resolution 267/2003, approved by the Ministry of Health of Argentina, lack a biomedical equipment management system. These systems are unavailable on the market and their high development costs can only be afforded by large hospitals. Consequently, there is an unsolved need for adequate records on the equipment at medical facilities, making management more difficult and creating unnecessary risks and expenses.In order to meet this need, a software program, called *Maguss*, has been developed for small and medium-sized medical facilities. It is a web-based software featuring low implementation costs, minimum hardware requirements, remote access from various mobile devices, and increased safety. The need for this type of systems was confirmed through its implementation and the excellent results obtained in two hospitals located in Córdoba city. At the same time, areas of improvement were pinpointed and steps are being taken, in collaboration with the Medical Informatics Department of the School of Exact, Physical and Natural Sciences, to remedy them.

1. Introduction

Maguss is an application software for managing medical equipment at healthcare facilities. This tool is designed to aid in the development of clinical engineering by making it possible to track the equipment within hospitals, clinics and other medical facilities in an orderly fashion. It is a free software product developed mainly with such tools as HTML, CSS, Java Script, IDE, MySQL databases, version 5.0, and a Linux server with PHP 5.3.

The main tasks performed by this software include:

- Keeping a record of all the biomedical equipment items at a medical facility through a customizable technical data sheet.
- Searching for equipment items by serial number.
- Displaying lists of equipment items by equipment and department through a filtering system.

- Tracking simple and complex maintenance work done on all equipment items through their Work History.
- Setting alarms as reminders of scheduled activities or tasks and assigning them a specific priority level.
- Listing spares and accessories indicating their availability and compatibility with the equipment.
- Listing suppliers along with the equipment provided and other relevant information.
- Keeping notes of the condition and movement of equipment on and off site.
- Viewing, printing and sending some of the above mentioned information via e-mail in Adobe Reader .pdf and Microsoft Office Excel .xlsx and .xls formats. Excel may also be used to filter and show information in statistical graphics.

In addition, it is a web-based application; therefore,

- It requires Internet access to operate.
- It may be remotely accessed from any device.
- Each facility is identified with a user name and a password.
- It may be customized and tailored to the specific requirements of each medical facility without altering its basic structure.

As regards access, the following points must be noted:

- Several users may be created for each facility.
- Users are assigned one of three different access levels, each of which is associated with particular ways of manipulating information:
  - Administrator: This user is granted full access and may view, enter, and modify information. He or she may also create new users of any of the three types mentioned in this section.
  - User: This user is allowed intermediate access and may view, enter, and modify information. However, he or she may not create other users.
  - Supervisor: This user is granted restricted access and may only view information. He or she may neither carry out any action that modifies the database nor create any new users.

2. *About the software*

2.1. Access to the software

To access the website at *http://www.magussingenieria.com.ar*, users must enter the user name and password provided by the software creators in the login window shown in Figure 1.

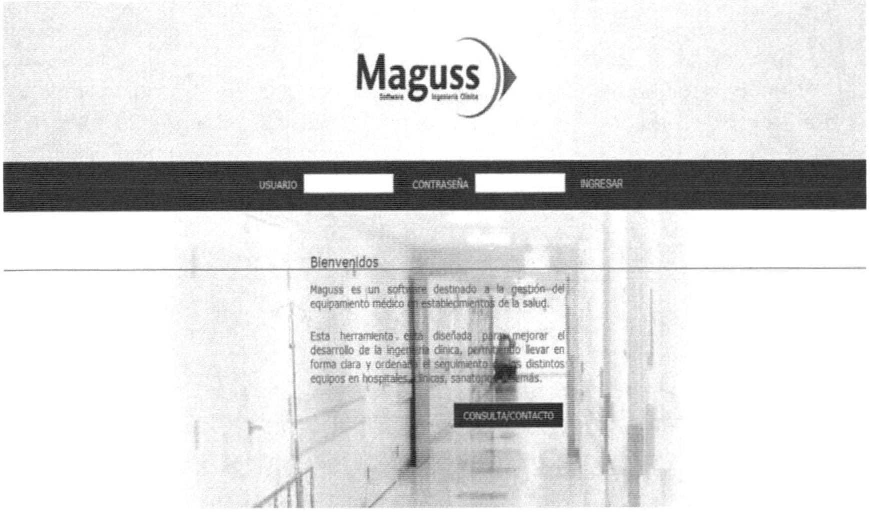

Figure 1: Login window to access the software at *http://www.magussingenieria.com.ar*

2.2. *Home*

To maintain consistency in user interfaces, the home page and all the other pages in this tool have a similar look and feel. They feature the Work Menu and the name of both the medical facility and the user, as shown in Figure 2.

In turn, the options on the Work Menu are "Manual" (Manual), "Contact" (Contacto), "Search" (Buscar), "Alarms" (Alarmas), "Add Equipment" (Registrar equipo), "Notes" (Notas), "Inoperative Equipment" (Equipos dados de baja), "Spares and Accessories" (Repuestos y accesorios) and "Suppliers" (Proveedores).

Figure 2: Home page. In this case no alarms were set; otherwise, they are displayed in this window.

2.3. *Adding equipment*

In order to add an equipment item to the database, a form must be filled in with some basic information. A distinctive feature of the software is that more specific information may be included through its customizable data sheet if the user considers it necessary, adding versatility to the software at this stage.

In addition, an image, such as a photograph, and a file, for example a manual, may be uploaded for each equipment item.

Once said form is filled in, if the user wishes to enter a similar item, he or she must click on the option "Duplicate" (Cargar y copiar). This opens a form containing the information on the item previously added, except for those details that are unique, such as the serial number and the internal code.

2.4. *Alarm setting*

Alarms are set on the data sheet page, where they are also deleted. When setting alarms, users must specify the reminder date, the frequency, the task, and the priority level as indicated by the colors red, orange, and blue. More than one alarm may be set per equipment item.

The alarms are displayed once the user signs in on the home page and are listed by date. They may be switched off by clicking on the button "Task performed" (Tarea realizada) on the same page.

2.5. *Alarm entries on the home page*

Alarm entries are listed chronologically on the home page, as shown in Figure 3, the one on top being for the current date, followed by those from previous days. If there are different alarm entries for the same day, these are listed in order of priority.

A tick is shown beside each alarm entry after clicking in the appropriate box to indicate the task is being carried out. Clicking on "Task performed" allows users to be notified of the completion of the task. The alarm is thus deactivated and the button now shows "Task performed by (user name)" [Tarea realizada por (nombre de usuario)]. This notification remains active throughout the day.

Figure 3: Alarm entries displayed on the home page.

2.6. Search page

Two search types are offered by the software, as shown in Figure 4.

On the one hand, an equipment item may be searched for by its serial number or internal code.

On the other hand, the software offers equipment lists filtered by equipment items and departments. Therefore, lists may be filtered by the following parameters:

- All the equipment at the medical facility.
- All the equipment in a department.
- All the equipment of the same type at the medical facility.
- All the equipment of the same type in a department.

Figure 6 shows a list of all the equipment at a medical facility.

If the user clicks on an equipment item on the resulting lists, its data sheet is displayed, as illustrated in Figure 5.

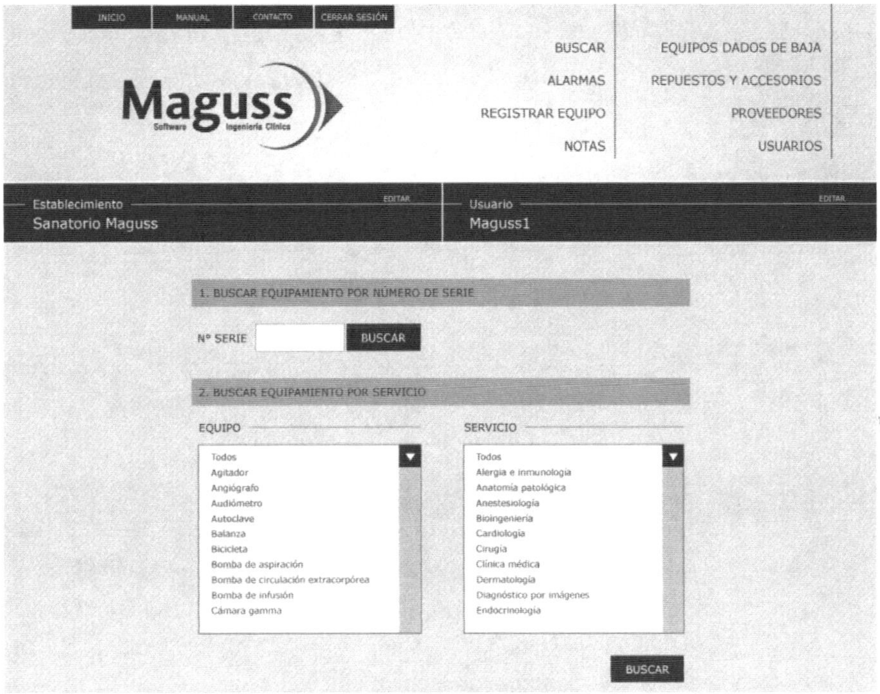

Figure 4: Search systems.

2.7. Data sheets

Figure 5 shows the data sheet of an equipment item with its photograph and all the relevant information about it. For further details, a file, like a manual, may be uploaded in this window.

Data sheets also include the Work History, showing the maintenance work done on the equipment.

This type of record presents the date and type of work performed on each item as well as a description of and the person responsible for it, making it possible to track repair and maintenance in an orderly fashion.

If the user wishes to enter an equipment item with features common to items already in the system, the shared information can be copied by clicking on the "Copy" (Copiar) button. In this way, entering several similar items becomes a user-friendly task.

Equipment with an inoperative status is deleted from the database of functional equipment and recorded separately.

It is possible to open the data sheet in Adobe Reader .pdf and Microsoft Office Excel .xlsx or .xls formats to print it and send it via e-mail.

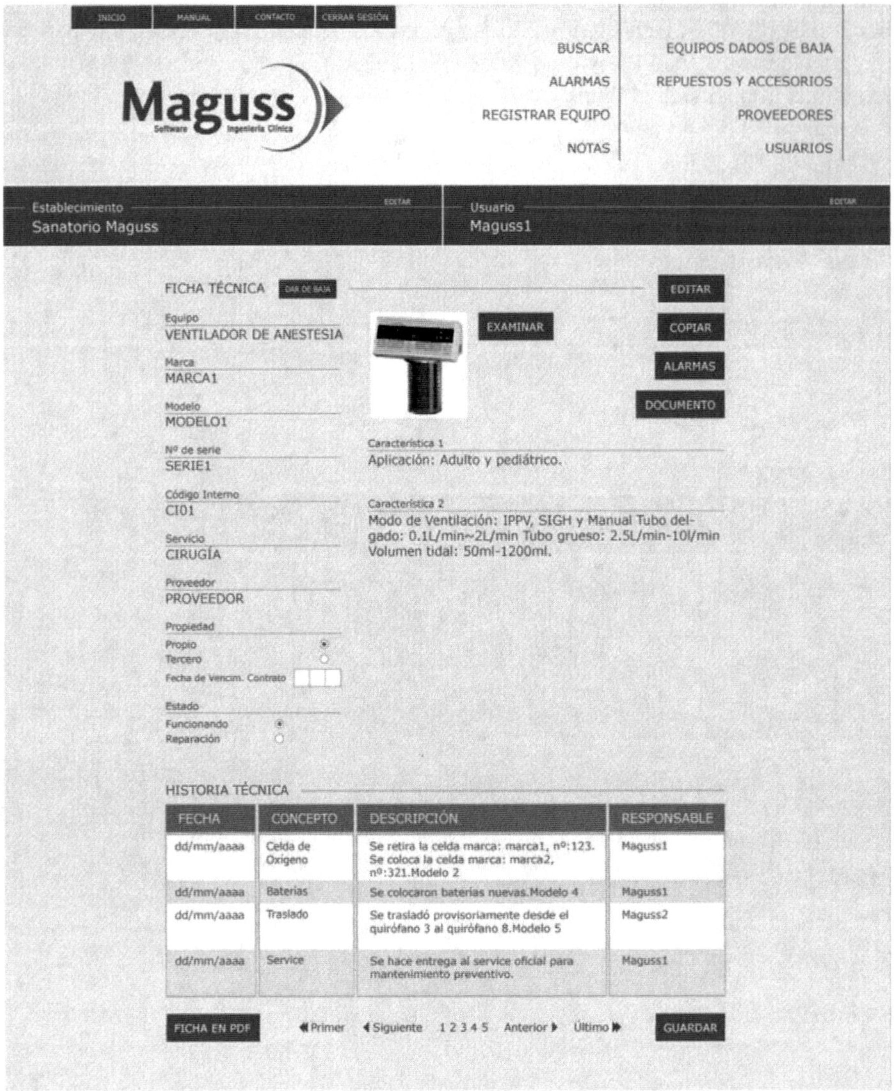

Figure 5: Data sheet. The figure above shows the data sheet of an anesthesia ventilator including the product details and its Work History.

2.8. Lists

The different software lists, such as that of equipment, suppliers, spares and accessories, and inoperative equipment, may be exported to Microsoft Office Excel .xls or .xlsx formats, allowing users to filter and present information, for instance, on a pie chart showing the functional equipment and that under repair. In turn, lists in said formats may be printed and sent via e-mail.

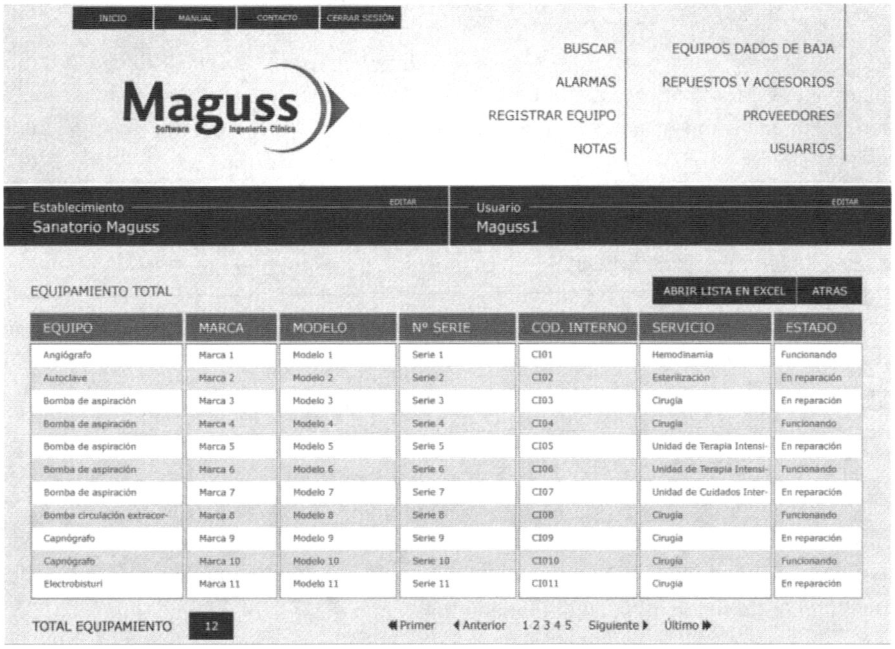

Figure 6: Software list. All the equipment at a medical facility.

2.9. Inoperative equipment

The software allows users to track inoperative equipment. The inoperative equipment list is ordered chronologically.

The reason for taking equipment out of service may be indicated in the Work History, where it is kept regardless of the equipment status.

When the equipment is back in running order, its status may be changed by accessing its data sheet.

More than one equipment item may be deleted at a time from the database.

The list may be opened in Microsoft Office Excel .xls or .xlsx formats to filter, print, and send the information via e-mail.

2.10. Spares and accessories

To add spares or accessories to the database, a form must be filled in by clicking on the "New" (Nuevo) button. These are then ordered alphabetically.

The list of spares and accessories also indicates the equipment with which they are compatible.

More than one spare or accessory may be deleted at a time from the database.

The list may be opened in Microsoft Office Excel .xls or .xlsx formats to filter, print, and send the information via e-mail.

2.11. Suppliers

To add a supplier to the database, a form must be filled in by clicking on the "New" button.
Suppliers are then ordered alphabetically by name or department. To do this, the user must click on the "Name" (Nombre) or "Department" (Especialidad) buttons, respectively.

More than one supplier may be deleted at a time from the database.

The list may be opened in Microsoft Office Excel .xls or .xlsx formats to filter, print, and send the information via e-mail.

2.12. Notes

The software features three types of notes to record and notify users of the condition and movement of the equipment. These include Service Requests (Solicitud de servicio), Removal of Equipment Notes (Salida de equipamiento) and Maintenance Reports (Informe de servicio). They are designed to record the work cycle.

They contain fields for the user to complete and present a similar layout. They may be exported to Adobe Reader .pdf format by clicking on the "Print" (Imprimir) button. The printed version will display the letterhead of the organization.

2.12.1. Service Request. By this type of note, also called 'Work Order', the personnel of any department may inform the biomedical engineering department about the condition of an equipment item.

2.12.2. Removal of Equipment Note. This type of note records the removal of equipment for different reasons, including repair, service, and relocation, among others. The same note may be used to record the removal of more than one equipment item.

2.12.3. Maintenance Report. It is a technical report describing the maintenance work performed on the equipment by both staff members and third parties.

3. Implementation

In order to evaluate the software performance and obtain some feedback about it, it was implemented at a typical medical facility in Córdoba, Argentina. An agreement was reached with the Hospital de Niños de la Santísima Trinidad, a children's hospital located in Córdoba city, to implement the software in its Department of Surgery.

Currently, this tool is in operation both at said hospital and the Hospital Municipal de Urgencias, another local medical facility.

4. Results

The objectives pursued in developing the software were as follows: to improve equipment management, organize maintenance work, schedule tasks of different priority for specific dates and with a certain frequency, reduce corrective maintenance work, and build an inventory. Taking these into account, and after its implementation, the software was found a suitable tool.

5. Conclusion

In the course of this work, numerous ideas as to the design and the functionalities of a biomedical equipment management software were considered. However, the goal established was to design a user-friendly tool that would help manage equipment at medical facilities and, once its development was complete, make it available for use.

Although the software is currently being improved to some extent, it already fulfills the authors' expectations in the sense that it serves as the basis for a full-featured management software as it enables the creation of a database of the equipment including its Work History, movement, and spares and accessories, among others.

Medical event notification system using SMS technology

J Voos, G Riva, C Zerbini, C Centeno, C Olmos and E Gonzalez

Grupo Ingeniería Clínica – Universidad Tecnológica Nacional, Facultad Regional Córdoba, Maestro López esquina Cruz Roja, Ciudad Universitaria, X5016ZAA Córdoba, Argentina.

jvoos@sdct.frc.utn.edu.ar

Abstract. This paper presents a medical event notification system implementation using short message service (SMS) over the cellular network. The main objective is the announcement for vaccination campaigns and availability of a particular medical attention at health sites where these services are not regularly provided. In this way, it provides a tool that facilitates healthcare in remote locations without medical assistance infrastructure. Since the main requirement of these systems is to achieve its massive adoption, SMS service is used to guarantee reaching all recipients regardless of their mobile device technology.

1. Introduction

Actually in our country there are remote areas that do not have good access to healthcare without health facilities and medical specialists in all disciplines. For this reason, public and private organizations use mobile assistance systems to provide medical care to these regions, including services such as vaccination, medical checkup and studies, prevention activities and medical specialist attention. Considering this need, we propose a medical event notification system based on SMS messaging service, in order to provide a communication channel that guarantees the announcement to all recipients of a health initiative.

1.1. Current situation

The use of information systems and communication technologies to support the achievement of health objectives (eHealth) is an area currently being developed worldwide, and has minimal initiatives in our country. The development of this kind of projects is a priority area in our country, as described in the publication libro blanco de la prospectiva TIC: proyecto 2020 [1], from the Ministry of Science, Technology and Productive Innovation of Argentina. In eHealth there are other more specific areas such mobile health or mHealth, defined as a health practice supported by mobile devices, such as mobile phones, patient monitoring devices, personal digital assistants (PDAs), and other wireless devices.

The growth of mobile phone subscriptions and the new technologies into phones enable the use of these devices for health services. According to the second global survey on mHealth compiled by the World Health Organization (WHO) [2] in 2011 most of its member countries (83% of 120 countries) reported the existence of at least one health program using the cell phone network. This implies that a large number of countries are adopting mHealth initiatives to address health needs in different areas of application, as shown in Figure 1.

Journal of Physics: Conference Series **477** (2013) 012015 doi:10.1088/1742-6596/477/1/012015

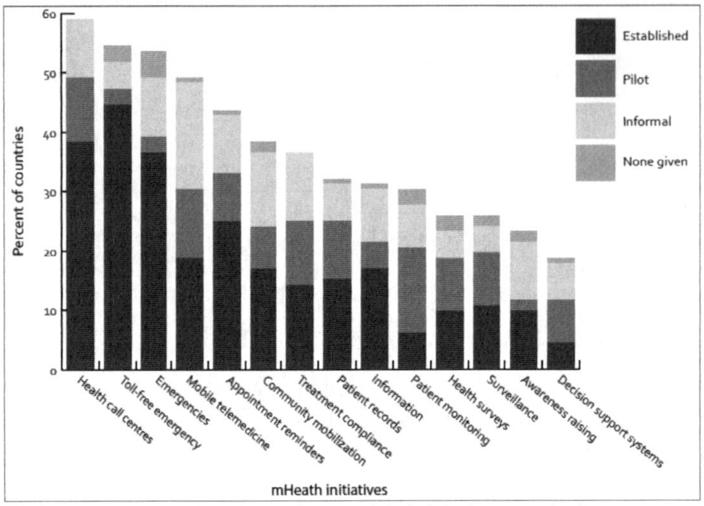

Figure 1. Adoption of mHealth initiatives and phases.

In Argentina there are telemedicine initiatives developed by private and public companies using cellphones, but require phones with special features like smartphones with internet connection to run. For this project, the SMS service was selected instead of an Internet connection using GPRS, due its wide coverage in our country and the possibility to receive/send messages from any cell phone. While the cost of GPRS service has declined considerably today in our country, the coverage provided for this service is still concentrated in big population cities reducing the geographic scope of mHealth initiatives, conditioning ultimately their social impact.

1.2. Proposal

Considering the potential of mobile technologies for telemedicine applications and prioritizing the massive adoption for low cost technologies, we propose the design and implementation of a medical event notification system based on SMS messaging service.

Additionally, the reduced data used on the SMS service messages enables delayed transmission according to the timeframe when the potential patient has signal; making efficient use of an irregular cellular network signals, a typical scenario for regions with low population density where these medical campaigns are implemented.

2. Development

2.1. Overview

The team involved in this initiative has experience on previous telemedicine projects including systems development and biomedical equipment for signals acquisition design; using the HL7 protocol and the cellular network as transmission channel [3] [4]. Following this telemedicine line of research, the existing platform will be upgraded developing new modules for sending SMS messages, as shown in Figure 2.

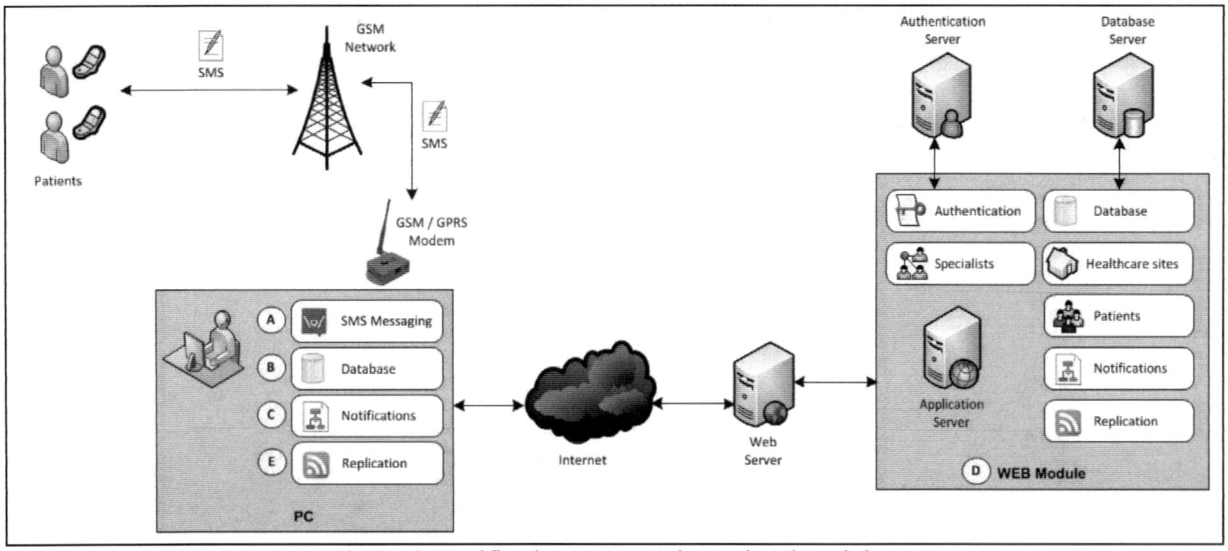

Figure 2. Notification system - Operational modules.

The project includes a software for send/receive SMS messages running on a PC with a modem (A) that allows it execution in remote sites without Internet connection. To select the recipients for a particular medical event, a database (B) is deployed to allow people selection according to different criteria (e.g. age group, sex, place of residence, illness, etc.) in response to the needs gathered in the place where the medical event will be made.

A web module is developed (D) for system access through Internet in order to let notifications management by specialists or other persons associated with a particular campaign, allowing also track responses to SMS messages sent, thanks to a replication software component (E) installed on the PC where the modem is installed.

Additionally, based on the settings per each notification, the system will send another SMS to the patient with the details of the assigned appointment time (C), in order to organize the medical attention and prevent having all the attendees going at the same time for a medical service.

2.2. Operational modules

Based on design showed in Figure 2, the system contains the following modules:

2.2.1. SMS messaging module. This module is implemented by one system running locally on a PC. After analyzing various open source solutions, FrontlineSMS messaging system [5] was selected due the integration facilities with other existing software components and the easy to use by end users without technical skills, as shown in Figure 3. The system was successfully implemented in other mHealth initiatives worldwide and has a support team ready to help based on their previous experiences on this kind of projects.

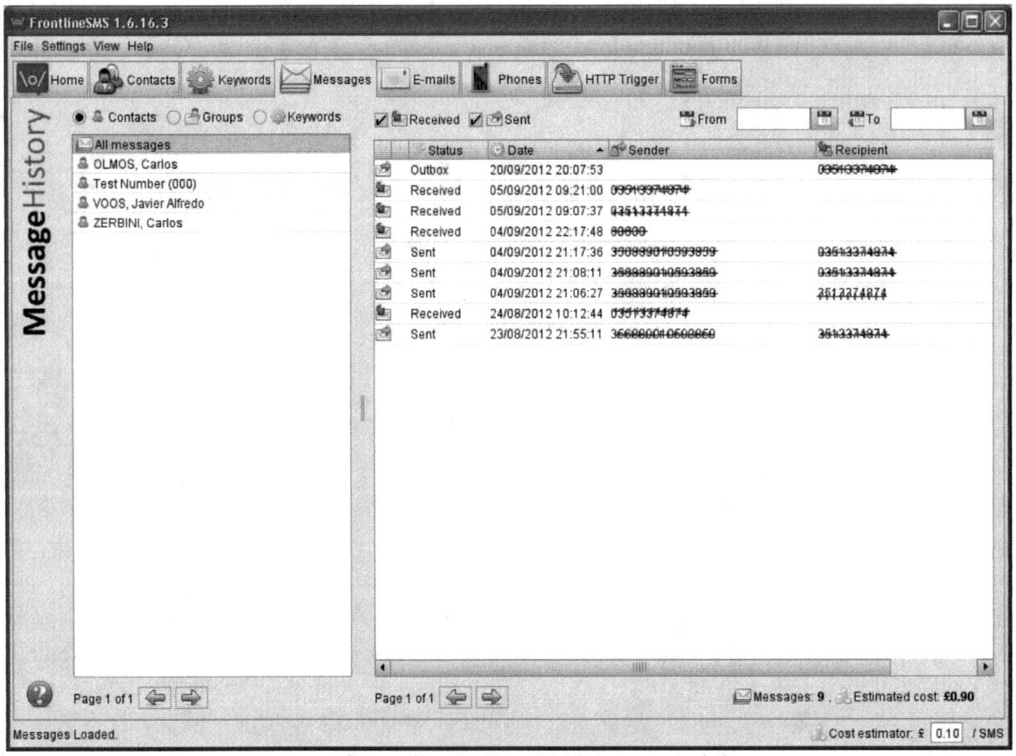

Figure 3. FrontlineSMS – SMS messages inbox.

2.2.2. Database module. To store patients and specialists information, healthcare sites and notification settings. It also contains the data related to messages and responses, in order to know in advance the patient participation for a particular campaign. The data model is described in Figure 4.

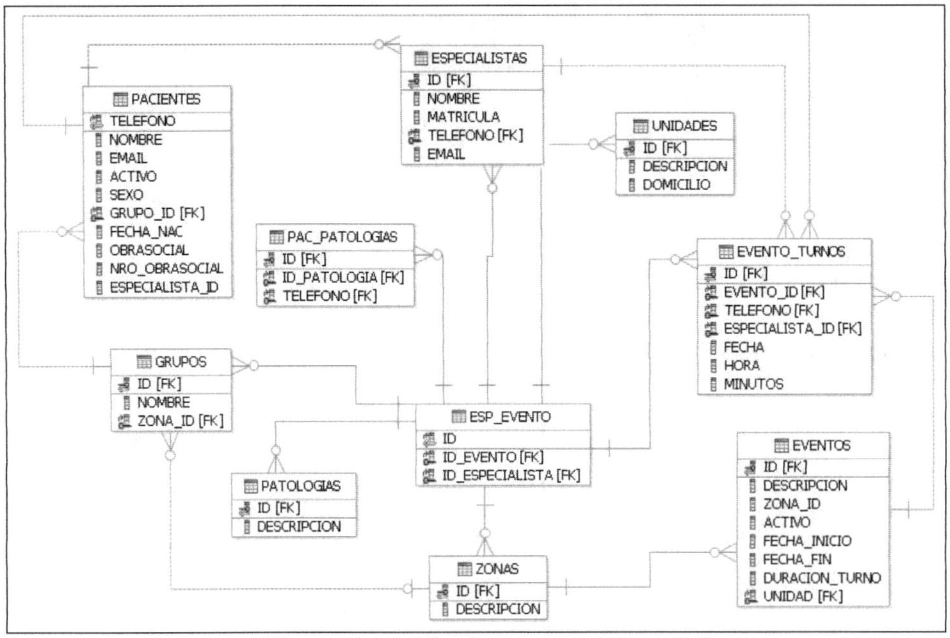

Figure 4. Notification system data model.

2.2.3. Notification management module. To provide the different notification settings related to the health initiative. Additionally, processes the responses sent to each notification and generate reports associated with the results for a particular campaign.

2.2.4. Web module. To offer a website for the database module maintenance, including new medical event creation to configure the appointment duration associated with this particular notification. Additionally it is feasible to visualize responses to the notifications sent and results for the health initiative.

2.2.5. Replication module. To replicate the information stored on the PC to the website and vice versa, in order to have the same information in both places.

3. Conclusion

This project aims to contribute the adoption and implementation of telemedicine projects, in particular mobile health provision, in our country. The main objective is applied low-cost services and execution feasibility in every place where a cell phone has enough signal to exchange SMS messages. This is especially suitable to our country social context: large geographical areas with low population density having sporadic and weak signal in remote areas. Currently we are working on this project implementation at public hospitals together with the Ministry of Health, a department for the Córdoba province government, Argentina.

There is a mature mobile phone technology and it is massively adopted by the population, reaching remote areas with low-cost services and high impact. The proposed system has the advantage of minimum hardware requirements, could be easy integrated with other existing telemedicine platform and the notification recipients can use any cell phone capable of receiving SMS messages for its execution.

The system can effectively assist to the medical specialists, allowing them to perform their activities in a more organized way, contributing to increase the medical campaigns productivity. Furthermore, the required training for potential patients for this system use is minimal since only involves receive/send SMS, a common activity in the daily cell phone use.

References

[1] *Libro blanco de la prospectiva tic : proyecto 2020.* 2009 1a Ed. Buenos Aires: Ministerio de Ciencia, Tecnología e Innovación Productiva. *ISBN* **978** 987 1632 00 8
http://www.mincyt.gov.ar/multimedia/archivo/archivos/Libro_Prospectiva_TIC_2020.pdf

[2] mHealth: New horizons for health through mobile technologies: second global survey on eHealth. 2011 World Health Organization. *ISBN* **978** 92 4 156425 0
http://www.who.int/goe/publications/goe_mhealth_web.pdf

[3] Voos J, Riva G, Zerbini, C, Centeno, C, Gonzalez, E. 2010 Custom HL7 V3 message provider using web services security features *PubMed ID* 21097078
http://ieeexplore.ieee.org/xpl/articleDetails.jsp?arnumber=5627666

[4] Voos J, Riva G, Zerbini, C, Centeno, C, Gonzalez, E. 2011 Medical Data Transmission Using Cell Phone Networks *J. Phys.: Conf. Ser.* **332** 012051
http://iopscience.iop.org/1742-6596/332/1/012051

[5] FrontlineSMS Messaging System
http://www.frontlinesms.com/

Characterization of multispectral aerial images of sugarcane

A Solano[1], G Schneider[1], A Kemerer[2] A Hadad[1]
[1]Facultad de Ingeniería, Universidad Nacional de Entre Ríos, Argentina
Grupo de Investigación en Inteligencia Artificial, Oro Verde, Entre Ríos, Argentina
[2]Grupo de Recursos Naturales y Factores Abióticos de la EEA Paraná del INTA
Cátedras de Climatología Agrícola y SIG de la FCA UNER Oro Verde, Entre Ríos, Argentina
E-mail: aguchosolano@gmail.com, gerschneider@gmail.com, akemerer@parana.inta.gov.ar, hadad@santafe-conicet.gov.ar

Abstract. In this paper we present a implementation and characterization the status of sugarcane plantations based on the analysis of multispectral aerial images. Currently there are no precise techniques to estimate objectively the cane area fall or overturned, and this causes significant losses in crop productivity and industrialization. For the realization of this work was made a dataset benchmark images, and implemented a software from which were obtained indicators related to agronomic phenomenon, and analyzes of the data generated. In addition was used Principal Component Analysis to visualize and integrate statistical texture features. The results indicate the statistical features used characterize partially the sugar cane phenomena and suggest include another texture focus to complement this feature set, previous to built an cane identification process.

1. Introduction

The cultivation of sugar cane in Argentina is found principally localized in the northwest region of the country (99 %), where represents, in agreement with [1], one of the productive more important activities, and in scarce extension in the littoral region (1 %). The productive systems present different technological levels according to the system of crop used, with systems from manuals to totally mechanized, although, at present more than 70% of the total volume it harvests of way mechanized. One of the more critical aspects of the process of crop of the sugar cane are the losses that it is produced and have decided as passable levels about 2,5%. The presence of fallen cane is one of the factors with bigger influence. Incidence of superior overturning to 20% decides increments in the losses that go from the 4 to 6,4%, according to express [2].

The presence of fallen sugar cane (CC) immediately of the crop brings at important losses in the collection and in the sugar mill. In the phenomenon of CC for bankrupt person take part in random aspects and unpredictable. It is normal that the sugar cane plantation in a productive square reach a height mediates, but presents variations of this variable in different sectors of the field. This must be because of the differential conditions of the earth that presents different levels of fertility, humidity, compressing, etcetera. For other side, each plant of the sugar cane plantation is constituted in a different way, for which will manifest in unique form the different external stimulus.
When the plant is young and has developed under the best conditions, reaches certain height and for its own weight it begins to warp, moving close to break once its breakage tension is surpassed product of the wind or environmental phenomena as the hail. The CC appears with a great space heterogeneity.

Once produces the phenomenon, the earth can short remain, or on the contrary can be produced a product with new green color of the appearance of new green sheets, with phototropic growth. This supreme heterogeneity in the presence of an external observation. The above-mentioned factors make evident the great complexity in the phenomenon in studies, which does it difficult of analyzing and quantize.

The strange matter, called trash, when enters to the sugar mill represents an important loss of efficiency in the process of manufacture of the sugar, assure [3]. The challenge of industrial engineering takes root in exploring the alternatives to control the harmful effects that these tailings or trash and the present dust in the cane have in the equipments, processes and operative performances of the factory. That is to say that the quality of the cane affects in direct form the industrial performance and the quality of the obtained sugar [4]. In turn the quality of the raw material (cane) can be affected for changing aspects of the agriculture of the sugar cane, such as the introduction of new cultivars, climatic variations, the use of ripening chemists, changes in the cultural practices and in the systems of crop or the appearance in the cultivations of illnesses or infections. As for the types of crop, it observes that the contents of trash and dust in cane harvested in form mechanized are appreciably lower that in the semi-mechanized, being more than a 2 to a 5,7 %. At the same time, the values of fiber (vegetable remainder) goes up from 14 to 18 % approximately, due to the presence of major quantity of vegetable parts for a less efficient crop [5]. This means that the industries become calm to the producers a significant part of the weight of the trash of the entered raw material for scales to cane value, as long as the form of muestreo and analysis of this parameter is generally inadequate.

To this you must incorporate the other additional costs that produces this strange product in the factory properly happiness, in any way whatever:

1 Wear in the mills for abrasion and a progressive loss in the capacitance of extraction of the cane juice.
2 Increase significant of the amount of sloth and with it of the losses of sugar in the same thing.
3 Increase of the consumption of chemicals for the treatment of the juices.
4 Increase of the color in the juice and subsequent in the sugar with it who to avoid it reduce the cycles of crystallization by decreasing the industrial efficiency.
5 Wear of equipments, pipeses and dazed.
6 Diminution of the can calorific of the waste pulp and consequently the efficiency in the burnt forest of the same, with an increase of the consumption of and of the emissions of boiler.
7 Other.

To decrease to the minimum possible the percentage of trash in the cane, several mills installed washing tables of cane, where they remove the content of dust by means of big quantities of water (approximate report of 3 m^3 of water by ton. of cane). This supposes saving of money, but adds problems like bigger costs for the required power, the necessary water and the additional manual labor, join to the new discussions with the producers since some studies confirm that in the wash of the cane loses a percentage of the performance for extracting sugar. Moreover, the technique requires availability for service of terrain to install settling tanks of sand and blunders to re-use the water and does not return the so to cause them by increasing the contamination. A promising strategy to achieve the reduction of this impact take root in achieving a bigger operative efficiency in the tasks of crop of the cane in the field, previous stage to the entrance to the factories, and fundamentally in sectors with broken cane that increase notably the percentage of trash. This is an area of intensive manual labor where for its particular properties, considers [5], have not achieved to introduce technological solutions that reduce its negative impact in the production.

At present, the quantization of the fall square productive sugar cane is an expensive process and ineficcient. It requires the sending of evaluators to the terrain and they must explore big surfaces to obtain a representative sampling that permit an estimation. The displacement around the field produce damage, require many man hours of work and generate estimations with a high degree of uncertainty. In turn, the presence of space variations in the productivity of the cane is very wide [6] decided variations of until 50 tn/has to level of lot of production. It is spectable that this owner, as well as variations in the quality of the cane observe to diverse scales, in a similar way to which has demonstrated for other variables they fall in the production, as the properties of earth [7], which raises objections establish a distance of unique sampling and reverberates negatively in the costs.

Other strategy is carrying out flights over the cultivations with observant specialists who estimates the percentage of fallen cane according to their experience of it observed from the air. This strategy is subjective and dependent of the available specialists in each area of interest.

By keeping in mind these antecedents it is made evident that has not observed report of some methodology with certainly measurable certainty to decide the affected area, in order to help to plan of better way the process of crop and the associates costs of the productive involucrate actors.

In this sense, the national institute of agricultural technology of Argentina, INTA, has approached the problematic, by carrying out works as of aerial photography of high resolution [8,9]. This data source works as a whole to advance with new technological strategies to approach the problem. In this work it presents the information developed substructure and the analysis of information and agronomic results.

From the information viewpoint mounted an application to carry out the labelling of images, as of which develops a pipeline type structure with capacitances of analysis of variables and classification of regions.

2. Methodology

It counted on a set of images obtained for the national institute of agricultural technology (INTA) of Argentina from a Sky Arrow 650 airplane TCNS ERA, whose system of capture is composed of a Global Positioning System, a multi-spectral chamber Geospatial MS4100 and a system of control and storage of data. The chamber employed permits the acquisition of photograms in three bands of the electromagnetic specter: green(530-580 nms.), red (650-685 nms.) and near infrared (770-830 nms.) with a resolution of image of 1920 x 1075 pixels. The GPS provides the position, attitude and height quickly in synchronism with the acquisition of each photogram.

The flight carried out on 5th May 2008, moment of the cycle of the cultivation where the presence of fall of the cane is clearly evident. The flights were carried out in the solar midday when all is said and done with a homogeneous illumination in the surface. The fly it is designed for an altitude of 1200 m. by resulting a size of pixel of 0.7 ms. There were obtained 540 photograms that were assembled by means of the reconnaissance anchoring knitted in consecutive photograms generating a mosaic.

With the supervision of an agricultural engineer it was carried out the labelling of the images in the classes of interest, using the application Label Me [10] of the Massachusetts Institute of Technology

(M.I.T.), who was in a local server to facilitate the access to the data and the incorporation to a pipeline of processing. The figure 1 shows the process of labelling who consisted in delimiting the images with polygonal continents of representative portions of the cane classes: standing cane (CP), fallen in form of patch (CCP), fallen cane in big areas (CCA), fallen cane with phototropism (CCF) and in excess or areas of exposition earth (C).

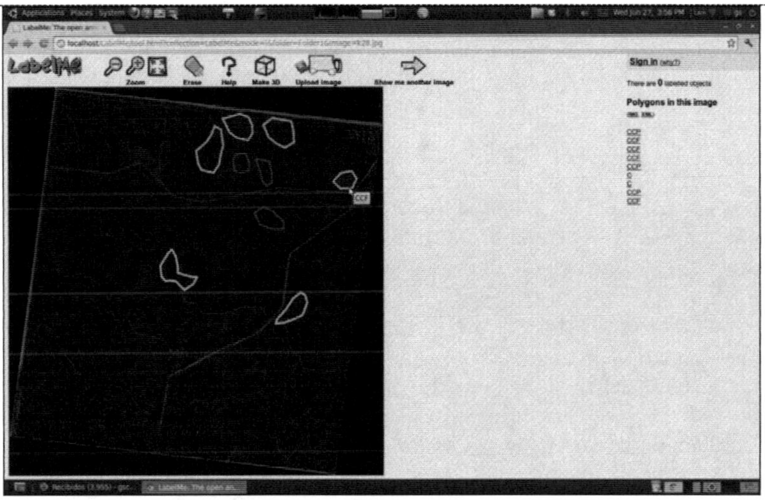

Figure 1. Capture of screen of an example image and of the process of labelling of the same thing with LabelMe

It was developed in Java a software capable of acquire the information with form of metadata generated in the labelling, and of cut and process automatically the portions of obtained images from the dataset, in order to obtain representative examples of each class. It incorporates in this stage a pre-processing of normalization in brightness of the images that is carried out by calculating the brightness mediates by canal of all polygonal labellings, and by fixing arbitrarily as parameter for the normalization a 120 % of this value, which enlarged the dynamic range without producing saturation in the histogram.

Given the characteristics of the agronomic phenomenon, carried out a boarding from the viewpoint of the textures of the images, by defining the following characteristics or indicators of interest:

- Standard deviation of the histogram of the sample by canal
- Entropy mediates of the histogram of the sample by canal
- Value mediates of the pixels of the sample by canal (mentioned as spectral signature)
- Index of vegetation of difference normalized (NDVI) [11]

The extraction of characteristics was carried out in an iterative process of the Batch type. The volume of information generated was stored in a Postgre database, of mode of can access concurrent immediately of the analysis to the whole information and it obtains the by using the potentialities of a language of SQL consultation.

By attending to the problem of the complexity of the data, minimized negative effects of the desbalance [12] by means of the generation of sets with similar number of samples for class, and obtained random of the dataset as of the SQL query:

```
DROP VIEW rocked;
CREATE VIEW balanced AS
( SELECT e.id FROM tablaAnalisisEspectro WHERE e.tamVentana=tamVent AND e
e.clase= "CP" ORDER BY RANDOM () LIMIT CantMuestras ) UNION
SELECT e.id FROM tablaAnalisisEspectro WHERE e.tamVentana=tamVent AND e
e.clase= "CCA" ORDER BY RANDOM () LIMIT CantMuestras ) UNION
( SELECT e.id FROM tablaAnalisisEspectro WHERE e.tamVentana=tamVent AND e
e.clase= "CCP" ORDER BY RANDOM () LIMIT CantMuestras ) UNION
( SELECT e.id FROM tablaAnalisisEspectro WHERE e.tamVentana=tamVent AND e
e.clase= "CCF" ORDER BY RANDOM () LIMIT CantMuestras ) UNION
( SELECT e.id FROM tablaAnalisisEspectro WHERE e.tamVentana=tamVent AND e
e.clase= ORDER BY RANDOM "c" () LIMIT CantMuestras );
```

Figure 2. SQL balanced query

To lead with dimensional problem was used Principal Component Analysis (PCA). PCA is a statistical technique whose purpose is to condense the information from a large set of correlated variables in fewer variables, without losing most of the variability present in the data set [13].

This method consists of transforming the original attributes or variables x1, x2,...., Xm of the examples into another set of attributes f1, f2,.... Fp, where p <m. This process can be geometrically seen as a change of shafts in the representation (projection).

The new attributes are derived as a linear combination of the variables in the data set, with "weights" chosen in such a way that the major components are not mutually correlated. Besides, the first fi are those containing most of the variability present in the original data.

Mathematically, if we assume m = p, the transformation process can be posed as follows.

$$\overline{f_{(mx1)}} = \begin{bmatrix} \overline{a_1}^T \\ \overline{a_2}^T \\ . \\ . \\ \overline{a_m}^T \end{bmatrix} .\overline{x_{(mx1)}} = A^T{}_{(mxm)} .\overline{x_{(mx1)}}$$

To find the vectors of matrix A, so that multiplying A by each example vector results in a new vector in the new dimensions space, variance is assumed to be higher for the first attributes than for the last ones. It can be shown that the result is obtained by first calculating the following estimate of the S covariance matrix:

$$S_{(mxm)} = \frac{1}{n-1} \sum_{i=1}^{n} (u_i)_{(mx1)} (u_i)^T{}_{(1xm)}$$

Where n is the number of examples in the data set, m is the number of attributes and ui difference vectors are calculated by subtracting each instance xi with the average attributes of all examples.

Once S covariance matrix is built, vectors of matrix A are obtained by calculating the eigenvectors, ordered from highest to lowest according to the extent of the eigenvalues associated with each eigenvector.

3. Results

The indicators of interest evaluated resulted, in general, promissory for the process of characterization of the fallen cane, to exception of the NDVI. This index presented low values in the case of the fallen cane where the cultivation reaches more biomass and the value must be bigger [14-16]. This happens because the band IR wins the variations in structure of the cultivation become a partner of the biomass and the index of area to folio. When the cultivation falls the structure seems modified, decreases, that does go down the IR and then the NDVI takes more low values. Also there is a confused effect of the contribution of reflectance of the earth in the fallen cane that decreases the NDVI [17,18]. Therefore, this index was not incorporate in the later analysis.
In the case of the values observed of deviation of the histogram it is observed for the cane class not fall (CP) a running towards valuing inferiors in the three canals analyzed (figures 3). It considers oneself to the standard deviation of the histogram as an indicating candidate of the cane class standing.

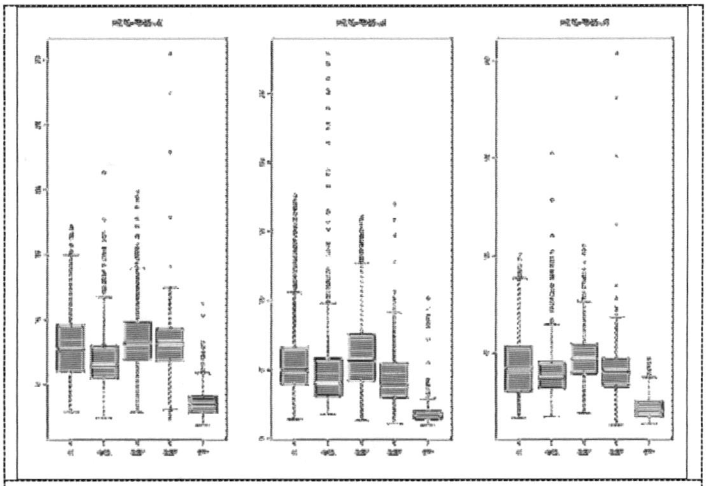

Figure 3. Standard deviation of the histogram of the contained samples in the polygons that agree the dataset, for each class and windows of V= 40 size px in the three analyzed canals.

The distribution of the entropy presented more high values for the classes of fallen cane (CCA, CCF and CCP) and exposition earth concern to the cane class standing, with valuing different media between classes. Therefore, it considers the entropy of the histogram a promissory indicator for the identification of cane standing. As for the spectral signature, although cavity is observed in the distributions between classes, for the IR canal a displacement is observed towards valuing superiors for cane standing. The spectral signature for the IR canal appears as promissory indicator to identify cane standing, for which will be considerate as one of the variables to incorporate.

Given that the proposed variables have similar behavior and overlap between classes, there was joined by PCA the indicators and reduced the dimensionality of the feature set in order to get a lesser number of possible descriptors.

By applying PCA was obtained that 96.6% of the variability in the data was explained by the first 3 principal components. However it was observed that in the space generated by the main components only three classes are distinguished: CP, C and CC. The latter groups the types of cane CCA, CCP and CCF. Said distribution is shown in Figure 4.

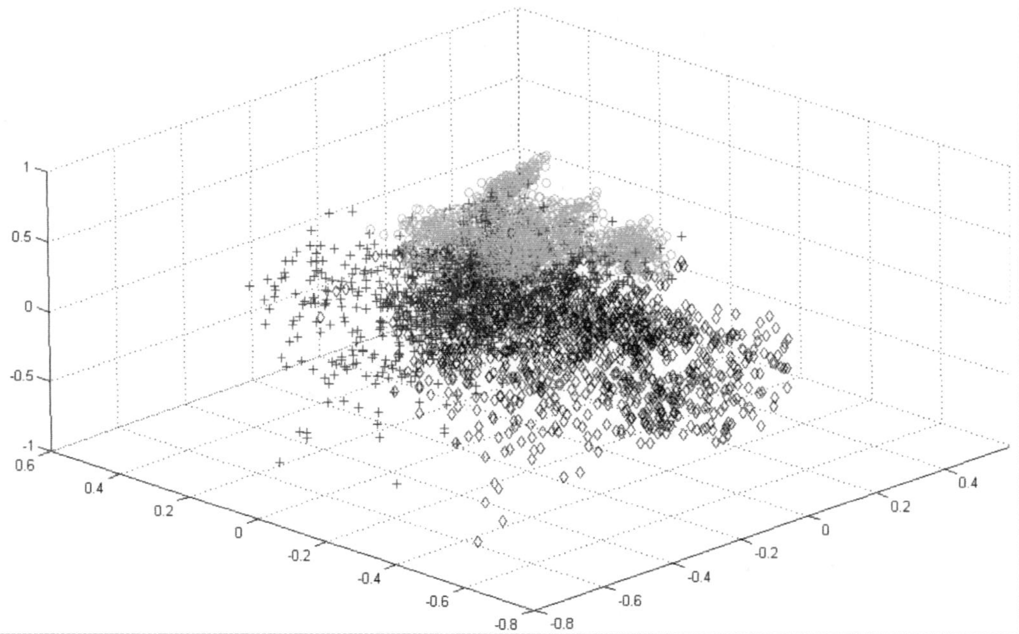

Figure 4. Distribution of the types of Cane in the Principal Components space.
(+)CC; (o)CP ; (diamond)C

4. Conclusions

It was compiled a reference dataset and a database for the study of the cultivation of sugar cane in its different states, from which the software created could provide information on the usefulness of the proposed indicators and to estimate the surface of sugarcane fall.
As to characterization of the cane types there is a need to combine several statistics associated with the texture of the image to address the problem of identification of sugar cane. In turn this combination by PCA can distinguish between walking cane and rod drop, resulting insufficient to differentiate most of the reeds falls. The only CC that can be identified is referred to as C (run) which corresponds to the zone of lower fertility. While it is a partial result has its value from the agricultural point of view have identified these areas of lower performance.
Furthermore it is noted that the approach of textures from the statistical viewpoint is insufficient for complete detection of canes falls, so it must include spectral or multiscale approach to the problem. As future work is planned to incorporate new descriptors in order to build a classifier oriented segmentation of such imagery.

5. References

[1] Pérez D, Fandos C, Mazzone L, Soria F, Scandalaris P and Scandalaris J. 2005 *Caña de azúcar en Tucumán y Argentina: evolución de algunos aspectos económicos y productivos en la campaña 2004* [on line]. Reporte agroindustrial. Estadísticas y márgenes de cultivos tucumanos. Vol. 2, Boletín N° 6, (ene.). Tucumán (Argentina): Estación Experimental Agroindustrial Obispo Colombres, EEAOC. p 1-6. <http://www.eeaoc.org.ar/upload/publicaciones/archivos/189/20120316170830000000.pdf> [consulted: 02/04/2013]

[2] Rodríguez R.A, Sopena, R.A, Saleme P.M and Vicini L.E 2010 *Pérdidas durante la Cosecha del Cultivo de Caña de Azúcar: Evaluaciones 2009 en la Provincia de Tucumán – Argentina* Informes Técnicos del proyecto Precop, No. 1. Famaillá (Tucumán, Argentina): INTA EEA Famaillá, N°1, 18 p. ISSN: 1852-9399.

[3] Tonatto J, Romero E.R., Leggio Neme M.F., Scandalaris J, Alonso J, Digonzelli P, Alonso L and Casen S 2005 *Importancia de la calidad de la materia prima en la productividad de la agroindustria azucarera* Gacetilla Agroindustrial de la EEAOC, No. 67, Tucumán (Argentina): EEAOC. 13 p.

[4] Larrahondo J.E. 1995 *Calidad en la Caña de Azúcar* CENICAÑA. El cultivo de la caña en la zona azucarera de Colombia, Cali, CENICAÑA. p.349-350.

[5] Sustaita G 2005 *Modelo estratégico para la industria azucarera regional.* Tesis Final MBA, San Miguel deTucumán (Tucumán, Argentina): Universidad Católica del Norte Santo Tomás de Aquino, Fundación del Tucumán y Pontificia Universidad Católica de Valparaíso.

[6] Viccini L 2007 *Determinación de la variación de rendimiento cultural de caña de azúcar en Tucumán, Argentina, posicionada con GPS* Actualización Técnica N° 8 (jun.). Manfredi (Córdoba, Argentina): Proyecto Nacional Agricultura de Precisión, INTA, 18 p.

[7] Anderson D.L., Portier K.M., Obreza T.A., Collins M.E. and Pitts D.J. 1999 *Tree regression analysis to determine effects of soil variability on sugarcane yields* Soil Science Society of America Journal, Vol. 63, No. 3, Madison (WI, USA), Soil Science Society of America, p. 592–600, ISSN: 0361-5995.

[8] Kemerer A., Melchiori A, Albarenque S and Melchiori R 2010 Utilización de fotografías aéreas multiespectrales para caracterizar la variabilidad espacial en la producción de caña de azúcar 9° Curso de Agricultura de Precisión y 4ª Expo de Máquinas Precisas (14-16/07/2010), Manfredi (Córdoba, Argentina): INTA - EEA Manfredi. Material del 9° Curso de Agricultura de Precisión, p. 177-182, 2010

[9] Melchiori A, Kemerer A, Melchiori R and Bellomo M 2009 *Utilización de fotografías aéreas multiespectrales de alta resolución para la cuantificación de caña de azúcar caída* I Congreso Argentino de Agroinformática (24-25/08/2009), Mar del Plata (Argentina): Sociedad Argentina de Informática y el Instituto Nacional de Tecnología Agropecuaria, INTA.

[10] Russel B.C., Torralba A, Murphy K.P. and Freeman W.T. 2008 *LabelMe: a database and web-based tool for image annotation* International Journal of Computer Vision, Vol. 77, No. 1-3 (may.). New York (USA): Springer US. p. 157-173, ISSN: 0920-5691.

[11] Rouse J.W., Haas R.H., Schell J.A. and Deering D.W. 1973 *Monitoring vegetation systems in the great plains with ERTS* Third ERTS Symposium (10-14/12/1973), Washington D.C. (USA): NASA. Proceedings of Third ERTS Symposium Vol. 1., p. 309-317.

[12] Toribio P, Rodriguez B.G. and Alejo R 2009 *Complejidad de los datos en las Redes Neuronales Artificiales: Estado de la cuestión* 7° Congreso Internacional de Cómputo en Optimización y Software, CICOS 2009 (17-20/11/2009), México (México): UAEM. Marco Antonio Cruz-Chávez, M.A. & Zavala-Díaz, J.C. (eds.). Memorias del 7mo Congreso de Cómputo CICOS 2009. ISBN(e) 978-607-00-1970-8, México (México): UAEM. p. 229-235. <http://campusv.uaem.mx/cicos/imagenes/memorias/7mocicos2009/Articulos/24%20%20Complejidad%20de%20los%20Datos%20en%20Redes.pdf> [consulted: 02/04/2013]

[13] Castells F, Laguna P, Sornmo L, Bollmann A and Roig JM 2007 *Principal Component Analysis in ECG Signal Processing* EURASIP Journal on Advances in Signal Processing. Volume 2007, Article ID 74580, 21 pages. doi:10.1155/2007/74580

[14] Flowers M.W.R., Heiniger R, Tarleton B and Meijer A 2003 *Field Validation of a Remote Sensing Technique for Early Nitrogen Application Decisions in Wheat* Agronomy Journal, Vol. 95, No. 1. Madison (WI, USA): American Society of Agronomy, p 167–176. ISSN: 0002-1962.

[15] Raun W.R., Solie J.B., Stone M.L., Martin K.L., Freeman K.W. Mullen R.W., Zhang H., Chepers J.S. and Johnson G.V. 2005 *Optical Sensor-Based Algorithm for Crop Nitrogen Fertilization* [online]Communications in Soil Science and Plant Analysis, Vol. 36, No. 19-20. Johannesburg (South Africa): Taylor & Francis, Inc. p. 2759-2781. ISSN: 0010-3624, EISSN: 1532-2416.

[16] Melchiori R.J.M., Caviglia O.P., Bianchini A.A., Faccendini N, Albarenque S and Raun W 2006 *Wheat Yield Prediction by Using an Active Sensor in the Northern Argentinean Pampas* ASA-CSSA-SSSA Annual meeting (12-16/11/2006), Indianapolis (Indiana, USA): American Society of Agronomy, ASA - Crop Science Society of America, CSSA - Soil Science Society of America, SSSA. Proceedings of ASA-CSSA-SSSA Annual Meeting 2006. Madison (WI, USA): ASA-CSSA-SSSA.

[17] Qi J, Chehbouni A, Huete A, Ferry Y and Sorooshian S 1994 *A modified soil adjusted vegetation index* Remote Sensing of Environment, Vol. 48, No. 2 (may.). Philidelphia (PA, USA): Elsevier Inc. p. 119-126, ISSN: 0034-4257. <http://www.sciencedirect.com/science/article/pii/0034425794901341> [consult: 01/04/2013]

[18] Rondeaux G, Steven M. and Baret F 1996 *Optimization of soil-adjusted vegetation indices* Remote Sensing of Environment, Vol. 55, No. 2 (feb.). Philadelphia (PA, USA): Elsevier Inc. p. 95-107. ISSN: 0034-4257.

9th Argentinean Bioengineering Society Congress (SABI 2013) IOP Publishing
Journal of Physics: Conference Series **477** (2013) 012017 doi:10.1088/1742-6596/477/1/012017

Temporal Assessment Model for Monitoring Critical Patients

A. Hadad, D. Evin, B. Drozdowicz
Artificial Intelligence Group, Faculty of Engineering, National University of Entre Ríos.
E-mail: hadad@santafe-conicet.gov.ar

Abstract: This paper gives an integrated view of developing a model for an assessment systems and monitoring tasks of critical patients. This model was designed according to the Temporal Abstraction dimensions and state interpretation levels, particularly the monitored physiological state. Monitoring arrhythmias was used as reference case for models development. A model based on the algorithm Smith-Waterman with the matrix substitution modified, was developed for the assessment state interpretation model, for this application. The modification was oriented to give more influence to the principal state for the temporal pattern represented. The assessment state space model, allow to obtain additional temporal information about a possible evolution, not previously available in classic monitors.

1. Introduction

One of the aims in the development of intelligent systems for monitoring clinical patients is to improve the common performance of traditional monitors by integrating multiple sources of observations and medical knowledge in the diagnosis process and patients´ evolution.

The assessment of critical patients requires a continuous representation of temporal features of the processes being monitored, which is derived from a great amount of data. This leads to the need of automatic mechanisms for the representation process, so that it can be more easily interpreted by users, improving the decision-making process speed and reliability and thus providing better assistance to patients.

There are three main tasks in the intelligent assessment of patients: data acquisition, data processing (producing reduced descriptions of collected signals) and diagnosis. The objective of this last task is to detect potential pathological states, as early as possible, and to characterize the situation by offering explanations based on passages from previous states **[1]**.

One of the ways of characterizing the inference methods used for the diagnosis task is to analyze how they operate in data and temporal dimensions. The level of data or interpretation is related to the data type used by the inference method, whereas the temporal abstraction dimension indicates if such method interprets individual data points or sequences of data points ordered in time. Figure 1 shows these dimensions and the corresponding inference methods **[2]**.

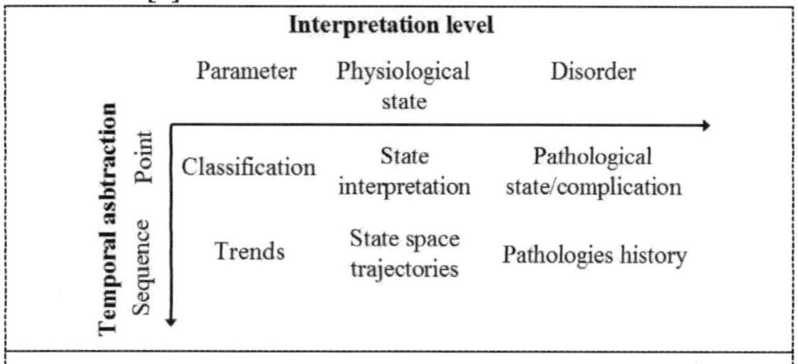

Figure 1. Dimensions and inference methods

On the one extreme of the interpretation spectrum, there are inference methods which interpret individual parameters values. In this level, if the system just considers points of isolated data, the inference operation is called classification, and it typically consists of a transformation of numbers into symbols. If the system interprets sequences of data, the inference method can be referred to as based on trends detection.

The following level of abstraction in the data dimension is based on states. The inference methods that operate at this level synthesize interpretations of physiological and physio-pathological states from the interpretation of parameter values. If they are extended to the degree of temporal state sequences, they reach the degree of trajectory in the state space. This level allows modeling the sequence of states followed by pathological cases.

The last level of abstraction in the interpretation dimension is that of disorders, which can be defined as a collection of physio-pathological states at a given time or as a trajectory of vectors in the state space over time. Traditional diagnosis Expert Systems operate at this level, but they suggest some limitations. For instance, a system that reasons in the three levels will be more robust, since it will be able to synthesize the physiological interpretations of parameter values and physiological states into interpretations of pathologies and complications **[2]**.

In order to model the temporal patterns from the data in the different levels of interpretation, a desirable representation of the time series is first required **[3]**.

As previously mentioned, a convenient methodology to obtain an appropriate and meaningful compact representation of temporal data is related to the construction of Temporal Abstractions (TAs) **[4]**. TA is a methodology of Artificial Intelligence characterized by leading a quantitative representation of the temporary data to an intervals-based qualitative representation. In each interval, a specific pattern is observed in the data. This pattern represents a summary of the original data and can be used to derive traits that characterize the dynamics of the system under observation. The algorithms that take a time series as input, generating a series of intervals, which correspond to periods in which a specific pattern is present in the input, are referred to as AT mechanisms **[5-7]**. In this work, they will be related to the cardiac arrhythmias problem.

The cardiac arrhythmias could be detected from
electrocardiogram (ECG). This signal represents the sequence of polarization (negative and positive) of atrias and ventricles. The electrical stimulus in a healthy heart is born in the nodule sinus atrial (a small muscular tissue in the right atria) and gives origin to a depolarization wave on both auricles (P wave). When this wave arrives at the site where right atrial and right ventricle join, it must go through another specialized structure: the atrioventricular nodule. This nodule has an electrical conduction speed that is slower than that of the other conduction tissue, causing a delay of the wave and a flat line on the electrocardiographic record.

In the atrioventricular nodule, the His beam (a muscular fiber beam with greater conductivity than other muscular fibers) extends through the wall that connects both ventricles and ends in fiber ramifications (Purkinje fibers) connected to the rest of the ventricular tissue. During the conduction period of the electrical impulse, from the His beam to the last cells of the ventricular myocardium, the so-called QRS wave is registered. Then, ventricles remain in an idle state (segment ST), after which they are re-polarized, which is reflected in the presence of the T wave. Besides the aforementioned waves, there is usually an additional wave called U wave, whose origin is not clearly known. When a fault in the heart rate occurs, regularity, the generation or conduction of the electrical impulses and/or the sequence of events do not follow the normal pattern previously described, and thus the arrhythmia occurs, which is very common in critical patients **[8]**.

Among the arrhythmias caused by the abnormal generation of the cardiac impulse, there are those that occur by an increase in automaticity (for example sinusal tachycardia, bradycardia sinusal, auricular tachycardia, etc.) or by the firing activity (syndrome of long QT, toxicity by digoxina, etc.). Conduction disorders of impulse include atrioventricular tachycardia or atrioventricular-nodules, atrial fluttering, atrial fibrilation, ventricular tachycardia, ventricular fibrilation, blockade atrioventricular and sinus atrial and branch blockade. It is possible to group these arrhythmias according to the degree of risk they represent for patients´ lives. A first group comprises tachycardia and ventricular fibrilation, which jeopardize patients´ lives and require immediate action by the personnel to revert the state. On the other hand, there is a set of other arrhythmias, which even though they do not represent an imminent threat of death for patients, they might require therapy to avoid further problems **[9]**.

One of the main reasons for which patients with myocardium infarct are admitted in intensive care units (or coronary care unit) is the possibility of monitoring them permanently and thus preventing malignant arrhythmias like ventricular fibrilation. In addition to the need for an intelligent supervision system that detects indicators of this type of arrhythmias, it is also interesting, that the intelligent system, to inform users of the presence of the second type of arrhythmias, which in some cases offer early indications of a possible malignant arrhythmia. For example, premature ventricular contractions (PVC) are often warning signals of the possible occurrence of malignant arrhythmias. Although PVC in ECGs of normal patients can be registered, several events per minute are considered pathological. A high rate of PVC may be due, among other causes, to poor coronary irrigation, specific medication effects and deficit of sanguineous oxygenation. This must be detected, especially in patients with infarcts **[9]**.

To reach a level of representation that allow for the search of temporal patterns, as the sequence of PVCs, it is first necessary to address the task of identifying V beats. These beats are normally mixed with normal heartbeats, which from the point of view of a classification model pose a problem with unbalanced classes. This imbalance between data sets for each class usually presents more "normal" examples than "abnormal". This problem becomes more serious when trying to make a multiple classification, because the classes containing fewer examples are disadvantaged by more than one class with more examples.

All approaches improve the classification rate to the original classification problems analyzed in each work, which coincides with the findings presented in **[10]**, which suggest similar effects on classification rates. On the other hand **[11]** questioned the incidence of examples imbalance in the degradation of the classification rate, arguing similar degradations in non-imbalanced problems.

In the temporal sequence dimension, this problem is reflected in the fact that pathological states are temporarily mingled with sequences of normal states with different lengths. This leads to the need of distinguishing the importance beat classes over the time sequence.

This paper will specifically address the time representation and modeling of the second type of arrhythmia, which offer early warning, representing states trajectory. For the representation of the trajectory in the space of arrhythmia states, a representation of the PVC time pattern is exemplified. It was previously described according to the application domain data through the Smith-Waterman algorithm with its substitution matrix being adjusted to the requirements of this application.

Database overview Section describes the signals and data being used. State-Space Assessment Model Section discusses the proposed model, based on the considerations made for the level of interpretation of the physiological status. Finally, we present the results and conclusions on the data used for the model application.

2. Database overview

2.1. Temporal Abstraction for Arrhythmia analysis

In the context of beats rate detection, temporal abstraction was conducted at three levels: at first, basic abstractions, and at the other two levels, complex abstractions were carried out. Figure 2 shows a diagram of these tasks.

The first level consists of the features-extracting module, which processes ECG samples to obtain a set of features that characterize a segment of the signal. The aim of the abstraction made by the features-extracting module is to facilitate the task of the classifying module. Even though it is possible to build a classifier using ECG samples, better results will be usually obtained using a number of features that is smaller than the number of samples in that interval **[12]**. It is also possible to use the outcome of the first abstraction to perform other diagnostic tasks such as detecting periods of ischemia in patients.

The second level of TA is performed by classifier I. This module uses the set of features of each segment given by the basic abstractions obtained in the previous stage to extract information regarding the type of heartbeat, either normal or arrhythmic, introduced by a patient.

The final level consists of classifying module II performing an abstraction based on sequences of beat types to extract information as regards temporal patterns at a higher level scale, which may provide evidence of poor oxygenation and malignant arrhythmias.

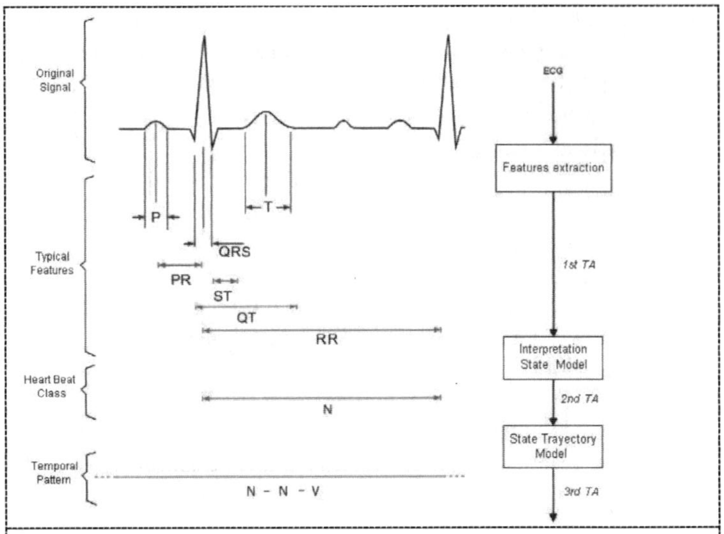

Figure 2. Temporal Abstraction Models

2.2. Signal database

ECG signals used in this work are part of the database of arrhythmia MIT-BIH [13]. This database contains 48 registers of 2 electrocardiographic derivations of 30 minutes each. These signals were filtered between 0.1 and 100 Hz before being digitized at 360 Hz, with a resolution of 11 bits and a range of + 5 mV.

This database consists of two sets of records: those ranging from 100 to 124 inclusive, which contain a set of representative samples of morphologies of both the waves and usual artifacts in clinical practice routines. The second set includes records ranging from 200 to 234 and were chosen to represent complex and unusual cases, such as juncture ventricular and complex supraventricular arrhythmias, as well as conduction abnormalities, which are generally more difficult to be distinguished either by the human experts or by automatic detectors.

In addition, each record has a file of annotations for the type of beat that is manually labelled by cardiologists. This information is vital to the classifying module because it works in a supervised fashion. Among the entries, there are 15 types of different beats. In this paper, they are grouped into 5 classes according to AAMI recommendations (AAMI, 1998): N class: normal beats; S class: supraventricular ectopic beats; V class: ventricular ectopic beats; F class: fusion between normal beats and ventricular ectopic beats; Q class: other types of beats, including odd ones.

DS comprises the following records: 109_A, 112_A, 115_A, 118_A, 121_A, 122_A, 124_A, 212_A, 220_A, 230_A, 233_A, 100_A, 103_A, 108_A, 111_A, 113_A, 114_B, 116_A, 117_A, 123_A, 231_A, 234_A.

3. State-Space Assessment Model

One way to characterize a sequence of states (in this case beats) is the search for similar sequences in databases as well as additional available data to contextualize them (age, sex, etc..).

When there are patterns of beats sequences, arrhythmias are interspersed among normal beats. Its quantities vary according to the patient, which means that there are variations to be considered during the search for similar patterns. With this in mind and with the intention of finding significant similarities in the sequence of beats that include some form of arrhythmia, a classification or scoring model is used during the process of seeking similarities. For this, the Smith-Waterman algorithm (based on dynamic programming) was used for local alignments of nucleotide sequences [14], with the substitution matrix adjusted for this application.

Taking into account the importance of the V beats for PVCs, defined qualify, with one higher magnitude order than other beats, the V-beat coincidence. The lowest positive value was assigned to the coincidence of normal heartbeats, given that this condition varies greatly according to each patient and has no clinical importance for this type of problem. Negative classifications are associated with non-matching types of beats and act as a penalty. As previously mentioned, this is implemented in a substitution matrix, which is used during the search process for similar sequences. In addition, “spaces” are incorporated in the sequence during comparison. These spaces allow a better alignment of sequences to find similarities, thus enabling a variable number of normal beats between arrhythmias.

The idea of the search process is to align complete sequences through optimal linear alignments according to an additional global score. This results from considering the qualifications of the substitution matrix and from penalization of incorporating spaces during the search. Partial qualifications of the alignment are stored in matrix F according to the following definition:

$$F(i,j) = \max\begin{cases} 0, \\ F(i-1, j-1) + s(x_i, y_j), \\ F(i-1, j) - d, \\ F(i, j-1) - d, \end{cases}$$

where d is the penalty for the included spaces, "x" and "y" are the sequences that are being compared, F (0.0) = 0, and s is the substitution matrix. Then, the maximum value of matrix F is sought in the last row and column, and from this element of the matrix the alignment is recovered seeking the previous maximum score in the previous row and column.

For instance, let us suppose that the following sequence of beats must be compared: *x*=NNVNNVNNP and *y*=NVNNPNNVNNNP. By applying the algorithm, the following alignment is obtained:

NN**V**NN---**V**NN**P**-

-N**V**NN**P**NN**V**NNN**P**

It can be seen that some spaces have been introduced in the alignment (represented with dashes), giving priority to the coincidence of beat V, which has the highest rating by coincidence in the substitution matrix. It is also intended to align heartbeat P, but since it has less weight, it admits some difference in the sequence.
The above example is intended to show the effects of the algorithm application to the comparison of two sequences. However, in a more realistic situation, it might be of clinical interest to find a sequence in particular within another longer one. To illustrate this, sequence x previously presented with sequence y extended in both directions by normal beats (e):

ye = NNNNNNNNNNNNNNNNNNNNNNN**V**NN**P**NN**V**NNN**P**NNNNNNNNNNNNNNNNNNNNNNNNNNNNNNNN

By applying the algorithm in a similar way, the following alignment is obtained:

---------------------NN**V**NN---**V**N-N**P**---------------------------------

NNNNNNNNNNNNNNNNNNNNNNN**V**NN**P**NN**V**NNN-**P**NNNNNNNNNNNNNNNNNNNNNNNNNNNNNNNN

Here, it is noted that the same algorithm maximizes the search score in the area where the similar temporal pattern is (Figure 3).

Journal of Physics: Conference Series **477** (2013) 012017 doi:10.1088/1742-6596/477/1/012017

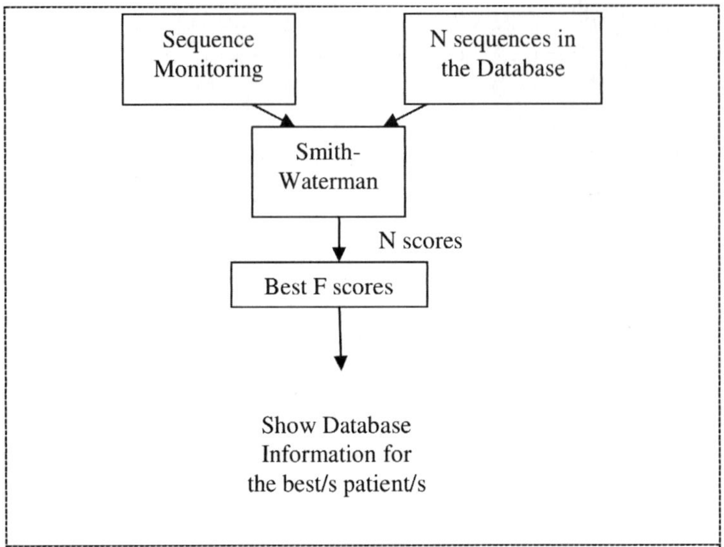

Figure 3. Activity Diagram for monitoring sequence analysis

4. Results

State-Space Assessment Model

Based on classifiers for normal beats, V-beats and other pathological beats, it is possible to compare a sequence x that is being monitored. To give greater importance to PVCs, the model put to the fullest qualification matches of the V beats in the temporal pattern. For this, a stretch (2.5 minutes) of a sequence (109_A record) is taken from the Database and compared with other nine full sequences through the Smith-Waterman algorithm.

Table 1. Results of sequence comparison.

Patient	Score (normalized)	Age	Sex
109_A	Current case	87	Female
112_A	1	64	Male
115_A	0.8758	54	Male
118_A	0.9876	-	Male
121_A	0.8758	39	Female
124_A	0.9123	69	Male
212_A	0.8758	51	Male
220_A	0.9441	77	Male
230_A	0.8758	87	Female

Sequence comparison to the monitored case
NNNNNNNNNNNNNNNNN**VP**NNNNNNNNNNNNNNNNNNNNNNNNNNNNNNNNN**V**NNNNNNNNNNNNNNNNNNNNNNNNNNNN
NN**V**NNNNNNNNNNNNNNNNNNNNNNNNNNNNNNNNNNNNNNN**V**NNNNNNNNNNNNNNNNNNNNNNNNNNN

Compact Representation: 17N**VP**33N**V**30N**V**39N**V**27N

Compared with the other nine patients, similarity qualifications (score standardized according to the highest score) and their associated data (age and sex for this example) are obtained.

Sequence Modified during the algorithm (by means of introduced dashes)
----NNNNNNNNNNNNNNNNN**VP**NNNNNNNNNNNNNNNNNNNNNNNNNNNNNNNNN**V**NNNNNNNNNNNN
NNNNNNNNNNNNNNNNNN**V**NNNNNNNNNNNNNNNNNNNNNNNNNNNNNNNNNNNNNNN**V**NNNNNNNNNNNNNNNNNNNNNNN
NNNN

Compact Representation: (4)17N**VP**33N**V**30N**V**39N**V**27N

More similar sequence (record 112_A)
N**V**NNNNNNNNNNNNNNNNNNNNNNNNNNNNNNNNNNNNNNNNNNNNNNNNNNNNNN**V**NNNNNNNNNNNNNNNNNNNNN**V**NN
NNNNNNNNNNNNNNNNNNNNNNNNNNNNNNNNNNNNNNNNNNNNNN**V**NNNNNNNNNNNNNNNNNNNNNNNNNNN

Compact Representation: NV54NV21NV48NV27N

In the example, the highest score was taken. A highest scores might have been taken and then contextualise the resulting group by grouping them by age, sex or other convenient clinical criteria. It should be stressed that the example can be extended to include more non-temporal data (weight, height, pre-clinical classifications, etc.). It is also possible to apply this model to other types of state sequences, such as stability / instability states a critical patient might undergo.
Then in the case of Database which provides the most similar temporal pattern, the associated data are available as indicated in Table 1, and the temporal sequence that comes after the temporal pattern found in case 112_A, which is the one that best assimilates the sought pattern. Thus it is possible to provide information on how that case evolved and other available data such as background, prior diagnoses, and so on. For this example, there is no other non-associated temporal information (in addition to age and sex). But there is information on the evolution of the case found by the model as similar in terms of sequence (Figure 4).

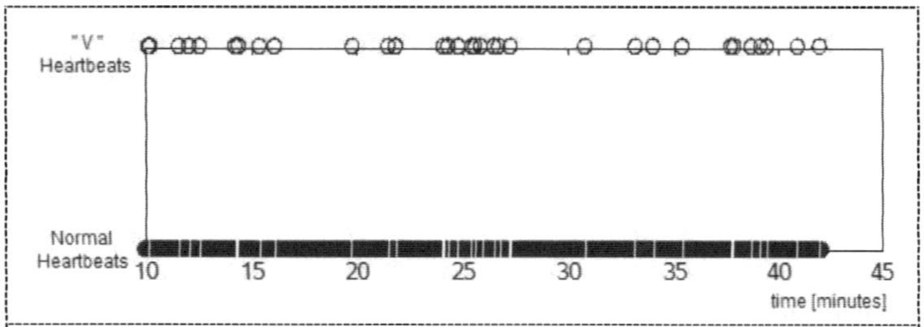

Figure 4. Sequence arrhythmias after found temporal pattern (112_A record, time in minutes)

In this particular case, there is a presence of arrhythmias class V in the forty minutes that follow the most similar case, which is additional information on the evolution of the monitored case.

Conclusions

The approach proposed in this paper allows a disaggregated analysis by granularities information provided by a monitor in a critical care unit, allowing higher quality information that facilitates the task of forecasting the evolution of patients.

As for the states space trajectory assessment model, this provides a vision in perspective about the evolution from a search considering temporal and non-temporal aspects. A greater availability of previously monitored cases would improve the quality of the database used to obtain conclusions.

The posed contextualisation through a clustering algorithm allows for selecting the most similar case, including information on non-temporal aspects, which are relevant to set a forecast in the area of critical care. This model form the basis of the operation logic for a system to support decision-making in the clinical setting, aimed at the representation of temporal patterns associated with early warning signs of critical issues, and enhancing the quality of the available information for decision making.

The temporal abstraction approach used for the development and implementation of monitoring models can be extended to other aspects of monitoring, including information on invasive blood pressure, oxygen saturation, heart rate and breathing, leading to the analysis of the changing stable / unstable and intermediate states. For this, is necessary to generate an indicator of stability from the available data, and then analyze the existing patterns during monitoring.

References

[1] Passariello, Mora, Carrault, Le Pichon (1993). Intelligent patient monitoring and management systems: a review. IEEE Eng Med Biol;12(4):23–33.

[2] Uckun S. (1993). Intelligent systems in patient monitoring and therapy management. A survey of research projects. Int J Clin Monit Comput 1994;11:241–53. Intelligent Systems in Patient Monitoring and Therapy Management.

[3] Höppner, F. (2001). Discovery of temporal patterns-learning rules about the qualitative behaviour of time series. In: Proceedings of the Fifth PPKDD, LNAI, 192-203.

[4] Shahar, Y. (1997): A framework for knowledge-based temporal abstraction. Art. Intelligence 90 79-133.

[5] Sacchi L., Bellazi R., Larizza C., Porreca R., Magni P. (2005) Learning Rules with Complex Temporal Patterns in Biomedical Domains. Book Series Lecture Notes in Computer Science. Publisher Springer Berlin / Heidelberg ISSN 0302-9743,

[6] Stacey M., McGregor C. (2007) Temporal abstraction in intelligent clinical data analysis: A survey. Artificial Intelligence in Medicine 39, 1—24.

[7] Hadad A., Evin D., Drozdowicz B. and Chiotti O. .Temporal abstraction for the analysis of intensive care information. Journal of Physics: Conference Series. Volume 90, 2007. http://www.iop.org/EJ/toc/1742-6596/90/1

[8] Artucio, Pereira (1991). Cardiac arrhythmias in critically ill patients: epidemiologic study. Critical Care Medicine 19(6):843-4.

[9] Hall, Schmidt, Wood (2005). Principles of Critical Care, 3Ed. Mc Graw Hill.

[10] Maloof Marcus A. (2003). Learning When Data Sets are Imbalanced and When Costs are Unequal and Unknown. Workshop on Learning from Imbalanced Datasets II, ICML, Washington DC.

[11] Japkowicz Natalie. Class Imbalances: Are we Focusing on the Right Issue?. Workshop on Learning from Imbalanced Datasets II, ICML, Washington DC, 2003.

[12] Chazal, O'Dwyer, Reilly (2004). Automatic Classification of Heartbeats using ECG Morphology and Heartbeat Interval Features. IEEE Transactions on Biomedical Engineering. Vol 51, N°7.

[13]Moody, Mark (1997). MIT-BIH Arrhythmia Database. Available: http://ecg.mit.edu/dbinfo.html

[14] Durbin R., S. Hedí, A. Krogh, G. Mitchinson (1998). Biological Sequence Analysis. Probabilistic Models of Proteins and Nucleic Acids. Cambridge University Press. ISBN 0-521-62041-4.

Secure reporting features for a low cost multiparametric PC-based patient simulator

P.P. Escobar[1]; M. Formica[2]; J.M. Massa[3];E. Arguiñarena[4]

[1] INTELYMEC, Fac. de Ingeniería, Universidad Nacional del Centro. Avda Del Valle 5737, Olavarría, Argentina.

[2] Fac. de Ingeniería, Bioingeniería, Universidad Nacional de Entre Ríos. Ruta 10, Km 11, Oro Verde, Paraná, E.R. Arg.

[3] INTIA, Facultad de Ciencias Exactas, Universidad Nacional del Centro. Campus Arroyo Seco, Tandil, Bs.As. Arg.

[4] PLADEMA, Fac. de Ciencias Exactas, Universidad Nacional del Centro. Campus Arroyo Seco, Tandil, Bs.As. Arg.

E-mail: pescobar@fio.unicen.edu.ar

Abstract. Preventive maintenance of biomedical devices is of particular concern to avoid failures and breakdowns, especially for critical care and life supporting devices. Signs of potential failures are not always visible and so, they should be inspected and tested periodically using the appropriate technology. Patient simulators have become one of the essential tools for any biomedical department on every health institution, but its cost is still high to be afforded by everyone. The Labview interactive virtual instrument SIMPAC, a simple PC-based patient simulator, showed to be a low cost powerful tool for simulation of healthy and pathological conditions using signals repositories of different biomedical devices such as EEG, ECG, etc., and also to record the output signals of devices like pacemakers, defibrillators, etc. Here we present an enhancement of the tool which consists of adding a secure reporting feature using XML for presenting tests results, applying digital signature for granting report integrity, assuring this way that reports can not be altered by intentional or accidental manipulation of the data contained on it. In addition, SIMPAC brings a low cost solution for teaching and practical purposes.

1. INTRODUCTION

The main purpose of preventive (and predictive) maintenance is to detect problems before they get develop. Nevertheless, certain kind of problems can not be detected by preventive procedures because they may occur suddenly. Maintenance of biomedical devices and the risk assessment due to its use are under the responsibility of the biomedical departments inside every health institutions. These professionals must set up a preventive and predictive maintenance (PPM) plan for medical equipment to guarantee the conservation of its adequate performances and requirements to be used safely on patients every time they are needed [1].

Maintenance programs should be designed aiming to detect minor problems before they result in major system failures, to reduce hazardous risk for users and patients, to minimize the equipment dead time, to avoid excessive repairing costs and also to verify that devices are compliant with codes, standards and regulations [2].

Biomedical devices are classified into risk categories, hence, a maintenance plan for each kind of device should be set up, considering they can fail from time to time with many diverse causes of failure including their own different nature, complexity, normal use, mechanical damage, user abuse, random component failure and obviously, aging [3].

Normally, when a new medical device is acquired and added to the inventory, we have to select an adequate maintenance schedule for it [4]. The election of the proper frequency for periodical inspections and performance tests is a key issue of the maintenance plan. High frequencies decrease the cost/effectiveness relationship and may also degrade device longevity. On the other hand, low frequencies may affect reliability, accuracy and safety [5].

Nowadays, the rapid advance of technologies is affecting to medical equipment maintenance, and traditional routines seem no longer enough to ensure that equipment is getting the best possible maintenance [6]. The wide variety of medical specialties set a hard challenge for biomedical departments, who have to catch up with new technologies and to incorporate new instruments to give a response according to necessities. A whole new set of analyzers are now essential tools for testing and analyzing modern medical devices. Some good examples are electrical safety analyzers, electrosurgery analyzers, patient simulators, X-ray testing devices, defibrillators analyzers and others.

Simulators and analyzers have a high cost, especially for hospitals and health centres in developing countries, situation that often turns them into prohibitive and makes impossible to carry out the proper maintenance, because of the lack of adequate instrumental. The "SIMPAC", a PC-based multipurpose simulator and analyzer is a low cost solution implemented with virtual instrumentation using *Labview™ v8.0* development suite and other resources available at the university labs, with a very modest investment on technology [7]. Figure 1 show a general overview of this simulator which was described in a previous work. Further details can be found following the references.

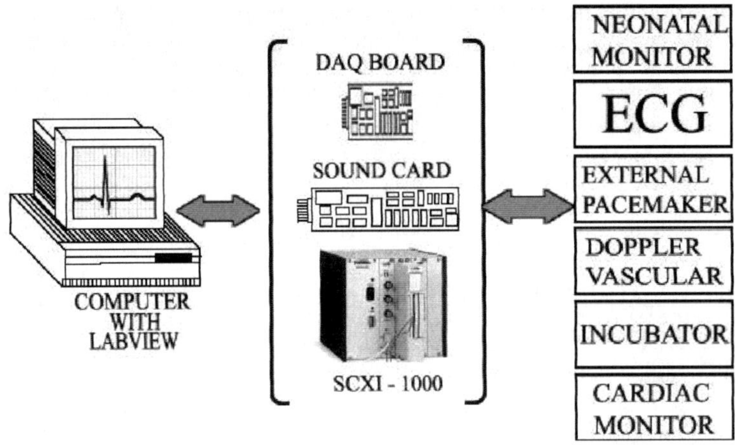

Figure 1: SIMPAC general system overview.

This versatile and economic tool can simulate many diverse physiological conditions, normal and pathological, in order to measure the output range values and verify that they meet manufacturer's specifications [8,9]. It also can function as an analyzer, recording the output signals given by specific biomedical devices like defibrillator, pacemakers, bi-directional vascular Doppler, etc.

The latest versions of the mentioned devices can connect to a PC through a serial port using RS-232 communication protocol or through a USB port. This connectivity capability allows for exporting acquired data to a PC giving us the possibility to generate reports from the result of each test performed, organizing the maintenance documentation process and giving a management tool for statistics, time/cost analysis and legal protection issues.

In this article we present an enhancement of the SIMPAC v1.0, adding a secure reporting feature using XML format for generating and archiving test results [10]. Security is here provided by means of a digital signature implementation [11,12,7] over the generated XML file, granting that the document will remain unaltered, avoiding intentional or accidental modification of any data contained on it. These structured XML reports can be integrated to any existing maintenance database and they can also be queried when we need to search something in particular.

2. DESIGN

2.1 SIMPAC overview

The PC-based simulator was designed and built in order to achieve a versatile tool for testing and calibrating biomedical devices, suitable for both practical and educational purposes. We chose *Labview™ v8.0* for the programming environment because it is available for several computer platforms such as Linux, Windows, Palm PC, etc, [14] and because its excellent connectivity with the outside world through DAQ modules, I/O and networks protocols which make it a great tool to perform computer-based experiments like acquisition, control, instrumentation and quick engineering solutions development [15, 16].

We considered the design of a virtual instrument (VI) with a user interface and full capability to interact with six different medical devices available at the lab as educational modules: a *bi-directional vascular Doppler* (frequency and beat noises); an *incubator* (room and skin temperatures); an *external peacemaker* (frequency and amplitude); an *electrocardiograph;* a *neonatal cardiac monitor* and a *cardiac monitor* (signal shape and frequency).

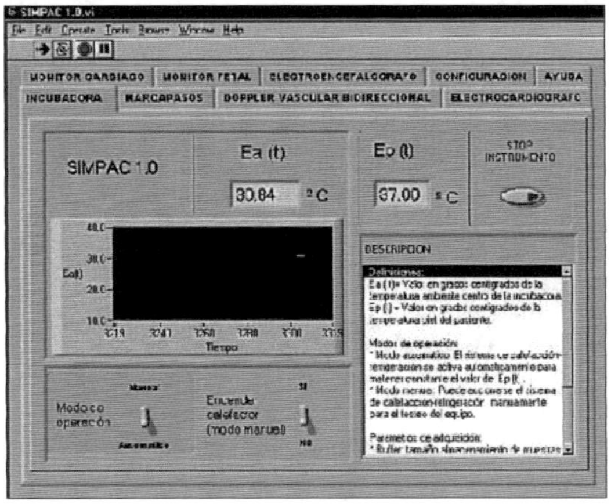

Figure 2: The SIMPAC v1.2 Virtual Instrument GUI.

At first, we chose a physical model for each device selecting the basic parameters to be controlled from the user interface. Then, a mathematical model was built to simulate different signal conditions. Signals generated by the mathematical model were stored in an excel sheet or a text file,

Journal of Physics: Conference Series **477** (2013) 012018 doi:10.1088/1742-6596/477/1/012018

creating this way a repository of signals that allowed us for testing devices, and then re-define the model if necessary [17].

Later, mathematical models were replaced by a repository of signals acquired by different methods. The most recent version provides two main modalities: simulation and testing. Simulation allows for sending signals from the computer to a device, and testing allows for acquiring signals from a device. All signals were allocated in a signal repository on the hard disk and used for simulation without the need of a patient, and also for testing and calibrating the devices.

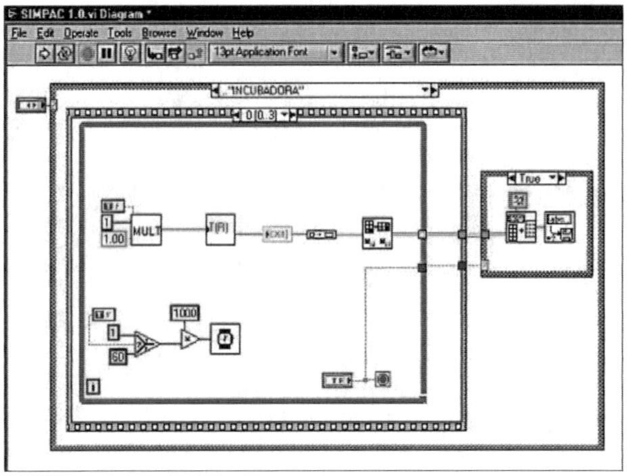

Figure 3: Design view of the incubator Sub VI.

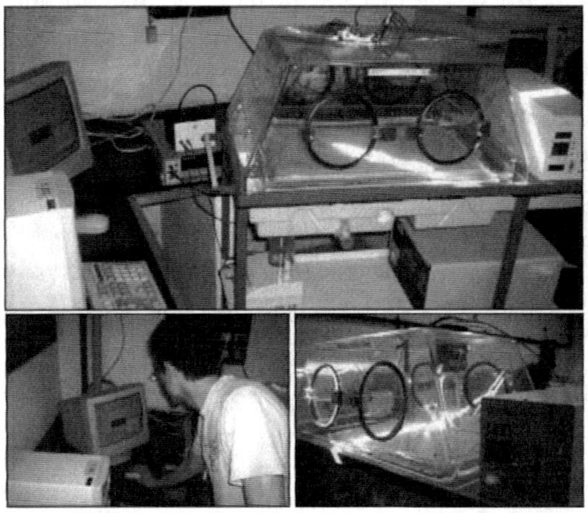

Figure 4: Working with a transport incubator Sub VI.

2.2 Considerations about maintenance documentation

Any PPM program must be effective in detecting faults and should reduce the number of repairs, without consuming much time for the task [6]. It is very important to build a failure history of a biomedical device, registering all relevant information obtained from every inspection.

Manual record keeping systems are still widely used, but computerized systems have great advantages and may add a lot of useful features for information processing and analysis. They allow for producing reports to identify equipment with higher repair costs, to support repair versus replacement decisions and to document potential user errors or equipment abuse [13], and to register functional ratings of a device in order to define whether it is working properly or not, according to manufacturer's specifications [14] or, when it should be replaced. They also provide documentation in case of a malpractice suit regarding to medical equipment maintenance; avoiding responsibilities proving to authorities that involved equipment was properly maintained.

2.3 The structure of a report

The structure of a report should be tailored for each device, because the signals we want to record are not the same for all devices. However, there are some common information items for all devices, like task responsible, time consumption, costs related to repairing or PPM inspection. A complete documentation process consists of three main parts: device ratings report (acquired by performing a test), device information (stored in an inventory database) and device's historical record (Fig.2).

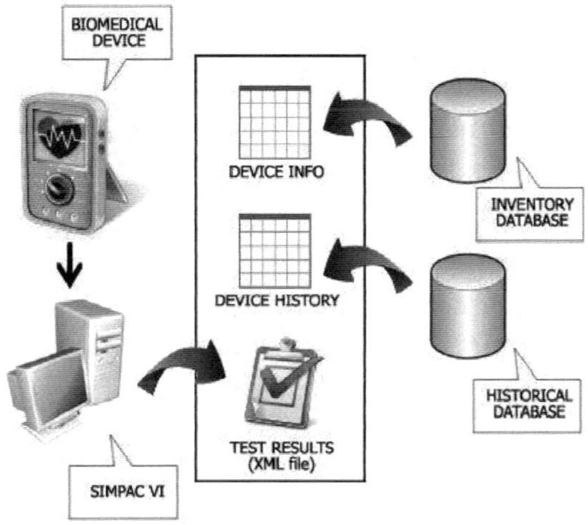

Figure 5: A complete documentation process scheme.

We focus here on the ratings report. The other two parts do not have a direct relationship with the SIMPAC virtual instrument and they will not be detailed extensively here. Information provided in the table named as "Device info" is shown briefly in Table 1.

Field Name	Data Type
Device Name	Text
Device ID	Number
Manufacturer	Text
Model	Text
Serial	Text
Location	Text
...	...
Age	Number

Table 1. Example of a Device Info table extracted from the Inventory DB.

Inventory database contains information about each device in the hospital (location, status, etc.) Historical database contains information about each preventive, predictive or corrective task performed on a device. There are other relevant tables involved in the process of managing and documenting information, for example, a table for each device listing its specific ratings for every considered parameter. This table will be useful in the future when checking tests results against these ratings, in order to determine how far the performance of a device from the manufacturer specifications is.

3. RESULTS

3.1 The XML reporting feature

When a PPM test is performed on a device, ratings are recorded and stored by the SIMPAC virtual instrument by reading a specific port where device is connected to. Data is saved to an Excel file (.xls) and then converted into an XML file in a process that is transparent for the user.

The conversion is made by a means of an application coded in Delphi 7, which uses a XML component with capabilities to create, to modify and to parse an XML file. This component can also import data from different databases and generate tree views of the content. The application is compiled and run inside the virtual instrument right after acquisition is over. In previous attempts we tried to directly save information in XML format using the "save as" option of Excel, but doing this we obtained an output file with a lot of unnecessary code.

XML is the 'Extensible Markup Language', universal language for data on the Web but not limited to. XML looks much like HTML. However, whereas HTML describes the layout of a page (for data displaying purposes) XML is concerned with semantically defining and describing data. An XML file can be thought as a mini database where tags or mark-ups in the document delineate the records and fields. The text between the tags is the data. An XML parser can perform operations like retrieving, updating, and inserting data into an XML file. Figure 3 show a XML report file structure [18]. The conversion process consist was successful every time we ran the virtual instrument to perform a test on a device. There was no problem or errors reported during conversions.

3.2 Security through digital signature

A digital signature of a document implies the use of the author/emitter private key and a unidirectional mathematical function, the Hash Algorithm [19], which calculates a code derived from the message, to extract a "fingerprint" of the information that have to be exchanged, shared or stored.

```
<?xml version="1.0"?>
<testresult>
  <device>
    <name>Defibrillator</name>
    <manuf>E&M</manuf>
    <model>C-12-S</model>
...
  </device>
  <test>
    <param1>energy</param1>
    <param2>chargingtime</param2>
...
  </test>
  <responsible>John Doe</responsible>
  <date>11-7-2004</date>
...
</testresult>
```

Example of an XML file tag structure.

The receiver uses the author/emitter public key to run the same operation on the document and comparing both hashes. If they match, the receiver can feel sure that the document was sent by the author/emitter and also it is unaltered. The relevant matter here is if the content of the document is altered, verification of signature will fail [20]. The digital signature schema is implemented using PGP software for key pairs generation and signing documents.

In this case, the author of a report is always the biomedical technician, who uses the SIMPAC patient simulator to perform a PPM task on a device generating automatically an XML report file. Once the report is generated, the technician digitally signs the XML document and then, he can store it locally on the PC hard disk or send it over the hospital network (maybe internet) to the biomedical engineers who are in charge of the documentation process and managing the biomedical technology department.

In this case, engineers (receivers) have to check the digital signature as it was explained above. If the comparison is successful, authenticity, non repudiation, integrity and confidentiality are assured. On the contrary, if it fails, the document was altered by someone and we can not trust on the information contained on it [21 22].

This is a key issue, because the PPM task reports act as documents in case of legal problems regarding to medical equipment maintenance [14]. If the device is functioning according to all requirements and specifications, a report is used for documentation purposes and the device is returned back to its original location inside the hospital. If the report shows that device is malfunctioning, it has to be retired from the service to be repaired and it can not be manipulated by medical or nursing staff and also, it must not be used on patients.

4. CONCLUSIONS

The SIMPAC PC-based patient simulator is a low cost, very useful tool, developed entirely as an original work. It allows for testing different kind of devices connected to the virtual instrument, which can work in two modalities: simulation and testing. Both modalities were accomplished successfully.

Labview shown to be a powerful tool and an excellent resource for building human-machine interfaces, with great coding languages integration capabilities [23].

The patient simulator is not fully developed yet, but this partial success has encouraged us to continue this way in order to improve its capabilities. The secondary goal or this work, education and training, was fully achieved, considering that basic and clinical science concepts were easily integrated with electronics to improve critical data analysis and to visualize real world dynamic processes in classes to enhance students' knowledge by means of a "hand on" experience.

The addition of reporting features enhances the simulator by providing a tool for secure documentation processes. Moreover, the XML file format chosen for the reports gives a semantic, machine-readable content, which is useful for query, search, replace and add data within a file.

The digital signature schema provides the security over the entire documentation process granting the integrity of information, the authenticity of both sender/receiver and non repudiation for authors identification.

This work is still under development and further results will be available soon.

5. References

[1] Amicucci GL. *Effectiveness indicator of periodic checks on medical devices*. Health Technology Assessment International. Meeting (2nd : 2005 : Rome, Italy). *Ital J Public Health.* 2005; 2: 236.

[2] R.Burak Arslan, Yekta Ulgen. *Smart-IPM: an adaptive tool for the preventive maintenance of medical equipment*. Institute of Biomedical Engineering, Bogazici University, 80815, Bebek, Istanbul, Turkey. 2001.

[3] John Webster, "Biomedical equipment maintenance", in *Encyclopedia of Medical Devices and Instrumentation*, pp.1227-1236. 1998.

[4] Bronzino, Joseph, *"Technology management", in Management of Medical Technology. A primer for clinical engineers*, Chapter 3, Butterworth-Heinemann pub., pp.67-110, 1992. 1995.

[5] Amoore, J.N.; Stay, D.G; Mather, J. *"Scheduled inspection of electromedical equipment. Planning and assessment"*, Journal of Medical Engineering & Technology, Volume 19, Issue 6 November 1995 , pages 211 – 218.

[6] A. Khalaf. *Evidence-Based Maintenance For Medical Equipment*. IFMBE Proceedings of World Congress on Medical Physics and Biomedical Engineering. Volume 25/7: Diagnostic and Therapeutic Instrumentation, Clinical Engineering, pages 804-807. Olaf Dössel and Wolfgang C. Schlegel. Springer Berlin Heidelberg.

[7] Escobar, Pedro; Formica, M. "Basis for the implementation of digital signature in Argentine's health environment". J. Phys.: Conf. Ser. 90 (2007) 012002 (6pp). ISSN (print): 1742-6588; ISSN (online): 1742-6596. Available online: http://www.iop.org/EJ/toc/1742-6596/90/1

[8] Del Aguila, C. 1994. *Electromedicina*. Ed. Hispano Americana.

[9] Bronzino, J.D. (Editor in chief), 1995. *The Biomedical Engineering* Handbook. CRC Press.

[10] W3C XML Standard. http://www.w3.org/TR/xmlenc-core/

[11] Menezes A, van Oorschot P, Vanstone S, *Handbook of Applyed Cryptography*, CRC Press, 1996

[12]Rivest, R., Shamir, A. and Adleman, L., *A Method for Obtaining Digital Signatures and Public Key.* Cryptosystems Communications of the ACM, 21(1978), pp. 120-126.

[13] Taylor, *Kevin* MScEng, PEng; Jackson, Stephen MA, A *Medical Equipment Replacement Score System.* Journal of Clinical Engineering: January/March 2005 - Volume 30 - Issue 1 - pp 37-41.

[14] Wang, Binseng Sc.D., C.C.E.; Levenson, Alan B.Sc. *Equipment Inclusion Criteria- A New Interpretation of JCAHO's Medical Equipment Management Standard.* Journal of Clinical Engineering. Volume 25, issue 1, February 2000. Aspen Publishers, Inc.

[15] Wells, K.K.; Travis, J. 1997. *Labview for everyone-Graphical Programming Made Even Easier*. Prentice Hall.

[16] Escobar, P.; Formica, M.; Acosta, G.; Massa, J. *"Design of a PC-based Patient Simulator for testing and calibration of electromedical devices using Labview"*. Proceedings of BIODEVICES 2008 - First International Conference on Biomedical Electronics and Devices. Volume 2, pp: 135-139.

[17] Dorf, R.C., Bishop, R.H. 1998. *Modern Ccontrol Systems*. Addison Wesley Longman.

[18] Delphi 7 XML component. http://delphi.about.com/library/bluc/text/uc050601a.htm

[19] RSA Data Security, *"Understanding PKI"*. (www.rsa.com) 1999.

[20] Lucena López, M.J. "*RSA: Criptografía y seguridad en computadores*" (3a ed). Univ. de Jaén. España. 1996.

[21] Salomaa, Arto. "*Public-key cryptography*". Berlin; New York: Springer-Verlag, cop. 1990.-- X, 245 p. : il. ; 25 cm.-- av ; 23.

[22] Van Leeuwen J.(editor), *Handbook of Theoretical Computer Science*, Elsevier Science Publishers, 1990 (Rivest R., Capítulo 13 : Cryptography)

[23] Wang, J.Y.-Z. *LabView in Engineering Laboratory Courses.*Frontiers in Education, 2003. FIE 2003. 33rd Annual Volume 2, Issue , 5-8 Nov. 2003 Page(s): F2E - F13 Vol.2

New approach to evaluate Acoustic Pollution in hospital environments

J.M. Olivera[1], L.A. Rocha[1,2], E. Ruiz[1], V.I. Rotger[1] and M.C.Herrera[1]
[1] Gabinete de Tecnologías Médicas, Departamento de Bioingeniería, FACET-UNT
[2] Instituto de Maternidad y Ginecología "Nuestra Señora de las Mercedes", Tucumán

E-mail: jolivera@herrera.unt.edu.ar

Abstract. There are many different services within a hospital. This means different types of noise which can be considered as acoustic pollution. Knowing that preterm infants exposed to high amounts of noise in the NICU are at a much higher risk because of their neurologic immaturity and physiologic instability, that excessive levels of noise also affect the persons and it can also impede some studies on patients, it was proposed to evaluate the Sound Pressure Level in some services of the Instituto de Maternidad, Tucumán, Argentina. During SABI 2011, this workgroup presented their first results in this matter. In this paper, there were evaluated the new Level III NICU –with new facilities in a different physical space in the hospital-, the laundry service and the attenuation factor (AF) of the cabinet cover in a neonatal incubator . The measurements were performed with two type II sonometer -CENTER 322, Cirrus CR832C) and it was also used an incubator analyzer (FLUKE INCU) for the incubator. The average values obtained were of 64.3 dBA for the NICU, and 80.8 dBA for the laundry room; the AF was between 2.88-3.09 and 3.43 –both under different conditions.

1. Introduction

During SABI 2011, this workgroup presented their first results in this matter. Those results were based in data collected in Hospital Maternidad de Tucumán in compliance with standards and legal regulations [1]. During 2012 there were important changes in the hospital facilities, creating a new Level III NICU, and new complementary resolutions to the National Law of Health and Safe at Work No 19587 were issued [2, 3, 4].

The Instituto de Maternidad is a public institution considered as Reference, in its specialty, in the northwest of Argentina. It is classified "Level 3" according to the Organization and Functioning of Neonatology and NICUs services guidelines of the National Ministry of Health –R.M. 306/02- and is the last step before to derive, the most complex cases, to Hospital Juan Garrahan in Buenos Aires, Argentina. This hospital has a family-centered policy, this fact reflects a shift from the traditional focus on the biomedical aspects of a child's condition to a concern with seeing the child in context of their family and recognizing the primacy of family in the child's life. This philosophy has a legal and ethic context in which the childs and its families are the real owners of this institutions, each family is considered unique and they will be the experts on the child's abilities and needs. The Tucuman Public Health System has a birth rate of about 25000 per year; half of them are attended in the Hospital Maternidad and proximately 1000 pregnant women per day are attended in all services. This hospital has 35 baby incubators, 25 infant radiant warmers, 4 transport incubators and 2 Advanced Life

Support ambulances. For adult patients it has 147 beds –93 for the Obstetric service with its own infant beds, 24 in the service of gynecology, 23 for High-Intermediate Risk, 4 in the service of oncology and 3 for Differential Cojoint Hospitalization-, and it also has 60 hospitalization units for neonatology.

It´s well known that Preventive Maintenance (PM) is a set of tasks performed by personnel for the purpose of maintaining equipment and facilities in satisfactory operating condition by providing systematic inspection to avoid failures either before they occur or before they develop into a major one, according to selected guidelines or standards [5]. The professional that takes care of these policies in a health institution is the Clinical Engineer. The American College of Clinical Engineering (ACCE) defines the Clinical Engineer as *A professional who supports and advances patient care by applying engineering and managerial skills to healthcare technology.*"-ACCE Definition, 1992-. As clinical medicine has become increasingly dependent on more sophisticated technologies and the complex equipment associated with it, the clinical engineer, as the name implies, has become the bridge between modern medicine and equally modern engineering [6].

Today, all over the world, Noise is considered as a modern plague. In Hospital Environments there are many sounds including those from beepers, alarms, machines, air-conditioning, staff activities, conversations, and medical equipment, among other sources. These can be severely irritating and sometimes harmful to patients, depending on their current conditions like age, hearing ability, medication intake, cultural background, and some psychological variables [7, 8]. Therefore, the acoustic pollution in healthcare environments is complex and requires a careful, strategic design and continuous control. This fact implies new healthcare design guidelines and there are many studies trying to identify strategies to improve this matter; therefore, any health institution must perform a regular control of the acoustic pollution in different indoor environments and it should be included in any PM schedule.

But, how do we define noise? We will consider as Noise any unwanted sound that exceeds the standards recommendations beyond any physical consideration. The unit used to measure Sound or Noise is decibel (dB), it is a logarithmic unit that indicates the ratio of a physical quantity relative to a specified reference level. It is well known that human hearing is less sensitive at very low and very high frequencies and requires an increase of about 10 dB before the sound subjectively appears to be twice as loud even when an increase of 6 dB represents a doubling of the sound pressure. The smallest change that can be perceived is normally about 3 dB. This is a direct consequence of the human frequency response.

When a noise evaluation is to be performed, the first step is to identify the type of noise in order to select the instrument –sound level meter-, noise parameters and the duration of the measurement [9]. The sound level meters, also known as sonometers, can acquire the sound level pressure and perform several different types of processing. The signal may pass through an electronic weighting network whose sensitivity varies with frequency in the same way as the human ear. There are three different internationally standardized characteristics termed "A", "B" and "C" weightings. The most common frequency weighting in current use is "A-weighting", it provides results often denoted as dB(A) which conforms to the response of the human ear. The "C-weighting" curve is applied when very loud or very low-frequency sounds are present. In some cases, Regulations may require an objective measurement of tonal content as well. In practice, this will be performed with an octave or one-third octave band filter attached or integrated into the instrument; in this case, they are called Octave Band Analyzers [9, 10].

What about the measurement report? It is common to report only dB values. This makes the report difficult to understand and turns it into a source of information of little or no use at all. Standards and recommended practices are a great help to make a measurement report, the level of detail must be consistent with its purpose and it is also important to write it in an easy-to-understand, readable style. ISO 1996-1:2003 states the minimal information to be recorded and also gives some recommendations. In Argentina, during January 2012 it was issued the Res. 85/2012 complementary to the National Law of Health and Safe at Work No 19587. This new resolution plus Res. 295/2003, also complementary, states the entire legal context for Noise and Vibration. In 2003 the maximum Equivalent Continuous Sound Pressure Level (LAeq), for an 8 h working-day or 48 h working-week, was reduced by the Res. 295/2003 from 90 dB to 85 dB, according to international standards. In other words, this means to recognize that the ear is affected at levels four times lower than the ones accepted years ago. This resolution also gives a theoretical framework leading to classify the type of noise before making any data collection. The measurement protocol and its report form, including an explanation of every item, are stated in Res. 85/2012; many of the suggestions, gives by ISO 1996-1:2003, were taken into account –although there is no item for photos, a good practice is to complement the report with a set of them describing the working conditions- [2, 3, 4, 11].

In this context, the objective of this paper is to measure Sound Pressure Levels at some services of the Instituto de Maternidad y Ginecología Nuestra Señora de las Mercedes, Tucumán, Argentina in compliance with the appropriate standards and regulations. In particular, in this preliminary study, the selected services were the new level 3 neonatal intensive care unit and the laundry. Also, it will be evaluated the Attenuation Factor for the cabinet cover of a infant incubator.

Why were these services selected?

Neonatal Intensive Care Unit (NICU): Numerous studies have documented the continuous noise exposure of infants associated with NICUs. Noise levels are often related to the new modalities of respiratory therapy such as high-frequency oscillatory ventilation and high-frequency jet ventilation, devices alarms, visitors (patient's family), among other sources. The exposure of medically fragile and low birthweight infants to NICU noise has been related to anoxia and bradycardia and has a negative impact on blood pressure, heart rate, perfusion, oxygen saturation and cerebral blood flow. Consequently, preterm infants exposed to high amounts of noise in the NICU are at a much higher risk because of their neurologic immaturity and physiologic instability. The Committee on Environmental Health of the American Academy of Pediatrics recommend a maximum noise pressure of 45 db(A) in daytime and 35 dB(A) at night [12, 13]. During 2012, new facilities were inaugurated; therefore, the objective is to compare those results obtained in 2009-2011 with those obtained in March 2013 at the new facilities. In this case, the report will be made following the ISO 1996-1:2003 and IRAM 4113-2:2010 [11, 14]. Figure 1 shows the floor plan of the old LEVEL III NICU while Figure 2 shows the floor plan of the new LEVEL III and LEVEL II facilities.

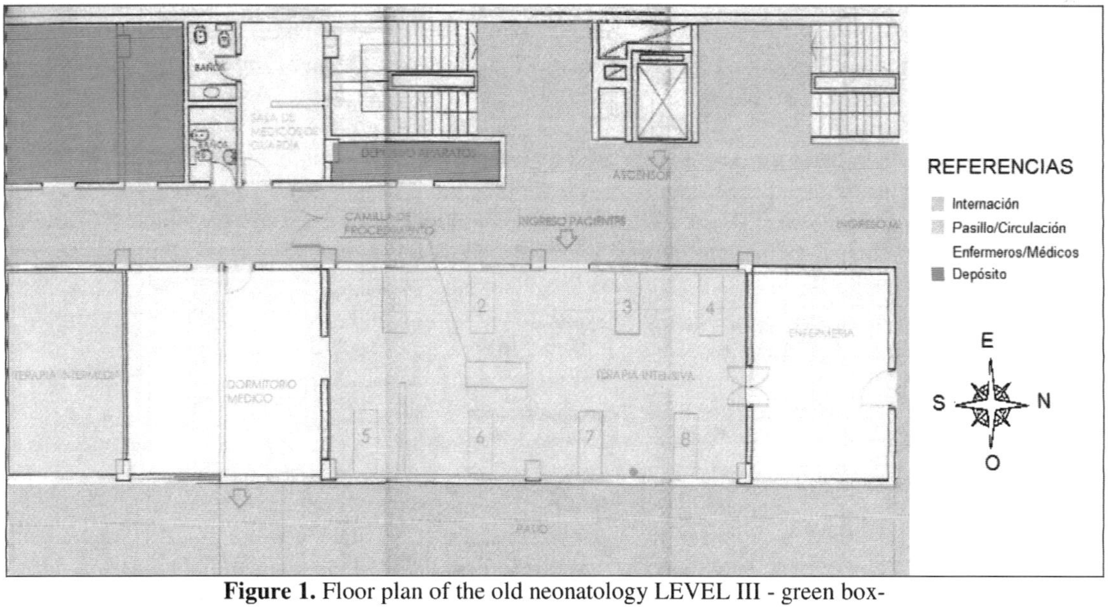

Figure 1. Floor plan of the old neonatology LEVEL III - green box-

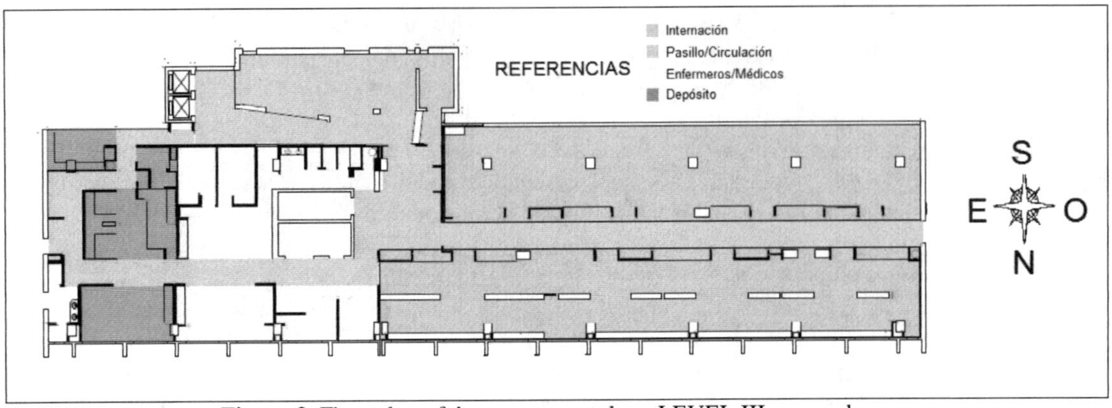

Figure 2. Floor plan of the new neonatology LEVEL III - green box-

Laundry: This is a typical case of noise at work and an occupational point of view can be used for its analysis. Occupational Safety and Health (OSH) can be important for moral, legal, and financial reasons. We should note that OSH can also reduce employee injury and illness related costs, including medical care, sick leave and disability benefit costs. OSH may involve interactions among many subject areas, including occupational medicine, occupational hygiene, public health, safety engineering, industrial engineering, among other disciplines. In Argentina, the standards and laws to be applied are the IRAM 4113-1:2009, IRAM 4113-2:2010, IRAM 4079:2006, National Law of Health and Safe at Work No 19587 and its complementary resolutions Res. 295/2003 and Res. 85/2012 [2, 3, 4, 14, 15, 16]. Occupational noise-induced hearing loss is one of the causes responsible for numerous occupational ailments and injuries. When the loss or reduction of hearing ability is present, there is a very high probability to be in presence of an irreversible process. Unfortunately, there are no low-cost solutions to reduce this problem; therefore, this task should be included in a MP schedule.

Neonatal Incubator: The objective is to evaluate the noise attenuation factor for the cabinet cover. The parameters to be measured are sound pressure, inside and outside the baby compartment.

This is a major challenge because of the complex conditions proper of level III NICUs. The IEC 60601-2-19:2009 defines the Compartment as an *environmentally-controlled enclosure intended to contain an infant and with transparent section(s) which allows for viewing of the infant.* The subclause 201.9.6.2.1.101-Sound level within the compartment- states that *it is unwise to keep preterm infants in an environment with ambient sound level above 50 dBA and with alarm impulses as high as 80dBA* [17]. All the infant incubators in the neonatology service are in compliance with IEC 60601-2-19:2009; but it should be noted that other institutions may have infant incubators in compliance with older standards than those from 2009 which accepted sound level, within the compartment, < 60 dBA.

2. Materials and Methods

The environment sound pressure level was measured with two type II sonometers. One of them was a CENTER 322 with data-logger, certified on June 2012. It was calibrated with a Sound Level Calibrator (TES-1356) at 94dB according with the manufacturer´s instructions. The instrument autonomy is about 36 h when it is powered by a 9V alkaline battery.

The sonometer parameters were set to:

- Level Range: Auto 30-120 dB; Frequency Weighting: A; Time Weighting: Fast
- Sampling time was programmed depending on the environment under evaluation

The recorded data were downloaded to a notebook, via an USB-RS232 cable adapter managed by the TestLink SE322 software, for Sound Level Meter, provided by the manufacture's sonometer. Furthermore, they can be also saved as a "csv" file format to be processed in any spreadsheet.

It was also used a sonometer Cirrus CR:832C Class 2 with 1:1 & 1:3 Octave Band Filters, certified on February 2012. It was calibrated with a Acoustic Calibrator (CR:514) according with the manufacturer´s instructions. The instrument autonomy is about 36 h when it is powered by a 9V alkaline battery.

The sonometer parameters were set to:

- Level Range: 30-100 dB
- Frequency Weighting: A
- Time Weighting: Fast
- Sampling time was programmed depending on the environment under evaluation

Neonatal Intensive Care Unit (NICU): The sound level pressure was recorded during 24 h, the sampling time was set to 5 sec and the sonometer was placed as part of the life support system of a baby incubator, as closely as possible of the incubator, in order to get data of the nearby environment under working conditions. Figure 3 shows a general view of the corridor, between the two hospitalization sectors, and the north beds area - light blue and green box in figure 2 floor plan-. Figure 4 shows the sonometer as part of the life support set. Figure 5 show a general view of the south beds area. It was asked to the personnel "to ignore the presence of the new instrument" in the service and it was explained that it does not record voices or conversations

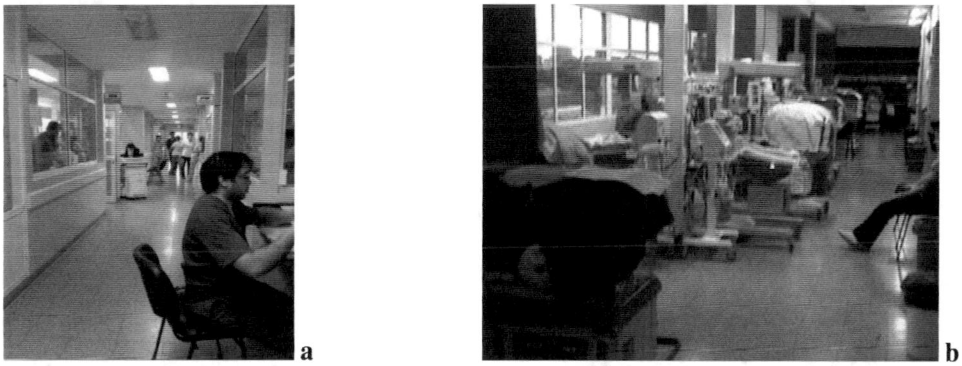

Figure 3. Level III NICU. General View: a- corridor between two hospitalization sectors; b- north beds area. Light blue and green box in figure 2.

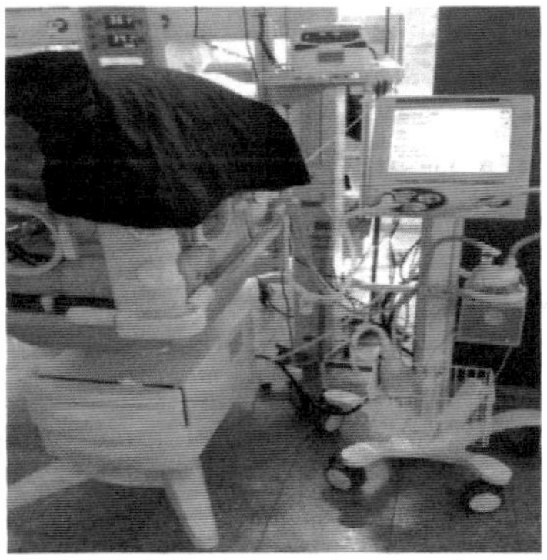

Figure 4. Sonometer as part of the life support set

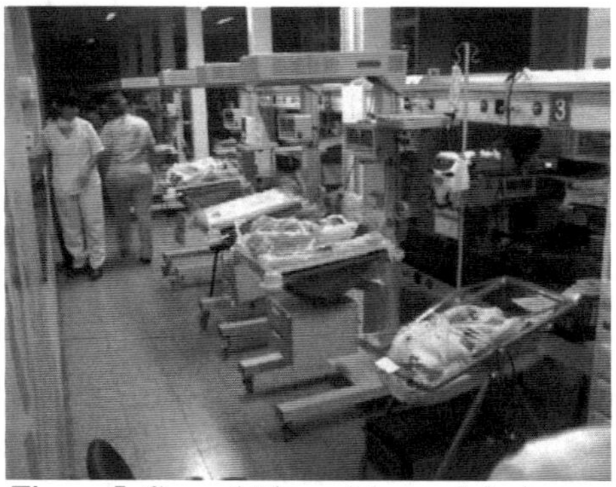

Figure 5. General View of the south beds area

Laundry: The sound level pressure was recorded using the CENTER 322 sonometer, during 5 min and the sampling time was set to 1 sec. It was asked to the personnel "to ignore the presence of the instrument" in the service and it was guaranteed that the device does not record voices or conversations. The sonometer was placed in the centre of the engine room, inside a foam rubber bag to avoid vibrations, with the microphone pointing to the noise sources. Figure 6 shows the engine room; Figure 7 shows the sonometer inside the foam rubber bag.

Journal of Physics: Conference Series **477** (2013) 012019 doi:10.1088/1742-6596/477/1/012019

Figure 6. Two pictures of the engine room of the laundry service

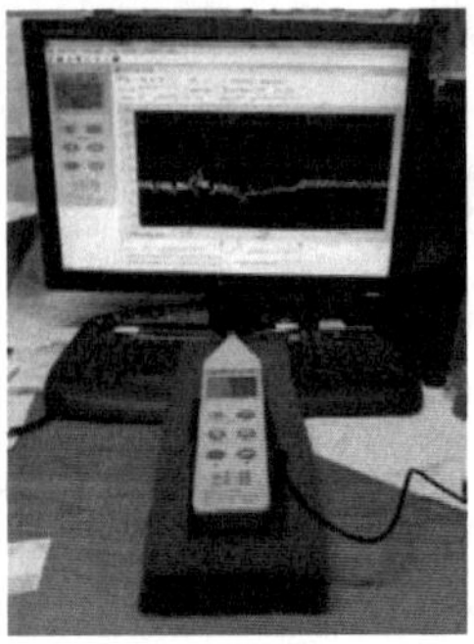

Figure 7. Sonometer inside the foam rubber bag

Neonatal Incubator: This test was performed in two different places. A first set of tests were performed inside a room placed at the old LEVEL III NICU under harder environment noise conditions than usual working conditions; at the outside; there was personnel, remodelling the existing facilities, using power tools like hammer drills, sanders, grinders, etc. Figure 8 shows the incubator under test and Figure 9 shows a view of the area under remodelling.

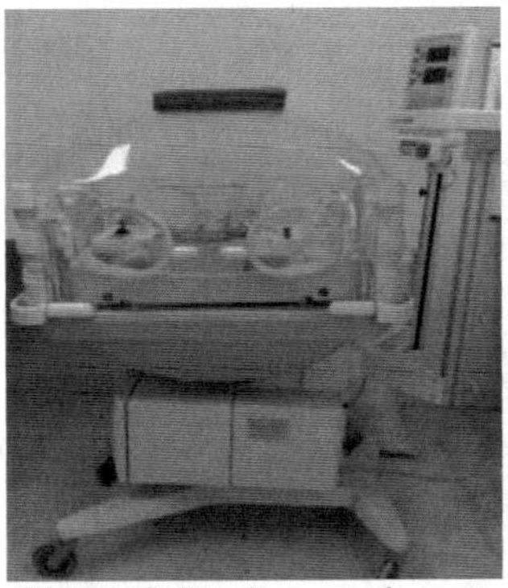

Figure 8. Incubator under test

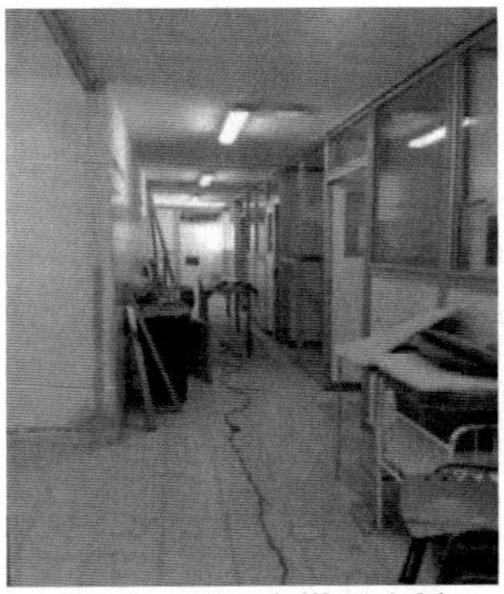

Figure 9. Area under remodelling (old neo facilities)

A second set of tests were performed inside the new neonatology facilities, at the corridor between both beds area shows in Figure 2 (light blue box) under normal conditions. In both cases, the ambient temperature was 24°C.

In order to evaluate the noise attenuation factor -for the cabinet cover- and the interior noise, there were used both sonometers, previously described, plus an incubator analyzer (Fluke Incubator Analyzer-INCU). The INCU is a portable device designed to verify the proper operation and environment of infant incubators recording important parameters of the incubator such as airflow, sound level, temperature (four individual measurement probes), and relative humidity. It can operate stand-alone and is programmed by a PC via an RS232 port; after the test is completed, the user uploads the data collected into the PC for display and analysis. The recorded data can be saved in a file or printed in a report. The INCU parameters were set to: Interval Time: 1 h 20 min with a Sample interval: 1 min.

The first test was performed without any external coverlet and the incubator was "ON" –Figures 8 and 9-. In the second test, the incubator was covered with an external coverlet, normally used to limit ambient light transmission into the inside –Figures 10 a and 10b-. In both cases, it was compared the INCU collected data –inside- against CENTER 322 data –outside-. The third test was performed, with and without different coverlets, while the incubator was OFF; the noise was recorded with the Cirrus sonometer –inside- and the Center 322 –outside-.

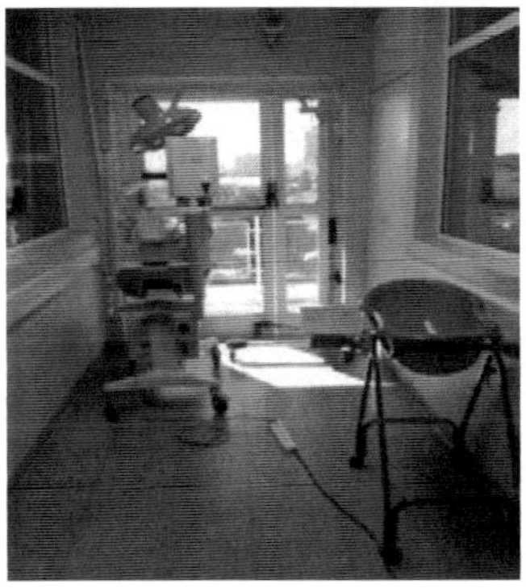
a

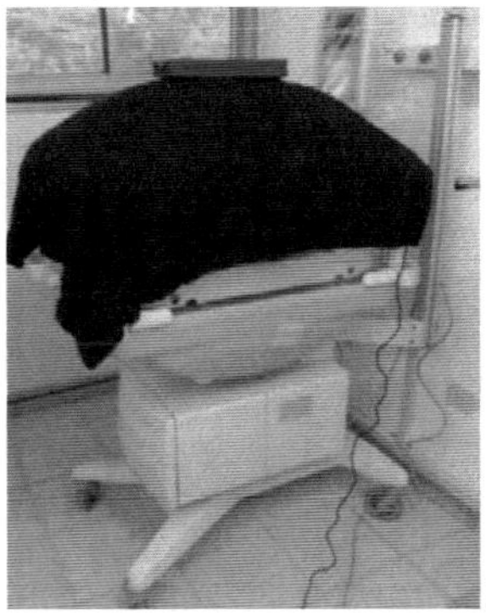
b

Figure 10. a- Corridor where the tests were performed¸ b- Incubator with external coverlet

3. Results

Neonatal Intensive Care Unit (NICU): It is well known that the sound pressure levels in level III NICUs are higher than in the other levels I & II. Previous studies, performed in the same service in 2009 and 2011, showed an average noise of 64.8 dBA and 63,6 respectively [1, 18]. In this case, we try to carry out the evaluation under the worst possible conditions. Figure 11 graphs the recorded data during 24 h.

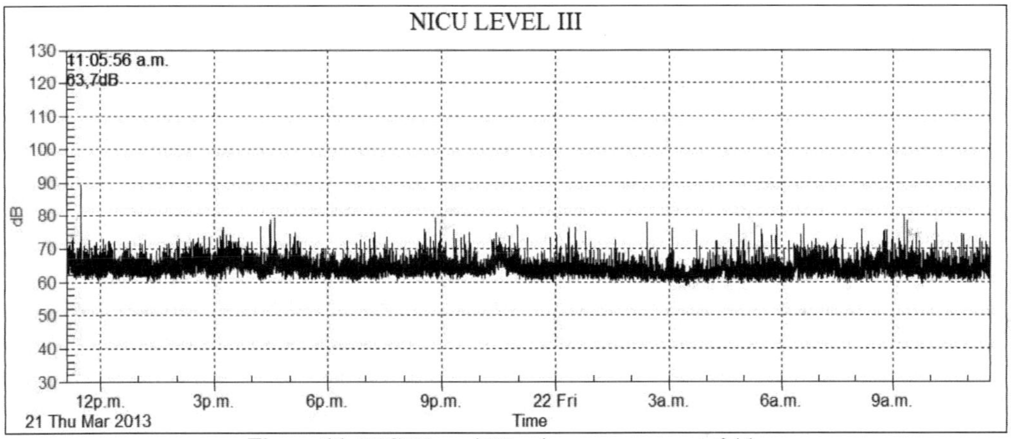

Figure 11. NICU Level III noise measurement -24 h-

The statistics values obtained are the following (24 h format):
Start Time:21/03/2013 11:05:56 a.m. End Time:22/03/2013 11:32:26 a.m
Sampling Rate:5 DataNo:17618
Avg.:64,3
Maximum:89,2@21/03/2013 11:29:31 a.m.
Minimum:59,0@22/03/2013 03:27:21 a.m.

If we compare these results with those obtained in 2009 and 2011, even though the present values were recorded at the new facilities, it is possible to conclude that ambient noise are similar but they should be improved if we take into account the American Academy of Pediatrics recommendations. But, when these results are compared to other real values, obtained in other institutions in our country and all over the world, it can be said that this is an acceptable result [19, 20, 21, 22].

Laundry: This is the second time that this type of evaluation is performed in this service and the personnel were pleased for being taken into account. Figure 12 graphs de recorded data.

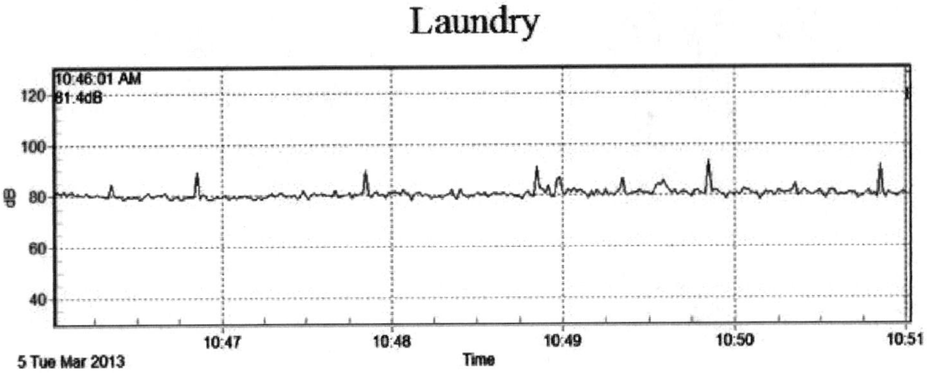

Figure 12. Laundry noise measurement

The statistics values obtained are the following (24 h format):
Start Time: 03/05/2013 10:46:01 AM End Time: 03/05/2013 10:51 AM
Sampling Rate: 1 sec Data: 300
Maximum: 93.6@3/5/2013 10:49:51 AM
Minimum: 78.2@3/5/2013 10:48:07 AM
Average Value: 80.8 dBA

These values are similar to those obtained in previous measurements [1]. Even when, the average value seems to be high, is lower than the maximum fixed by the law which is 85 dBA.

Neonatal Incubator: The goal of this test is to check if the cabinet cover has any noise attenuation factor. The actual model of incubators at level III NICU has a technical specification of "own noise" less or equal to 50 dBA, but there is no specification for cover attenuation factor. This matter is still under analysis by many groups [21, 22, 23, 24]. Preliminary results show that the attenuation factor for the ambient noise is 1.15, for an external noise of 64 dBA [1].
The first test was performed inside a room placed at the old LEVEL III NICU –under remodelling- in harder environment noise conditions than usual working conditions. A second test was performed at the corridor between the beds area. In both cases, the incubator was "ON". The values obtained are showed in Table 1.

Incubator is ON	Area under remodelling		Corridor	
	Inside	Outside	Inside	Outside
Leq dBA	59.5	59.5	63.5	62.3
L min dBA	55.5	42	59.7	54.5
L max dBA	73.6	78.9	70.3	74.9
Sampling rate	60 s	60 s	60 s	60 s
Interval Time	3 h	3 h	3 h	3 h

Table 1.

Both tests have many uncontrolled external variables but the attenuation factor is similar to the preliminary obtained in 2011. These results are very useful when we analyzed them; in the first test the fact to be noted are the minimum values, we can assumed that the L min for the inside is the own internal noise of the incubator, therefore it should be scheduled for a preventive maintenance. The values measured in the second test surprised us because we measured Leq inside > Leq outside. Fortunately, other group reported described similar results that suggest a reverberance effect in the interior space of the closed incubator due to lack of effective acoustic damping. In addition, it is apparent that the physical transmission loss of the plexiglass of the enclosure hood is negligible and this is supported by additional test results [21].

Based on these results, we decided to performed other two tests with the same incubator but in "OFF" state. The first one was performed while there were personnel talking "normally" closely to the incubator –normal conversation-, the second one was performed with music -closely to the incubator-; in both cases with and without an external coverlet. We assumed that the spectral frequency is different. These results are shown in Table 2.

Incubator is OFF	Without coverlet						External coverlet					
	Normal conversation			Music			Normal conversation			Music		
	Inside	Outside	AF	Inside	Outside	AF	Inside	Outside	AF	Inside	Outside	AF
Leq [dBA]	49.8	59	2.88	54.9	64.7	3.09	47.8	58.5	3.43	53.7	64.4	3.43
L min [dBA]	41.5	54.6		41	54.9		40.5	54		40.7	53.9	
L max [dBA]	77.4	76.6		82.4	77.1		75.2	75.6		81.3	80.4	
Sampling rate [sec]	1	1		1	1		1	1		1	1	
Interval Time [min]	10	10		10	10		10	10		10	10	

Table 2. AF= attenuation factor

In this analysis we have to take into account some facts: a- a human ear requires an increase of about 10 dB before the sound subjectively appears to be twice as loud; b- an increase of 6 dB represents a doubling of the sound pressure; c- AF=1 means that (P outside = P inside); AF=2 means (P outside = 2* P inside) –this means 6 dB-; AF=3 means (P outside = 3* P inside) –this means 9.5 dB-. A baby cannot tell us if he can perceive or not any variation in loudness; therefore, we will consider an acceptable value if AF ≥2 and a good AF when this value is ≥3

4. Conclusion

Environmental noise data is often used in legal proceedings. In fact, we should recommend collecting data with the assumption that someday it may have to be examined in court testimony. Therefore it is of outmost importance that the conditions of the measurement be carefully documented in a formal measurement report. A comprehensive and carefully documented formal has the best chance in any legal proceedings or noise control negotiations settled out of court. All reports should be documented according to Res. 85/2012 -complementary to the National Law of Health and Safe at Work No 19587- and with ISO 1996-1:2003 recommendations. It is also important to write the report in an easy-to-understand, readable style [2, 3, 4, 11].

In the case of neonatal incubators, it is necessary to perform complementary tests, with octave analysis, to determine the frequency components in order to discriminate its own internal noise and the contribution of the external life support set. In 2012, a group of the Vibration and Acoustics Laboratories of Virginia Tech and the Department of Internal Medicine Eastern Virginia Medical School, obtained results that would indicate a significant level of low frequency sound particularly when devices such as ventilator pumps are employed [21]. In 2013, other group described a multidisciplinary design project, conducted in an academic setting, reflecting a systems-oriented, human-centered philosophy in the design of neonatal incubator technologies, proposing some tips to enhance the incubator performance [24].

We do not conclude anything new if we said that a PM schedule is a major task in any health institution, but we believe that a very significant contribution is to include in the schedule a regular control of the acoustic pollution all over the health institution.

Acknowledgments

Authors want to deeply thank to the BME students Verónica Menichetti and Daniel Guillermo Estrada, for their collaboration in collecting the data and to Gustavo Wain EE for his support in IEC standards. Also thank to all the hospital personnel in the person of the hospital director Rossana Chala MD and Sebastián Díaz Carrasco BME. This work was supported in part by grants from the UNT Research Council (CIUNT), 26/E422 Research Program and INSIBIO-CONICET.

References

[1] Olivera JM, Rocha LA, Rotger VI, Ruiz EV and Herrera MC 2011 Acoustic pollution in hospital environments *Anales del XVIII Congreso Argentino de Bioingeniería SABI 2011 y VII Jornadas de Ingeniería Clínica and Journal of Physics: Conference Series* **1:**1-10

[2] Ley Nacional de Higiene y Seguridad en el trabajo No 19587

[3] Especificaciones técnicas sobre ergonomía y levantamiento manual de cargas, y sobre radiaciones *Resolución Nro 295:2003 modificatoria de Anexo V de la Ley 19587*

[4] Protocolo para la Medición del nivel de Ruido en el Ambiente Laboral *Resolución Nro 85:2012 complementaria a la Ley 19587*

[5] Industrial Accident Prevention Association 2007 A Health and Safety Guideline for Your Workplace: Preventive Maintenance IAPA guidelines *www.iapa.ca*

[6] American College of Clinical Engineering *http://www.accenet.org*

[7] Ceilings & Interior Systems Construction Association (CISCA) 2010 White Paper: Acoustics in Health Care Environments *http://www.cisca.org*

[8] Goines L & Hagler L 2007 Noise Pollution: A Modern Plague *Southern Medical Journal* 100(3):287-294

[9] Brüel&Kjær 2001 Environmental Noise–booklet *Brüel&Kjær Sound & Vibration Measurement*

[10] Tim South 2004 Managing Noise and Vibration at Work *Elsevier Butterworth-Heinemann*

[11] ISO 1996-1:2003 Acoustics - Description, measurement and assessment of environmental noise - Part 1: Basic quantities and assessment procedures

[12] Levy G, Woolston D, Browne J 2003 Mean Noise Amounts in Level II vs Level III Neonatal Intensive Care Units *NEONATAL NETWORK* **22:2** 33-38

[13] Committee on environmental health 1997 Noise: A Hazard for the Fetus and Newborn Pediatrics *American Academy of Pediatrics* **100(4):**724-7

[14] IRAM 4113-2:2010 Acoustics -Description, measurement and assesment of environmental noise. Determination of environmental noise levels -guidelines-

[15] IRAM 4113-1: 2009 -Acoustics. Description, measurement and assessment of environmental noise. Basic quantities and assessment procedures-

[16] IRAM 4079:2006 -Noises. Maximum admissible levels at occupational environments to prevent hearing impairment. Relationship between noise exposure and hearing threshold permanent displacement-

[17] IEC 60601-2-19:2009 -Particular requirements for the basic safety and essential performance of infant incubators-

[18] Brandán R, Halloy N, Sanchez MA, Sappia LD, Sueldo J, Rocha LA, Herrera MC, Rotger VI y Olivera JM 2009 Contaminación Acústica en salas de neonatología *Anales del XVII Congreso Argentino de Bioingeniería y VI Jornadas de Ingeniería Clínica, SABI 2009* **1:**1-4

[19] Alvarez Abril A, Boschi C, Gomez M 2007 Revisión de ruido en unidades de cuidados intensivos neonatales–Análisis regional *Anales del XVI Congreso Argentino de Bioingeniería y V Jornadas de Ingeniería Clínica, SABI 2007* **1:**1-4

[20] Werner AF (MD) Evaluación del ruido ambiental en las Unidades de Cuidados Intensivos de Recién Nacidos 2008 Anales *del VI Congreso Iberoamericano de Acústica-FIA 2008* **1:**1-11

[21] Chris Fullera et al 2012 Measurement and active-passive control of noise in neonatal incubators *INTER-NOISE and NOISE-CON 2012 Congress and Conference Proceedings* pp. 6723-6734(12) ISSN 0736-2935

[22] Moreira Pinheiro E et al 2011 Noise at the Neonatal Intensive Care Unit and inside the incubator Rev. Latino-Am. Enfermagem 19(5):1214-21

[23] Wubben SW et al 2011 The sound of operation and the acoustic attenuation of the Ohmeda Medical Giraffe OmniBed *Noise and Health Journal* **13:50** 37-4

[24] Ferris TK, Shepley MM 2013 Journal of Perinatology 2013 *The design of neonatal incubators: a systems-oriented, human-centered approach* Suppl 1:S24-31 doi: 10.1038/jp.2013.11.

Journal of Physics: Conference Series **477** (2013) 012020 doi:10.1088/1742-6596/477/1/012020

Simulation of a non-invasive glucometer based on a microwave resonator sensor

Santiago Pimentel[1], Pablo Daniel Agüero[1], Alejandro José Uriz[2], Juan Carlos Bonadero[1], Mónica Liberatori[1] Jorge Castiñeira Moreira[2]

[1] Laboratorio de Comunicaciones - Facultad de Ingeniería - Universidad Nacional de Mar del Plata

[2] CONICET / Laboratorio de Comunicaciones - Facultad de Ingeniería - Universidad Nacional de Mar del Plata

E-mail: {spimentel,casti}@fi.mdp.edu.ar

Abstract. In this paper a simulation of a microwave resonator sensor for constructing a non-invasive blood glucose meter is presented. A relationship between changes of the dielectric permittivity of the blood and the frequency response of S parameters of the sensor is observed. This can lead to a measuring procedure in which the glucose level present has a correlation with the value of the frequency resonance of the sensor. The test bank consists of a planar spiral microwave resonator over which the individual under test places his/her finger. This modifies the initial frequency resonance of the resonator because of the change produced in the measuring procedure over the dielectric permittivity of the resonator. Simulations show a correlation between dielectric permittivity blood changes, and changes in the value of the frequency resonance, in the frequency response of S parameters of the resonator.

1. Introduction

Diabetes mellitus [1] is a common disease that causes significant levels of mortality and morbidity. It is a serious disease that has now reached epidemic proportions and the prevalence rates are expected to go even higher in the future [2]. In 1995 the number of people suffering this disease was estimated to be 135 million, rising to be 154 million in 2000. It is expected that in 2025 this number will be 300 million, with a higher increasing rate in developed countries [4]. In general terms we can identify two types of diabetes. Type 1 diabetes [3] is the most serious type and affects $5 - 10\%$ of the diabetic population. It is a disease that develops during childhood or adolescence and is characterized by a severe deficiency of insulin secretion, resulting from atrophy of the islets of Langerhans, and causes hyperglycemia and eventually ketoacidosis [11], [13].

Type 2 diabetes, also known as diabetes mellitus [14], affects $80 - 95\%$ of the diabetic population, and it is present mostly in adults and more frequently obese people. The disease generally deals with an inefficient use of insulin. A lack of proper treatment increases the rate of mortality due to diabetes mellitus. Type 1 diabetics have to provide themselves with insulin by injection or pumps, whereas most of type 2 diabetics can control their blood glucose levels by a proper diet, exercise, loss of weight and oral medication. Type 1 diabetics usually test their blood glucose levels once or twice a day, but a proper treatment of this condition would require around

6 to 7 tests a day. This is however quite bothering due to the puncturing of fingers for obtaining the blood sample of classic glucometers. This is why a non invasive glucometer would be of great help to these people, to have a better control of their insulin and glucose levels. A sensor based on a microwave resonator is selected here as the main part of a non invasive glucometer. This leads to an introduction of important parameters of a given microwave resonator like microwaves, dielectric permittivity, S parameters, blood characterization, and mainly what is understood as a sensor.

This paper is organized as follows. Section 2 describes the basis of measuring blood glucose levels by using a microwave resonator sensor. Section 3 depicts the designed scheme for simulations. Section 4 presents results of simulations. Finally section 5 is devoted to conclusions and further work.

2. Measurement of glucose in blood

In a recent paper [5], Hofmann et. al. analyses the behaviour of microwave based non-invasive measurements in glucose concentrations in blood using in-vitro samples, for 40 cases of study. They found a clear correlation between microwave measurements of dielectric properties and a concentration in solutions and emulsions. They also identify a high sensitivity of the proposed techniques that is also sensitive to errors. In a previous work of the authors [6], They also use open electromagnetic waveguides for non-invasive glucose monitoring, finding also a correlation between the glucose concentration variation and the phase of the measured parameter.

Based on this evidence, but considering that the above results were obtained over in-vitro samples, the present paper approaches by simulation, the behaviour of a microwave resonator sensor, that takes into account the additional complications that appear in a non-invasive measure of glucose concentrations in blood. This is done by considering the effect of the different tissues associated to a measurement done over of a human finger.

In order to provide an introduction of the technique that is studied, some of the related concepts are briefly described below.

Microwaves are electromagnetic waves that are characterized for being in the frequency range between $300MHz$ to $300GHz$. The corresponding wavelengths are between $1m$ to $1mm$. Microwaves are utilized because their wavelengths are such that the signal can easily penetrate most of the materials [7].

Materials can be characterized in terms of their electrical properties by constants, one of which is the so called dielectric permittivity, a number that measures the capacity of the material to polarize its particles in the presence of an external Electric Field.

Since all materials have always a dielectric constant that is equal or larger than that of the free space, we can make use of the so called relative dielectric constant [10], a number that indicates how many times larger is the dielectric constant of that material with respect to that of the free space. The dielectric constant is a parameter that can be experimentally measured. By measuring dielectric properties of a given material, we can indirectly measure other properties of that material related to their molecular structures. A given electric network of two ports ca be suitably characterized by its S parameters [9]. They are a measure of the relation between reflected and incident waves generated when the electric network is connected to a transmission line characterized by a given impedance, called the characteristic impedance of the transmission line.

Among the classic four S parameters of a two port electric network, parameters S_{11} (relation between the reflected and incident waves at the input port) and S_{21} (relation between the transmitted wave at the output port, and the incident wave at the input port) [11] were utilized. These two parameters contain valuable information to characterize a given microwave resonator.

Microwave sensors [12] are suitable for measuring blood characteristics because they operate at high frequencies (small wavelengths) and can easily pass trough skin and fat layers of the human body. Resonance techniques have been widely used because their capacity of providing a measure with precision. In our case, we have used a planar resonator that can easily be placed on the human body surface in contact with the skin. A resonator sensor acts when electric fields interact with the dielectric that is part of the sensor, and energy is coupled to it. This coupled energy generates a peak response at the resonating frequency of the sensor, where the peak depends on the relative permittivity of the material under study [13]. Changes in the dielectric permittivity of the material modifies the frequency response generating peaks or valleys in it.

Since blood is a complex substance in which cells, proteins, hormones, glucose and other particles are dissolved in water, its permittivity depends on the frequency, and is also influenced by the other parameters. Bloods permittivity is influenced by each of its constituents. A microwave sensor that has a frequency response at multiple frequencies can potentially aisle the response of one of the parameters in spite of the change in the other parameters [15].

Dielectric permittivity of the different layers of the body can have noticeable variations as a function of the frequency [16].

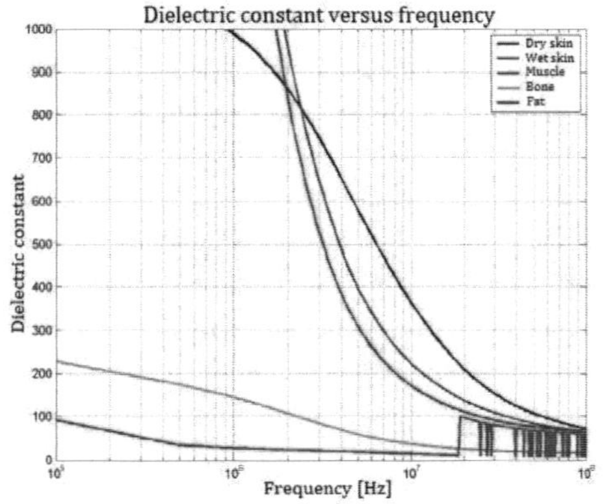

Figure 1. Relative Dielectric constant of different biologic tissues as a function of the frequency [17].

Since dielectric permittivity is a function of the frequency, the second order Debye Equation 1 is taken into account to model this behaviour [18].

$$\varepsilon(\hat{\omega}) = \varepsilon_\infty + \sum_{n}^{2} \frac{\Delta\varepsilon_n}{1 + j\omega\tau n} \tag{1}$$

Glucose sensor has to be placed on an area with the least amount of fat in order to maximize the transmission of waves th rough the different layers.

At a frequency of $1GHz$ a given wave passing through skin and fat layers is transmitted in a 50% of its intensity, whereas if the same wave passes trough skin and muscle layers it can be transmitted in a 93% of its intensity. Thus, the sensor would be in touch with more blood if it is placed on muscle than on fat tissues. The more blood is measured, the more accurate will be the measure. The sensor has to be able to properly detect changes in the dielectric permittivity. In our experiment, the individual places his/her finger on the surface of the planar resonator generating changes in its parameters, and this way the measuring procedure makes use of a body part with relative low content of fat and high content of blood that allows us to perform a measure with high level of the transmission of energy. With the use of this sort of sensors, diabetics, especially type 1 diabetics, could perform several measures a day in a non invasive way, making a better control of the glucose blood levels and improving the treatment and their life quality.

Journal of Physics: Conference Series **477** (2013) 012020 doi:10.1088/1742-6596/477/1/012020

3. Simulations

A first step in the development of a microwave resonator as the main part of a non invasive glucometer is to perform several simulation using software for electromagnetic simulation. The selected software is the CST Microwave Studio [20], which can accurately predict the behaviour of the microwave resonator. A simplified model of the different tissues of the body together with the second order Debye equation [18] was implemented using CST.

Figure 2. Layers model of a finger

The layer model seen in Fig. 2 is constituted of a dried skin layer of $0.015mm$, a wet skin layer of $0.3mm$, a fat layer of $0.2mm$, a blood layer of $0.5mm$ and a muscle layer of $0.2mm$. This simplified layer model models the finger of the individual placed on the planar resonator surface.

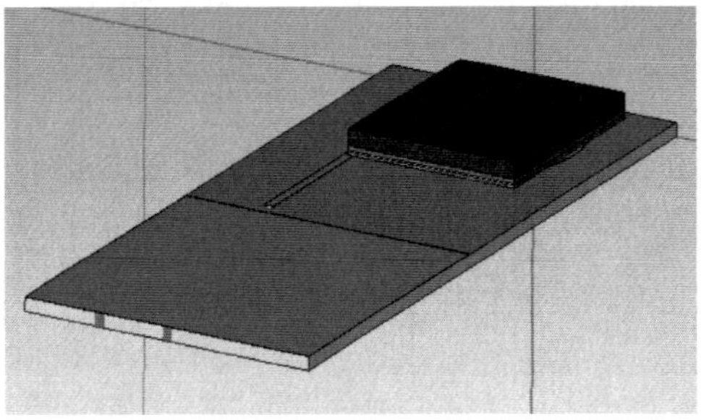

Figure 3. Layer model in contact with the resonator, as simulated using CST Microwave Studio.

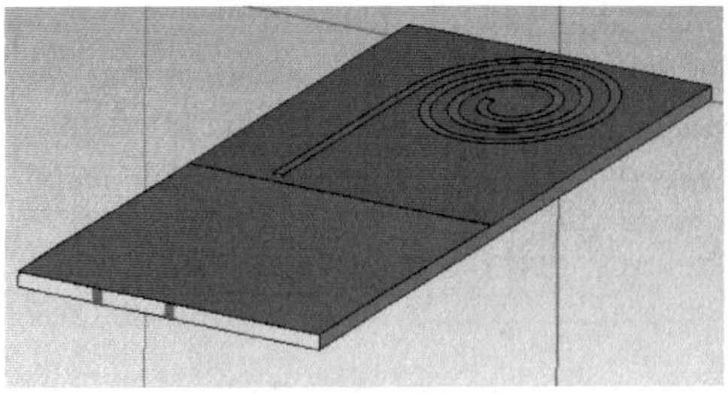

Figure 4. Spiral sensor simulated using CST Microwave Studio.

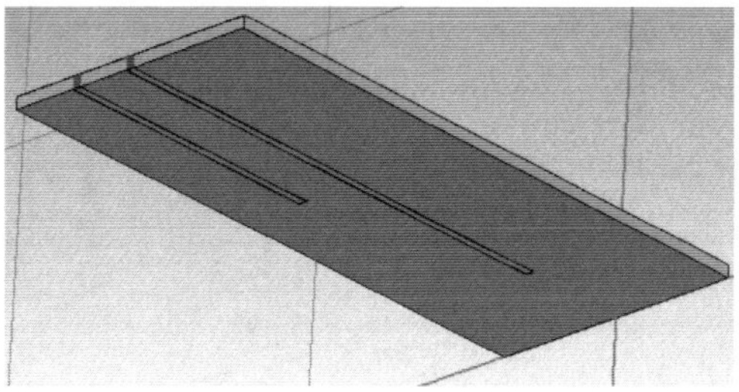

Figure 5. Bottom view of the spiral sensor.

The board on which the resonator is constructed is of a material called Rogers3010 [19], known as RT/Duroid, whose relative dielectric constant is 10.2. This material has lower losses at microwave frequencies than other materials available like the so called FR4. The boards length is $83.067mm$, its width is $36.39mm$ and its thickness is $1.6mm$. Copper layer has a thickness of $0.035mm$, and the width of the microstrip is $1.1582mm$. This width and thickness of the microstrip makes its characteristic impedance be equal to 50Ω, so that the system is matched. The spiral form of the resonator aims to maximize the contact between the fingers surface and the resonator surface. The microstrip transmission line starts at the input port develops its spiral form and ends at the bottom of the board as the output port. This makes the structure be a resonator. Its behaviour can be understood as a wired antenna, thus behaving as an inductance, and copper at the top and bottom of the board separated by the dielectric material acts as a capacitance. Therefore at several given frequencies a resonance is observed, appearing as peaks or valleys of the frequency response of the S parameters of the resonator.

4. Obtained Results

The layer model is placed on the microwave sensor surface, a voltage pulse is applied at the input of the resonator, and the frequency response of parameters S_{11} and S_{21} is evaluated for different values of the bloods layer dielectric permittivity.

Transient solver tool of CST Microwave Studio [21] is set by using as boundary conditions the mode Open Add Space, and by defining a suitable resolution in dB in order to obtain a given precision in the frequency response of the S parameters under measure. The frequency range of the simulation is $0Hz$ to $3Ghz$.

Since the second order Debye model is used, blood permittivity frequency response is evaluated with three fixed values of the parameters $\varepsilon_\infty = 2,5$, $\Delta\varepsilon_1 = 9$ and $\Delta\varepsilon_2 = 35$ [18]

Then these parameters are modified individually to determine their influence on the frequency response. Since variations over $\Delta\varepsilon_2$ and ε_∞ are no significant over the frequency response, then $\Delta\varepsilon_1$ changes are only evaluated.

Figures 6 and 7 show the influence on the frequency response of parameters and when the value of parameter $\Delta\varepsilon_1$ is varied.

Simulations show variations in discrete values of parameters $\Delta\varepsilon_1$ in the range of 9 to 90.

Figures 6 and 7 show that simulation of changes in the dielectric permittivity of blood results into changes of the frequency of resonance of the microwave resonator. Peaks and valleys become more abrupt, change in frequency or both. By means of a given algorithm, implemented using programmable electronics, we can take into account this behaviour in order to do a correlation between the observed changes in the resonance response of S parameters and a given value of the glucose present in blood, as one of the main factor that produce the change in its dielectric permittivity.

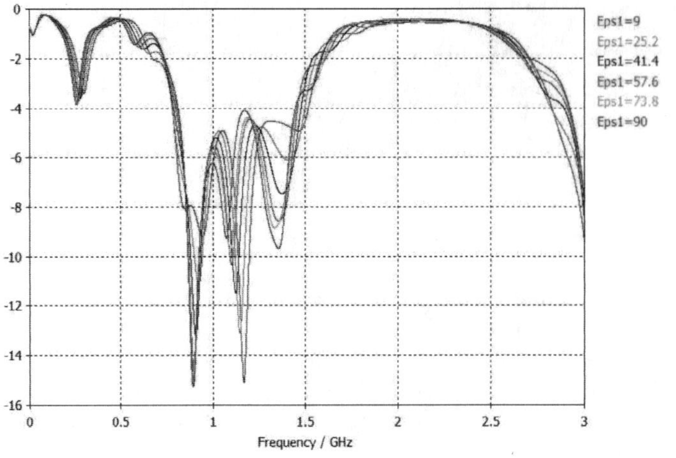

Figure 6. Frequency response of S parameters as a function of variations of $\Delta\varepsilon_1$

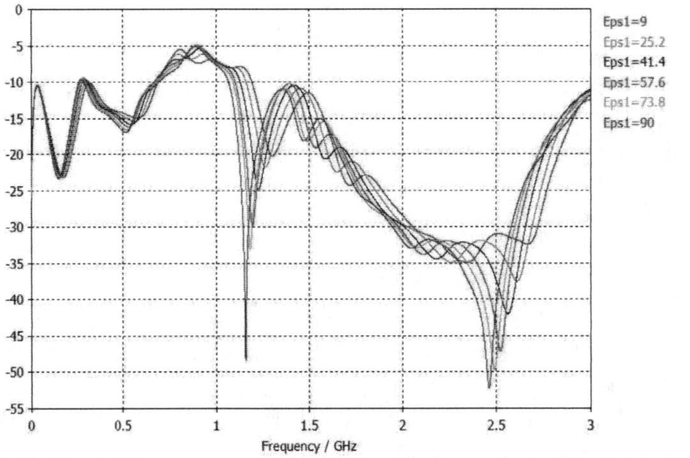

Figure 7. Frequency response of S parameters as a function of variations of $\Delta\varepsilon_1$

5. Conclusions

A microwave resonator sensor constructed as a spiral microstrip line has been simulated to show significant changes in the frequency response of its S parameters due to changes in the permittivity of the material under test. These simulations show that is possible to measure physiological changes produced by changes in the glucose blood levels when the material under test interacts with a microwave resonator.

The correlation between the blood dielectric permittivity changes (produced by changes in glucose levels in blood) and changes in the frequency response of S parameters of the structure (as shifts in the value of the frequency for which the resonance is present before and after the measure) appears as further work.

Probable difficulties to make a suitable calibration of the measure with respect to the user profile could be expected. In order to avoid the dependence of the measure on the specific characteristics of the system with respect to the individual under test (that is to say, all the involved parameters apart from the blood layer in the proposed sensor can adopt different values depending on the individual under test) . A possible solution to this problem can be the implementation of a prototype of the glucometer that can be used to take enough number of tests with individuals of known blood glucose levels in order to obtain statistical information, able to establish the correlation between blood glucose levels and frequency response changes or values.

Acknowledgment

The authors would like to thank to CONICET, National University of Mar del Plata and University FASTA for the funding received through projects *Secuencias caóticas digitales en procesamiento y encriptado de señales* and *Codificación y Control de Errores*, respectively.

References

[1] Geosalud: http://www.geosalud.com/diabetesmellitus/
[2] World Health Organization: http://www.who.int/es/
[3] Superintendencia de Salud: http://www.supersalud.gob.cl/difusion/572/w3-article-580.html
[4] T. Smith, *Our Deadly Diabetes Deception*, Nexus Magazine, vol 11, June July 2004.
[5] M. Hofmann, G. Fischer, R. Weigel, D. Kissinger, *Microwave-based noninvasive concentration measurements for biomedical applications.* IEEE Transactions on Microwave Theory and Techniques 61(5), pp. 2195-2204, 2013.
[6] M. Hofmann, M. Bloss, R. Weigel, G. Fischer, D. Kissinger, *Non-invasive glucose monitoring using open electromagnetic waveguides*, European Microwave Week 2012: "Space for Microwaves", EuMW 2012, Conference Proceedings - 42nd European Microwave Conference, EuMC 2012 , pp.546-549, 2012.
[7] L.C. Shen and J.A. Kong, *Applied Electromagnetism*, 3 Ed Boston, PWS Pub. Co., pp. 606, 1995.
[8] R. Resnick, D. Halliday, K. Krane, Física, vol 2, CECSA, 2010.
[9] J.M. Miranda, J.L. Sebastián, M. Sierra, J. Margineda, *Ingeniería de Microondas. Técnicas Experimentales*, Prentice Hall, pp. 317-320, 2002.
[10] R.L. Boylestad, *Introducción al Análisis de Circuitos*, Prentice Hall, 10 Ed, pp. 379-383, 2004.
[11] F. Sischka, *Basics of S-Parameters, Part 1* , Agilent Technologies Characterization Handbook, 2006.
[12] I.D. Robertson, *Microstriplines*, 2008.
[13] K. Saeed, A.C. Guyetts, I.C. Hunter, R.D. Pollard, *Microstrip Resonator Technique for Measuring Dielectric Permittivity of Liquid Solvents and for Solution Sensing*, IEEE/MTT-S International Microwave Symposium 2007, pp. 1185-1188, 2007.
[14] Asociación Latinoamericana de Diabetes, *Guías ALAD de Diagnóstico, Control y Tratamiento de la Diabetes Mellitus Tipo 2*, http://www.alad-latinoamerica.org/phocadownload/guias alad.pdf
[15] FDA News, *FDA Approves GlucoWatch Device for Children with Diabetes*, 2002.
[16] A. Rosen, M.A. Stuchly, A.V. Vorst, *Applications of RF/Microwaves in Medicine*, IEEE Trans. Microwave Theory Tech., vol 50, pp. 963-974, 2002.
[17] A.J. GarcaCollado, G.J. MolinaCuberos, M.J. Núñez, *Transferencia energética en un modelo multicapa de tejido humano sometido a irradiación electromagnética de alta frecuencia*, Revista Mexicana de Física. 55 (1). pp.17, 2009.
[18] S. Gabriel, R.W. Lau, C. Gabriel, *The Dielectric Properties of Biological Tissues: III. Parametric Models for the Dielectric Spectrum of Tissues*, Physics in Medicine and Biology, vol 41, pp. 2271-2293, 1996.
[19] Rogers Corporation: http://www.rogerscorp.com
[20] Computer Simulation Technology: http://www.cst.com/
[21] CST Microwave Studio - Transient Solver. 2010.

Journal of Physics: Conference Series **477** (2013) 012021 doi:10.1088/1742-6596/477/1/012021

Automatic Segmentation of Exudates in Ocular Images using Ensembles of Aperture Filters and Logistic Regression

Marco Benalcázar[1,2,3], Marcel Brun[3], and Virginia Ballarin[3]

[1] Consejo Nacional de Investigaciones Científicas y Técnicas (CONICET), Argentina.
[2] Secretaría Nacional de Educación Superior, Ciencia, Tecnología e Innovación (SENESCYT), Ecuador.
[3] Grupo de Procesamiento Digital de Imágenes, Universidad Nacional de Mar del Plata, Mar del Plata, Argentina.

E-mail: marco_benalcazar@hotmail.com

Abstract. Hard and soft exudates are the main signs of diabetic macular edema (DME). The segmentation of both kinds of exudates generates valuable information not only for the diagnosis of DME, but also for treatment, which helps to avoid vision loss and blindness. In this paper, we propose a new algorithm for the automatic segmentation of exudates in ocular fundus images. The proposed algorithm is based on ensembles of aperture filters that detect exudate candidates and remove major blood vessels from the processed images. Then, logistic regression is used to classify each candidate as either exudate or non-exudate based on a vector of 31 features that characterize each potensial lesion. Finally, we tested the performance of the proposed algorithm using the images in the public HEI-MED database.

1. Introduction

Exudates are one the main signs of diabetic macular edema (DME), which occurs when the retina swells as a complication of diabetic retinopathy. There are two kinds of exudates, hard and soft. Hard exudates are composed of lipid and protein deposits that are leaked from the bloodstream into the retina. Soft exudates are superficial retinal microinfarcts. Both hard and soft exudates are direct consequences of the obstruction of the ocular blood vessels due to elevated blood sugar levels. The presence of exudates within the central macula, where the majority of photoreceptors are concentrated, can cause vision loss or even blindness in diabetic patients. In fundus ocular images of the retina, exudates appear as yellowish structures with well defined edges and variable shapes [1-4].

Diabetic retinopathy, and consequently DME, can be inhibited by early diagnosis and appropriate treatment. However, since diabetic retinopathy is asymptomatic in early stages, many people remain undiagnosed until they start losing their vision due to the increasing amount of hard and soft exudates and other pathologies that accompany diabetic retinopathy such as mycroaneurysms and new blood vessels. In the ideal case, a mass screening of all diabetic patients, even of those experiencing no vision issues, would help to diagnose DME early enough for optimal treatment [1]. At this time, such an undertaking would be too work and time intensive because each image must be analyzed by a specialist, making the early diagnosis of this pathology difficult. On the other hand, for those cases when diabetic retinopathy has been already diagnosed, the detection of exudates in fundus ocular images generates valuable information for treatment planning. Namely, the position of exudates is important for laser treatment, which is the photocoagulation of the lesions to stop their progress and prevent blindness [3].

Several issues must be addressed when applying computer aided methods for the segmentation of exudates in fundus ocular images. First, hard and soft exudates have similar attributes to those of the optic disk in terms of color, brightness, contrast, and texture. Second, hard exudates have well defined structures; whereas, soft exudates have unclear boundaries. Third, in fundus ocular images, variations in illumination, contrast, and color in regions within the images (intra-image variability) as well as among the images (inter-image variability) occur. Intra-image variations are caused by differences in light diffusion inside the eye, lesions, and variations in reflectivity and thickness of the retina. Inter-image variations are caused by differences in the acquisition conditions of the images (e.g., illumination and acquisition angle), materials (e.g., cameras), and the ethnic group from which the screened patients come [1,3].

Because of the aforementioned issues, the automatic segmentation of exudates is still an open field of research, especially when we want to develop algorithms that perform well in a broad spectrum of application. In the present work, we combine tools from digital image processing, pattern recognition, and machine learning to propose a new algorithm for the automatic segmentation of both hard and soft exudates. It is important to mention that in this work, we make no distinction between hard and soft exudates. One reason is because differentiating between them is problematic even for specialized ophthalmologists. Additionally, both kinds of exudates are direct consequences of diabetic retinopathy, and their differentiation poses no significant clinical advantage for the diagnosis of DME [1-4]. Therefore, hard and soft exudates will be referred to simply as exudates from now on.

This paper is organized as follows. In Section 2, we describe both the source of the images as well as the methodology used in this work. In Section 3, we present the proposed algorithm for automatically segmenting exudates. Section 4 contains the results obtained and compares them with the results presented in recent works published in the scientific literature. Finally, in Section 5, we summarize the main contributions of this work and outline future possible developments.

2. Images and Methodology

2.1. Images
We used the public Hamilton Eye Institute Macular Edema database (HEI-MED) [2]. This database is composed of 169 fundus ocular images (color images) with their associated ground truth images (binary images). These images, color and binary, are formed by 2196x1958 pixels. The binary images correspond to manual segmenetations of the regions containing exudates. Small lesions were clustered together for those cases in which the individual ones were not clearly discernible. The database contains images from a mixture of ethnicities and stages of diabetic retinopathy and includes images without any pathology. For this work, we randomly divided the 169 images into 3 subsets: TrS_1, TrS_2 and TtS. TrS_1 and TrS_2 are composed of 30 images each and are used for the training stage of the proposed algorithm. TtS contains the reamining 109 images, which were used for testing.

2.2. Methodology
In this subsection, we explain some theoretical aspects of the methods that compose the different steps of the proposed algorithm.

2.2.1. Color to Grayscale Transformation

Let the function $G:\mathbf{E}\rightarrow L$ be a *grayscale image*. $\mathbf{E}$ is a nonempty subset of $\mathbf{Z}^2$, and $L=[0,l]$, with $l\in\mathbf{Z}^+$, represents a gray-level interval. For usual applications we consider $l=255$. The set of all possible images from $\mathbf{E}$ to L is denoted by $L^{\mathbf{E}}$. A *color image* is modeled as a function $\overline{H}:\mathbf{E}\rightarrow[L_R\times L_G\times L_B]$, usually $L_R=L_G=L_B=[0,l]$, such that $\overline{H}=(H_R,H_G,H_B)$, with $H_R,H_G,H_B\in L^{\mathbf{E}}$, where H_R, H_B and H_G are the channels of the image $\overline{H}$. In the particular case of RGB images, H_R, H_B and H_G are the red, green and blue channels, respectively. The set of possible color images from $\mathbf{E}$ to $[L_R\times L_G\times L_B]$ is denoted by $[L_R\times L_G\times L_B]^{\mathbf{E}}$. A *pixel* t of the image $\overline{H}$ is a coordinate in $\mathbf{E}$ of the form $t=(x,y)$ that takes the triplet $\overline{H}(t)=(H_R(t),H_G(t),H_B(t))$, where $H_R(t),H_G(t),H_B(t)\in[0,l]$.

Given an RGB image $\overline{H}$ and an arbitrary point $t\in\overline{H}$, the function $\Omega:[L_R\times L_G\times L_B]^{\mathbf{E}}\rightarrow L^{\mathbf{E}}$ is a transformation of RGB into grayscale images, such that

$$G(t)=\Omega(\overline{H}(t))=round\left(\alpha_1 H_R(t)+\alpha_2 H_G(t)+\alpha_3 H_B(t)\right), \tag{1}$$

where $\alpha_1+\alpha_2+\alpha_3=1$, and $round(\bullet)$ rounds to nearest integer.

2.2.2. Aperture Filters

Formally, let a *spatial window* $W=\{w_1,w_2,...,w_n\}$, with $w_i\in\mathbf{Z}^2$ and $i=1,2,...,n$, be a small subset of $\mathbf{Z}^2$, centered at the origin of $\mathbf{E}:(0,0)\in W$. The number of points of W is denoted by $n=|W|$. Moreover, let consider a *range window* $K=[-k,k]$, with $k\in\mathbf{Z}^+$ and $k<<l$, be an interval that represents the number of input gray-levels for filtering. Finally, let an *aperture* $W\times K$ be the Cartesian product of a spatial window $W=\{w_1,w_2,...,w_n\}$ and a range window $K=[-k,k]$. Therefore, given the grayscale image O and the aperture $W\times K$, an *aperture configuration* (also called *observation*) is a function $\mathbf{v}:W\rightarrow K$, so that $\mathbf{v}=(v_1,...,v_n)$ with $n=|W|$. We obtain an aperture configuration $\mathbf{v}$ from the image O by applying the following two steps: 1. The *spatial translation* of O by t: $O_{-t}=O(t'-t)$, being $t,t'\in\mathbf{E}$, and then the observation of the pixel values of O_{-t} within the spatial window W. This step returns the vector $\mathbf{u}_{O(t)}=\left(O(w_1-t),...,O(w_n-t)\right)$. For simplicity of notation, we will refer to vector $\mathbf{u}_{O(t)}$ with $\mathbf{u}$ from now on. 2. The *range translation* of $\mathbf{u}$ by a scalar $z\in\mathbf{Z}^+$: $\mathbf{u}_{-z}=(O(w_1-t)-z,...,O(w_n-t)-z)$, followed by the projection of points of $\mathbf{u}_{-z}$ outside $K=[-k,k]$ to either $+k$ or $-k$. Formally, and denoting the projection operation by $(\bullet)^*$, each point $v_i\in\mathbf{v}$, with $i=1,2,...,n$, is obtained from a given observation $\mathbf{u}=(u_1,...,u_n)$ as follows:

$$v_i=(u_i)^*=\min(\max(-k,u_i-z),k). \tag{2}$$

Given the definitions above, let an *aperture filter* $\mathbf{\Psi}:L^{\mathbf{E}}\rightarrow\{0,1\}^{\mathbf{E}}$ be a function that maps grayscale images into binary images. That is, given the grayscale image O, then $I=\mathbf{\Psi}(O)$ is a binary image so that $I:\mathbf{E}\rightarrow\{0,1\}$. Aperture filters are characterized by functions of the form $\psi:K^W\rightarrow\{0,1\}$, called *characteristic functions*, so that for each $t\in\mathbf{E}$:

$$I(t)=\mathbf{\Psi}(O)(t)=\psi\left((O(w_1-t)-z)^*,...,(O(w_n-t)-z)^*\right)=\psi(\mathbf{v}_{O(t)}), \tag{3}$$

where z is a function $z: L^W \to L$ that allows us to place the aperture $W \times K$ vertically in the pixel $t \in O$. In this work, we chose z to be $z = median(u_1,...,u_n)$.

Moreover, in the language of pattern recognition, the characteristic function ψ can be seen as a *classifier* that maps observations $\mathbf{v} \in \{\mathbf{v}_1,...,\mathbf{v}_m\}$ to labels from the set $\{0,1\}$. Based on this assumption, the segmentation of a grayscale image O is reduced to a classification of its pixels within $W \times K$. Finally, aperture filters fulfill three important properties: 1. *spatial translation invariance*: $\boldsymbol{\Psi}(O_t) = \boldsymbol{\Psi}(O)_t$ for each pixel $t \in O$; 2. *range translation invariance*: $\boldsymbol{\Psi}(O(t)+z) = \boldsymbol{\Psi}(O(t)) + z$ for any $z \in \mathbf{Z}^+$; and 3. *local definition* within $W \times K$: $\boldsymbol{\Psi}(O)(t) = \boldsymbol{\Psi}(O \cap (W \times K))(t)$ [5,6].

2.2.3. *Statiscal Design of Aperture Filters*

In the section above, we reduced the segmentation of an image to a classification of its pixels. Based on this fact, the statistical design of aperture filters is reduced to the well known problem of designing classifiers in the field of pattern recognition. In this context, we model the images to be processed as random realizations of the two processes ***O*** and ***I***, which are stochastic, discrete, stationary, and joint. Process ***O*** produces observed images, and process ***I*** produces ideal images [6,7]. *Observed images* represent the problem to be solved; whereas, *ideal images* represent the desired output of the processing.

Thus, given a pair of images (*O*,*I*) drawn from the processes (***O***,***I***), the goal of the statistical design is to find an aperture filter $\boldsymbol{\Psi}$ so that $\boldsymbol{\Psi}(O)$ is as close as possible to I. If the measure of closeness between $\boldsymbol{\Psi}(O)$ and *I* is the mean square error (MSE), then the cost function to be minimized is

$$\varepsilon(\boldsymbol{\Psi}) = MSE = E[(1/|\mathbf{E}|)\sum_{t\in\mathbf{E}}(I(t) - \boldsymbol{\Psi}(O)(t))^2], \tag{4}$$

where *E* is the expectation operator and t is an arbitrary point of $\mathbf{E}$, which is possible because (***O***,***I***) are stationary processes. When $\boldsymbol{\Psi}$ is characterized by ψ; the processes are stationary; and the observation $\mathbf{v} = (v_1,...,v_n)$ is an *n*-upla of random variables, v_i, then equation 4 yields

$$\varepsilon(\boldsymbol{\Psi}) = \varepsilon(\psi) = \sum_{\{\mathbf{v}:\, \mathbf{v}\in O\cap W(t)\times K\}} \Pr(I(t) \neq \psi(\mathbf{v})|\mathbf{v})\Pr(\mathbf{v}), \tag{5}$$

where $\Pr(I(t) \neq \psi(\mathbf{v})|\mathbf{v})$ represents the probability of misclassifying an observation $\mathbf{v}$ seen by $W \times K$ placed at $t \in \mathbf{E}$, and $\Pr(\mathbf{v})$ represents the probability of the occurrence of $\mathbf{v}$. Therefore, the statistical design of an aperture filter $\boldsymbol{\Psi}$ reduces to find its optimal characteristic function ψ_{opt} (classifier), which minimizes equation 5 through the following *decision rule*:

$$\psi_{opt}(\mathbf{v}) = \begin{cases} 1 & \text{if } \Pr(I(t) = 1|\mathbf{v}) \geq \tau \\ 0 & \text{otherwise} \end{cases}. \tag{6}$$

In the equation above, if $\tau = 0.5$, then ψ_{opt} is the Bayes classifier. However, for practical applications, we select the value for the threshold τ based on a tradeoff between the false positive and false negative rates of ψ_{opt}.

Finally, the size $|\mathrm{H}|$ of the space of hypothesis, H, for the statistical design of aperture filters grows exponentially as a function of both the size $|W|$ of the spatial window W and the size $|K|$ of the range window K, according to $|\mathrm{H}| = 2^{|K|^{|W|}}$.

2.2.4. Automatic Design of Aperture Filters using Logistic Regression

From equation 6, we can conclude that the statistical design of an aperture filter $\mathbf{\Psi}$ consists of estimating the value of $\Pr(I(t) = 1|\mathbf{v})$ for each observation $\mathbf{v}$ that can be seen by $W \times K$. In this work we model the probability $\Pr(I(t) = 1|\mathbf{v})$, which we will refer to simply as $\Pr(I = 1|\mathbf{v})$ from now on, with the following *logistic function*:

$$\Pr(I = 1|\mathbf{v},\boldsymbol{\beta}) = 1/(1+\exp(-f(\mathbf{v},\boldsymbol{\beta}))), \tag{7}$$

where $f: K^W \to \Re$ is definded by $f(\mathbf{v},\boldsymbol{\beta}) = \beta_0 + \beta_1 v_1 + \ldots + \beta_n v_n$. Furthermore, we used $\Pr(I = 1|\mathbf{v},\boldsymbol{\beta})$ to denote the dependence of $\Pr(I = 1|\mathbf{v})$ on $\boldsymbol{\beta}$. In the language of machine learning, equation 7 is called *logistic regression* [9,10]. We chose logistic regression because it is a low complex model that can be easily trained, avoids *overfitting*, and has good capabilities of prediction [10].

Using logistic regression to estimate the value of $\Pr(I = 1|\mathbf{v})$ for each observation $\mathbf{v}$ that can be seen by $W \times K$ poses a constraint in the space of search H of the optimal aperture filter $\mathbf{\Psi}_{opt}$. In fact, the new space of search H', being $\mathrm{H}' \subset \mathrm{H}$, is composed by only linear classifiers. Although the use of constraints yields a suboptimal filter $\mathbf{\Psi}_{subopt}$ with increased error compared to the unconstrained optimal operator, $\varepsilon(\mathbf{\Psi}_{subopt}) > \varepsilon(\mathbf{\Psi}_{opt})$, using constraints produces lower estimation error. In general, a constraint is beneficial if the reduction in the error of estimation is larger than the error induced by the constraint [5-7].

Finally, by using logistic regression, the design of aperture filters is reduced to the task of estimating the vector $\boldsymbol{\beta} = (\beta_0,\ldots,\beta_n)$ by maximizing the likelyhood $\Pr(\mathcal{D}|\boldsymbol{\beta})$ given by equation 8. This task is performed using a set of examples $\mathcal{D} = \{(\mathbf{v}_i, freq\,(I=0,\mathbf{v}_i), freq\,(I=1,\mathbf{v}_i))\}$, where $i = 1,2,\ldots,m$, and $freq(I=0,\mathbf{v}_i)$ is the frequency with which the label 0 is seen given the observation $\mathbf{v}_i$ (a similar rule applies for $freq(I=1,\mathbf{v}_i)$). Observation $\mathbf{v}_i$ and frequencies $freq(I=0,\mathbf{v}_i)$ and $freq(I=1,\mathbf{v}_i)$ are obtained by translating $W \times K$ through the images in the training set $\mathbf{D} = \{(O_i, I_i)\}$, being $i = 1,2,\ldots,p$.

$$\Pr(\mathcal{D}|\boldsymbol{\beta}) = \prod_{i=1}^{m} \Pr(I=0|\mathbf{v}_i,\boldsymbol{\beta})^{freq\,(I=0,\mathbf{v}_i)} \Pr(I=1|\mathbf{v}_i,\boldsymbol{\beta})^{freq\,(I=1,\mathbf{v}_i)}. \tag{8}$$

The maximization of the equation above is equivalent to minimizing $-ln(\Pr(\mathcal{D}|\boldsymbol{\beta}))$ [10].

2.2.5. Aperture Filter Design with Imbalanced Datasets

Imbalanced datasets are characterized by a high number of examples from one class (majority class), and only few, or even no examples, from the other class (rare class). Assuming that the majority class is labeled with 1, an imbalanced dataset is characterized by $\left(\sum_{i=1}^{m} freq\,(I=1,\mathbf{v}_i)\right) >> \left(\sum_{i=1}^{m} freq\,(I=0,\mathbf{v}_i)\right)$. If we train a classifier using imbalanced datasets, then the classifier will usually underestimate the conditional probabilities of the rare class, because they tend to be biased towards the majority class. Several approaches have been proposed to address this issue, including undersampling examples from the majority class, or oversampling the examples from the rare class [11]. One important drawback of the former approach is the elimination of some examples from the majority class. In this work, we increased artificially (oversample) the frequency $freq(I=1,\mathbf{v}_i)$ of each example $\mathbf{v}_i \in \mathcal{D}$ from the minority class using the following expression:

$$freq_{balanced}(I=1,\mathbf{v}_i) = round(freq(I=1,\mathbf{v}_i)(\sum_{i=1}^{m} freq(I=0,\mathbf{v}_i)/\sum_{i=1}^{m} freq(I=1,\mathbf{v}_i))), \quad (9)$$

where $freq_{balanced}(I=1,\mathbf{v}_i)$ is the new frequency with which the example $\mathbf{v}_i$ influences the maximization of equation 8.

2.2.6. *Ensemble of Aperture Filters*

Even though we are using a low complex model for classification (i.e., logistic regression), the estimation of $\boldsymbol{\beta}$ using all the triplets $(\mathbf{v}, freq(I=0,\mathbf{v}), freq(I=1,\mathbf{v})) \in \mathcal{D}$ is highly time and memory consuming. To circumvent this issue, we used an ensemble of aperture filters [8]. Under the ensemble approach, instead of designing a single aperture filter using all the m triplets from the set $\mathcal{D}$, we designed as many aperture filters $\boldsymbol{\Psi}_1,\ldots,\boldsymbol{\Psi}_p$ (called base aperture filters) as number of training pairs of images we had in set $\mathbf{D}$, where $\boldsymbol{\Psi}_i$ was designed using the pair $(O_i, I_i) \in \mathbf{D}$, being $i \in \{1,2,\ldots,p\}$. Then, to process a new image O', we weighted and then combined the predictions of all the p base aperture filters using the Shannon entropy. Formally, the prediction of an ensemble of aperture filters, $\boldsymbol{\Psi}_E$, is given by a weighted sum of the predictions of the base aperture filters $\boldsymbol{\Psi}_1,\ldots,\boldsymbol{\Psi}_p$:

$$\Pr(I=1|\mathbf{v})_E = \sum_{i=1}^{p} Pe_i \Pr(I=1|\mathbf{v})_i \,. \quad (10)$$

In the equation above, $\Pr(I=1|\mathbf{v})_E$ is the probability predicted by the ensemble $\boldsymbol{\Psi}_E$ that the observation $\mathbf{v}$ will be assigned label 1. $\Pr(I=1|\mathbf{v})_i$ is the probability predicted by the base aperture filter $\boldsymbol{\Psi}_i$ that the observation $\mathbf{v}$ will be assigned label 1. The weight Pe_i controls the influence of the prediction of $\boldsymbol{\Psi}_i$ in the final prediction of the ensemble and is computed by

$$Pe_i(\mathbf{v}) = (1 - H(\psi(\mathbf{v})_i))/(p - \sum_{j=1}^{p} H(\psi(\mathbf{v})_j)) \cdot \quad (11)$$

In the equation above, $H(\psi(\mathbf{v})_i) = -\sum_{j=\{0,1\}} \Pr(I=j|\mathbf{v})_i \, Log_2(\Pr(I=j|\mathbf{v})_i)$ is the Shannon entropy of the prediction of $\boldsymbol{\Psi}_i$ regarding the observation $\mathbf{v}$.

3. Proposed Algorithm

In this section, we describe the algorithm we propose to segment automatically exudates in fundus ocular images.

a) Color to Grayscale Transformation: In this step, we transformed the color ocular image $\overline{H} \in [L_R \times L_G \times L_B]^{\mathbf{E}}$ (Figure 1a) to a grayscale image $H_1 \in L^{\mathbf{E}}$ (Figure 1b), with 256 gray-levels, using equation 1 with the weights $\alpha_1 = 0.07$, $\alpha_2 = 0.9$ and $\alpha_3 = 0.03$. We chose these weights because it was observed that the green channel H_G of the image $\overline{H}$ has the highest contrast between exudates and the other components of the eye fundus. The red, H_R, and blue, H_B, channels have low contrast and some level of noise [1,8].

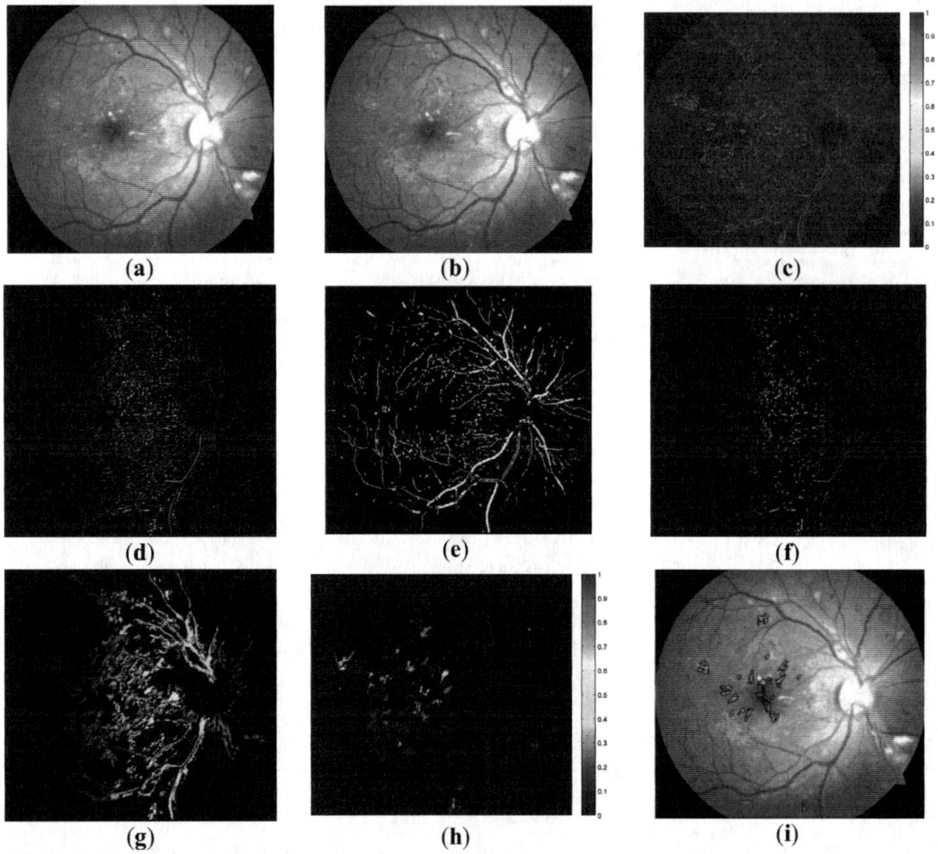

Figure 1: Example of the automatic segmentation of exudates: (**a**) Original image (**b**) Grayscale image (**c**) Lesion probability map (**d**) Exudate candidate markers (Each marker is shown by a different color) (**e**) Image with blood vessels automatically segmented (**f**) Image without blood vessels and noise (**g**) Exudate candidates (Each candidate is shown by a different color) (**h**) Exudate candidate probability map (**i**) Edges of the segmented exudates superimposed on the original image

b) Exudate Candidate Marker Detection: This step starts with the generation of a lesion probability map, in which we assigned a probaility of being part of an exudate region to each pixel $t \in H_1$ (Figure 1c). Each pixel from the image H_1 was processed using an ensemble of aperture filters (Section 2.2.6). In this case, we used the aperture $W \times K$, where W is a 15x15-squared spatial window and $K = [10{,}10]$ is a range window. The base aperture filters were trained using the training set TrS_1 (Section 2.1) and the procedure described in Section 2.2.4. In the ocular images the number of pixels that belong to exudates is, in general, much less than the number of background pixels. For this reason, before designing the base aperture filters, we balanced the training triplets using the procedure described in Section 2.2.5. Then, we applied a threshold at $\tau = 0.2$ and grouped pixels from the lesion probability map using a connectivity of 8, obtaining a binary image $H_2 \in \{0,1\}^{\mathrm{E}}$ composed of clusters of pixels that belong to exudate candidate markers (Figure 1d).

c) Blood Vessels and Noise Removal: In this step, we started by removing the clusters from the binary image H_2 that are composed of pixels that belong to major blood vessels using an algorithm developed in our previous work [8]. The threshold we used to classify each pixel from image H_1 as either blood vessel or background was $\tau = 0.5$ (Figure 1e). Then, we defined small clusters, those with fewer than 100 pixels, as noise and removed them to obtain the binary image H_3 (Figure 1f).

d) Exudate Candidate Detection: This step consists of obtaining a binary image H_4 composed by clusters of pixels that are potential exudates (Figure 1g). Image H_4 is obtained by morphological reconstruction [12] of the binary image H_3 (marker) from the binary image obtained by thresholding the lesion probability map from step b at $\tau = 0.1$ (mask).

e) Feature Extraction and Exudate Candidate Classification: Image H_4 contains spurious clusters of pixels as consequence of false markers detected in step b. In this step, we classified each cluster from image H_4 as either exudate or non-exudate by using logistic regression. The parameters of the logistic regression were estimated maximizing equation 8 (Section 2.2.4) using the training pairs $(\mathbf{x}_i, y_i)$, being that $i = 1,2,...,q$. q denotes the total number of potential exudates detected in images from the training set TrS_2. Vector $\mathbf{x}_i$ is composed by 31 features, and y_i takes label 1 if the region covered by the exudate candidate in the ideal image belongs to an exudate or otherwise label 0.

The 31 features that compose the vector $\mathbf{x}_i$ represent the main characteristics of each exudate candidate as follows:

- *shape* (extracted from image H_4): area, perimeter, eccentricity, major and minor axis length;
- *texture* (extracted from image H_1): edge strength (mean pixel value within the candidate border detected by a morphological gradient using a disk-shaped structuring element of radius 1), mean and standard deviation of the pixels within the candidate, and local pixel contrast (mean/standard deviation pixel value within the candidate subtracted from the mean/standard deviation pixel value from a 6-pixels-border-wide around the potential lesion); and
- *color* (extracted from image $\overline{H}$): median, mean and standard deviation from the LUV (U and V channels), RGB (all the three channels), and HSV (H and S channels) color spaces.

Finally, we applied the trained logistic regression classifier to each lesion of the image H_4, assigning to each a probability of being an exudate (Figure 1h). Candidates having a probability equal to or greater than the threshold $\tau = 0.7$ were classified as exudates (Figure 1i).

4. Results and Discussion

In this section, we present the results obtained by applying the proposed algorithm to each of the 109 images that comprise the testing set TtS (Section 2.1). The ultimate goal of the automatic segmentation of exudates is the diagnosis of the DME condition. Therefore, we evaluated the performance of our algorithm based on its ability to classify patients as either having or not DME. We defined a patient with DME as one in which at least one exudate was detected. In Figure 2, we show the receiver operating characteristic (ROC) curve [10] obtained, where $Sensitivity = TP/(TP+FN)$ and $Specificity = TN/(TN+FP)$. *TP* is the number of patients correctly classified as having DME. *FN* is the number of patients that our algorithm incorrectly classified as being healthy. *TN* is the number of patients correctly classified as healthy. *FP* is the number of patients wrongly classified as having DME. The area under the ROC curve (AUC) is 0.77. This value and the shape of the ROC curve indicate that our algorithm tends to detect exudates in some images that belong to healthy patients.

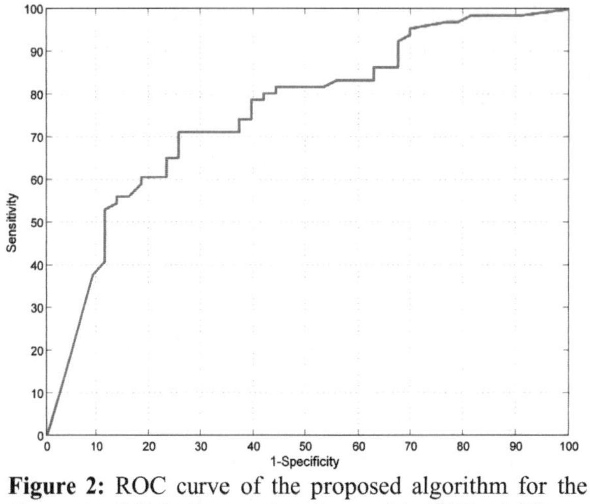

Figure 2: ROC curve of the proposed algorithm for the diagnosis of diabetic macular edema with AUC = 0.77.

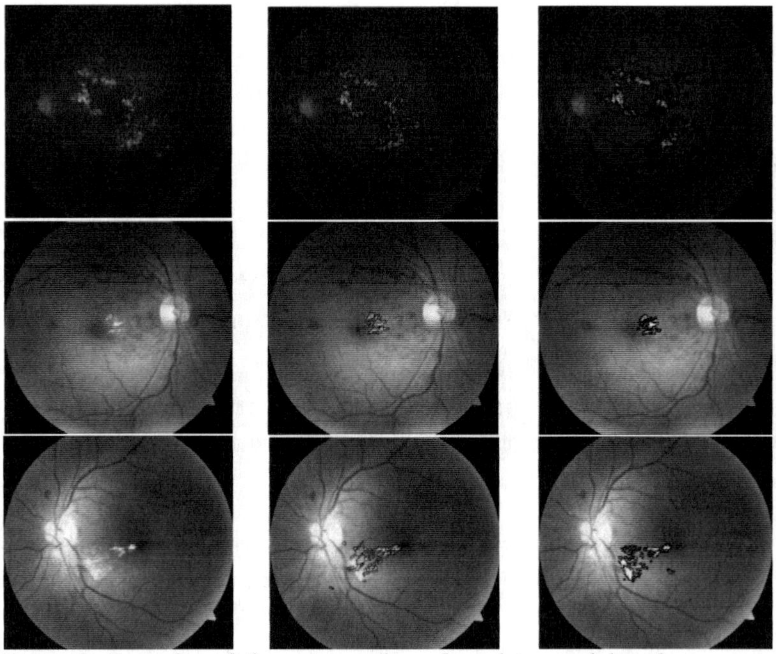

Figure 3: Examples of the segmentation of exudates applying the proposed approach to images from the HEI-MED public database: Each row contains a different test image. In the first column are the original images. The second column shows the results obtained by applying the proposed algorithm. The third column contains the ground truth images of the exudates.

In the scientific literature several algorithms have been proposed to address automatic segmentation of exudates. However, in most of the cases the performance evaluation is done using either private databases with only a few images or public databases that have no ground truth images showing the borders of the exudates. Therefore, to perform a benchmarking between the different proposed algorithms is very difficult. To our knowledge, we only found that [2] and [3] evaluated their algorithms for the diagnosis of DME using the HEI-MED database. They obtained values for the AUC of 0.81 and 0.82, respectively. Although these algorithms perform better than the proposed algorithm in diagnosing DME, they do not perform exudate segmentation. Therefore, one potential issue regarding these works can be the diagnosis of DME based on false positive exudates. Conversely, our algorithm provides information on the borders and positions of the exudates (Figure 3).

5. Concluding Remarks

In this work, we proposed a new approach for the automatic segmentation of exudates. The information generated by our approach is useful not only for the diagnosis of DME, but also for treatment and blindness prevention. The proposed method is based on ensembles of aperture filters that were used for both detecting exudate candidates and removing parts of the major blood vessels from the processed images. Finally, a logistic regression classifier was used to assign a probability of being an exudate to each candidate detected.

We also treated the aperture filter design as a pattern recognition problem. Using this approach, we reduced the aperture filter design to a classifier design by basing it on logistic regression. Additionally, the practical complexity issue that occurs when designing aperture filters with large training sets of images was circumvented by using an ensemble method. The ensemble method allowed us to combine the predictions of several base aperture filters. Each base aperture filter was designed using only one training pair of images.

Lastly, the results obtained in this paper and other publications demonstrate that automatic methods can be used to assist in the diagnosis of DME. However, we still need to improve the performance of such methods. In the case of our particular approach, future work includes the development of new methods for designing aperture filters to process color images directly without a color to grayscale transformation.

6. References

[1] Winder R, Morrow P, McRitchie I, Bailie J, Hart P, 2009, *Algorithms for digital image processing in diabetic retinopathy*, Computerized medical imaging **33** 608–622.

[2] Giancardo L, Meriaudeau F, Karnowski T, Li Y, Tobin Jr, Chaum E, 2011, *Automatic retina exudates segmentation without a manually labelled training set*, International symposium on biomedical imaging, 1396–1400.

[3] Giancardo L, Meriaudeau F, Karnowski T, Li Y, Garg S, Tobin Jr K, Chaum E, 2012, *Exudate-based diabetic macular edema detection in fundus images using publicly available datasets*, Medical image analysis **16** 216–226.

[4] Youssef D and Solouma H, 2012, *Accurate detection of blood vessels improves the detection of exudates in color fundus images*, Computer methods and programs in biomedicine **108** 1052–1061.

[5] Hirata R, Brun M, Barrera J, Dougherty E, 2006, *Advances in nonlinear signal and image processing*, Chapter 2, Hindawi.

[6] Hirata R, Dougherty E, Barrera J, 2000, *Aperture filters*, Signal processing **80** 697–721.

[7] Benalcázar M, Brun M, Ballarin V, 2013, *Artificial neural networks applied to statistical design of window operators*, Pattern recognition letters **34** 970–979.

[8] Benalcázar M, Brun M, Ballarin V, 2012, *Segmentación de vasos sanguíneos en angiografías retinales usando ensambles de filtros aperture*, III Simposio internacional de innovación y tecnología (SIIT), 125–129.

[9] Duda R, Hart P, Stork D, 2001, *Pattern classification*, Wiley.

[10] Murphy k, 2012, *Machine learning: a probabilistic perspective (adaptive computation and machine learning series*, Masachusets Institute of Technology.

[11] Weiss G, 2004, *Mining with rarity: a unifying framework*, ACM SIGKDD Explorations Newsletter **6** 7–19.

[12] Vincent L, 1993, *Morphological grayscale reconstruction in image analysis: applications and efficient algorithms*, IEEE Transactions on image processing **2** 176–201.

SVD inversion for the bi-dimensional Conical Radon Transform

Javier Cebeiro and Marcela Morvidone

Universidad Nacional de San Martín, Av. 25 de Mayo y Francia, B1650HMP Gral. San Martín, Buenos Aires, Argentina

E-mail: jcebeiro@unsam.edu.ar

Abstract.
In nuclear medicine, the Conical Radon Transforms family includes $2D$ and $3D$ models for scatter radiation imaging. The underlying idea is taking advantage of Comptom effect in order to provide image reconstruction without rotating the detector system. Several techniques currently applied for the classical Radon transform may also be useful in this new context. Nevertheless, since the physical model is rather different, it is necessary to assess the performance of each method. In previous works, we have studied inversion by filtered-backprojection and adaptive algebraic reconstruction techniques. In this paper we study the performance of a method based in the factorization of the projection matrix in order to invert the projections. Particularly, we test our algorithms on the bi-dimensional Conical Radon Transform, a simple member of the Conical Radon family. We obtain useful results on the performance of the three methods which should be valuable in the more realistic $3D$ context.

1. Introduction

Although gamma radiation is required as means of investigation thanks to its privileged properties of penetration, it suffers severely from Compton scattering phenomenon which degrades the quality of images acquired by the detectors. The Compton scattering radiation is due to the presence of electric charges in the propagation medium. This event causes a lose of the energy of the incident radiation and also a deviation of its trajectory characterized by an angle ω with respect to its initial direction angle. It is probably the most important cause of degradation in gamma-ray imaging.

A new imaging principle that takes advantage of Comptom scattering to perform reconstructions is based on the fact that the energy measured in a detected photon contains information about its direction after a first-order Comptom scattering. The Conical Radon Transform family is a mathematical model of this new principle for camera gamma imaging [1–5]. Image reconstruction from its projections is guaranteed since the Conical Radon Transforms are analytically invertible [6]. This model provides a theoretical basis for the development of a nonmoving camera gamma with a fixed collimator.

Some aspects of the bi-dimensional Conical Radon Transform ($\mathbb{TV}$) have been already studied and both, analytical [1, 7, 8] and algebraic [9] inversion have been obtained. In particular, it was shown that the $\mathbb{TV}$ can be inverted by a filtered back-projection technique (FBP) completely adapted to the geometry of the problem. The FBP method is widely used in applications of the classical Radon transform as it provides the most efficient analytical algorithm for inversion.

However, this method performs well when many projections are available but the image quality is significantly reduced when a limited number of projections is used. It was shown in [6] that an iterative algebraic reconstruction technique (ART) is superior when the number of projections is limited. Another drawback of FBP for the $\mathbb{TV}$ case is the upper V-shaped artifact in the image probably due to the discretization grid. Reconstructions carried out with ART exhibit the same artifact suggesting that the problem is linked to the model itself. We focus here on an alternative reconstruction technique, the Singular Value Decomposition (SVD) [10, 11] which has been previously used in related problems of image reconstruction from projections [12–14]. In particular, we study the inversion of the $\mathbb{TV}$ transform using SVD, and we compare its reconstruction performance with our previous results.

2. The bi-dimensional Conical Radon Transform

In this context, the object under study is thought as a flat three-dimensional object, i.e., of very low thickness, in which a non uniform radioactivity source distribution exists. A collimated linear detector, set parallel to the plane of the object, absorbs the photons coming from a perpendicular direction, after Compton scattering. The situation is represented in Figure 1.

The source of radiations is represented by a nonnegative integrable function $f(x, y)$ with bounded support. Let E_0 be the energy of primary photons emitted by the object. We consider the total number of photons arriving on the detector with an energy $E \leq E_0$ resulting from a single Compton scattering. Calling **D** a detecting site on the x axis, the collision points **M** are on a line perpendicular to the detector at point **D**. When the detector is fixed to detect exactly one photon energy, the scattering angle is fixed and the point sources **N** which will contribute to the measures lie on two lines that form an angle ω with the line **MD**. The photon flux density measured at a detecting site **D** is the sum of scattered radiation flux densities outgoing from the set of scattering sites **M** lying along the axis of the collimator at **D**.

Then, the data acquisition may be formulated as follows:

$$\tilde{g}(\xi, \omega) = K(\omega) \int_l^\infty \frac{d\eta}{\eta} \int_0^\infty \frac{dr}{r} [f(\xi + r\sin\omega, \eta + r\cos\omega) + f(\xi - r\sin\omega, \eta + r\cos\omega)] \quad (1)$$

where ξ, η, l and $r \in \mathbb{R}$ and $0 < \omega < \pi$. Here $\tilde{g}(\xi, \omega)$ is the measured photon flux density at **D** under a scattering angle ω, using the Cartesian coordinates of Figure 1. For ease of notation, we included all physical factors resulting from Compton scattering into one variable $K(\omega)$. This variable contains the square of the classical electron radius, the average electron density, and the Klein-Nishina scattering probability function.

The inner integral with $\eta = 0$ is called the $\mathbb{TV}$ transform.

$$g(\xi, \omega) = [T\mathbb{V}f](\xi, \omega) = K(\omega) \int_0^\infty \frac{dr}{r} [f(\xi + r\sin\omega, r\cos\omega) + f(\xi - r\sin\omega, r\cos\omega)] \quad (2)$$

So the $\mathbb{TV}$ is a particular case of (1) where **D** and **M** are coincident. While this assumption leads to a less realistic model since the detection and collision sites are coincident, the model is mathematically simple and useful to study issues concerning the discretization without being obscured by other problems appearing in (1) such as multiple integrals or divergences ($r - 0$ case) that make necessary to regularize the integral by applying further mathematical procedures.

The value of $g(\xi, \omega)$ gives the detected photon flux density at a detector labeled by $\xi \in \mathbb{R}$ from angle $0 \leq \omega < \pi/2$. This transformation has an analytic inverse [2] which rests on the action of a cosine-Fourier transform. We have also established the inversion by FBP in [7, 8]. In what follows, we will eventually refer to $g(\xi, \omega)$ as the *projections*.

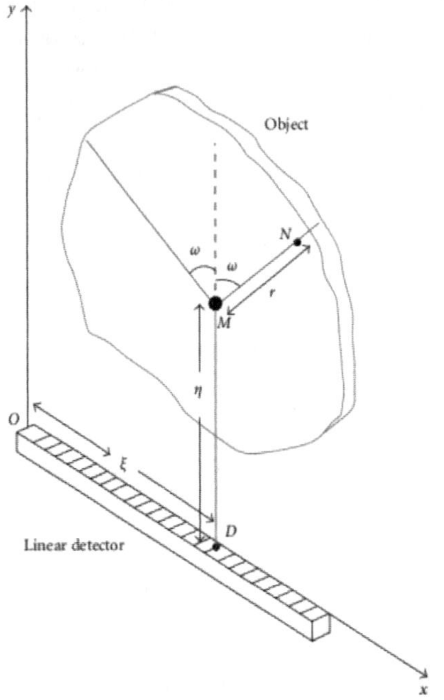

Figure 1: Experimental setup and definition of variables

2.1. The Point Spread function

A very useful formulation of the $T\mathbb{V}$ transform reads as follows

$$g(\xi,\omega)=\int\int dxdy\,\mathrm{PSF}_{\mathbb{V}}(\xi,\omega|x,y)f(x,y) \tag{3}$$

where the Point Spread function $\mathrm{PSF}_{\mathbb{V}}$ is the impulse response of the operator $T\mathbb{V}$, i.e., the image of a source point

$$f(x,y)=\delta(x-x_0)\delta(y-y_0). \tag{4}$$

Some easy calculations show that

$$\mathrm{PSF}_{\mathbb{V}}(\xi,\omega|x_0,y_0)=K(\omega)\frac{1}{y_0}\left[\delta(x_0-\xi+y_0\tan\omega)\delta(x_0-\xi-y_0\tan\omega)\right]. \tag{5}$$

The $PSF_{\mathbb{V}}$ is also called the kernel of the $T\mathbb{V}$ transform.

3. Analytic and algebraic reconstruction

There are different ways to recover the image of the source object from projections. We focus here on two of the most relevant approaches, analytic techniques, which we mention succinctly, and algebraic techniques, which are our main interest in this work.

3.1. Analytic methods

Analytical methods are methods of reconstruction in the ideal case of continuous variables. They rely on individual analytical relations between $f(x,y)$ and its projections $g(\xi,\omega)$. Detailed implementations for FBP in the $T\mathbb{V}$ case can be found elsewhere [1, 7, 8].

3.2. Algebraic methods
In this approach, the projection formula is discretized leading to a system of linear equations. Then, we search for $f(x, y)$ in a space of finite dimension $n = N \times N$, as a linear combination of basis functions Φ_i:

$$f(x, y) = \sum_{i=0}^{n} f_i \Phi_i(x, y).$$

where Φ_i are the characteristic function of the pixels, defined by:

$$\Phi_i(x, y) = \begin{cases} 1, & \text{if } (x, y) = \text{ pixel } i \\ 0, & \text{otherwise} \end{cases}$$

Now, the value of the projection at one point is

$$g_j = g(\xi_l, \omega_k) = [T\mathbb{V}f](\xi_l, \omega_k) = \sum_{i=1}^{n} f_i \, [T\mathbb{V}\Phi_i](\xi_l, \omega_k) = \sum_{i=1}^{n} A_{ji} f_i,$$

where g_j is the value of the projection with angle ω_k at detector ξ_l.

Then, the matrix formulation of the image formation process reads

$$A\mathbf{f} = \mathbf{g} \tag{6}$$

Here $\mathbf{g}$ is the projection vector (observed data), each of its components is a projection value, and its size is $m \times 1 = DP \times 1$, i.e., the number of detectors D times the number of angles P; $\mathbf{f}$ is the image vector, each component is the intensity of a pixel of the source, its size n is equal to the number of pixels $N \times N$, and finally A is the projection matrix, of size $m \times n = DP \times N^2$. This algebraic formulation of the problem is, in some sense, the discrete relative of equation (3). Since the algebraic problem is sometimes ill-posed (noise usually puts in evidence ill-posedeness of the problem), a straightforward inversion of matrix A is not always possible. Algebraic methods provide reconstruction using either iterative [6, 9, 14–17] or single step [14] algorithms. In scattered radiation problems, we have already tested the adaptive algebraic reconstruction technique with a random permutation scheme (RPS-AAAR) [9]. We now present an alternative inversion method, often used in the context of ill-posed problems, the SVD.

3.2.1. Singular value decomposition The SVD [10, 11] is a non iterative technique that enables algebraic inversion for the problem involving a matrix $A_{m \times n}$ with $m \geq n$. Singular values are the square roots of the eigenvalues of the square matrix $A^t A$. SVD allows factorization for a matrix in the form:

$$A = U_{m \times m} \times S_{m \times n} \times V^t_{n \times n} \tag{7}$$

where U and V are orthogonal matrices whose columns are eigenvectors of AA^t and $A^t A$ respectively and S is a diagonal matrix containing the singular values of A. This factorization allows us to write the pseudo inverse $A^\dagger$ for A:

$$A^\dagger = V \times S^{-1} \times U^t \tag{8}$$

and thus the inversion problem is solved as:

$$\mathbf{f} = (V \times S^{-1} \times U^t)\mathbf{g} \tag{9}$$

SVD allows accurate inversion and provides a practical way to solve the problem of instabilities due to limited numerical resolution by truncating the smallest singular values [10].

Table 1: Projection matrices for different angular discretizations with $N = 64$ and $D = 64$.

Number of angles (P)	rows	columns	Condition number PSF	R
295	18880	4096	$1.82\ 10^4$	—
64	4096	4096	$3.20\ 10^{19}$	$3.53\ 10^{14}$
32	2048	4096	$2.92\ 10^4$	$5.33\ 10^6$
16	1024	4096	$5.82\ 10^3$	$1.73\ 10^5$

4. Numerical results

In this work, all matrices, projections and reconstructions were implemented in a machine with an $\epsilon = 2.2204\ 10^{-16}$. This parameter conditions the problem numerically. We compare reconstructions made with SVD with those previously obtained using FBP and RPS-AART.

4.1. Projection matrices

We considered two different ways of generating the projection matrix $[T\mathbb{V}\Phi_i](\xi_l, \omega_k) = A_{ji}$

(i) by discretizing the analytical formula of the kernel of the $T\mathbb{V}$ transform from equation (5). Each column of the projection matrix is the discretization of $PSF_{\mathbb{V}}(\xi, \omega | x_0, y_0)$ for a given point (x_0, y_0). We call this matrix PSF.

(ii) by using an algorithm that performs the numerical calculation of the $T\mathbb{V}$ defined by (2) of every basis element Φ_i, $i = 1, ..., n$. Each column of A is the T$\mathbb{V}$ of a basis element. We call this matrix R.

In both cases, matrix dimension depends on the discretization of spacial and angular parameters: if ξ in (2) is discretized by D values and ω by P values, the size of A is $DP \times N^2$.

We considered 64×64 pixel images and different number of angles (P) between 16 and 295. The number of detectors was fixed to $D = N = 64$. For the case (ii), where the projection matrix R is generated by transforming each element of the basis, we use a $dr = 0.01$ in the discretization of equation (2). Table 1 summarizes the properties and dimensions of the projection matrices. Notice that, for the 295 angles case, only the PSF was calculated because the computational cost of the R matrix is very high. Figures 2 and 3 show the SVD spectra of these matrices, *i.e.*, the diagonal elements of matrix S in (7).

Our data $g(\xi, \omega)$, are the projections of a 64×64 Shepp-Logan phantom simulated either by direct transformation using (2) or by multiplying the projection matrix PSF times the image in vector form.

4.2. Reconstructions

We compare results using three different reconstruction techniques: FBP and 20 iterations RPS-AART (both presented in previous works and included here for comparison) and SVD. We used different angular resolutions (P = 16, 32 and 64), for both types of projection matrices, R and PSF (Figure 4). Figure 5 shows the three reconstructions for a fixed angular resolution ($P = 295$ angles), using PSF matrix in the case of algebraic methods. Notice that there is no R matrix for $P = 295$ because its calculation takes too much time. Two figures of merit were considered in order to assess the results [9]: mean square error (MSE) and correlation. Table 2 shows their values for all the reconstructions. Notice that the SVD reconstruction for the PSF with $P = 64$ is very poor: from Table 2 we see that the MSE is almost 3 and Figure 5 m) shows the reconstructed image. Nevertheless, truncating the SVD at the 4046th singular

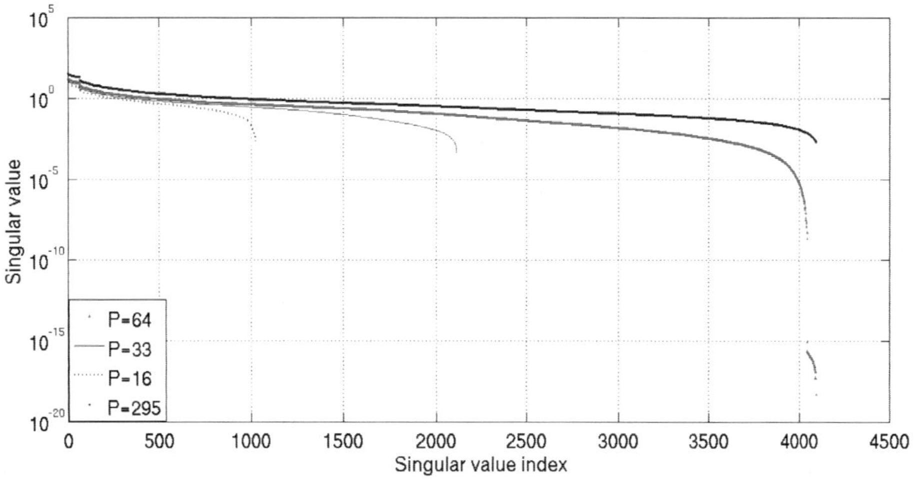

Figure 2: Singular value spectra of the PSF matrix using different number of angles P

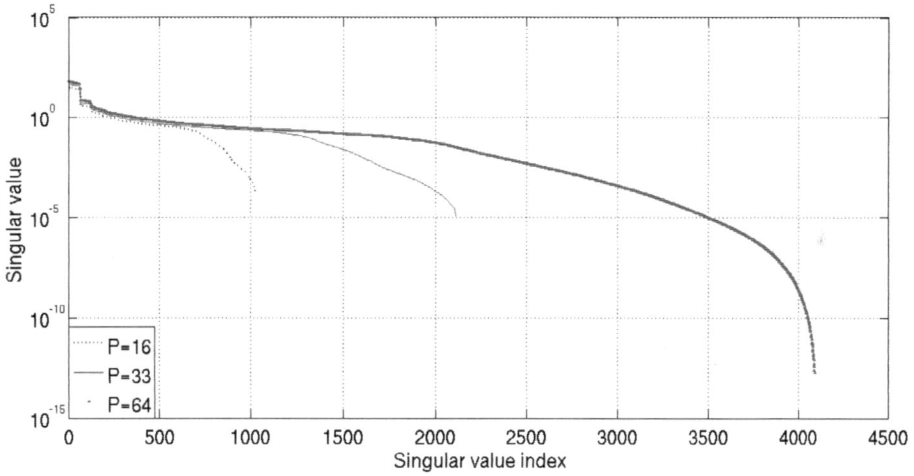

Figure 3: Singular value spectra of the R matrix using different number of angles P

value, the following figures of merit are obtained: MSE = 0.1730 and Corr = 0.9862, and the reconstructed image is shown in Figure 6 a). Figures 6 b)c)d) show the MSE as a function of the number of singular values kept after regularization. For example, a truncation index of 500 means that the first 500 singular values were used in the reconstruction, while the rest was discarded. Figure 6 b) illustrates the MSE of a well-conditioned matrix in the sense of [10] (SVD decomposition of a square R matrix for a 32×32 case with $D = P = 32$ and condition number $= 9.56\ 10^{11}$) where a $MSE = 0$ is achieved without truncation. Figure 6 c) shows the MSE for four truncated SVD reconstructions with a small number of angular projections for both types of matrices, the truncation index T moves between 1 and DP. Figure 6 d) shows the MSE for the SVD reconstruction using the PSF ($P = 64$) and truncating at different levels between 3840 and 4096 (256 last singular values).

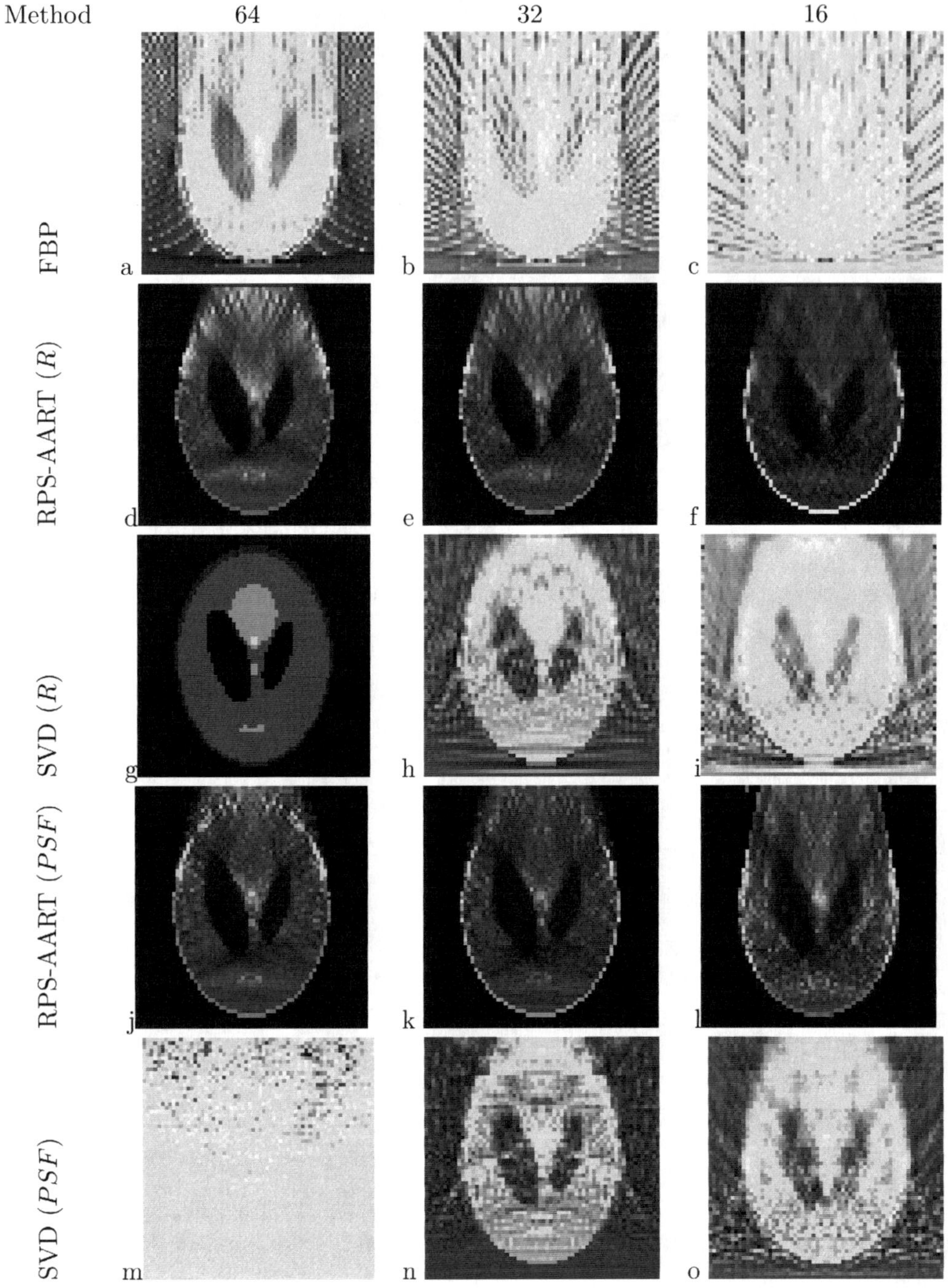

Figure 4: Shepp-Logan phantom reconstructions with different angular resolutions (number of angles $P = 16, 32$ and 64) and methods. Projections were generated using equation (2) (a-i) and equation (6) with $A = PSF$ (j-o).

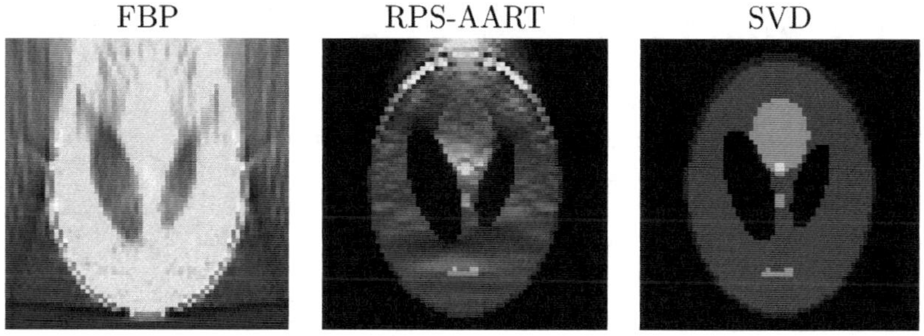

Figure 5: The three methods with $P = 295$ angles.

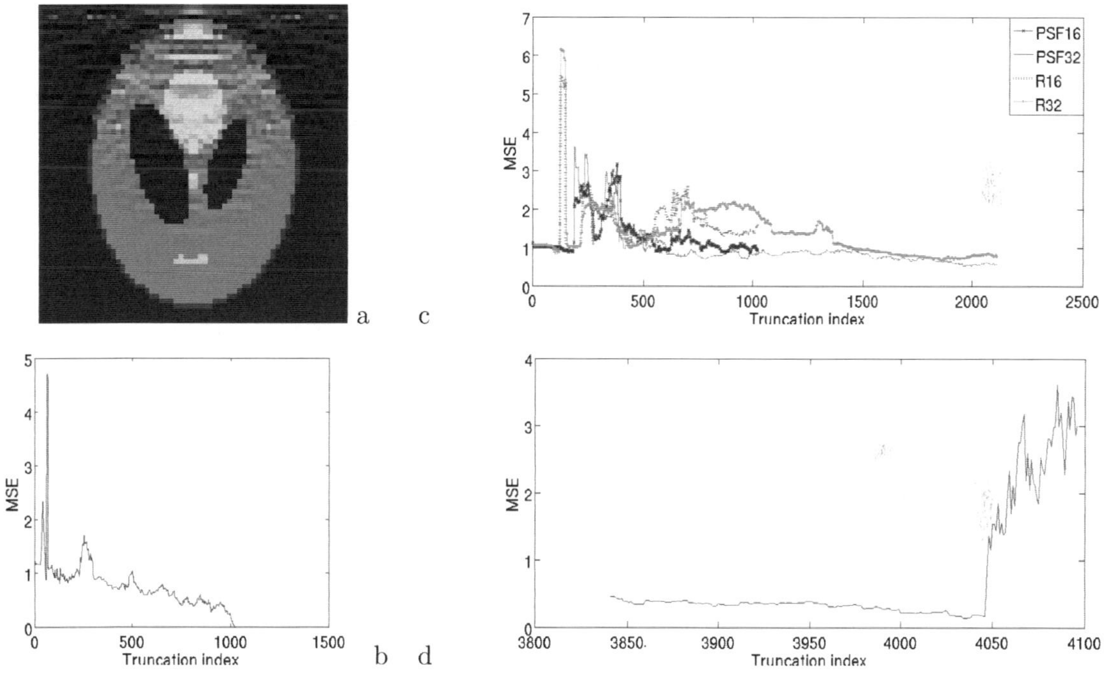

Figure 6: a) Regularized SVD reconstruction for PSF with $P = 64$ and truncation index $T = 4046$. b) MSE for 1024 different reconstructions of a 32×32 phantom ($D = P = 32$) covering all truncation indexes between 1 and 1024 with final $MSE = 0$. c) MSE for PSF and R reconstructions with $P < N$. d) MSE for reconstructions truncating at different levels between 3840 and 4096 (256 last singular values) with $P = 64$.

5. Discussion

The problem of inverting the $T\mathbb{V}$ transform using a singular value decomposition is numerically conditioned by the machine epsilon, $\epsilon = 2.2204\ 10^{-16}$ in our case. The condition numbers of the matrices we used are shown in Table 1. In Figure 4 m) it can be seen that the reconstruction with the PSF matrix for the $P = 64$ case is not stable. This fact is explained by the high condition number of the matrix. The SVD spectra for this case exhibits the smallest singular values ($< 10^{-12}$) and an important gap, more than five orders of magnitude, between consecutive singular values at the 4046th value (Figure 2). On the other hand, truncating the smallest singular values improves significantly the result. The gap indicates the optimal truncation index for regularization. We have an ill-conditioned matrix [10] with a well-determined optimal truncation index ($T = 4046$) making it possible to regularize the solution. This fact is confirmed by the graph showing the MSE for reconstructions using different number of singular values (Figure 6 d)) where it can be seen that for indexes higher than 4046, the MSE suddenly increases. Graphical and numerical results indicate that the SVD method (with or without truncation, depending on the condition number) clearly outperforms FBP and RPS-AART when enough angular projections are available. This is explained by the SVD factorization general hypothesis which states that the number of rows in the matrix should be greater than or equal to the number of columns: since the number of detectors was kept constant ($D = 64$), the number of angular projections determines the final dimension of the matrix. On the other hand, results are better for the RPS-AART when a limited number of projections is used.

Table 2: MSE and correlation for each reconstruction technique without truncation in SVD

			FBP	RPS-AART		SVD	
	Angular resolution [rad]	Number of angles (P)		PSF	R	PSF	R
MSE	0.005	295	0.98	0.11	—	0.00	—
	0.023	64	1.10	0.27	0.28	2.99	0.00
	0.050	32	1.73	0.32	0.32	0.58	0.78
	0.100	16	2.44	0.34	0.45	0.93	1.56
Corr	0.005	295	0.68	0.93	—	1.00	—
	0.023	64	0.66	0.80	0.82	0.04	1.00
	0.050	32	0.54	0.80	0.79	0.79	0.79
	0.100	16	0.39	0.75	0.70	0.69	0.66

6. Conclusion

We tested the SVD technique for inverting the TV-transform. The method outperforms RPS-AART and FBP when the number of projections is enough to ensure the $m > n$ case, required in the SVD theory. With a low number of projections, RPS-AART is superior. Matrices were well-conditioned with the exception of one case. The instability can be overcome by a proper truncation of the smallest singular values: the SVD spectra shows that, although the matrix is ill-conditioned, the optimal truncation index is well-determined. Since our findings can be extended to other Conical Radon Transform family members, the results obtained provide a useful basis for an inversion problem including $2D$ and $3D$ reconstructions. Some concern appears when analyzing a possible $3D$ implementation of the reconstruction problem since its dimensions could make the SVD factorization computationally expensive. Additional work should be done in order to determinate the prospect of using this technique in that context.

References

[1] Morvidone M, Nguyen M K, Truong T T and Zaidi H 2010 *International Journal of Biomedical Imaging, doi:10.1155/2010/208179* (Special Issue on Mathematical Methods for Images and Surfaces)
[2] Nguyen M K and Truong T T 2006 *Imagerie par Rayonnement Gamma Diffusé* (France: Hermes Science Publishing)
[3] Nguyen M K, Truong T T, Bui H D and Delarbre J L 2004 *Inverse Problems in Science and Engineering* **12** 225

[4] Nguyen M K and Truong T T 2002 *Inverse Problems* **18** 265
[5] Delarbre J L 2005 *Imagerie d'émission gamma : Nouvelles méthodes d'exploitation du rayonnement diffusé* Ph.D. thesis Université de Cergy-Pontoise
[6] Herman G and Mayer L 1993 *IEEE Trans. on Med. Imaging* **12** 600
[7] Morvidone M, Nguyen M K, Truong T T and Zaidi H 2010 *Proc.* 17th *IEEE Int. Conf. on Image Processing* (Hong Kong: IEEE) p 629
[8] Morvidone M, Nguyen M K, Truong T T and Zaidi H 2011 *International Journal of Biomedical Imaging, doi:10.1155/2011/913893*
[9] Cebeiro J, Morvidone M and Rubio D 2013 *Proc.* 4th *Congress on Industrial Computational and Applied Mathematics* (Buenos Aires: ASAMACI)
[10] Hansen C 1986 *The truncated SVD as a method for regularization* (Stanford University Stanford, CA, USA: Stanford University)
[11] Burden R and Faires D 2010 (Boston, MA, USA: Brooks/Cole CENGAGE Learning)
[12] Selivanov V and Lecomte R 2001 *IEEE, Trans. Nucl. Sci.* **48** 761
[13] Shim Y S and Cho Z H 1981 *IEEE Transactions on Acoustics, Speech and Signal Processing* **29** 904
[14] Driol C 2008 *Imagerie par rayonnement Gamma diffusé à haute sensibilité* Ph.D. thesis Université de Cergy-Pontoise
[15] Guan H and Gordon R 1994 *Phys. Med. Biol.* 2005
[16] Lu W and Yin F F 2004 *Medical Physics* **12** 3222
[17] Andersen A H and Kak A C 1984 *Ultrasonic Imaging* 81

Journal of Physics: Conference Series **477** (2013) 012023 doi:10.1088/1742-6596/477/1/012023

Distal Femur Allograft Selection Using a Spectral Shape Descriptor

F E Milano[1,2]*, L E Ritacco[2], G L Farfalli[3], L A Aponte-Tinao[3] and M Risk[1,4]

[1] Instituto Tecnológico de Buenos Aires, Buenos Aires, Argentina
[2] Unidad de Planeamiento y Navegación Virtual, Hospital Italiano de Buenos Aires, Buenos Aires, Argentina
[3] Departamento de Oncología Ortopédica, Hospital Italiano de Buenos Aires, Buenos Aires, Argentina
[4] Consejo Nacional de Investigaciones Científicas y Técnicas - CONICET, Argentina

E-mail: fmilano@ieee.org

Abstract. The automatic selection of bone allografts in a virtual bone databank is an active line of research. This work presents a new approach to solve this problem, based on a recently published image descriptor, called Volumetric Heat Kernel Signature. This descriptor is used to compare the size and shape of three dimensional thresholded computed tomography volumes. This approach is compared to a published method that uses the Iterative Closest Points algorithm to register a segmented search template to different candidates present in the databank. Statistical testing of the agreement between the two methods show that both approaches render similar results in the relevant clinical context. The proposed method avoids incorrect registrations due to local minima and does not require lengthly manual image segmentation and positioning before its use. The new method is conceptually simple and its mathematical basis is sound.

1. Introduction

In this work we study the problem of automatic selection of allografts of the distal femur in a bone databank composed of thresholded Computed Tomography (CT) images. The aim of this work is to develop an algorithm to automate repetitive measurement tasks executed in long bone tissue banks, as described in a previous paper by Ritacco et al. [1], and to compare it to a previous automatic method published by Sleiman et al. [2].

The measurements obtained in distal fresh frozen femora stored in a donor tissue bank serve as a general size filter for allograft selection. The selection of a replacement bone for a patient under reconstructive surgery is based on different biomechanical compatibility criteria. The first and most basic criterion is the similarity in extent between host and donor bones; in this case the sizes of the bones are evaluated using three linear measurements, each of them between two well-defined anatomical landmarks. These are the transepicondyle distance (A), the anterior-posterior distance in the medial condyle (B), and the anterior-posterior distance in the lateral condyle (C). These landmarks, as shown in Figure 1, can be accurately determined by a trained surgeon from visual inspection of a rough cortical bone threshold based segmentation of a CT

*This work was partially financed by a CONICET doctoral scholarship.

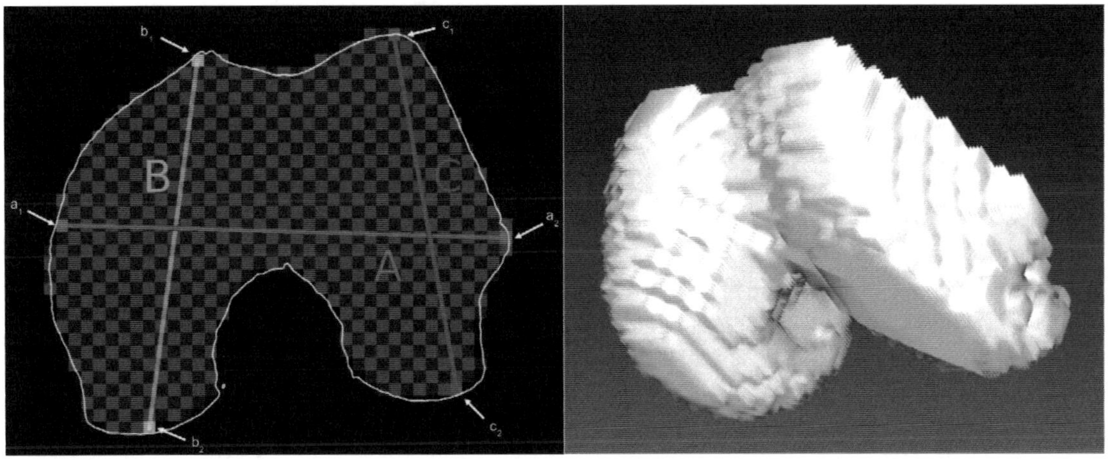

Figure 1: (Left) Main distances used to evaluate the size of a distal femur: transepicondyle distance (A), the anterior-posterior distance in the medial condyle (B), and the anterior-posterior distance in the lateral condyle (C). (Right) A volumetric rendering showing the 3D morphology of the distal femur.

scan. The second criterion is the degree of morphological similarity of the bone surfaces. Both criteria and the virtual bone databank workflow are reported by Ritacco et al. [3]. The main aim of this paper is to describe and to evaluate a novel technique that would serve to select an allograft fullfilling both criteria in the context explained above.

1.1. Related work

There are a number of different methods for bone databank matching reported in the literature. These methods vary greatly according to the intended application and anatomical area. Paul et al. [4] report a method for hemipelvic allograft selection based on a rigid three dimensional (3D) registration. The registration is performed directly on the CT volumes, using a gradient descent approach to minimize an euclidean metric. The different registrations are validated by visual inspection. Seiler et al. [5] implement a polyaffine registration process to recover six anatomical landmarks on the distal femur that serve as endpoints for the ABC measures. This strategy yields an acceptable estimation of the allograft size, but it does not address the morphology similarity issue. Sleiman et al. [2] describe a surface registration method based on the iterative closest points (ICP) algorithm [6]. The registration error is measured as mean surface distance and as Hausdorff distance. An advantage of this strategy is that both the size and morphology of the bone are taken into account in the error measure. However, this approach has the disadvantage of not being robust enough, since the method used produces registrations where the metrics fall in local minima, but the visual inspection of the result determines an incorrect alignment of the bones. In this work we compare the performance achieved by the mentioned method with the new proposed method.

1.2. Contribution

Most of the methods cited above require the existence of an acceptable image segmentation process. This is not always the case since quality segmentation is usually achieved through a long and tedious manual process which is executed by a knowledgeable expert in the domain of anatomical interest. In this work we present as a contribution the application of a spectral based descriptor called Volumetric Heat Kernel Signature (VHKS) to automatically select an adequate distal femur replacement in a bone databank of thresholded CT volumetric scans. VHKS were recently introduced by Raviv et al. [7] and yields useful properties from the viewpoint of a feature descriptor. VHKS are naturally invariant to isometric transforms of a volume. It has been shown in practice that VHKS are also robust to mild non isometric deformations as those commonly found in natural geometric shapes. This is usually the case for distal femora shape variations between subjects. Since VHKS is not scale invariant, it can be used to evaluate both the size and morphology criteria.

2. Materials and Methods

2.1. Experimental Data and Preprocessing

The experimental data is composed of ten left femora CT volumetric images scanned on a Toshiba Aquilion CT scanner, with a resolution of 0.877 mm and slice increments of 1 mm. The images were thresholded to segment the cortical bone. The segmented images were resampled to a resolution of 5 mm in each dimension, to obtain cubic voxels. The resampled images were cropped to keep only the lower extremity of the femora. Manual segmentations produced by an expert medical surgeon were used to generate surface models, represented as triangular meshes, of each distal femur.

2.2. Heat kernel based descriptor

In recent years different spectral methods for shape characterization have been published. Among these methods, the *Heat Kernel Signature* (HKS) introduced by Sun et al. [8] uses the heat diffusion process on a surface to describe the local and global geometry of each point in the surface with a feature vector. The *Volumetric Heat Kernel Signature* (VHKS) [7] is derived from the HKS and it is computed on a volume embedded in a three dimensional space.

Heat distribution in a domain is governed by the heat diffusion equation,

$$\left(\alpha\Delta_X + \frac{\partial}{\partial t}\right) u = 0 \tag{1}$$

where α is the thermal diffusivity and Δ_X is the Laplace operator. The solution $u(x,t)$ of the heat equation, given adequate initial and boundary conditions, describes the amount of heat on the point x in time t. The solution of the heat equation with an impulse on the surface as the initial condition, $u_0(x) = \delta(x-z)$, is called the heat kernel and can be written as

$$K_t(x,y) = \sum_{l=0}^{\infty} e^{-\lambda_l t}\Phi_l(x)\Phi_l(y). \tag{2}$$

where λ_l, Φ_l are the eigenvalues and eigenfunctions of the Laplace operator with Neumann boundary conditions [7]. A shape descriptor, the VHKS, is derived by the application of a spatial restriction to the heat kernel

$$h(x,t) = K_t(x,x). \tag{3}$$

If the VHKS of each voxel x of the volume is sampled logarithmically in time ($t = \gamma^\tau$), then the discrete function

$$h_{x,\tau} = h(x, \gamma^\tau) \tag{4}$$

is obtained. A vector $\mathbf{h}_x$ with the first n samples of $h_{x,\tau}$ is used as a n-dimensional voxel descriptor, where τ_1, ..., τ_n are different time-scales. The whole volume is described as a normalized sum of the descriptors

$$\mathbf{h}_X = \frac{\sum_X \mathbf{h}_x}{\#X} \tag{5}$$

being $\#X$ the number of voxels in X.

The VHKS is invariant to isometric transforms and it is multi-scale, capturing both local features at small t values and global features at large t values. The VHKS is also an informative descriptor, that is, under mild conditions, if two volumes have equal volumetric heat kernel signatures, then these volumes are isometric. However, the VHKS is sensitive to volume scale [8].

The VHKS is computed using a finite differences scheme to evaluate the parabolic partial differential equation that describes the distribution of heat in a body. For each image in the experimental set, the VHKS is calculated for each voxel, starting with an impulse initial condition. Since the heat kernel solution decreases exponentially, each component is normalized against its sum along all the VHKS; also the signature is sampled uniformly at logarithmic time. In this way, the local and global components of the VHKS contribute equally.

2.3. Search algorithm

The search algorithm accepts as inputs the VHKS descriptor vector for the template distal femur and the list of descriptor vectors for the bone databank. In this context, the euclidean distance between vectors $\mathbf{h}_X$ and $\mathbf{h}_Y$ is used as an index of size and morphology difference between the volumes X and Y. The distal femur volume in the databank whose VHKS vector minimizes this euclidean distance is the best match for the template that is being looked for.

2.4. Iterative closest points

The method proposed in this work is compared to the iterative closest points (ICP) registration process. The mean surface distance between two registered surfaces is the metric used to rank the possible matching candidates. Before executing the ICP registration, most of the surfaces were manually positioned to avoid, to a certain extent, the bad alignments due to local minima.

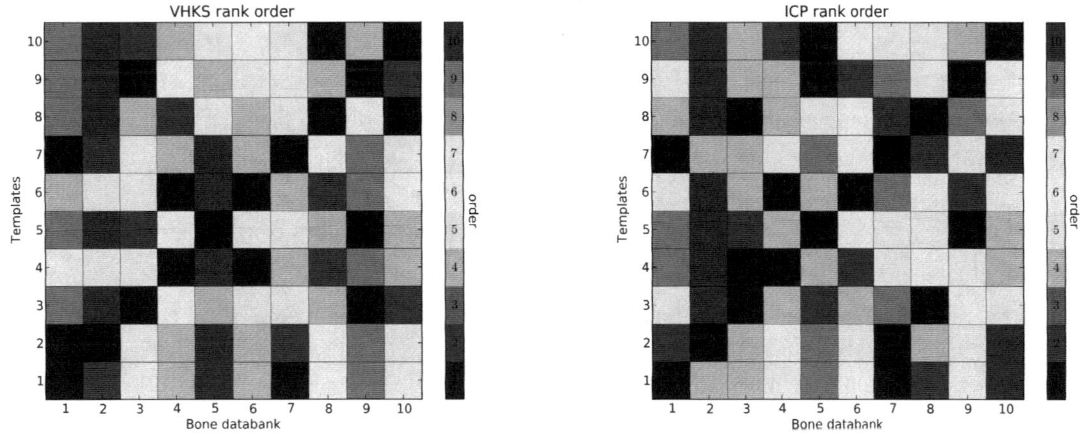

Figure 2: Each row in the matrix represents a ranking for the case that corresponds to that row. Each column represents a candidate from the bone databank. The blue diagonal serves as control: the best match is always of a template with itself. (Left) Rankings for the VHKS method. (Right) Rankings for the ICP method.

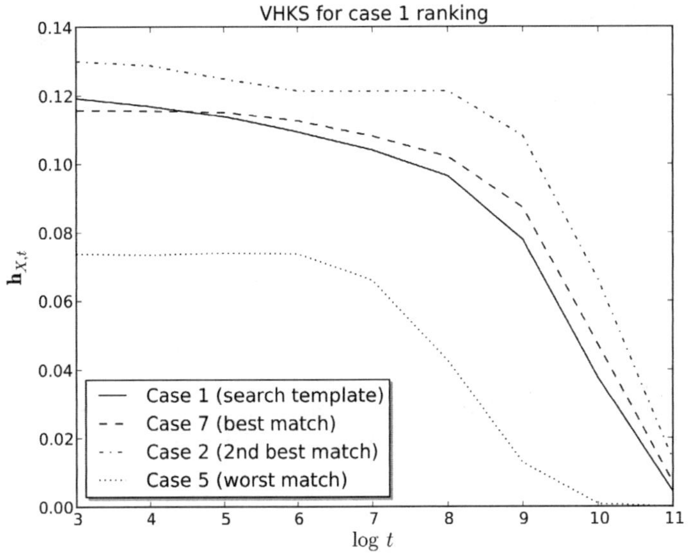

Figure 3: Plots of $\mathbf{h}_X$ for different volumes X, representing matches found in the ranking generated by the search process for template number 1.

Without this preprocessing step, an elevated number of cases would have been registered in an incorrect way.

2.5. Experimental scheme

The agreement between the two methods is measured for the whole experimental set using a leave-one-out scheme:

(i) Choose a case to use as template, and leave it out from the experimental set.

(ii) Choose a case to use as candidate, and exclude it from the bone databank.

(iii) Calculate the euclidean distance between VHKS template vector and candidate vector (for ICP, calculate the mean surface distance between the registered surfaces). Store this distance.

(iv) If there are still volumes in the experimental set, go back to (ii).

(v) Use the stored distance to rank candidates, from lower to higher magnitudes.

(vi) If there are still volumes not used as template, then restore the bone databank and go back to (i).

The process described above produces a total of 100 comparisons and 10 rankings. This scheme is applied to both the VHKS and ICP methods. These rankings are shown in two matrices. Each matrix row is a ranking for certain template; its cells are color coded according to the selection order of the candidate in that column (Figure 2).

2.6. Statistical analysis

Cohen's weighted kappa coefficient is used to measure the inter-rater agreement between the two evaluated methods [9]. The weights are linear and increase indicating the seriousness of the disagreement. For example, weights in the main diagonal of the table of occurrences are 0, weights in the sub and super-diagonal are 1, weights in the diagonals right below and above the sub and super-diagonal are 2, and so on.

The statististical analysis and the implementation of the Cohen's weighted kappa coefficient were carried out in the R environment [10].

3. Results

Cohen's weighted kappa coefficient is $\kappa = 0.45$, 95% CI $[0.22, 0.68]$, a moderate agreement according to Landis et al. [11]. If the occurrence table is restricted to the three best matches, then the coefficient is $\kappa = 0.77$, 95% CI $[0.34, 1.21]$, a substantial agreement. Color coded matrices showing the different rankings for both evaluated methods are shown in Figure 2. Each row represents a search process. Each column represents a match candidate. The cell color shows the selection order proposed for the template case in its row and the candidate case in its column.

A sample of fours VHKS vectors showing the template case 1, the two best matches and the worst match is shown in Figure 3.

4. Discussion

The new method proposed in this paper for distal femur automatic allograft selection, produces acceptable results in a short list of best matches. This method is compared to a published method based in the ICP algorithm. The importance of this comparison between ICP and VHKS lies in the fact that the second method is an intrinsic measure of the shape of a volume, whose meaning is not as intuitive as the closest points distances used by the ICP algorithm.

The agreement of the two methods, to a certain point in the application domain, is a step forward accepting the VHKS as a reliable method to select allografts of appropriate size and morphology. Although the general agreement between both methods is classified as "moderate", in terms of the statistical analysis, however the medical relevance of both methods is in the first two or three best matches. These are the selection possibilities the physician is presented with; given this short list he or she will choose the best allograft according to his or her medical expertise. In this scenario the new method may help to obtain the best selection, when restricted to the three best matches, both methods present a "substantial" agreement.

Future developments will refine two aspects of the proposed method. The first one, and most important, is a manual validation of the rankings. In this work the comparison with the ICP based method serves only as a bronze standard. The second aspect is that the low resolution of the grid generated to compute the VHKS using finite differences may show discrepancies in the volume shape that would not be so with finer resolutions. A parallel programming implementation may be able to compute larger problems with a fine-grained grid and with a better performance.

A suitable application of this method may be to select bone allografts in other anatomical regions of the body, like the pelvis. This line of work will be continued in the future.

References

[1] Ritacco L E, Seiler C, Farfalli G L, Nolte L, Reyes M, Muscolo D L and Tinao L A 2012 *Cell and tissue banking* ISSN 1573-6814 URL `http://www.ncbi.nlm.nih.gov/pubmed/22484825`

[2] Sleiman H B, Ritacco L and Aponte-Tinao L 2011 *Annals of biomedical* 1–9 URL `http://www.springerlink.com/index/3067551845734004.pdf`

[3] Ritacco L E, Farfalli G L, Milano F E, Ayerza M A, Muscolo D L and Aponte-Tinao L 2013 *Sarcoma* **2013** 524395 ISSN 1357-714X URL `http://www.ncbi.nlm.nih.gov/pubmed/23690733`

[4] Paul L, Docquier P L, Cartiaux O, Cornu O, Delloye C and Banse X 2010 *Acta orthopaedica* **81** 252–7 ISSN 1745-3682 URL `http://www.ncbi.nlm.nih.gov/pubmed/20175643`

[5] Seiler C, Pennec X, Ritacco L and Reyes M 2011 *Proceedings of SPIE* vol 7962 p 79620G

[6] Wyawahare M, Patil D and Abhyankar H 2009 *International Journal of Signal Processing, Image Processing and Pattern Recognition* **2** 1–5 URL http://www.sersc.org/journals/IJSIP/vol2_no3/2.pdf
[7] Raviv D, Bronstein M and Bronstein A 2010 *Proceedings of the ACM* 3–8 URL http://dl.acm.org/citation.cfm?id=1877817
[8] Jian S, Maks O and Leonidas G 2009 *Eurographics Symposium on Geometry Processing 2009* **28**
[9] Cohen J 1968 *Psychological Bulletin* **70** 213–220 ISSN 00332909 URL http://psycnet.apa.org/journals/bul/70/4/213/
[10] R Core Team 2013 *R: A Language and Environment for Statistical Computing* R Foundation for Statistical Computing Vienna, Austria URL http://www.R-project.org/
[11] Landis J R and Koch G G 2012 *Society* **33** 159–174 ISSN 0006341X URL http://www.jstor.org/stable/2529310

High intensity region segmentation in MR imaging of multiple sclerosis

F Rodrigo[1], M Filipuzzi[1], R Isoardi[2], M Noceti[2] and JP Graffigna[1]
1 Facultad de Ingeniería, Universidad Nacional de San Juan, San Juan, Argentina
2 Fundación Escuela de Medicina Nuclear (FUESMEN), Mendoza, Argentina

E-mail: jgraffig@gateme.unsj.edu.ar, roberto@fuesmen.edu.ar

Abstract. Numerous pathologies are often manifest in Magnetic Resonance Imaging (MRI) as hyperintense or bright regions as compared to normal tissue. It is of particular interest to develop an algorithm to detect, identify and define those Regions of Interest (ROI) when analyzing MRI studies, particularly for lesions of Multiple Sclerosis (MS). The objective of this study is to analyze those parameters which optimize segmentation of the areas of interest. To establish which areas should be considered as hyperintense regions, we developed a database (DB), with studies of patients diagnosed with MS. This disease causes axonal demyelination and it is expressed as bright regions in PD, T2 and FLAIR MRI sequences. Thus, with more than 4300 hyperintense regions validated by an expert physician, an algorithm was developed to detect such spots, approximating the results the expert obtained. Alongside these hyperintense lesion regions, it also detected bone regions with high intensity levels, similar to the intensity of the lesions, but with other features that allow a good differentiation.The algorithm will then detect ROIs with similar intensity levels and performs classification through data mining techniques.

1. Introduction

Multiple Sclerosis is a disease of the central nervous system that affects more than 2.5 million people all over the world. There is no known cure for the disease so far; however, there are therapies that may slow its progression.

The neural regions where myelin is lost (plaques or lesions) appear as hyperintense regions in MRI studies, particularly on T2, PD and FLAIR sequences [1]. In MS, these lesions appear at different times and locations of the brain and spinal cord. Once identified, a growth in size or temporal spread of the lesions may occur. And so, it is very important to delimit all the lesions if it is meant to evaluate the development of the disease or the results of the treatment drugs [2].

However, the human ability to distinguish subtle variations in the images is quite limited. Another important fact is that delineating all lesions in a particular study is a tedious and time consuming task. This is where Digital Image Processing has a wide range of applications.

While MRI studies are widely used for the diagnosis of neurodegenerative diseases and monitoring their evolution, conventional methodology fails to detect the underlying disease process, and consequently, the correlation between the clinical findings and conventional MRI is very limited [3]. This discrepancy clinical/radiological is probably due to the difficulty of the analysis and detection of specific major pathologic patterns associated with each disease.

Physicians reporting studies must be trained based on certain conditions and perform a number of annual reports. Double reading of reports increases the sensitivity of lesion detection minimum 7% but at a high time cost [4].

From the observation of the evolution of hyperintense regions it can be analysed the development of the abnormalities caused by MS, which in turn serve to monitor the disease and assess the effectiveness of various drugs.

Therefore, the aim of this work was to develop an automatic algorithm based on statistical parameters that could delimit the regions acceptably, in comparison with a DB delimitated by a specialist, after a validation process.This segmentation program might then be used with different artificial intelligence techniques in order to develop an automatic detector of MS lesions in MRI.

Tissue segmentation is one of the processing phases in medical imaging and defines different regions within images or volumes obtained from the different diagnostic techniques [5]. This process can be applied for the diagnosis or treatment of various diseases by detecting and tracking the injured areas.

Magnetic Resonance Imaging (MRI) is a method widely used in the analysis of neurological pathologies. To accomplish this, it is important to acquire different sequences that allow one to enhance particular characteristics of each tissue to detect pathologic regions. Among the sequences used,it can be mentioned: proton density, T1, T2, FLAIR, fMRI, Perfusion, Diffusion, DTI, etc. [6]. In the acquired images, the regions under evaluation are often presented as hyper or hypo-intense regions (see figure 1).

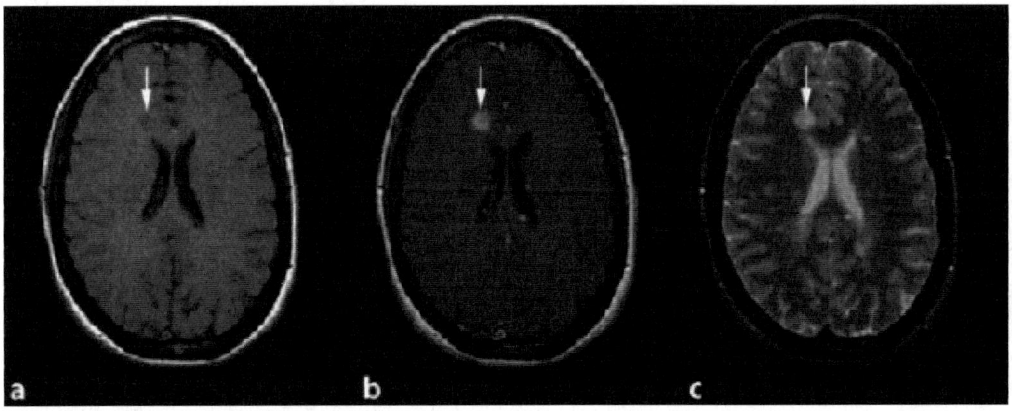

Figure 1. Axial images of sequences T1(a), T1 with contrast (b) and T2(c) of a patient with MS. In T1 sequences de lesion are defined as hypo-intense regions. While in T1 with Gd the lesions are shown as hyper-intense (courtesy of Sahraian M A and Radue E W 2008).

There are segmentation techniques based on the multi-sequence analyzing seeking the delimitation of just the MS lesions avoiding the reduction of false-positive [7], [8]. Other methods are based on fuzzy logic techniques intended to improve the detection of small lesions [9]. Akselrod-Ballin et al [5] propose a method of segmentation and classification using two data sets: PD-T1-T2 for 25 patients and FLAIR for 16 patients. Examples of MS lesions are shown in figure 2. Based on the proposed technique, they indicate that the sequence FLAIR has slight improvements in sensitivity and specificity. However, most of these studies use databases with a small number of studies, making it difficult to generalize these techniques.

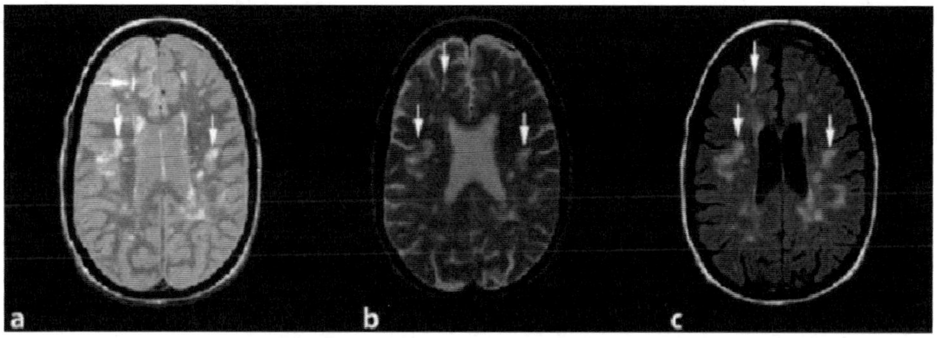

Figure 2. Axial images of sequences PD(a), T2(b) and FLAIR(c). It can be seen a better definition of the lesions in the last image (courtesy of Sahraian M A and Radue E W 2008).

For MS patients, FLAIR sequence shows the lesions as hyper-intense regions. In addition to these structures, there are other hyper-intense regions related with injuries of different etiology or bone structures. This paper aims to apply segmentation tools of hyper-intense regions.This algorithm is then used to classify them in lesions of MS or other structures,see figure 3.

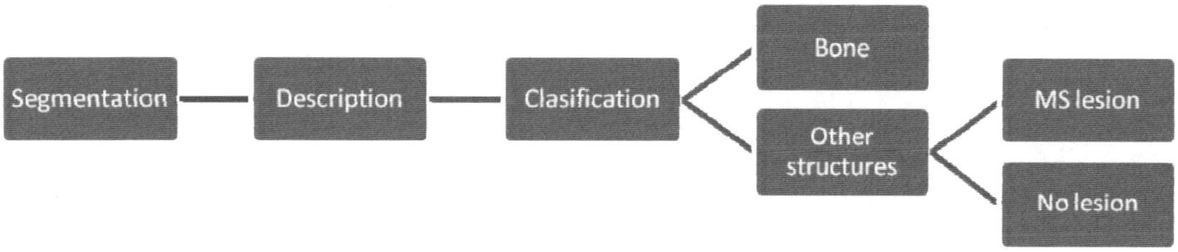

Figure 3. Steps of a usual Computer-Aided Diagnosis System.

This paper presents a segmentation technique based on adaptive thresholds and growth regions. Proposed techniques have highlighted a better sensitivity over the specificity, proposing for classification stages discrimination between segmented volumes corresponding to MS lesion or other structures.

2. Methods and Materials

2.1 Data Base

The database is composed of 100 studies of 68 patients with MS, acquired in a 0.5-T magnetic resonator develop by own project research [12]. These studies have multiple sequences, within which, in the Axial FLAIR series was conducted a process of delimiting regions of injury. This arduous process was carried out by a specialist in neuroimaging, which indicated 2 points for each lesion: the first inside the injury and the second in the periphery. This allowed subsequently delimit ROIs by growing seed point, to the gray level of the border point. The marks were recorded on an overlay under DICOM standards and all were validated by the same professional.

2.2 Automatic detection of hyperintense regions

Once defined the regions of injury, it was necessary to develop a system for automatic detection of regions with the same relative intensity of bright or gray level, which can be used, for example, in systems of Computer-Aided Diagnosis (CAD) for further classification process [10]. For this, it was used a Region Growing Technique.

It was necessary to solve two specific problems: the location of the seed points of growth and the amplitude of growth of these seeds.

Identifying regions of hyperintensity (seed points) was performed by dynamic threshold binarization according to the result of Equation 1, where P1 is a constant.

$$Detection\ Threshold = \frac{Median^2}{Mean} * P1 \tag{1}$$

The average value of the image is modified by the amount of extreme samples (hyperintensities), while the median value of the image has slight response to this variation. The greater amount of hyperintensities in the image, the greater the average and therefore the detection threshold will be reduced in order to detect every one of them. The result of this first binarization is a black and white image, where white regions correspond to the brighter pixels of the hyperintense areas in the original image. To perform the growth of the ROIs, using the centroid of these regions as seed points, a new gray level is determined by Equation 2, where P2 is a constant.

$$Growing\ Threshold = \frac{Mean^2}{Median} * P2 \tag{2}$$

This equation resembles to the idea that sometimes hyperintense regions do not present a strong contrast to neighbor regions that are also bright, so while brighter the image, the gray level thresholding should be increased to ensure well defined ROIs and avoid not desirable connection.
To set the value of the constants P1 and P2, which limit seed points and ranges of growth, it were analyzed all database studies. For this, it was used an exhaustive method by scanning a range in the P1-P2 space. This range was established based on a medium value of detection and growing of the lesions, after an iterative process. All this permitted to define superposition of regions automatically marked and labeled by the physician in order to detect errors.
The overlay, composed only by the automatic detected lesions, was analyzed by XOR (exclusive-or) resulting in 1 when the values of the both binaries images, automatic process and validated DB, were different. This means that XOR operation result is related with errors in segmentation. Therefore, lesion areas marked by the professional (I_{prof}) that overlap automatically marked regions (I_{autom}) are deleted, resulting in zero. Hence, the amount of white pixels is considered as a quality measure, the lower the best. The differences image is defined as show in Equation 3.

$$I_{diff} = XOR(I_{prof}, I_{autom}) \tag{3}$$

That is, the result is a "white value or 1" when automatically marked regions do not correspond to those marked in the database due to poor growth, to an excessive growth, or that are inconsistent with lesion marks (for example, bone).
This defines whether automatically labeled regions are smaller or larger than those of the database, and so, expect a minimum area resulting of XOR output when regions are as close as possible.

3. Results

The proposed method was used to perform scanning in the parameter space (P1, P2) to analyze the resulting area of the XOR output and obtained the result shown in figure 4. The scan was made on 100 studies with 18images each, resulting in a total of 1800 images. It can be seen that the area increases as the parameter of growth increases because there are more variations with respect to the database. On the other hand, in the opposite direction, if the amount of seed points decreases, the number of ROIs detected will become smaller, so that the output area decreases, showing poor results.

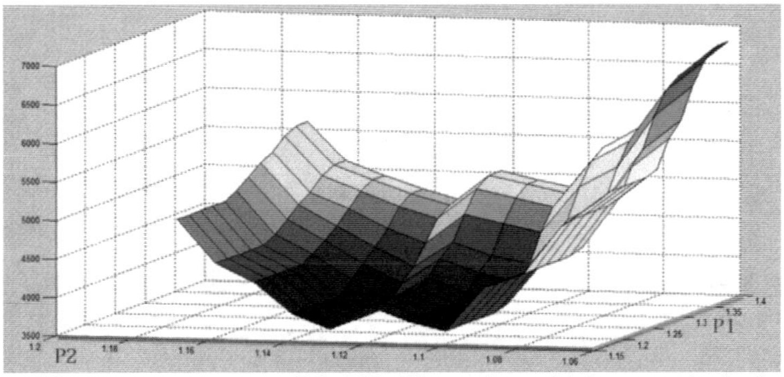

Figure 4. Resulting parameter space (P1, P2). The region with best results is shown in blue.

As can be seen, the image has an area of minimum values, which corresponds with the fewer error in the ROIs delimitation. As mentioned with regard to seed points, the best parameters values are those who result in a minimum value in the XOR-area surface. This is because all the lesion areas are indicated by a dot seed and the growth of these is consistent. On the other hand, if the "seed parameter" is set to low, there might be too many seed points, and the error would increase.

To evaluate better the response in this area and establish the best values for the parameters P1 and P2, a more extensive analysis was carried out in a minimum zone, using a smaller step to sweep both parameters. The final results are shown in the figure 5.

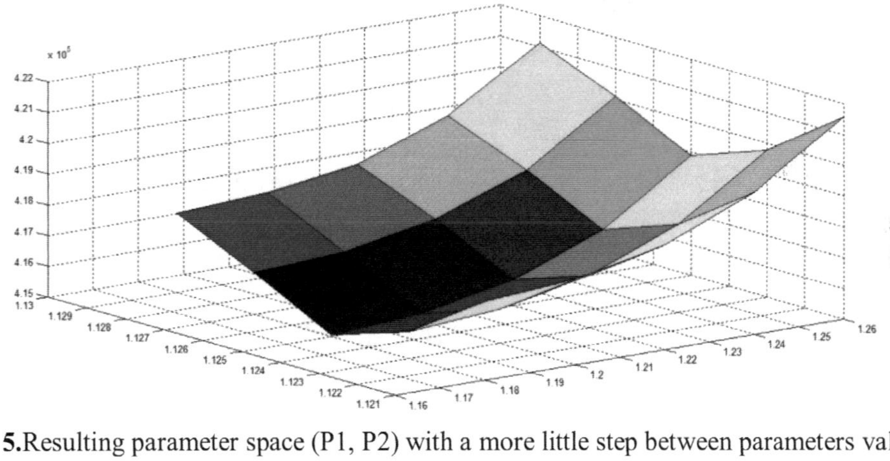

Figure 5. Resulting parameter space (P1, P2) with a more little step between parameters values.

With these values, the segmentation algorithm was tried out and validated against the data base in an automatic process based on verifying the pixels establish as ROI, using the optimum values of P1 and P2. This process gave the best result of a 91% of lesion segmentation, with an 8% of extra pixels and a 19% of fewer pixels, for the parameters P1=1.18 and P2=1.125.This means that all over the 100 studies the algorithm detects a 91% of the lesions but has the main problem in the growing parameter, which it seems to be too small, giving ROIs smaller than the ones of the data base. With the intention to improve this, a smaller value of the parameter was used but the results were worst, probably because an excessive delimitation. Nevertheless, a result of 91% is acceptable to start a real situation validation with the values found as optimal for P1 and P2 in equal acquisition conditions. This is, same equipment and image features.

These percentages correspond only to the lesions marked in the data base. The hyper-intensities due to bone and other structures are also delimitated but were not took into account for the validation, because the mayor objective is to evaluate the lesions segmentation. As mentioned above, after the segmentation, a classification process should categorize all the structures detected.

The best values of P1 and P2 were then used in the detector with excellent delimitation results (see figure 6). The bone was eliminated of the segmentation process by using a classification stage not described in this work, based in features detection of bone regions and neural network classification [13].

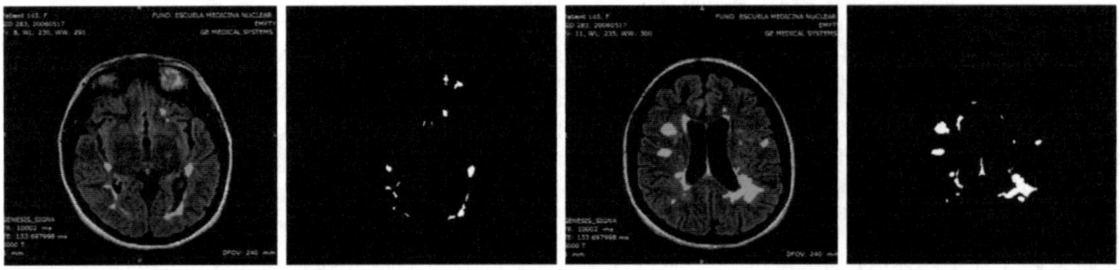

Figure 6. Specialist delimitations in pink next to automatic segmentation results on the right.

4. Conclusions

Detection of hyperintense regions in MRI studies is of great diagnostic value in clinical practice. Usually, this segmentation process is based on complex interpretation algorithms [11]. By systems that delimit these areas, similar to a medical specialist or maintaining similar parameters, it is possible to detect and classify hyperintense areas by specialized systems or obtain characteristics of these regions that might be used as clinical diagnostic support. For example, evaluate an increase in the size of lesions of a patient under a given treatment. The search of a better equation for the growing parameter is the main objective at the moment, basically trying to obtain bigger ROIs without detecting other adjacent structures.

Through this work we present an algorithm that detects hyperintense regions of MS, with a 91% validation but slightly more little lesions with respect to the data base, as a prelude to the classification of injury and other bright regions.

References

[1] Calabresi P 2007 Multiple sclerosis and demyelinating conditions of the central nervous *Saunders Elsevier* (chap 436)

[2] Sahraian M AandRadue E W 2008 MRI Atlas of MS Lesions *Springer Berlin Heidelberg* New York

[3] McDonald W I, Compston A and Edan G 2001 Recommended diagnostic criteria for multiple sclerosis: guidelines from the International Panel on the diagnosis of multiple sclerosis *Ann. Neurol. 50***1** 121–7

[4] Chan H P and Galhotra S 1987 Image Feature Analysis and Computer-Aided Diagnosis in Digital Radiography: Part 1 Automated Detection of Microcalcifications in Mammography *Medical Physics***14**

[5] AyeletAkselrod-Ballin, Galun M, Gomori J, Filippi M, Valsasina P, Basri R and Brandt A 2009 Automatic Segmentation and Classification of Multiple Sclerosis in Multichannel MRI *IEEE transactions on biomedical engineering*, **56**, 10

[6] Gonzalez R and Woods R 2007 Digital Image Processing *Pearson Prentice Hall* 3rd edition

[7] Hasanzadeh M andKasaei S 2007 Multispectral brain MRI segmentation using genetic fuzzy systems *9th Int. symposium on signal processing and its applications* pp. 1–4

[8] Wang P and Wang H L 2008 A modified FCM algorithm for brain image segmentation*Int. seminar on Biom. Informat. Eng.* pp. 26–29

[9] Aymerich E M, Montseny E, Sobrevilla P, Rovira A 2011 Detection of Hyperintense Regions on MR Brain Images using aMamdani Type Fuzzy Rule-Based System Application to the Detection of Small Multiple Sclerosis Lesions *IEEE International Conference on Fuzzy Systems* Taipei, Taiwan

[10] Marriott J J and Miyasaki J M 2010Therapeutics and Technology Assessment Subcommittee of the American Academy of Neurology *American Academy of Neurology* USA

Journal of Physics: Conference Series **477** (2013) 012024 doi:10.1088/1742-6596/477/1/012024

[11] Fenton JJ, Taplin SH, Carney P A and Abraham L 2007Influence of computer-aided detection on performance of screening mammography *N Engl. J Med.*
[12] Filipuzzi M, Rodrigo F, Graffigna JP, Isoardi R, Noceti M: Diseño e implementación de una Base de Datos de Esclerosis Múltiple e Interfaz Gráfica de Usuario. 3o Congreso Argentino de Informática y Salud. CAIS 2012. La Plata. Argentina. ISSN: 1853-1881 http://gateme.unsj.edu.ar/esclerosis
[13] Filipuzzi, Rodrigo, Laciar and Graffigna. (2012). Clasificación Automática de Lesiones de Esclerosis Múltiple Mediante Métodos de Conjunto en Imágenes de Resonancia Magnética. Argencon 2012 IEEE. ISBN:987572076-3.

Journal of Physics: Conference Series **477** (2013) 012025 doi:10.1088/1742-6596/477/1/012025

Validation of algorithm used for location of electrodes in CT images.

J. Bustos[1], J. P. Graffigna, R. Isoardi, M. E. Gómez, R. Romo
Gabinete de Tecnología Médica, Facultad de Ingeniería.
Universidad Nacional de San Juan, San Juan, Argentina

E-mail: jgraffig@gateme.unsj.edu.ar

Abstract. It has been implement a noninvasive technique to detect and delineate the focus of electric discharge in patients with mono-focal epilepsy. For the detection of these sources it has used electroencephalogram (EEG) with 128 electrodes cap. With EEG data and electrodes position, it is possible locate this focus on MR volumes. The technique locates the electrodes on CT volumes using image processing algorithms to obtain descriptors of electrodes, as centroid, which determines its position in space. Finally these points are transformed into the coordinate space of MR through a registration for a better understanding by the physician. Due to the medical implications of this technique is of utmost importance to validate the results of the detection of electrodes coordinates. For that, this paper present a comparison between the actual values measured physically (measures including electrode size and spatial location) and the values obtained in the processing of CT and MR images.

1. Introducción

The neurophysiology requires different equipment and studies for increasingly complex diagnosis. The pathologies are approached from anatomical, functional and metabolic information from different modalities. The work is framed in the medical imaging of the brain and the goal is the validation of method devolped for electrode detection and localization. The correct location of electrodes allows the neurophysiologist to combine the information from the EEG with the images of the brain. Studies using Magnetic Resonance (MR), Computed Tomography (CT), Single Photon Emission Computed Tomography (SPECT), etc.. MR and EEG recordings are the most important studies and fusion of both improves significantly the diagnosis. This requires the precise location of EEG electrodes on the MR. In addition, this information uses automatic process to locate electrical focus. These techniques use mathematical algorithm as described in [8] and [9]. In [16] was made the acquisition of the electrodes using MR and CT to evaluate the distortions produced by the electrodes on the image. From the results of the MR with 128 electrodes it was concluded that, using tin electrodes, it is not possible to obtain useful information from the image. Therefore, we performed acquisition protocol for locating electrical focus consisting of three related studies: CT with electrodes cap (See Figure 1), diagnostic MR and EEG of 128 channels.Using morphological operations and other techniques, it‘s possible obtain a segmentation of electrodes and eliminate most of the other structures (See Figure 2). This process, associated with the description of the volumes allows properly isolation of all electrodes. Then, a fusion method is applied to transform electrodes coordenates to MR space, in order to obtain a study with electrodes information. Using other application, the neurophysiologist can obtain better diagnostic merging EEG data with MR study with electric activity of cortex.

Figure 1. 128-electrodes cap.

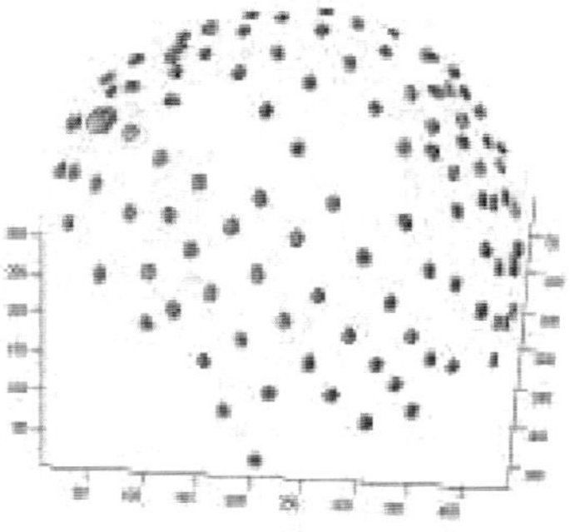

Figure 2. Detected electrodes.

All process must be validated to use in clinical application. Validation is the process whereby the software is shown to satisfy the needs of application with accuracy and other performance criteria. Validation of medical image algorithms will usually follow a sequence of measurements using computer-generated models (software phantoms), images of physical phantoms of accurately known construction and dimensions, and images of patients or volunteers. The process must demonstrate both high robustness and high accuracy [15].
In this paper it`s described a validation protocol for electrodes location method. For this, it'was performed a comparative study between a physical measurement and CT volume adquiered with phantom. Electrodes coordenates was obtained using image processing techniques described in a next section.

2. Materials and Method

To validate phantom was used polymethylmethacrylate (PMA) rectangular parallelepiped with 6 electrodes placed on their faces (see Figure 3). This object was scanned in different locations with CT64 Philips scanner. For the physical mcasurcmcnt uscd a scaler.

From the volume of CT centroids were obtained using electrodes GPL software in C + +.

We worked with the ITK library [11], the compiler MinGW and the development environment CodeBlock [12].

Figure 3: Phantom with electrodes on their faces.

A. Electrodes Location

The first step in the detection, after the acquisition, is the segmentation of electrodes. Due to the large difference between the CT number of the anatomical structures and the electrodes, it is possible to apply thresholding operations for the first segmentation. However, these techniques detected, together with the electrodes, wires and connectors make it difficult to detect the ROIs. Moreover, the electrodes have a hole through which enters the gel and are placed in different orientations.

By morphological operations we have tried to isolate the electrodes. Dilation, erosion, opening and closing in three dimensions have been used [1] [10]. However, preliminary results have not been positive. On the other hand, when we are seeking expansion of the electrode to fill holes, cables and connectors join in larger structures. Moreover, erosion removes the cables, but it generates strong distortions in the electrode.

In order to solve these disadvantages, we propose a hybrid operation that seeks to fill large structures (hole) and remove small objects (wires and connectors). The erosion operation on the volume V(x, y, z), using a structure element B with a cubic form with side n, can be defined as follows:

$$V_o(x,y,z)\begin{cases} =1, if \sum v_i = n^3, \forall\ v_i \in B_{(x,y,z)} \cup V \\ =0, if \sum v_i < n^3, \forall\ v_i \in B_{(x,y,z)} \cup V \end{cases} \quad (1)$$

where V_o (x,y,z) is the output value of the voxel at that point (x,y,z), $B_{(x,y,z)}$ is the structure element with center in (x,y,z) and n^3 is the number of voxels of B. That is, if the sum is equal to n^3, the central voxel in Vo is equal to 1; otherwise if the sum is less than n^3 take the value 0. This defines a threshold for erosion which will be n^3.

Dilation D on the same volume can be defined as:

$$V_o(x,y,z)\begin{cases} =1, if \sum v_i > 0, \forall\ v_i \in B_{(x,y,z)} \cup V \\ =0, if \sum v_i = 0, \forall\ v_i \in B_{(x,y,z)} \cup V \end{cases} \quad (2)$$

The hybrid operation proposal is as follows:

$$V_o(x,y,z)\begin{cases} =1, if \sum v_i > n^3/3, \forall\ v_i \in B_{(x,y,z)} \cup V \\ =0, if \sum v_i \le n^3/3, \forall\ v_i \in B_{(x,y,z)} \cup V \end{cases} \quad (3)$$

This function approaches to dilation when $B_{(x,y,z)} \cup V$ has many ones and to erosion when $B_{(x,y,z)} \cup V$ has many elements zeros. Thus, it is possible to fill the holes of the electrode and remove the wires and connectors.

B. Validation of detection electrodes

This process was analysing in two phases. First we compared a Physical size of the electrodes, and then we perform measurement of Physical location of the electrodes and distances between each one.

Validation of physical size consists of comparing electrode dimensions using physical meter and tomographic images. It is therefore necessary to measure with a scaler the actual size of the electrodes as shown in Figure 4.

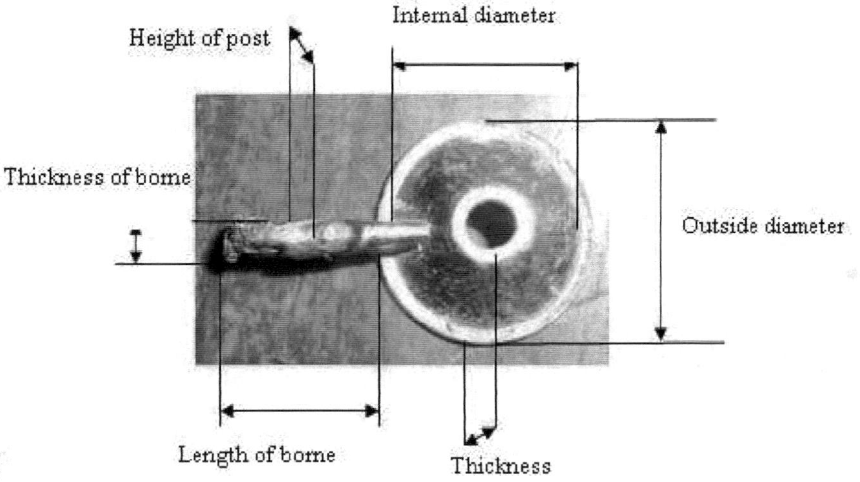

Figure. 4: Parameters measured on the electrode.

The CT scanner was need to adquire fanthom with electrodes and to obtain a measure of them (using descriptors). Images are acquired by performing three series which differ in the phantom position (see figure 5).

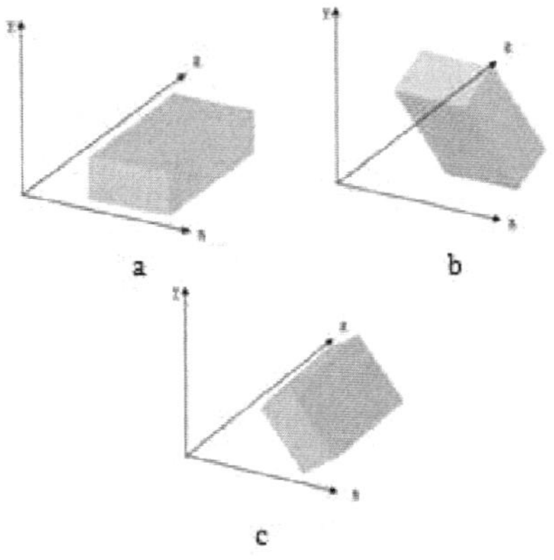

Figure 5: Layout the phantom with respect to the axes of the team. a. phantom aligned to the team. b. phantom tilted y, z. c. phantom tilted at x, y.

For obtain the electrode location in CT, we used the technique described previusly. Slice of original volume and render of electrodes are shown in figure 6. Once the images are acquired, we proceed to extract the necessary characteristics of the electrodes to be compared with actual physical dimensions. Using a software devolped with ITK/VTK, it was calculated descriptors of electrodes. Compartive analysis was made using length of axes [13] and oriented bounding box size [14].

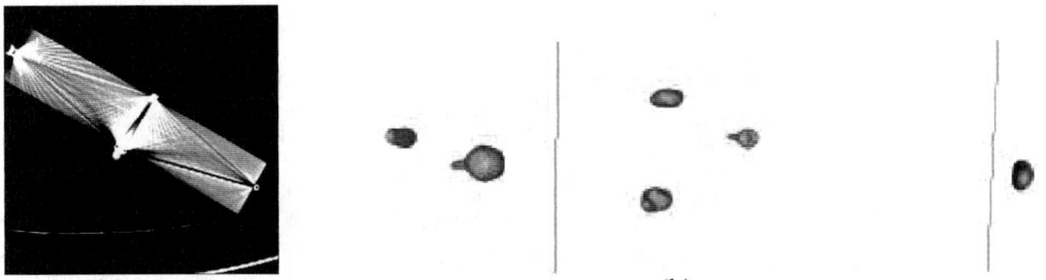

(a) (b)

Figure 6: Tomographic image of the electrodes and render of electrodes.

The validation for the electrode location was focused on measuring the physical distance between the opposite sides electrodes of the phantom, see figure 7, and compare these measurements with those obtained on images acquired previously. A second validation is to compare the distance between the centroid for each series acquired and assess whether the positioning of the phantom affects the calculation.

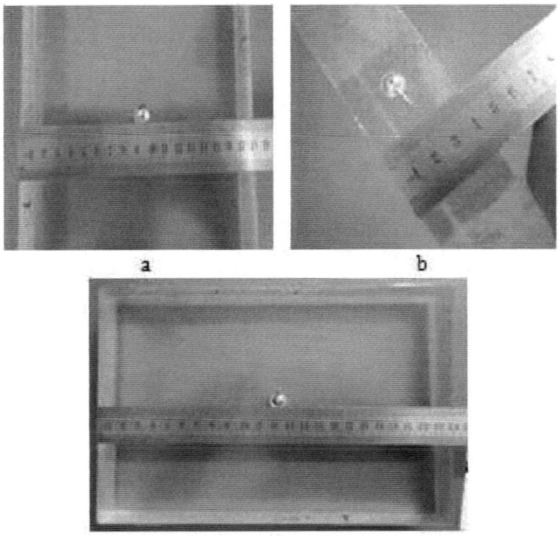

Figure 7: Measuring the dimensions of the phantom, to measure the width of the phantom, b. measuring the thickness of the phantom, c. measuring the length of the phantom.

For most accurate measurement we must consider phantom dimension and distance between surfaces and electrode centroids (See Figure 8). This requires finding the center of gravity mathematically electrode and the length phantom.

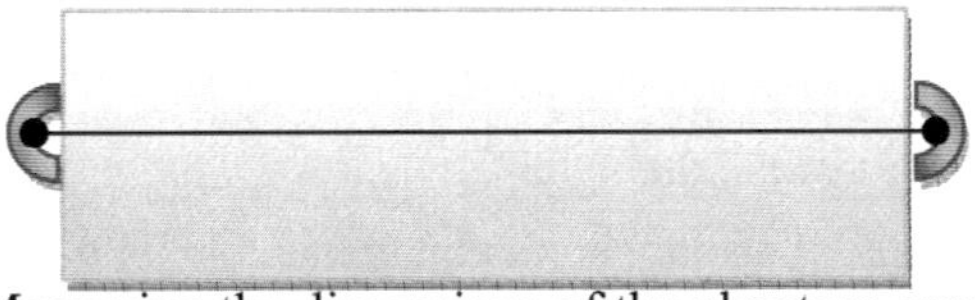

Figure 8: Measuring the dimensions of the phantom considering electrodes centroids

The centroid of the electrode is calculated by considering the electrode as a hemisphere whose base is the largest diameter of the electrode that rests on the surface of the phantom and which corresponds to the axes x, z, see figure 9.

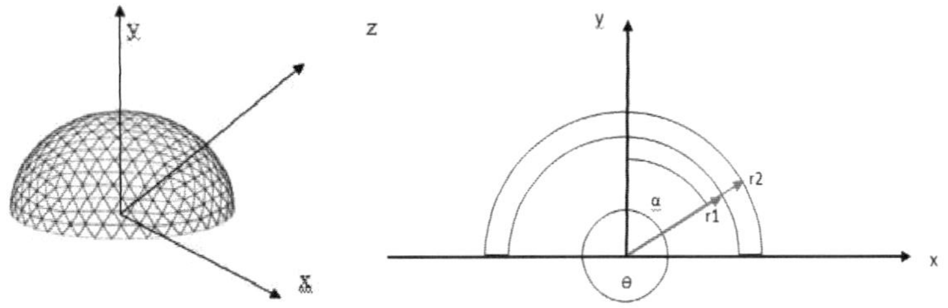

Figure 9: Representation of the electrode where the plane x, z corresponds to the surface of the phantom and its profile.

Therefore the displacement must be evaluated relative to the axis and by x = 0 and z = 0. Taking a profile, it was formuled throw following integral:

$$Y = \frac{\int_0^{360}\int_0^{90}\int_{r1}^{r2} r^3 * sen(\alpha) * \cos(\alpha) dr d\alpha d\theta}{\int_0^{360}\int_0^{90}\int_{r1}^{r2} r^2 * sen(\alpha) * dr d\alpha d\theta} \quad (8)$$

where r1 = 5mm is the outer radius of the electrode, r2 = 4.5mm is the inner radius of the electrode, α is the angle formed by the radius and the y axis, and θ is the angle around all electrode circumference.

In CT study, we used same process to obtain electrodes segmentation. To find the electrodes distance we developed a program that calculates the centroid (cen) of each of the electrodes and computes the Euclidean distance between the opposed pairs in the phantom. The following equation determines the calculation of this distance.

$$Distance = \sqrt{\left(cen_{xi} - cen_{xj}\right)^2 + \left(cen_{yi} - cen_{yj}\right)^2 + \left(cen_{zi} - cen_{zj}\right)^2} \tag{9}$$

where x, y, z are the coordinates of the electrodes i and j that are opposed.

3. Results

For validation of size of electrodes, it was compared actual physical size with tomographic images obtained. Actual physical dimensions of the electrodes were measured with the scaler. The values obtained are:

Table 1: Different measurements on the electrode in mm.

External electrode diameter	10 mm
Internal electrode diameter	9 mm
Electrode thickness	3 mm
Pin length	7 mm
Pin thickness	1.5 mm
Height pin	2.5 mm

Using phantom CT volume and described method, it was calculated descriptors for each electrode. With length of axis value and bounding box, it was defined length, width and diameter for electrode. Values were compared with the measures shown in Table 1. If the electrode is segmented accuraly, then we can assume that biggest axis correspond with length of electrode (10mm+7mm), the medium axis correspond to electrode diameter (10mm) and smaller axis is compatible to thickness (3mm). We used information from 18 electrodes obtained in three acquisitions. The results are shown in figures 10 and 11.

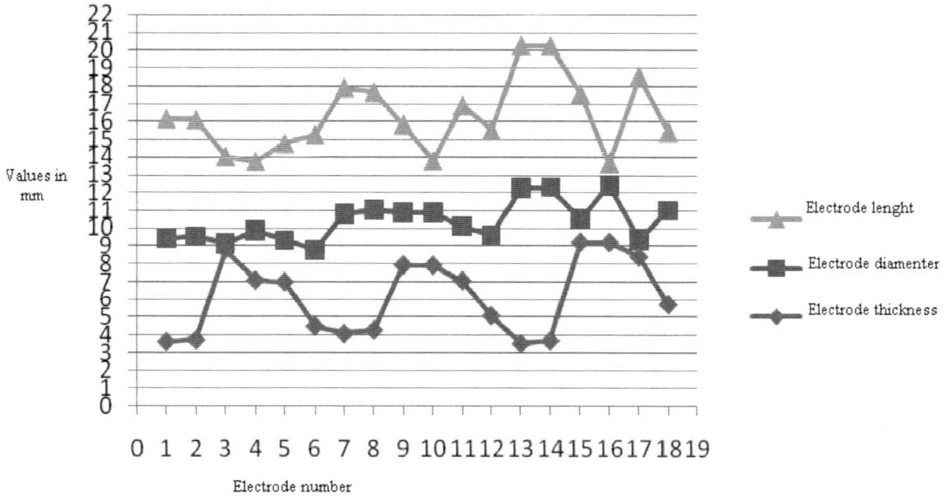

Figure 10: Variation of the length, diameter and thickness of the electrodes for the three sets acquired, which total 18 electrodes.

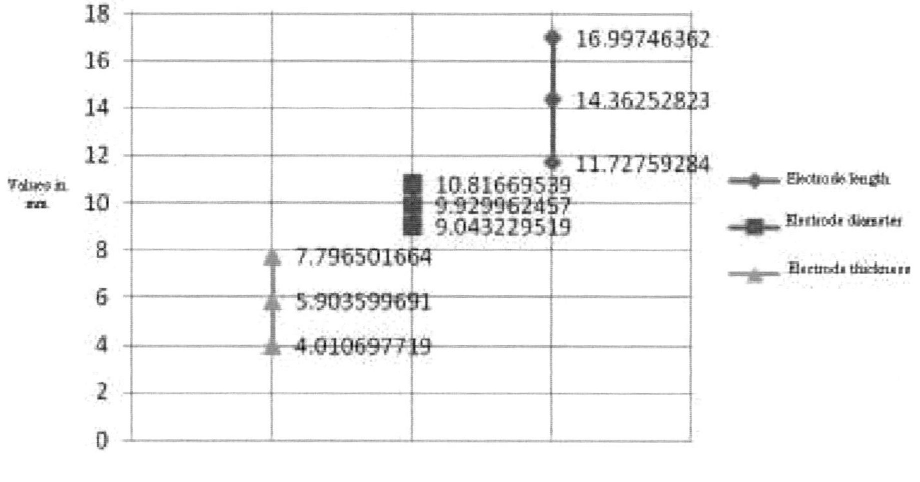

Figure 11: Mean values and standard deviation for length, diameter and thickness of the electrodes for the three series acquired.

It's possible appreciate that electrode diameter obtained by CT images are consistent with real value, but others have any differences associated to application of morphological operations. In length value we need to eliminate wire and thiny pins, then value is smaller than physical units. In other case, to eliminate hole in electrode, thickness is enlarged in this direction. However, most important value in EEG location is electrode diameter.

In second phase, we calculate distances between centroids of opposite electrodes. All measures were made using physical scale obtained from DICOM attributes. We calculated euclidean distances and compared with physical values. For these measures, we consider dimension of PMA and centroid obtained mathematically (this calculation add 1.3mm for each side). Table 2 show these values.

Table 2: Comparison of actual length values and processed

Measure	Actual Values	Processed Media Values	Standard Deviation
Large	242.6 mm	242.86 mm	0.214 mm
Width	183.6 mm	182.48 mm	0.048 mm
Thickness	42.6 mm	41.27 mm	0.107 mm

We can see that the standard deviation is minimal. Therefore, it is demonstrate that the distance values obtained by processing tomographic volumes substantially correspond with the actual physical measurements and allows the reliability of implemented method for locating the electrodes.

4. Conclusion

The paper presents a validation of method for detecting electrodes on CT volumes. This technique is associated to acquisition protocol for locating electrical focus and consists of three related studies: CT with electrodes, diagnostic MR and EEG of 128 channels. Therefore, a hybrid morphological operation permits segment electrodes and eliminate most of the other structures.

For development of algorithms, used ITK (Insight ToolKit) libraries provided all the classes needed for processing the images.

To validate the actual physical size of the electrodes, it was analyzed theoretical aspects associated to descriptors in images. With the use of a scaler, electrodes used in clinical application were measure to obtain theirs dimensions. Electrodes were acquired in CT volume using a phantom placed in different locations over PMA. Due to the metallic electrodes properties, it was appear star artifacts in images. The descriptors used to represent the size of electrodes were lengths of the axes and oriented bounding box size. These descriptors allowed a comparison with the actual measured values. Results obtained were satisfactory, mainly using diameter.

Validating actual physical location of the electrodes required for the calculation of the centroids and the Euclidean distances. Compared with the real physical values shown satisfactory results, which allows properly estimate the location of the electrodes. The coordinates of these electrodes permit in conjunction with EEG to determine the location of the epileptic discharge focus. At a later stage CT-MR registration will place the focus on the study of MR.

References

[1] Rafael C. Gonzales, Richard E. Woods, TRATAMIENTO DIGITAL DE IMAGENES, Addison-Wesley Iberoamericana, S.A, Delaware-Wilmington, 1996.
[2] Seyed M. Mirsattari, Donald H. Lee, Daniel Jones, Frank Bihari, John R. Ives. MRI compatible EEG electrode system for routine use in the epilepsy monitoring unit and intensive care unit. Published by Elsevier Ireland Ltd, 2004.
[3] Karen Mullinger, Stefan Debener, Ronald Coxon, Richard Bowtell. Effects of simultaneous EEG recording on MRI data quality at 1.5, 3 and 7 tesla. Published by Elsevier Ireland Ltd, 2007.

[4] Stephen B. Baumann, Douglas C. Noll. A modified electrode cap for EEG recordings in MRI scanners. Published by Elsevier Ireland Ltd, 1999.
[5] Hill, Derek L. G., Hajnal, Joseph V., Hawkes, David J. Medical Image Registration, CRC Press, New York, (2001).
[6] Bankman Isaac N. Handbook of Medical Imaging. Processing and Analysis. ACADEMIC PRESS, USA, (2000).
[7] Sonka, Milan, Fitzpatrick, J. Michael. Hanbook og Medical Imaging. Volume 2. Medical Image Processing and Analysis. SPIE Press, Washington USA, (2000).
[8] Ademoglu, A.; Demiralp, T.; Istefanopolus, Y.; Comu, S.; Baykan, B. "Epileptic source localization using wavelet prefiltering and MUSIC scanning", Proceedings of the 25th Annual International Conference of the IEEE Engineering in Medicine and Biology Society, 17-21 Sept. 2003, vol. 3, pp: 2366 – 2369.
[9] Wei W.; Wenyan J.; Hesheng L.; Xiaorong G.; Guojun Z.; Yuping W. "Localization of epileptic foci from preictal EEG data using standardized shrinking LORETA-FOCUSS algorithm", Proceedings of First International Conference on Neural Interface and Control, 26-28 May 2005, pp: 209–212.
[10] Azpiroz J., Medina V., Lerallut J-F: "Biomedical Image Processing" (in Spain)-Chapter 6: Mathematic Morphology. Universidad Autónoma Metropolitana. Mexico. 2000.
[11] Ibáñez L, Schroeder W, Ng L, Cates J: "The ITK Software Guide. Second Edition Updated for ITK version 2.4". Insight Software Consortium
November 21, 2005
[12] Björklund A et al.: "Code::Blocks Manual" Version 1.0. Code::Blocks Project. 2007.
[13] Rafael C. Gonzales, Richard E. Woods, "Digital Image Processing", Second Edition, Prentice-Hall, Inc. Upper Saddle River, New Jersey 07458, 2002.
[14] Dirk Padfield, James Miller, "A Label Geometry Image Filter for Multiple Object Measurement", National Alliance for Medical Image Computing (NAMIC), funded by the National Institutes of Health through the NIH Roadmap for Medical Research, NY, September 5, 2008.
[15] Joseph V. Hajnal, Derek L.G. Hill, David J. Hawkes, "Medical Image Registration", Series Editor Michael Neuman, "The BIOMEDICAL ENGINEERING Series", Boca Raton London New York Washington, D.C, http://www.crcpress.com, 2001.
[16] Graffigna JP, Gomez ME, Bustos JJ: Detection of EEG Electrodes in Brain Volumes. 32nd Annual International Conference of the IEEE Engineering in Medicine and Biology Society (EMBC2010), ISSN: 1557-170X, September 2010, pp. 5133-5136.

A development and implementation of a tinnitus treatment method

Alejandro José Uriz[1], Pablo Daniel Agüero[2], Juan C. Tulli[2], Jorge Castiñeira Moreira[1], Esteban L. González[2], Graciela Moscardi[3], Elber E. Sajama[2]

[1] CONICET / Laboratorio de Comunicaciones - Facultad de Ingeniería - Universidad Nacional de Mar del Plata

[2] Laboratorio de Comunicaciones - Facultad de Ingeniería - Universidad Nacional de Mar del Plata

[3] Facultad de Ciencias Médicas - Universidad FASTA

E-mail: {ajuriz,pdaguero}@fi.mdp.edu.ar

Abstract. Tinnitus is a physiological phenomenon where a person listens sounds which have not been generated by any external source. Today, many people suffer this condition. Although, in very few cases therapeutic methods completely eliminate tinnitus, it is possible to apply a variety of techniques to improve the quality of life of people with this condition. One of the most used methods to treat tinnitus consists of masking the tinnitus using an external sound. The main goal of this work is to present the development of a tinnitus treatment method, which optimizes the synthesized sounds in order to improve the life's quality of the user. Subjective tests and experimental results are used to analyze the performance of the method.

1. Introduction

Tinnitus [1] is a condition that affects an important percentage of the Society. It is defined as the condition where a person perceives a sound that has not been issued by any external source. This phenomenon is a hearing condition that can be chronic or not, and affects a large number of people. Tinnitus may be caused either by congenital reasons, stress or trauma of the hearing organ due to louder sounds. Recently, the number of people suffering from this condition has increased considerably due to improper use of portable playback of sounds, for example MP3 players.

While therapeutic methods in very few cases completely eliminate tinnitus, it is possible to apply a variety of techniques to improve the quality of life of people with this condition. In this group of therapies are included medication [2], [3], surgical methods [2] and relaxation sound therapy [4], [5], [6].

The most widely used approach for the treatment of tinnitus is the sound therapy [4], [5], [6]. Such techniques are based on the phenomenon of spectral masking [5], [7],[8]. In this phenomena, the brain in the presence of two spectrally adjacent sounds, can reduce the perception of one of the two sounds, and even eliminates one of them.

There are several types of therapies based on relaxation sounds. Firstly, there are assistive listening devices ([9], [10]) that allow users to amplify the environmental sounds with the aim of masking the tinnitus. This effect reduces the perception of the tinnitus and the individual is able to improve performance in their daily activities. On the other hand, it is possible to reduce the effect of tinnitus with sounds of interest to the user. This type of therapy consists of playing sounds (either music or sounds that are pleasant) so the brain focuses on them and consequently reduce the perception of tinnitus. This approach is widely used for the rest of the individual.

Finally, there is a group of techniques that have been developed in recent years whose goal is to mask the tinnitus [4], [5], [6]. For this aim, a colored noise generator is used to synthesize the signal which must partially or completely mask the tinnitus. This technique is used in high-end commercial assistive listening devices [9], [10]. While this approach improves the quality of life of individuals who suffer from this disease, there are cases where the noise generated to mask the tinnitus reduces the perception of other sounds. This is undesirable, because it could cause problems such as loss of intelligibility of some words, and also an alert sound could not be perceived. Therefore, it may cause a decrease in the independence of the user.

The first goal of this work is to improve the performance of tinnitus treatment techniques using spectral masking. In this way, a hearing impairment simulator was used [12] to simulate and study tinnitus. Based on obtained data, it is possible to optimize the sound synthesizer in order to mask only the tinnitus and not other sounds of interest for the user.

Algorithms are implemented in MATLAB and a DSP device from Microchip. The chosen device is the dsPIC33EP512MU810 [11].

Section 2 describes several techniques to reduce the perception of a tinnitus. Section 3 depicts the algorithm used to reduce the perception of a tinnitus, while Section 4 presents the most important features of the implemented device. Section 5 presents the experimental measures and subjective experiments carried out in order to validate the proposed algorithm and test the developed device. Finally, Section 6 summarizes the conclusions of the work and provides future research lines.

2. Tinnitus therapy using sounds

Tinnitus is a hearing condition that can be very rarely cured, even using medical treatments and surgical methods. In recent years methods that seek to improve the quality of life of people with tinnitus were developed. The main goal of methods based on sound therapies is to reduce the perception of tinnitus in the individual. There are various types of therapies based on sounds. In the following subsections three types of the most commonly used methods are described.

2.1. Tinnitus treatment using masking sounds

This kind of therapy can be divided into two subgroups: techniques based on zen music, and based on colored noise [4], [5], [6]. In the first subgroup of therapies are included techniques based on natural sounds such as the sound of rain o waterfalls, among others. These sounds can be recorded or, as in the case of some high-end assistive devices[9], also they may be generated using fractal based mathematical algorithms.

The second subgroup includes techniques based on colored noise. These use the spectral masking phenomenon between components[7],[8]. Masking is defined as the process taking place in the auditory system of a person, where a sound of a certain energy can reduce the perception of another sound which is spectrally adjacent to the first.

Because the masked components will not be perceived by the person, there are sound compression algorithms (such as the MPEG-I Audio Layer 3 [14]), which can compress the sound using only the spectral components that are not masked. There are three possible situations defined between the tinnitus and the sound generated for masking it. Firstly, if the sound does not change the perception of tinnitus, it is stated that it is not masking it. Moreover, if the sound level generated a partial alteration of tinnitus is said that the tinnitus is partially masked by the external stimulus. Finally, in the case where tinnitus is not perceived due to the external stimulus, it is said that it has been completely masked. Note that when applying these therapies, care should be taken that the external stimulus not causes damage to the auditory system of the person.

Therapies based on colored noise attempts to partially or fully mask the tinnitus using synthesized noise. Actually, this technique is included in high-end assistive listening devices[10], where the colored noise is obtained by filtering additive white gaussian noise (AWGN). In these devices, the filtering is done by mean of an equalizer with up to 16 subbands, which are also divided into four bands located between 100Hz-500Hz, 500Hz-1400Hz, 1400Hz-3500Hz and 3500Hz-8000Hz.

2.2. Therapies based on environmental sounds

An interesting option to reduce the effect of tinnitus in daily life of the user is to amplify existing sounds in the environment of the user[4]. This therapy reduces the difference of level between the sounds of the environment and the tinnitus, so it can be viewed as an improvement in the signal to noise ratio (SNR) of the signal perceived by the user. Such therapies should be applied using an assistive listening device that amplifies the signal so that ambient sounds at least partially mask tinnitus. This technique is currently implemented by some assistive listening devices and it is suitable for the user during their everyday activities. The main disadvantage of these techniques is produced when a high level of amplification of the environmental is needed. In this case, additional hearing impairments may appear in the long term.

2.3. Therapies based on sounds of interest for the user

Due to the nature of tinnitus, its perception can be reduced using sounds of interest for the user[4]. This therapy is based on how the brain perceives the tinnitus and generally is used in environments of relaxation of the user. They are very useful in cases where the tinnitus causes sleep disorders in the patient.

3. Proposed technique

3.1. Description of the proposed technique

Methods presented in Section 2 are widely used, but they are not methods exclusively developed for treating tinnitus. Then, these methods often fail due to a poor identification of the nature of the tinnitus that affects a patient. This is evident in the techniques presented in Subsections 2.2 and 2.3. The technique based on relaxation sounds, must take into account the energy of treated tinnitus in order to determine the level of the masker generated noise, in particular if white or colored noise it is used. As was previously described, this technique is widely used in hearing aid devices, and it can be seen that the spectral shape of the generated noise masks the tinnitus afflicting the user, but may mask other spectrally adjacent sounds.

Figure 1 shows the case of tinnitus composed by a pure tone located at $f = 1500$Hz. Continuous red line depicts the masking threshold of this tone. Continuous green line depicts the absolute threshold of hearing (ATH) of a young person without hearing problems in a quiet environment. Blue line is the spectrum of a composite signal by additive white gaussian noise filtered using a bandpass filter with cutoff frequencies of 1000Hz and 5700Hz. The absolute threshold of hearing of this condition is shown in striped black line. Figure 1 shows that by adding the filtered white noise, the resultant absolute threshold of hearing is elevated and individual components which are below this will not be perceived.

In this paper a method that optimizes generated colored noise is proposed to reduce this effect. In this approach a way to succesfully mask the tinnitus and that, moreover, improve the perception of other sounds is presented. The first step of the proposed method consists of determining the ATH of the patient by the presence of tinnitus. Then, based on the phenomenon of masking tones using noise [5], [7],[8], it is possible to determine the minimum level of noise as

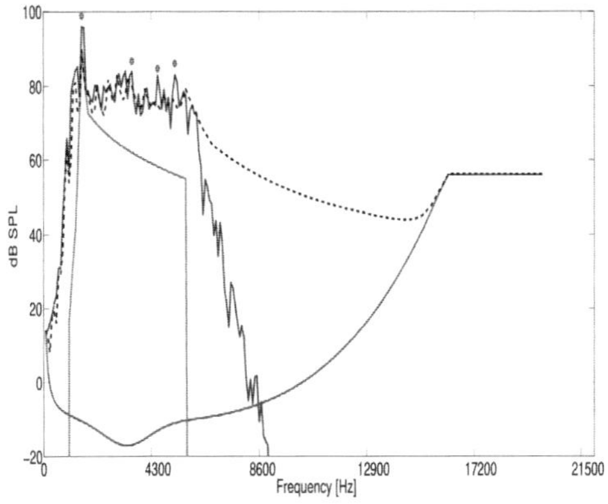

Figure 1. ATH variation generated by the colored noise used to mask tinnitus.

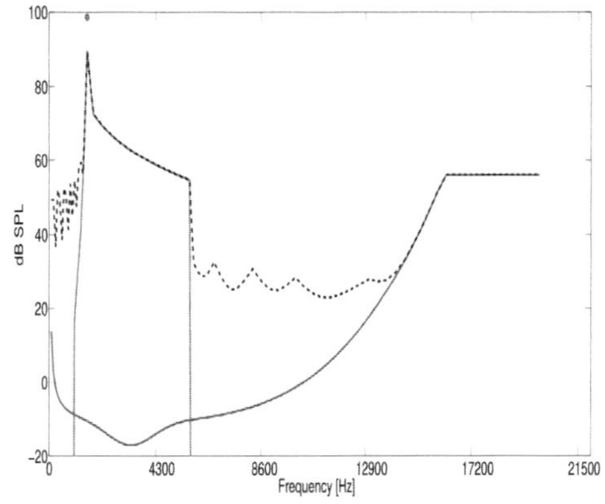

Figure 2. Masking threshold variation induced by tinnitus.

a function of frequency to mask the sound. Then, through this procedure, it is possible to set the noise level necessary to mask the tinnitus. Once obtained these specifications, they are used to process noise, and a bandpass filter is generated. This filter is used to fit the spectrum of white noise to mask the tinnitus. This step of the process is shown in Figure 2, which assumes a tinnitus tone at f = 1500Hz. Continuous red line represents the masked threshold of the tone. This curve indicates that components below it can not be perceived by the person suffering from such condition. In continuous green line, it is shown the ATH for a healthy young person. Finally, in striped black line is shown the curve of the ATH resulting of applying additive white gaussian noise filtered using tinnitus masking curve.

The curves of the resulting absolute thresholds of hearing are compared in Figures 1 and 2. It can be appreciated that the proposed method has a hearing threshold of lowest energy. Therefore, we can say that the proposed method allows the perception of more sounds that the reference method. The proposed system is presented in Figure 3, where it can be seen that in the first stage the diagnosis of the patient is used to estimate the masking threshold caused by tinnitus. Later, a bandpass filter is generated based on the previously obtained information, which is responsible for filtering the additive white gaussian noise. The filtered signal is used as a stimulus to the person who suffers the tinnitus.

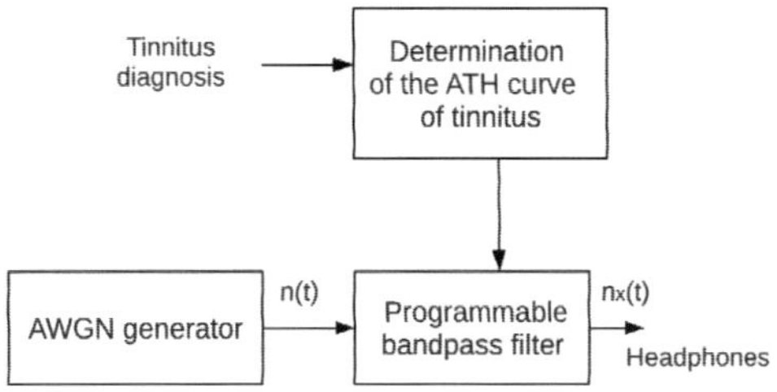

Figure 3. Block diagram of the proposed system.

The result of this method is a therapy optimized to the patient, which is appropriate for most of the cases. Furthermore, this technique has almost direct implementation in sound synthesizers and assistive listening devices. The main disadvantage of this technique is that, the tinnitus afflicting of the user must be known in detail to set the noise level to be synthesized.

3.2. Implementation of the proposed technique

The proposed technique was implemented in a hearing impairment simulator [12]. This tool was developed by the research group to simulate a particular tinnitus. Moreover, the simulator can be used to estimate the masking threshold. Then, by introducing to the system a particular tinnitus, two outputs are obtained: the audible signal which simulates how a person with a particular hearing condition perceives a tinnitus and the masking curve associated with this phenomenon.

The model of the ATH for a young person, in a quiet environment is obtained from expression 1.

$$S(f) = 3.64(\frac{f}{1000})^{-0.8} - 6.5e^{-0.6(\frac{f}{1000}-3.3)^2} + 10^{-3}(\frac{f}{1000})^4 \quad (1)$$

Another advantage of working with this simulator is the posibility to perform experiments with individuals with a healthy auditory system. Thus, it is possible to obtain a greater number of people for the experimental studies, which is critical for validating the results. The first step of the proposed method is to make the diagnosis of tinnitus by a specialist and enter the data into the simulator. In this way, the ATH for the patient is obtained using the simulator. In this way, the masking spectrum caused by each of the spectral components can be obtained. Once the masking produced by tinnitus is established, the level of noise necessary to mask such tonal component is determined. Then, it is used to filter additive gaussian white noise and colored noise is obtained.

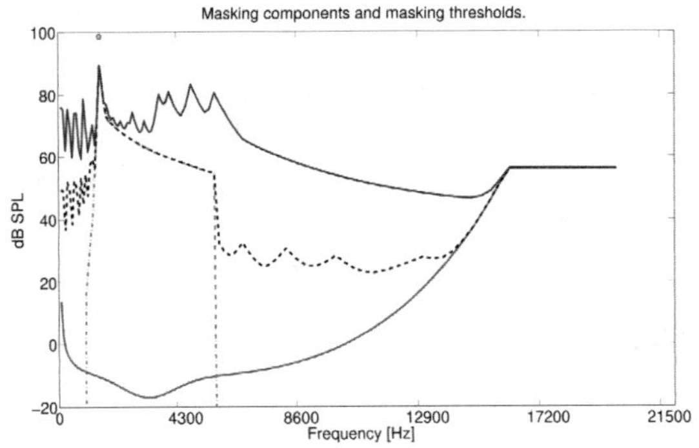

Figure 4. Masking thresholds obtained using noise filtered with a bandpass equal to the ATH of tinnitus and the proposed techniques. The proposed method allows a lower ATH than the reference method.

Once obtained the colored noise, the synthesized sound is used for the treatment of tinnitus. This may be done by a music player or an assistive listening device. The hearing impairment simulator [12] works with a spectral resolution of 32.25 Hz. It is possible to fit the spectrum of the filtered noise to the masking threshold with a precision better than many commercial assistive listening devices, which, as previously mentioned, only have up to 16 bands of equalization. These bands are in the range from 100Hz to 8kHz. Figure 4 shows in continuous blue line the ATH curve of the reference method, and in dashed black line the ATH curve of the proposed method. It can be seen that exists a diference between the ATH curves of the proposed and reference methods. Components placed in this range will be perceibed using the proposed method, while they will be masked if a therapy based on the baseline methods would be applied.

Thus, the synthesized sound is adjusted to the desired output, and it masks the tinnitus and reduces the masking of other external spectrally adjacent sounds to the tinnitus. A complete block diagram of the implemented system is shown in Figure 5, where the implemented system is divided into two stages: training and operation. During the training stage, tinnitus diagnosis data provided by the specialist is loaded to the simulator, which computes the threshold of masking of the tinnitus. Once the system was trained, the ATH obtained is used to reprogram the simulator, which in this case it is used as a bandpass filter. The output of the simulator is used to stimulate the user through headphones.

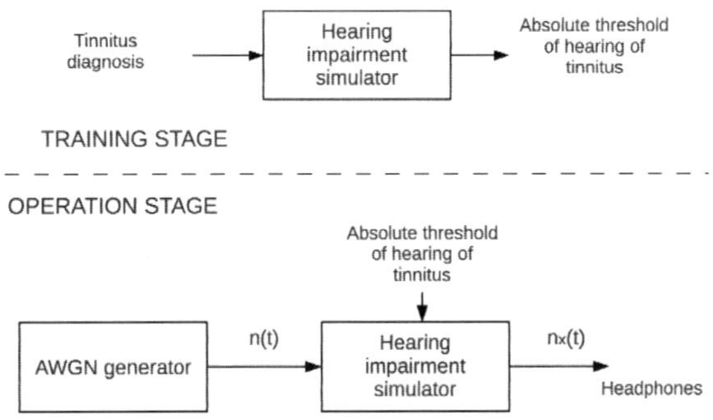

Figure 5. Block diagram of the implemented method.

4. Developed device

The main goal of the developed device is to acquire the sound, and add to this sound the synthesized noise. Thus, the synthesized signal contains the sum of the processed signal and generated noise. Then, the algorithm proposed in the previous section was implemented in a low-cost digital signal processor (DSP). The main goal of the proposed device is to mask the tinnitus using synthesized sounds. In the next subsections, the features of the most important components of the device are described.

4.1. Analog front-end

In order to reduce the size of the device and improve its performance, an audio codec is used for the analog front-end. The chosen codec is the PCM3052A[13] of Texas Instruments. This integrated circuit is a low cost, single-chip stereo audio codec. The most relevant features of this device are listed below:

- Integrated analog-to-digital converter (ADC) and digital-to-analog converter (DAC), with integrated antialiasing filter and output amplifier included, respectively.
- 24 bits data word length.
- Microphone input with preamplifier, stereo line-in input and stereo output.
- Programmable sampling rate between 16KHz and 96KHz.

Among other features, the system has an external line which is used to control the input source between the microphone amplifier and the stereo line input.

4.2. Digital Signal Processor

The chosen DSP is the dsPIC33EP512MU810 [11] from Microchip. This device has 52KB of data memory and 512KB of program memory, thus it is possible to use cross-compilers. Also, the chosen device has an Universal Serial Bus (USB) interface, which allows a communication between the device and other peripherals or a computer.

One of the greatest advantages of the dsPIC33EP512MU810 is its capability to run multiple tasks simultaneously, in this particular case those tasks are the transmission of a data segment and the processing of the previous one. This is accomplished by using the Direct Memory Access (DMA) module of the device [11], which independently operates from the main processor. In the proposed system, the signal is acquired using the codec, which sends the data to the dsPIC. Then, data is processed by the microcontroller and, finally, the processed data is sent back to the codec, which synthesizes the processed signal.

4.3. Implemented device

Due to the capability of the codec for input selection, the device has one microphone stereo input and a telecoil input. Also, it has a stereo output with a headphone amplifier. With the aim of configuring the hearing aid device with additional parameters, it has an USB interface, which can be connected to a computer. Finally, the implemented hearing aid has a SPI interface which can be used to connect a SD card with external configuration parameters, or to link the device with another hearing aid. Also, due to the amount of data and program memory, a set of very important features in a hearing aid can be implemented simultaneously.

A block diagram of the built system is presented in Figure 6.

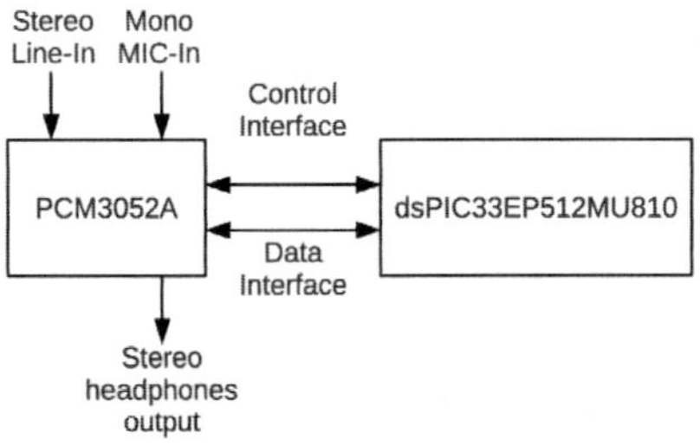

Figure 6. Block diagram of the implemented device.

Due to the packaging of the integrated circuits, the device has reduced dimensions (5cmx7cmx1,5cm) which is important for a portable device. Also, the device requires only one supply of 6V to operate and, current consumption is about 100mA. Figure shows the implemented device. A photography of the device is shown in Figure 7.

Figure 7. Developed Device.

5. Experiments

With the aim of compare the methods, a set of eight audios with a simulated tinnitus ($f = 1000Hz$) were generated. This frequency was chosen because is the most sensible region of the hearing range. Then, in each audio a tinnitus masking method was generated. The methods under study are listed below:

- **1**. Additive white gaussian (AWG) noise.
- **2**. AWG noise filtered using a bandpass filter parameterized with the tinnitus masking threshold (proposed method).
- **3**. No masking (reference).
- **4**. AWG noise filtered using a bandpass filter with $f_C = 1000Hz$ and $B_P = 625Hz$.
- **5**. AWG noise filtered using a bandpass filter with $f_C = 1000Hz$ and $B_P = 1750Hz$.
- **6**. AWG noise filtered using a bandpass filter with $f_C = 1000Hz$ and $B_P = 312Hz$.

Then, in order to analyze the performance of the methods in operational conditions, the masking methods were evaluated with additional prerecorded words. With this aim, the set of phonetically balanced words of Tato[15] were used. Those words are widely used in Argentina to perform hearing tests. Results of this experiment are shown in Table 1, where volunteers were asked about the intelligibility of each word (MOS-I), degree of tinnitus masking (MOS-M), and transcription of each word. This set of features are analyzed for the simulated configurations.

Subjective experiments were done in order to test the performance of the proposed algorithm. 15 volunteers (which accepted an informed consent) were asked to listen to the synthesized algorithms in random order. Listeners were asked to judge the tinnitus masking level of the sounds using a 5-point scale, from 1 (tinnitus is not masked) to 5 (tinnitus is completely masked). They also evaluated the intelligibility of the synthesized audios. In the same way, volunteers judged the intelligibility of the word contained into the audio using a 5-point scale, from 1 (word is unintelligible) to 5 (word is completely intelligible). Finally, a transcription of each perceived word was required.

Table 1 shows the results of the experiment. In the second column of Table 1 it can be seen that, except for the reference method, all the implemented methods present similar masking scores. In the reference method, because it does not apply any masking method, the score shows that the tinnitus is completely perceived. On the other hand, in the third column of Table 1, it can be seen that the proposed method and the reference method have similar scores of intelligibility, which are also higher than the other implemented methods. A similar conclusion may be obtained from the fourth column of the table. Then, it is possible to conclude that the proposed masking method has a similar performance to other implemented methods, and also, the intelligibility of spectrally adjacent sounds is similar to the reference method, in which no masking method was applied. In words sinthesized using the other masking methods listeners had generally problem to distinguish between m and n , and also between v and d sounds, which have components in the range of the generated noise.

In Figures 8 and 9, results of Table 1 are shown using boxplots. In Figure 8, it can be seen that, excluding the reference, all the implemented methods have similar scores in the masking test. On the other hand, Figure 9 shows results of the intelligibility test. It can be seen that the proposed method and the reference have a similar performance.

A set of experimental measurements were also made to test the performance of the system. The sampling frequency used in the experiment is 16.383kHz. In the first test, the processing time of each process and the acquisition time for a frame of 256 samples were measured.

Table 1. Results of masking and intelligibility mean opinion scores, and % of transcription of each method.

Method	MOS-M	MOS-I	Transcription
1- AWGN	3.80	3.45	77%
2- Proposed	**3.90**	**3.85**	86%
3- Reference	1.50	3.95	86%
4- AWGN filtered using a BPF ($f_C = 1000Hz$, $B_P = 625Hz$)	3.90	3.35	68%
5- AWGN filtered using a BPF ($f_C = 1000Hz$, $B_P = 1750Hz$)	3.85	3.25	63%
6- AWGN filtered using a BPF ($f_C = 1000Hz$, $B_P = 312Hz$)	4.05	3.30	50%

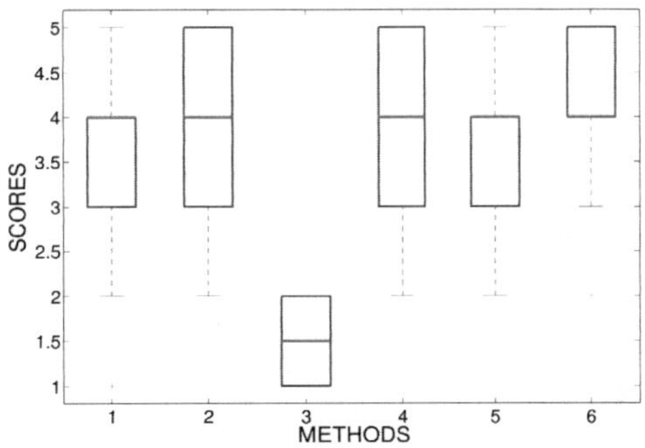

Figure 8. MOS-M boxplots of implemented methods.

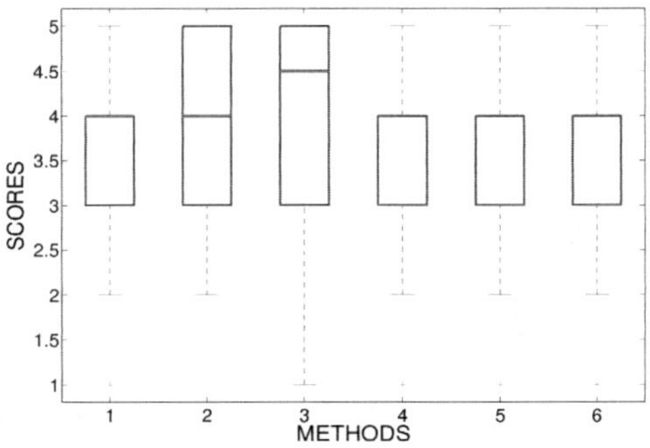

Figure 9. MOS-I boxplots of implemented methods.

The sum of communication ($61\mu s$), the noise synthesizer algorithm ($1.5ms$) and other processing times ($2.07ms$), is lower than the acquisition time($15.63ms$). Thus, the system operates in real-time.

The available DMA memory is totally used, and only 37% of RAM memory and 4% of ROM memory are used. Thus, there is a good deal of remnant data memory and program memory. Therefore, it is possible to implement additional functions in the device. Also, the proposed algorithm requires less data memory than others noise synthesizers used in hearing aid devices. It is possible to conclude that other tinnitus treatment methods can also be implemented in the device.

6. Conclusions

In this work, a new tinnitus treatment algorithm was presented. It was demonstrated that the proposed method, which masks the tinnitus through filtered noise using a bandpass filter which was set using the absolute threshold of hearing of the tinnitus, at least has the same masking performance of other widely used methods, and also, the proposed technique improves the perception of spectrally adjacent sounds to the tinnitus.

Additionally, a digital assistive listening device which uses digital signal processing techniques to obtain the most important features of a high-end commercial assistive device was presented.

Finally, it was demonstrated that it is possible to successfully implement the tinnitus treatment algorithm in the developed device. This is very important because this type of methods must be applied using a hearing aid device.

The proposed hearing aid system in a low scale of production has a material cost many times lower than the final cost of a commercial high-end hearing aid. Thus, the developed device could satisfy a need of the Society. As a further work, additional features will be implemented.

Acknowledgment
The authors would like to thank to CONICET, National University of Mar del Plata and University FASTA for the funding received through projects *Secuencias caóticas digitales en procesamiento y encriptado de señales*, *Análisis de señales destinados a mejorar la calidad de vida* and *Estudio de la relación entre un Sistema informático de Análisis acústico de la voz y la Escala de evaluación cualitativa RASATI*, respectively. The authors would like to thank the cooperation of the Phonoaudiologist Lila Rovedo for her assistance in the work.

References

[1] S. Holmes and N. D. Padgham, *Ringing in the Ears: Narrative Review of Tinnitus and Its Impact*, Biological Research for Nursing, Vol. 13, No. 1, pp. 97-108, 2011.
[2] J. J. Shea and J. R. Emmett, *The medical treatment of tinnitus*, The Journal of Laryngology and Otology. Supplement Vol 1981, No. 4, pp.130-138, 1981.
[3] A. B. Elgoyhen and B. Langguth, *Pharmacological approaches to the treatment of tinnitus*, Drug Discovery Today, Vol. 15, No. 7, pp. 300305, 2010.
[4] J. A. Henry, T. L. Zaugg, P. J. Myers and M. A. Schechter, *Using therapeutic sound with progressive audiologic tinnitus management*, Trends on amplification, Vol. 12, No. 3, pp.188-209, 2008.
[5] A. J. Schleuning and R. M. Johnson, *Use of masking for tinnitus*, The International Tinnitus Journal, Vol.3, No.1., pp. 25-29, 1997.
[6] H. Feldmann, *Masking of tinnitus - Historical remarks*. In Proceedings III International Tinnitus Seminar, pp. 210-213, 1987.
[7] H. Fletcher, *Auditory Patterns*, Modern Physics Vol. 12, No. 1, pp. 4765, 1940.
[8] T. Painter and A. Spanias, *Perceptual coding of digital audio*. In Proceedings of the IEEE, Vol. 88, No. 4, pp. 451513, 2000.
[9] Widex Inc.*http://www.widex.com/*, Lynge, Denmark.
[10] Siemens Hearing Instruments, *Introduction to tinnitus management*, http://hearing.siemens.com, 2008.
[11] Microchip Inc., *dsPIC33EP512MU810 Data Sheet, High Performance 16-bit Digital Signal Controllers*. http://www.microchip.com/, 2011.
[12] A. J. Uriz, P. D. Agüero, J. Castiñeira Moreira, J. C. Tulli and E. L. González, *Simulador de deficiencias auditivas*, Revista Iberoamericana de Automtica e Informtica e Industrial, Vol. 8 , No. 2 , pp. 52-62 , 2011.
[13] Texas Instruments,*PCM3052A Datasheet*, http://www.ti.com/, 2005.
[14] International Organization of Standardization, *ISO/IEC JTC1/SC29/WG111 MPEG IS13813(1994): Generic Coding of Moving Pictures and Associated Audio*, International Organization of Standardization. Ginebra, Suiza, 1994.
[15] G. De Sebastian, J.J. Badaroco and D.G. Postan, *Audiología Práctica*, Ed. Médica Panamericana, 1992.

Identification of drowsiness and alertness conditions by means of Spectral F-Test applied to pupillometric signals

V C G Soares[1], J K S Souza[1], G E Ginani[2], S Pompéia[2], C J Tierra-Criollo[1,3] and D B Melges[1]

[1] Graduate Program in Electrical Engineering - Federal University of Minas Gerais - Av. Antônio Carlos 6627, 31270-901, Belo Horizonte, MG, Brazil
[2] UNIFESP/ Psychobiology Department, Federal University of São Paulo, São Paulo, Brazil
[3] UFRJ/Biomedical Engineering Program, Federal University of Rio de Janeiro, Rio de Janeiro, Brazil

valcienesoares@yahoo.com.br, jkssbh@gmail.com, giuliano.ginani@gmail.com, spompeia@gmail.com, carjuliot@gmail.com, danilomelges@gmail.com

Abstract. The autonomous regulation of pupil size provides an objective physiological measure of alertness level. Pupil diameter decreases and its fluctuations increase during drowsiness. In this work, a statistical based technique in frequency domain, known as Spectral F-Test (SFT), was applied in order to compare the power of pupillometric signals measured in two conditions, wakefulness and sleepiness, from eleven volunteers. SFT was calculated based on Welch Periodograms of time series of pupil diameter (TSPD) signals. The median power for TSPD in two conditions and each volunteer at frequencies bellow 0.8 Hz were compared by means of the paired Wilcoxon signed-rank test. The percentages of volunteers for whom power during drowsiness were statistically higher than power during alertness achieved 100% for several frequencies bellow 0.2 Hz. For frequencies from 0.2 to 0.8 Hz, the percentages presented high variability, fluctuating from 40 to 90%. The Wilcoxon test indicated that median power of TSPD during drowsiness was higher than TSPD during alertness for frequencies bellow 0.2 Hz presented statistical difference. Hence, the SFT showed to be a suitable tool for drowsiness/alertness differentiation that could be applied to sleep studies.

1. Introduction

Multiple Sleep Latency Test (MSLT) is a current gold standard method to evaluate somnolence, but its main drawback is the expense of the test, which is based on polysomnography [1]. This procedure consists of recording physiological signals during sleep, such as the electroencephalogram (EEG), electro-oculogram (EOG), electromyogram (EMG), electrocardiogram (ECG), capnogram, oximetry and a video recording of the exam. It also requires patients to remain all day in laboratory, the presence of a specialized professional on sleep analysis during all exam, and an expensive infrastructure.

On the other hand, pupillometry has been used as an alternative method for objective quantification of sleep. This method consists of recording pupil diameter oscillations [1–3], which is a simple and non-invasive test, cheaper than MSLT.

The measurement of pupil parameters is clinically important, since pupil size indicates, apart from adjustments to different lighting conditions and visual accommodation [4], the presence of physical or mental dysfunctions such as: multiple sclerosis [5], autonomic neuropathy [6], depression [7], Alzheimer's and Parkinson disease [8], retinitis pigmentosa [9] and glaucoma [10]. In addition, pupil size increases with cognitive effort [11].

Pupil dilation also varies between conditions of drowsiness and alertness. During alertness, pupil diameter is relatively stable, while drowsiness leads to considerable pupil size fluctuation [12].

Pupil size can be registered by a camera, using techniques of image processing, resulting in a time series of pupil diameter (TSPD). Analytical tools are applied to the TSPD in order to obtain indicators of drowsiness.

Various pupillometry studies [3,13–15] have shown that as subjects become sleepy, the pupil diameter decreases and oscillates at lower frequency and higher amplitude. This parameter, known as hippus, was also investigated by other researchers [14]. Hence, these works suggested that pupil diameter could be used as a drowsiness indicator [1,3,14,15]. Particularly, the power of pupil oscillations in low frequency components has been used for sleep deprivation identification [14–16]. However, there is no consensus about the frequencies that better translate the drowsiness-to-alertness transition, and conversely.

Lüdtke et al. [15], Wilhelm et al. [2] and McLaren et al. [1] proposed indicators based on the signal power at frequencies up to 0.8 Hz in order to objectively assess the alertness level. These three studies used the same methodology by calculating the following parameters: power of pupillary oscillations (PPO) from 0.01 to 0.8 Hz, mean pupil diameter (MPD) and pupillary unrest index (PUI); the latter measures the pupil instability [1,2,14].

Lüdtke et al. [15] concluded that there is difference in PPO and PUI when these parameters are compared for alert and sleepy subjects, suggesting these variables would be useful to quantify drowsiness. After that, Wilhelm et al. [2] used the three parameters (PPO, MPD and PUI) to differentiate successfully healthy patients and hypersomniacs. Moreover, McLaren et al. [1] found correlation between Sleep Latency (MSLT) and the parameters PPO and PUI. However, any variable was able to substitute MSLT [1]. On the other hand, PPO and PUI are still the main parameters studied today [5,12,16–20]. Different from these parameters, an alternative approach using frequency components from the power spectrum density (PSD) of pupillary response has been suggested [20,21]. In these studies, three self-report questionnaires were used to measure sleepiness: the Stanford Sleepiness Score (SSS), Visual Analogue Scales (VAS) and the Epworth Sleepiness Scale (ESS). These questionnaires were associated with the power in four different frequency ranges of pupillary oscillations in order to discriminate between two groups of different sleepiness level.

In the present paper, a statistical based technique in frequency domain [22], known as Spectral F-Test (SFT), was applied in order to compare the power of pupillometric signals in two conditions: alertness and drowsiness. Statistical difference between median power in both conditions was assessed in order to establish the best frequencies to distinguish them.

2. Methods

2.1. Signal acquisition

The pupillogram was recorded from eleven male (aged from 21-33 years) volunteers without a history of neurological or psychiatric diseases, using the pupillometer developed at Biomedical Engineering Laboratory of Federal University of Minas Gerais (UFMG) [23]. The participants were awake and continuously supervised during the experiment, which was performed in Department of Psychobiology

(Federal University of São Paulo - UNIFESP). The pupillometry video was recorded (sampling rate: 120 samples per second) at 8 am after a night of sleep (wake session) and again after 24 h of continuous alertness (sleep deprived condition) for 10 minutes each (Table 1). During pupil size estimation, participants carried out a Psychomotor Vigilance Test (PVT) [24], which involves pressing a button every time they saw a visual stimulus generated by a LED stimulator.

Factors that could interfere with the drowsiness quantification such as caffeine and medication ingestion were used as exclusion criteria [23]. The temperature and light were controlled and the volunteers had access to external stimuli such as day time and day light. This research was approved by local ethics committee (UNIFESP; 0763/10) and all volunteers signed written informed consent [23].

Table 1. Signal characteristics

Section	Time	Day	Signal characteristics
1st	8:00	1st	Record after a normal sleep night
2nd	8:00	2nd	Record after 24 hours* of sleep deprivation

*Considering the Circadian Rhythm.

2.2. *Pre-processing*

After video acquisition, pupil size was estimated leading to time series of pupil diameter (TSPD). When blinks occurred (pupillary diameter = 0), this measure was adjusted by interpolation [23].

Considering that for the first two minutes of recording the volunteers were alert even for the 2th session (drowsiness condition), this segment was discarded. The TSPD was filtered by a 2 Hz low-pass, 4th order, zero phase, Butterworth filter.

2.3. *Spectral F-Test (SFT)*

Spectral F-Test (SFT) assess whether two sample spectra are from populations with identical power spectrum [25]. In this study, the SFT was calculated as the ratio between the power of filtered TSPD in drowsiness condition (2th session) by the power in alertness condition (1st session) [22]:

$$\phi(f) = \frac{\hat{P}_{yy}(f)}{\hat{P}_{xx}(f)} = \frac{\frac{1}{M_y}\sum_{i=1}^{M_y}|Y_i(f)|^2}{\frac{1}{M_x}\sum_{i=1}^{M_x}|X_i(f)|^2} \tag{1}$$

Where $\hat{P}_{yy}(f)$ and $\hat{P}_{xx}(f)$ are the periodograms of signal $y[k]$ and $x[k]$, respectively, f is frequency, "^" superscript denotes estimation, $X_i(f)$ and $Y_i(f)$ are respectively, the Fourier Transform of i-th epoch of $x[k]$ and $y[k]$, and M is the number of epochs.

Considering that $\hat{P}_{yy}(f)$ belongs to the same population that $\hat{P}_{xx}(f)$, one can establish the null hypothesis (H_0) of power equality. Assuming that TSPD, in both conditions, is a zero mean Gaussian random signal, it can be showed that the probability density function of the ratio in equation (1) tends asymptotically to an F-distribution with $2M_x$ and $2M_y$ degrees of freedom [25]. Hence, critical values for H_0 and a given significance level α [22], can be calculated as:

$$\phi_{crit} = F_{crit}(2M_x, 2M_y, \alpha) \tag{2}$$

Statistical difference between powers of distinct sessions is given when SFT estimation exceeds the critical value for a frequency $f(\phi(f) > \phi_{crit})$, which leads to the null hypothesis rejection of power equality.

2.4. Processing

After pre-processing, the Power Spectral Density (PSD) for signals in both conditions, $\hat{P}_{yy}(f)$ and $\hat{P}_{xx}(f)$, were calculated using Welch Periodogram for 8-minutes recordings, windowed in 2-minutes epochs (spectral resolution of 0.008 Hz) with 5% overlapping, leading to $M_x = M_y$ =61 epochs. The significance level α=0.05 was used in equation (2) for critical value calculation.

3. Results

Figure 1 shows pupillograms (TSPD) for subject #11 in two different conditions, after filtering process. The drowsiness signal clearly presents higher amplitude fluctuations than alertness one. PSD of TSPD for the same subject shows that the pupillogram power at lower frequencies, mainly below 0.2 Hz, increases during somnolence (Figure 2). This behavior described for both time and frequency for volunteer #11 was similar for other individuals.

The percentages of volunteers for whom power during drowsiness were statistically higher than the power during alertness, estimated by means of SFT for frequencies up to 0.8 Hz are presented in figure 3. It is worth noting, below 0.2 Hz, several frequencies presented percentages of 100%, whereas above this frequency, only two components (0.337 and 0.747 Hz) showed this rate. Percentages higher than 90% are also more common below 0.2 Hz.

In order to verify the frequency range that is more suitable to differentiate drowsiness from alertness condition, a paired Wilcoxon signed-rank test was applied to PSD for all volunteers. P-values for the test are showed in figure 4 that evidences significant differences between median powers for two conditions powers, particularly for the frequency range below 0.2 Hz (p-values lower than 0.01, except for 0.08 Hz). For frequencies over 0.2 Hz, the p-values presented high variability, with values often over 0.01.

4. Discussion

The phenomenon of pupillary fluctuations, known as “waves of fatigue”, was firstly described by Lowenstein et al. in 1963 [18 apud 1]. Later, other studies [3,13,14,15] involving pupillary activity became important references when they noted the increasing of pupillary oscillations at low frequencies.

Similar to Lüdtke et al. [15] work, in this study, power of pupil oscillations in low frequency range was used to differentiate two conditions: alertness and drowsiness for eleven volunteers. However, the power of time series of pupil diameter (TSPD) was compared by means of a statistical test, the Spectral F-Test (SFT), applied to two different conditions, for each volunteer. The SFT results showed that the 2th session (sleep deprivation) presented statistically significant higher power than the 1st in frequency range below 0.8 Hz for the majority of volunteers. These findings agree with other works that employed different parameters based on pupil oscillations power [1,2,14]. However, the paired Wilcoxon signed-rank test applied to PSD values of both conditions for all volunteers showed that frequencies up to 0.2 Hz presented significant differences in median power. Hence, frequencies below 0.2 Hz, instead of 0.5 Hz [12] or 0.8 Hz [1, 2,14], should be used to differentiate alert from sleepy subjects. In fact, these frequencies presented the highest percentages of volunteers for whom power during drowsiness were statistically higher than during alertness. These percentages were often above 90%. For frequencies from 0.2 to 0.8 Hz, the percentages presented high variability, fluctuating from 40 to 90%.

It should be borne in mind, however, that the participants in the present study were carrying out a cognitive task while the pupil size was measured. Hence, it is possible that cognitive effort under situations of different arousal levels led to changes observed here.

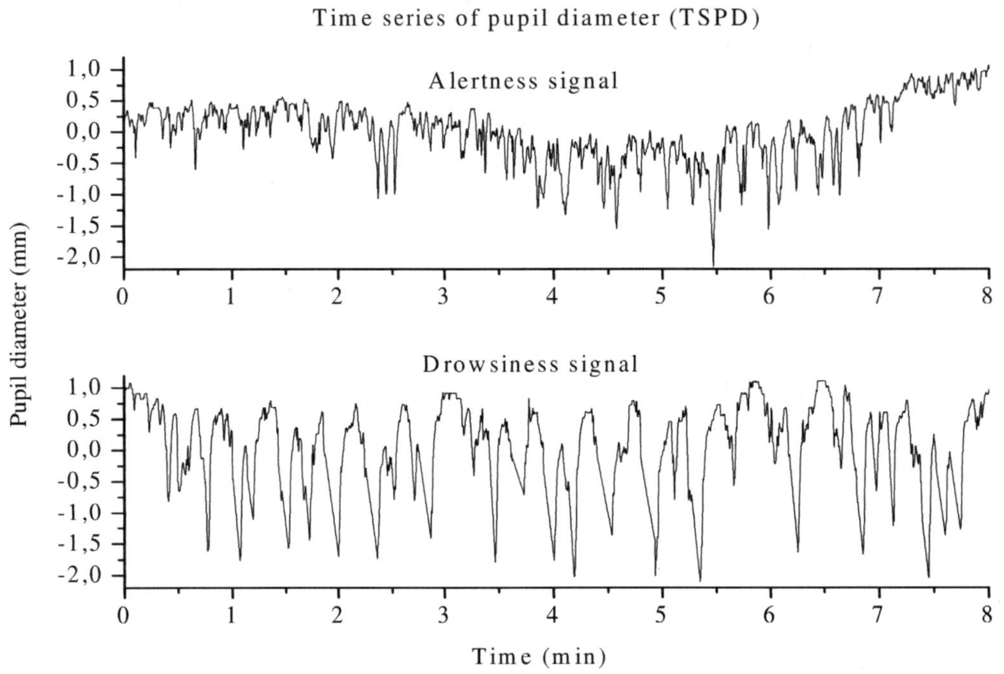

Figure 1. TSPD for subject #11 in two states: alertness and drowsiness.

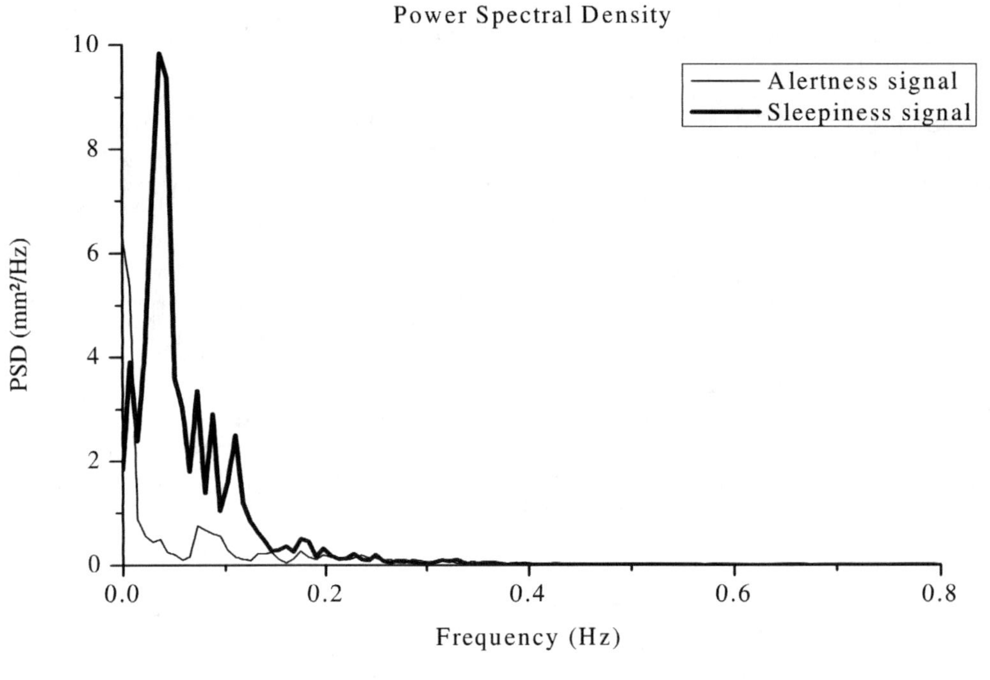

Figure 2. PSD for subject #11 at two states: alertness and sleepiness.

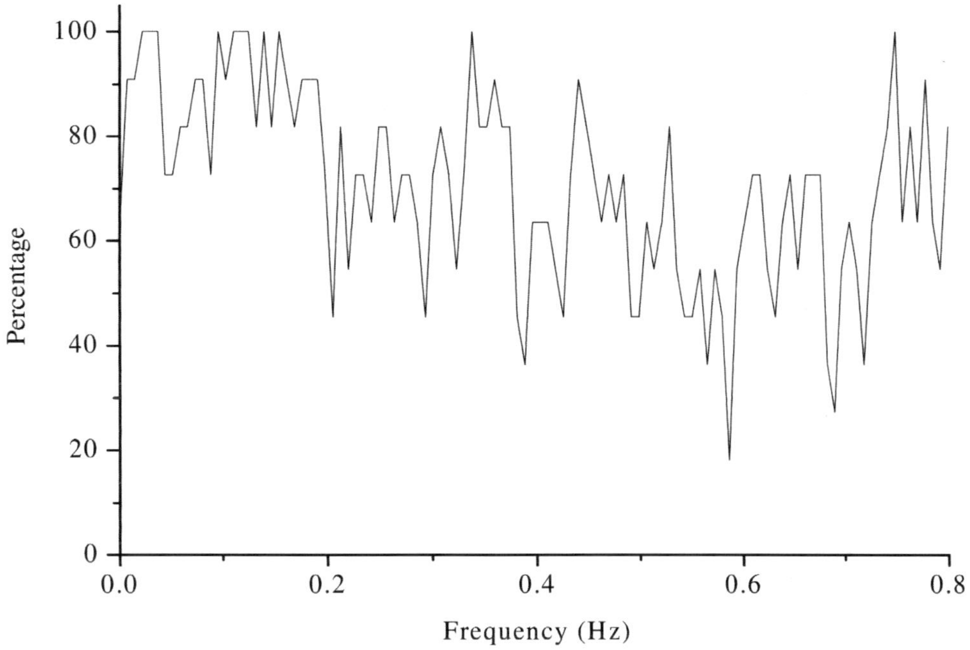

Figure 3. Percentages of volunteers for whom power during drowsiness were statistically higher than the power during alertness, estimated by means of SFT.

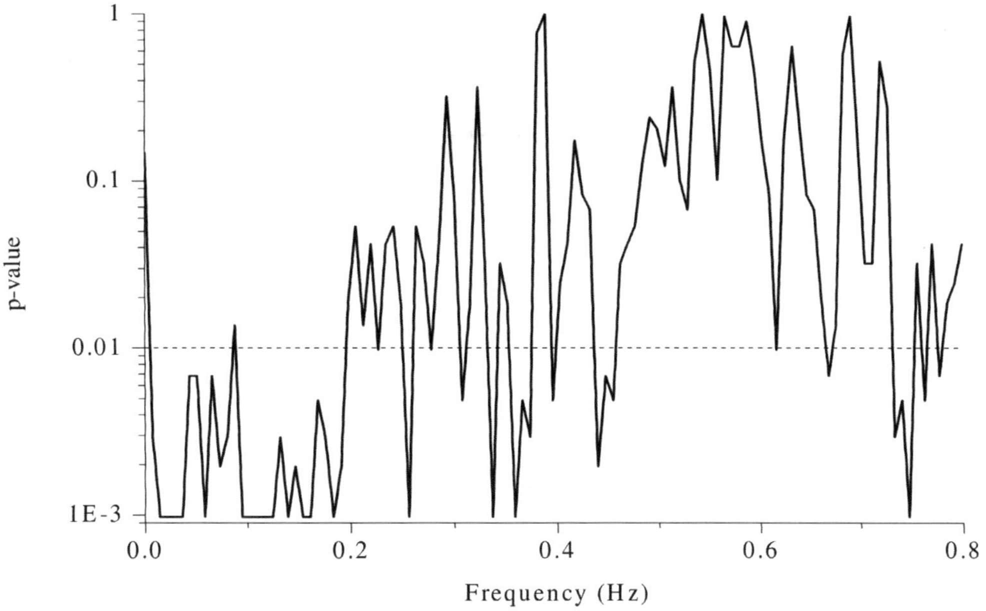

Figure 4. Result of Wilcoxon paired test indicating frequencies bellow 0.2 Hz presented statistical difference between alertness and drowsiness. Y-scale in logarithmic values.

5. Conclusion

This work proposed a novel approach to differentiate alertness from drowsiness condition based on a statistical test. Spectral F-Test showed to be a suitable technique for this purpose. Moreover, a more limited frequency band was identified as the range where the power changes occur. Thus, the best frequencies for evaluating drowsiness seem to be below 0.2 Hz. As conclusion, the analysis of pupillometric signal using SFT and suitable frequencies can be an important tool for sleep studies, since this technique allowed the discrimination of data from two sessions corresponding to different physiological states. Further, this could be applied to identify variations in pupil size associated with circadian rhythm (metabolic cycle involving organic processes that repeat every day, including sleep-wake cycle), by analyzing other sessions performed at different hours of the experiment.

Finally, future studies should focus on the dynamic time evolution of SFT in order to identify drowsiness-to-alertness transitions, and conversely.

Acknowledgments

To CAPES, FAPESP (process 2011/01286-0), CNPq and FAPEMIG for financial support.

References

[1] McLaren J W, Hauri P J, Lin S-C and Harris C D 2002 Pupillometry in clinically sleepy patients. *Sleep Med* **3** 347–52

[2] Wilhelm H, Lüdtke H and Wilhelm B 1998 Pupillographic sleepiness testing in hypersomniacs and normals. *Graef Arch Clin Exp* **236** 725–9

[3] Morad Y, Lemberg H, Yofe N and Dagan Y 2000 Pupillography as an objective indicator of fatigue. *Curr eye res* **21** 535–42

[4] Kandel E R, Schwartz J H and Jessell T M 2000 *Principals of neural science* (New York: McGraw-Hill)

[5] Frauscher B, Egg R, Brandauer E, Ulmer H, Berger T, Poewe W and Högl B 2005 Daytime sleepiness is not increased in mild to moderate multiple sclerosis: a pupillographic study. *Sleep Med* **6** 543–7

[6] Ferrari G L, Marques J L B, Gandhi R a, Heller S R, Schneider F K, Tesfaye S and Gamba H R 2010 Using dynamic pupillometry as a simple screening tool to detect autonomic neuropathy in patients with diabetes: a pilot study. *Biomed Eng Online* **9** 26

[7] Siegle G J, Steinhauer S R and Thase M E 2004 Pupillary assessment and computational modeling of the Stroop task in depression. *Int J Psychol* **52** 63–76

[8] Fotiou D F, Stergiou V, Tsiptsios D, Lithari C, Nakou M and Karlovasitou A 2009 Cholinergic deficiency in Alzheimer's and Parkinson's disease: evaluation with pupillometry. *Int J Psychol* **73** 143–9

[9] Kardon R, Anderson S C, Damarjian T G, Grace E M, Stone E and Kawasaki A 2011 Chromatic pupillometry in patients with retinitis pigmentosa. *Ophthalmology* **118** 376–81

[10] Link B, Jünemann A, Rix R, Sembritzki O, Brenning A, Korth M and Horn F K 2006 Pupillographic measurements with pattern stimulation: the pupil's response in normal subjects and first measurements in glaucoma patients. *Invest Ophth Vis Sci* **47** 4947–55

[11] Stanners R, Coulter M, Sweet A and Murphy P 1979 The pupillary response as an indicator of arousal and cognition *Motiv Emotion* **3**

[12] Wilhelm B J, Widmann A, Durst W, Heine C and Otto G 2009 Objective and quantitative analysis of daytime sleepiness in physicians after night duties *Int J Psychol* **72** 307–13

[13] Lowenstein O and Loewenfeld I E 1958 Electronic Pupillography: A New Instrument and Some Clinical Applications *Arch Ophthalmol* **59** 352–63

[14] McLaren J W, Erie J C and Brubaker R F 1992 Computerized analysis of pupillograms in studies of alertness. *Invest Ophth Vis Sci* **33** 671–6

[15] Lüdtke H, Wilhelm B, Adler M, Schaeffel F and Wilhelm H 1998 Mathematical procedures in data recording and processing of pupillary fatigue waves. *Vision Res* **38** 2889–96

[16] Souza J K S, Ginani G E, Pompéia S, Baron J and Criollo C J T 2012 Identificação de privação de sono utilizando pupilômetro *XXIII Congr Bras Eng Biomed* 5

[17] Warga M, Lüdtke H, Wilhelm H and Wilhelm B 2009 How do spontaneous pupillary oscillations in light relate to light intensity? *Vision Res* **49** 295–300

[18] Prasad B, Choi Y K, Weaver T E and Carley D W 2011 Pupillometric assessment of sleepiness in narcolepsy. *Frontiers in Psychiatry / Frontiers Research Foundation* **2** 35

[19] Eggert T, Sauter C, Popp R, Zeitlhofer J and Danker-Hopfe H 2012 The Pupillographic SleepinessTest in adults: effect of age, gender, and time of day on pupillometric variables. *Am J Hum Biol* **24** 820–8

[20] Nakayama M, Yamamoto K and Kobayashi F 2012 Estimation of sleepiness using pupillary response and its frequency components *Int J Bioinformatics Res Appl* **8** 342–65

[21] Nakayama M and Eye A 2008 Estimation of Sleepiness using Frequency Components of Pupillary Response *BioCAS* 357–60

[22] Melges D, Infantosi A and Miranda de Sá A 2012 Frequency-Domain Objective Response Detection Techniques Applied to Evoked Potentials: A Review *Applied Biological Engineering - Principles and Practice* (Rijeka, Croatia: InTech) pp 47–84

[23] Souza J K S, Ginani G, Pompéia S and Tierra-Criollo C 2013 Equipamento portátil para ensaios simultâneos de teste de vigilância psicomotora e pupilometria *Rev Bras Eng Biomed* **29** 97–109

[24] Dinges D F and Powell J W 1986 Microcomputer analyses of performance on a portable, simple visual RT task during sustained operations *Behav Res Meth Ins C* **17** 652–5

[25] Infantosi A F C, Miranda de Sá A M F L and Sft A T S F 2007 A Statistical Test for Evaluating the Event-Related Synchronization / Desynchronization and its Potential use in Brain-Computer-Interfaces *IV Latim Am C Biomed Eng* **2** 1122–6

Journal of Physics: Conference Series **477** (2013) 012028 doi:10.1088/1742-6596/477/1/012028

Objective detection of Steady State Visual Evoked Responses for different lighting conditions

A M Tannus[1], C J Tierra-Criollo[1,2], D B Melges[1]

[1]Graduate Program in Electrical Engineering - Federal University of Minas Gerais - Av. Antônio Carlos 6627, 31270-901, Belo Horizonte, MG, Brazil
[2]Biomedical Engineering Program - Federal University of Rio de Janeiro – Av. Horácio de Macedo 2030, 21941-914, Rio de Janeiro, RJ, Brazil

E-mail: alexandretannus@gmail.com; carjuliot@gmail.com;danilomelges@gmail.com

Abstract: This work aims at evaluating the performance of two Objective Response Detection (ORD) techniques: the Magnitude-Squared Coherence (MSC) and the Multiple Coherence (MC) for detecting Steady State Visual Evoked Response (SSVEP) for two different lighting conditions (scotopic and photopic). EEG signals were collected (using a wireless headset) from ten volunteers without history of neurological diseases during stimulation with white LED flickering at frequencies of 6, 7, 8, 9, 10 and 11 Hz. The experimental protocol was composed by 13 sessions in a dark room and 13 in a bright one. The frequency order was randomized. The MSC was applied to derivations [P7], [P8], [O1] and [O2], and MC to the combinations of leads [O1-O2], [O1-O2-P7], [O1-O2-P8] and [O1-O2-P7-P8]. Both estimates were calculated using $M = 11$ epochs. The detection rates for MSC presented an important decrease at photopic condition compared to the scotopic one for all derivations and all frequencies. The employment of MC to [O1-O2-P7-P8] allowed overcoming this limitation, leading to detection percentages of at least 80% for all frequencies (except 9 Hz). These results show that MC is a promising technique for building SSVEP Bran Computer Interfaces (BCI).

1. Introduction

Assistive technology development has grown in the last years, and it is mainly directed to people with severe motor impairment, caused by spinal cord injury, cranial trauma and degenerative neurological diseases such as multiple sclerosis [1], amyotrophic lateral sclerosis (ALS) [2] and primary lateral sclerosis [3]. In Brazil, according to IBGE (Geography and Statistics Brazilian Institute), there are around 13 million people who declare to have a motor disability [4]. From these people, 4.5 million present severe disabilities.

In this context, the brain-computer interface (BCI) is a class of assistive technology that allows these individuals to control devices or assess cognitive functions [5] employing the brain electrical activities (electroencephalogram – EEG), usually during some kind of stimulation (evoked potential - EP) [6]. Generally, the EP is commonly categorized in transient or steady-state. The EP is classified as transient when the response to a stimulus finishes before another stimulus is presented, and as steady-state, on the contrary.

The signals employed in a BCI can be obtained by invasive or non-invasive techniques. The invasive methods can require the use of needle or even surgically implanted electrodes [7]. These techniques include electrocorticogram (ECoG) and methods for recording action potentials (AP) and local field potentials (LFP) [7]. Among the non-invasive methods there are: the EEG, the magnetoencephalogram (MEG), and the functional magnetic resonance imaging (fMRI). However, the most used non-invasive techniques are the EEG and evoked potentials that have low-cost and simple-execution [8].

Particularly, the visual evoked potential (VEP) produces cortical response in the same frequency of the stimulation and, therefore, has been used for many different BCI applications, such as controlling a computer mouse [9], a car [10] and a robotic hand [11]. The VEP can be elicited by different types of stimuli, such as checkerboards, flashing images and flashing LEDs [12,13]. The latter is known to produce cortical response with high signal-to-noise ratio [14].

Statistics based techniques, known as Objective Response Detection (ORD), have been used for identify cortical responses to sensory stimulation, including auditory [15], somatosensory [16,17] and visual [18]. Among these techniques, the Magnitude-Squared Coherence (MSC) has presented promising results, as well as its multivariate version, the Multiple Coherence (MC).

Hence, this work aims at evaluating the performance of MSC and MC for detecting response to visual stimulation for two different lighting conditions (scotopic and photopic) and investigating whether there is habituation to the stimulus.

2. Material and methods

2.1. Magnitude Squared Coherence (MSC)

The MSC is the square module of the coherence function and represents the parcel of the squared mean value of the measured EEG signal $y[k]$ caused by stimulation $x[k]$ for a frequency f. A coherence estimator can be calculated based on the discrete windowed signal by [19]:

$$\hat{\kappa}^2(f) = \frac{\left|\sum_{i=1}^{M} Y_i(f)\right|^2}{M\sum_{i=1}^{M}\left|Y_i(f)\right|^2} \quad (1)$$

where $Y_i(f)$ is the Discrete Fourier Transform of the ith epoch of the EEG signal and M is the number of EEG epochs. When there is no response to stimulation, $\hat{\kappa}^2(f)$ tends to zero and, when there is consistent response, $\hat{\kappa}^2(f)$ tends to 1.

In order to infer about the presence/absence of the response, the critical value, which consists of a detection threshold for the MSC, can be calculated for M EEG epochs and the significance level α (false positive rate) by [20]:

$$\hat{\kappa}^2_{crit} = 1 - \alpha^{\frac{1}{M-1}} \quad (2)$$

When the estimate value exceeds the critical one ($\hat{\kappa}^2(f) > \hat{\kappa}^2_{crit}$), there is detection for such a frequency f.

2.2. Multiple Coherence (MC)

The multiple coherence estimate between a stimulus $x[k]$ and a set of N signals $y[k]$ is given by [21]

$$\hat{\kappa}^2_N(f) = \frac{V^H(f)\hat{S}_{yy}^{-1}V(f)}{M} \quad (3)$$

Where M is the number of epochs, H is the Hermitian operator and $V(f)$ and $\hat{S}_{yy}^{-1}$ are, respectively

$$V(f)=\left[\sum_{i=1}^{M} Y_{1i}^{*}(f) \quad \sum_{i=1}^{M} Y_{2i}^{*}(f) \quad \ldots \quad \sum_{i=1}^{M} Y_{Ni}^{*}(f)\right]^{T} \quad (4)$$

and

$$\hat{S}_{yy}^{-1}=\begin{bmatrix} \hat{S}_{y_1y_1}(f) & \hat{S}_{y_1y_2}(f) & \ldots & \hat{S}_{y_1y_N}(f) \\ \hat{S}_{y_2y_1}(f) & \hat{S}_{y_2y_2}(f) & \ldots & \hat{S}_{y_2y_N}(f) \\ & & & \\ \hat{S}_{y_Ny_1}(f) & \hat{S}_{y_Ny_2}(f) & \ldots & \hat{S}_{y_Ny_N}(f) \end{bmatrix} \quad (5)$$

where $Y_{ji}(f)$ is the Fourier Transform of the ith epoch of the signal $y[k]$, * denotes the complex conjugate and $\hat{S}_{y_py_q}(f)$ is the cross-spectrum of derivations p and q.

The critical value of MC for a significance level α, M EEG epochs and N derivations, can be calculated as:

$$\hat{\kappa}_{N\mathrm{crit}}^{2}=\frac{F_{\mathrm{crit}\alpha,2M,2(M-N)}}{F_{\mathrm{crit}\alpha,2M,2(M-N)}+(M-N)N^{-1}} \quad (6)$$

The detection for a frequency f is based on the same principle used for MSC, that is, when the estimated value is higher than the critical value ($\hat{\kappa}_{N}^{2}(f)>\hat{\kappa}_{N\mathrm{crit}}^{2}$).

2.3. EEG Data Acquisition
EEG signals were collected from 14 derivations, AF3, F7, F3, FC5, T7, P7, O1, O2, P8, T8, FC6, F4, F8, AF4, using the Emotiv Epoc^{TM} wireless headset (Emotiv). The derivations were filtered from 0.2 to 45 Hz, sampled at 128 Hz (14-bit resolution). EEG was acquired from 10 adult volunteers (7 men and 3 women, aged from 20 to 35 years old), without history of neurological diseases. This research was approved by the Ethics Committee of Federal University of Minas Gerais (UFMG) - protocol n. 11525613.4.0000.5149 - and all volunteer signed written informed consent.

2.4. Experimental Protocol
The volunteers were stimulated by white LEDs with the following frequencies: 6 Hz, 7 Hz, 8 Hz, 9 Hz, 10 Hz and 11 Hz. The order of the stimulation frequency was randomized. All frequencies were presented in two lighting conditions: in a dark room (luminance lower than 2 lux) and in a controlled – luminance room (around 80 lux). The stimulation was applied using equipment developed in the Biomedical Engineering Laboratory of UFMG [13].

The experimental protocol was composed by 26 sessions of two-minutes duration, 13 in a dark room and 13 in bright one, alternating spontaneous EEG and EEG during visual stimulation, as illustrated in Figure 1Figure 1. The first 13 sessions were performed in scotopic condition while the other 13 were performed in a photopic situation.

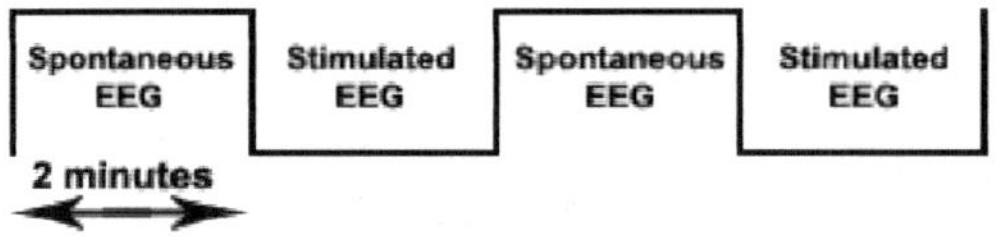

Figure 1. Experimental Protocol

2.5. Pre-processing
The EEG signals were digitally band-filtered from 2 to 40 Hz (Butterworth 6th order, zero-phase) and windowed in 4-second epochs with 50% overlapping. Then, an artefact rejection technique was applied to each window. This technique consists in calculating the standard deviation (SD) of a noisy-free EEG segment and remove EEG windows where 10% of any samples or 5% of consecutive samples exceeds ± 3 SD [17].

2.6. MSC and MC performance evaluation
The MSC and MC were employed to identify the stimulation response. $\hat{\kappa}^2(f)$ and $\hat{\kappa}_N^2(f)$ were estimated for $M = 11$ epochs. The critical values $\hat{\kappa}^2_{crit}$ and $\hat{\kappa}^2_{N\,crit}$ were calculated with α=0.05. The MSC was obtained for derivations [O1], [O2], [P7] and [P8], and the MC for the combinations: [O1-O2], [O1-O2-P7], [O1-O2-P8] and [O1-O2-P7-P8]. Based on the response identification for each frequency and all casuistry, the detection rates were obtained.

3. Results
The percentages of detection for derivations [O1], [O2], [P7] and [P8], in the scotopic condition, using MSC, are presented to 4 different sets of $M = 11$ EEG epochs in order to identify habituation effects (Figure 2). As it can be seen, no adaptation to the stimulation was evidenced for a specific frequency, i.e., there was no important decreasing in the detection rates along the experiment.

In the same figure, it is possible to note that the best detection rates occurred at 7 and 8 Hz, which achieved more than 80% for all sets in derivation [O2]. A 100% detection rate was reached in 8 Hz at [P8]. On the other hand, responses to 6 and 11 Hz presented the worst detection rate, presenting no more than 70% for all sets and all leads, except for the second set at [O2]. Frequencies of 9 and 10 Hz presented higher fluctuations in the percentages, achieving more than 80% in some sets and less than 60% in other for [O2], [P7] and [P8].

In the photopic condition, detection rates were lower than 60% for the majority of frequencies, independently of the derivation or the set considered (Figure 3). Detection rate of 100% was achieved only for 11 Hz at derivation [O2] and 10 Hz at [P7].

Detection rates for the MSC applied to the first set of 11-epochs are presented in Figure 4 for both scotopic and photopic conditions. As it can be seen, the percentages of detection decreased for the majority of frequencies in all leads, except for 10 Hz at [P7] and 11 Hz for [P8], [O1] and [O2]. However, the proportion test indicated significant difference only for [P7] at 7 and 9 Hz, and at [P8] from 6 to 9 Hz.

Figure 5 shows the percentages of detection for the Multiple Coherence applied to four different combinations of derivations, [O1-O2], [O1-O2-P7], [O1-O2-P8] and [O1-O2-P7-P8]. The 2-leads combination [O1-O2] presented rates of at least 80% at 7, 10 and 11 Hz for both conditions. For 3-leads combinations ([O1-O2-P7] and [O1-O2-P8]), higher rates than 90% were achieved at scotopic and photopic situations for 7, 8 and 10 Hz. Finally, [O1-O2-P7-P8] showed a detection rate of 100% at 8 and 10 Hz for both conditions. Moreover, only for 9 Hz, lower detection than 80% was observed in the photopic situation. However, the proportion test showed no significant difference between any detection rates, when the conditions are compared.

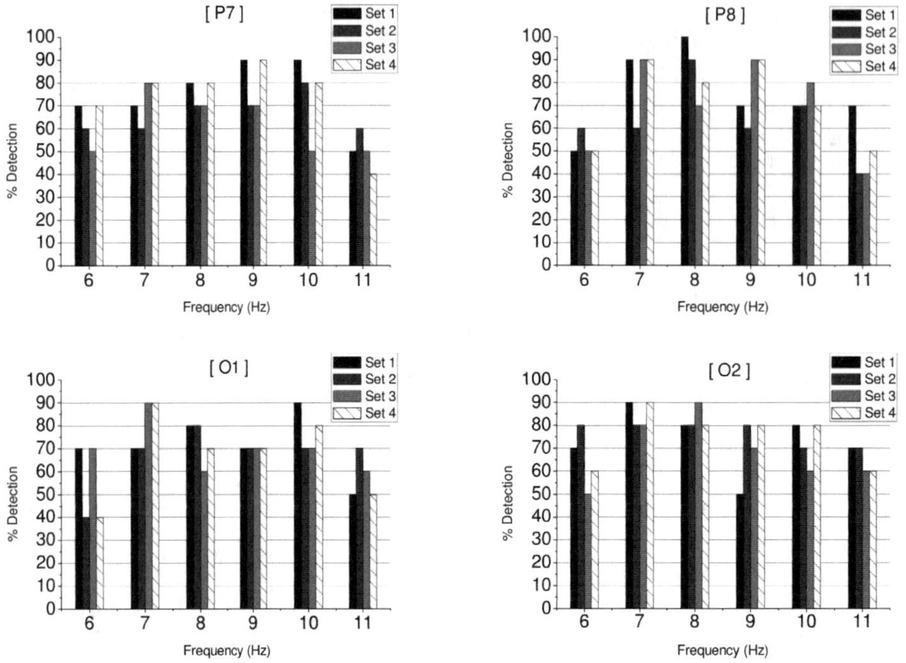

Figure 2. MSC detection rates at occipital and parietal derivations for $M = 11$ at scotopic condition.

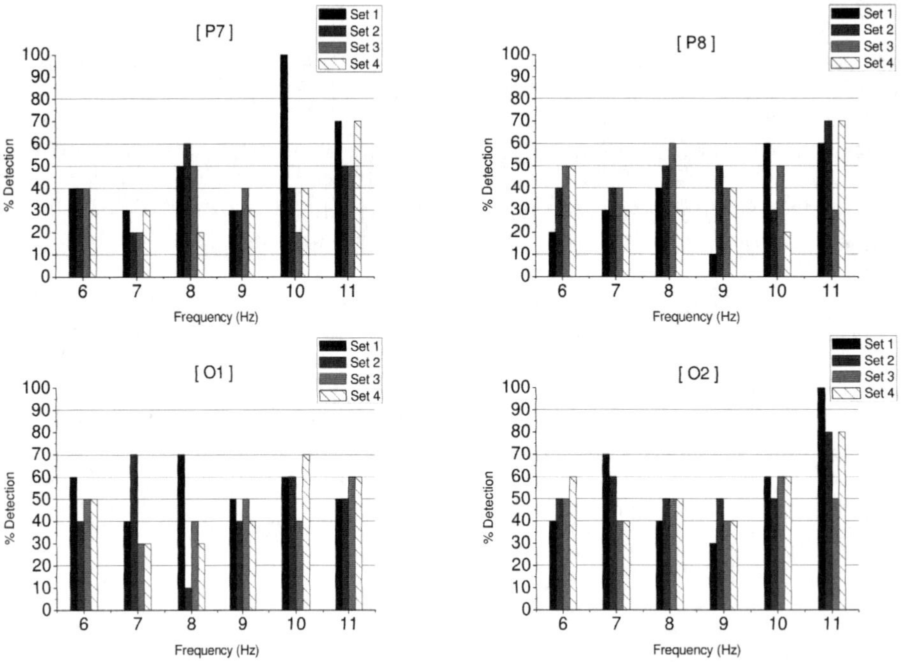

Figure 3. MSC detection rates at occipital and parietal derivations for $M = 11$ at photopic condition.

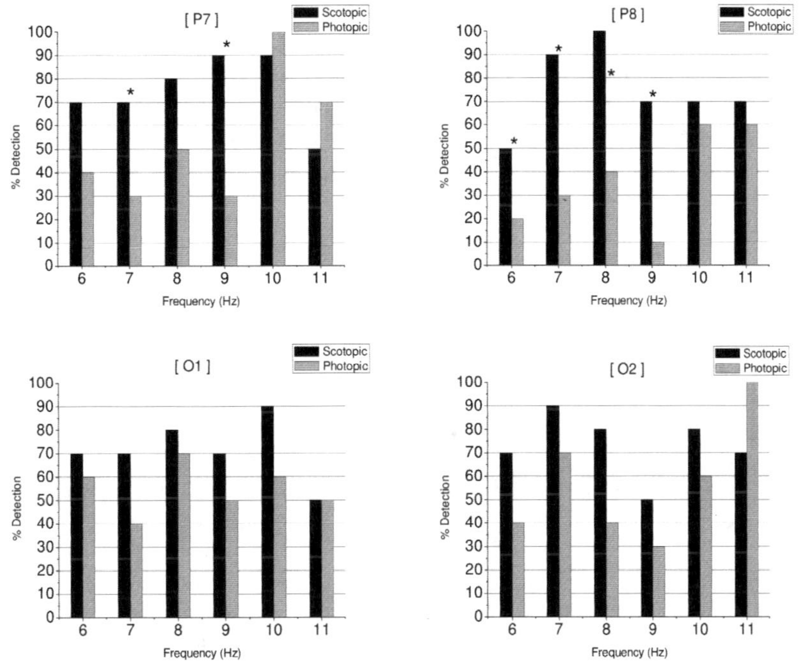

Figure 4. Comparison between detection rates (MSC) at scotopic and photopic conditions for the first set of M = 11 epochs at all derivations. * indicates significant difference between detection rates.

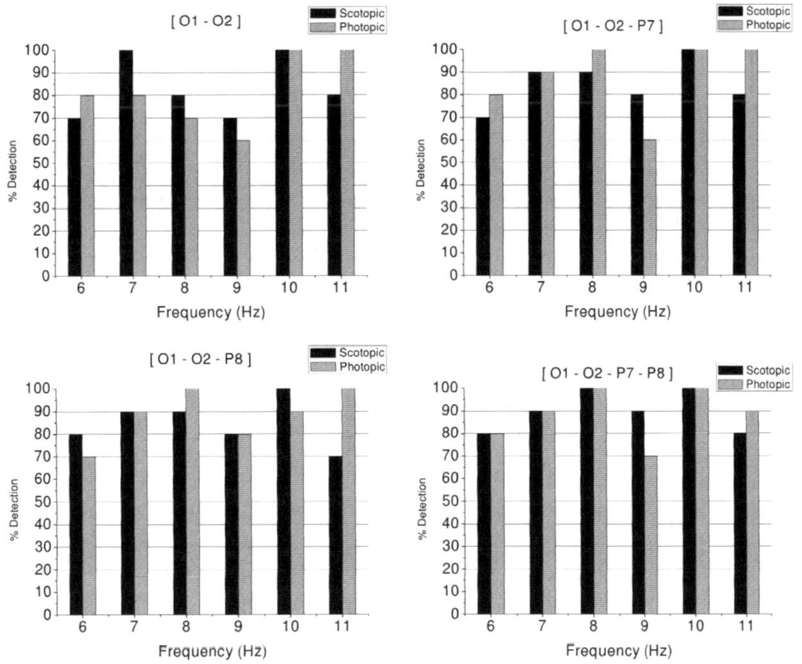

Figure 5. Comparison between detection rates (MC) at scotopic and photopic conditions for the first set of M = 11 epochs at all combinations of leads.

4. Discussion

In this work, the MSC and MC were applied to verify the best frequencies for a BCI based steady-state visual evoked potential for two different lighting conditions (scotopic and photopic).The first technique showed a very poor performance at photopic situation for all used frequencies. At scotopic condition, better detection rates were achieved, reaching 80% in [O2] for 7 and 8 Hz. These frequencies were already studied by Pinto [18], who reported a detection rate of at least 74% with MSC at them and also at 9 Hz. However, in that work, the author employed a clinical EEG amplifier, which has a better signal-to-noise ratio.

In order to improve the detection rates, a multiple coherence (MC) analysis was performed. Four combinations of leads were investigated, all including the occipital derivations [O1] and [O2]. Using this technique the results were better than for MSC, in both conditions, and the detection rates increased with the addition of new leads. A four leads combination led to the best detection rate, with percentages higher than 80% at all frequencies, except 9 Hz. The increase in the performance with new leads addition was also reported by Miranda de Sá *et al.* [19], who compared MSC and MC in the detection of 10 Hz SSVEP, by Zanotelli *et al.*[22] in the evaluation of auditory evoked responses and by Melges *et al.* [16] in an experiment with somatosensory stimulation and the same techniques.

For SSVEP BCI, other signal processing methods had been employed for feature extraction. The Wavelet Packet Transform (WPT) was used by Bian et al. [23] to develop a four-command BCI with stimulation frequencies of 7, 10, 15 and 20 Hz. This BCI showed an accuracy of 83%. Another technique studied was the Continuous Wavelet Transform (CWT), which was applied by Zhang et al. [24] to detect responses at 8.57, 10, 12 and 15 Hz. In this work, the authors compared CWT with six different mother-wavelet and a FFT based technique, reporting that the highest performance was achieved with a Complex Morlet CWT, which presented more than 90% accuracy. Wang [10] and Muller-Putz et al. [25] studied the SSVEP employing the power spectrum analysis. The first developed an electrical car control using 9, 11, 13, 15 and 17 Hz stimulation frequencies, obtaining an average accuracy higher than 85%. The latter implemented a hand orthesis control with commands associated with 6, 7, 8 and 13 Hz. The accuracy of this system varied between 44% and 88%.

Similar to our study, Braga et al. [26] applied Objective Response Detection technique, the Spectral F-Test (SFT), to identify response to stimulation at 6, 7 and 8 Hz, with average hit rate higher than 90%, calculated based on the time evolution analysis of SFT. On the other hand, in this study, some of the above mentioned frequencies were employed. However, the detection rates were used, instead of the accuracy rate, to compare the detection probability for two different lighting conditions.

Finally, the percentages of detection for different sets of EEG epochs revealed no habituation effect, since no important decrease in the rates was observed for each derivation.

5. Conclusion

Based on the results of detection rates for the MSC, 7 and 8 Hz were considered the best frequencies for obtaining SSVEP. Thus, these frequencies should be assigned to commands in a BCI that need fast and reliable response. As second choice, the frequency of 10 Hz, with slight lower rates, could be also used. However, for natural illumination environment, which presents luminance compatible with the photopic condition employed in this work, the use of MSC is not recommended since an important decrease in the detection percentages was observed.

The use of a multivariate ORD technique, Multiple Coherence (MC), allowed overcoming the MSC limitation, leading to better detection rates for scotopic and photopic situations. MC presented high detection probability for all frequencies, when parietal and occipital derivations were employed.

Finally, the time evolution analysis of the MC and the calculation of BCI accuracy have to be investigated in order to build a SSVEP BCI.

Acknowledgments
To CAPES, CNPq, FAPEMIG and PRPq/UFMG for the financial support.

References

[1] Ziemssen T 2009 Multiple sclerosis beyond EDSS: depression and fatigue. *J Neurol Sci* **277 Suppl** S37–41
[2] Gupta P K, Prabhakar S, Sharma S and Anand A 2012 A predictive model for amyotrophic lateral sclerosis (ALS) diagnosis. *J Neurol Sci* **312** 68–72
[3] Le Forestier N, Maisonobe T, Spelle L, Lesort A, Salachas F, Lacomblez L, Samson Y, Bouche P and Meininger V 2001 Primary lateral sclerosis: further clarification. *J Neurol Sci* **185** 95–100
[4] IBGE 2010 Tabelas de resultados Censo Demográfico 2010 Características gerais da população, religião e pessoas com deficiência **2010** Available on ftp://ftp.ibge.gov.br/Censos/Censo_Demografico_2010/Caracteristicas_Gerais_Religiao_Deficiencia/tab1_3.pdf Accessed on 18/05/2013
[5] Iversen I H, Ghanayim N, Kübler A, Neumann N, Birbaumer N and Kaiser J 2008 A brain-computer interface tool to assess cognitive functions in completely paralyzed patients with amyotrophic lateral sclerosis. *Clin Neurophysiol* **119** 2214–23
[6] Misulis K E 1994 *Spehlmann's Evoked Potential Primer* (Newton, MA: Butterworth-Heinemann)
[7] Moran D 2010 Evolution of brain-computer interface: action potentials, local field potentials and electrocorticograms. *Curr Opin Neurobiol* **20** 741–5
[8] Leuthardt E C, Schalk G, Roland J, Rouse A and Moran D W 2009 Evolution of brain-computer interfaces: going beyond classic motor physiology. *Neurosurg Focus* **27** E4
[9] Ming C M C, Xiaorong G X G, Shangkai G S G and Boliang W B W 2005 Stimulation frequency extraction in SSVEP-based brain-computer interface *First International Conference on Neural Interface and Control* (Wuhan: IEEE) pp 64–7
[10] Wang H 2010 Remote control of an electrical car with SSVEP-Based BCI *2010 IEEE International Conference on Information Theory and Information Security* (Beijing: IEEE) pp 837–40
[11] Ortner R, Allison B Z, Korisek G, Gaggl H and Pfurtscheller G 2011 An SSVEP BCI to control a hand orthosis for persons with tetraplegia. *IEEE Trans Neural Syst Rehabil Eng* **19** 1–5
[12] Vialatte F-B, Maurice M, Dauwels J and Cichocki A 2010 Steady-state visually evoked potentials: focus on essential paradigms and future perspectives. *Prog Neurobiol* **90** 418–38
[13] Pinto M A S, Souza J K S, Baron J and Tierra-Criollo C J 2011 A low-cost, portable, micro-controlled device for multi-channel LED visual stimulation. *J Neurosci Methods* **197** 82–91
[14] Wu Z, Lai Y, Xia Y, Wu D and Yao D 2008 Stimulator selection in SSVEP-based BCI. *Med Eng Phys* **30** 1079–88
[15] Cagy M, Infantosi A F C and Gemal A E 2000 Monitorização do Plano Anestésico por Técnicas Estatísticas no Domínio da Freqüência *Rev Bras Eng Biomed* **16** 95–107
[16] Melges D B, Miranda de Sá A M F L and Infantosi A F C 2012 Tibial nerve somatosensory evoked response detection using uni and multivariate coherence *Biomed Signal Process Control* **7** 215–20
[17] Simpson D M, Tierra-Criollo C J, Leite R T, Zayen E J and Infantosi a F 2000 Objective response detection in an electroencephalogram during somatosensory stimulation. *Ann Biomed Eng* **28** 691–8
[18] Pinto M A S 2011 *Estudo do potencial evocado visual em regime permanente baseado em LED para interface cérebro máquina* (UFMG)
[19] Miranda de Sá A M F L and Felix L B 2001 On the Detection of Visual Evoked Potential Responses by Using Multiple Coherence *23rd Annual EMBS International Conference* (Istanbul: IEEE) pp 2002–5

[20] Miranda de Sá A M F L and Infantosi A F C 2007 Evaluating the relationship of non-phase locked activities in the electroencephalogram during intermittent stimulation: a partial coherence-based approach. *Med Biol Eng Comput* **45** 635–42

[21] Miranda de Sá A M F L, Felix L B and Infantosi A F C 2004 A matrix-based algorithm for estimating multiple coherence of a periodic signal and its application to the multichannel EEG during sensory stimulation. *IEEE Trans Biomed Eng* **51** 1140–6

[22] Zanotelli T, Santos T S, Felix L B and Tierra-Criollo C J 2010 Deteção do potencial evocado auditivo em regime permanente utilizando coerência e coerência múltipla *XXII Congr Bras Eng Biomed* (Tiradentes) pp 1162–5

[23] Bian Y, Li H, Zhao L, Yang G and Geng L 2011 Research on Steady State Visual Evoked Potentials based on Wavelet Packet Technology for Brain-Computer Interface *Procedia Eng* **15** 2629–33

[24] Zhang Z, Li X and Deng Z 2010 A CWT-based SSVEP classification method for brain-computer interface system *Int Conf Intell Control Info Process* (Dalian: Ieee) pp 43–8

[25] Muller-Putz G and Pfurtscheller G 2008 Control of an electrical prosthesis with an SSVEP-based BCI *IEEE Trans Biomed Eng* **55** 361–4

[26] Braga V C C, Cerqueira F G G and Criollo C J T 2012 Interface Cérebro Máquina por Potencial Evocado Visual em Regime Permanente Wireless com Coleta Online *XXIII Congr Bras Eng Biomed* (Porto de Galinhas) pp 1–5

Comprehensive feature extraction for objective *Dynamic Gait Index* assessment of risk of falls in the elderly

M T Miralles[1,2,4], I Ghersi[1], R Vecchio[1], R Paterson[2], M Pérez Akly[3], M Ferrando[3], A Paterson[2] and F Álvarez[3]

[1] Pontificia Universidad Católica Argentina (LaBIS-Facultad de Ciencias Fisicomatemáticas e Ingeniería-UCA), Av. Alicia Moreau de Justo 1500, Ciudad de Buenos Aires, Argentina
[2] Universidad de Buenos Aires (CIDI-FADU-UBA), Intendente Güiraldes 2160, Pabellón III, Ciudad Universitaria, Ciudad de Buenos Aires, Argentina
[3] Unidad Asistencial Dr. César Milstein (Servicio de Neurología), La Rioja 951, Ciudad de Buenos Aires, Argentina

E-mail: mmiralles@gmail.com

Abstract. The evaluation and understanding of human balance and risk of falls, particularly in the elderly, is a growing field of study. Complementeing classic clinical tests (relying on comprehensive, although mostly qualitative assessments) with new devices that acquire precise patterns associated to the balance process, appears as a great oportunity in this area. Objective features can be obtained from these patterns, summarizing relevant data about balance and gait processes. This work presents results from the Dynamic Gait Index assessment over 42 subjects, while wearing a specially designed 3D-linear acceleration sensing device. With knowledge about the test, features related to step frequency, energy expenditure and changes in signal shape were defined and obtained from the three acceleration signals. The evolution of these features over time is of interest, so a moving-window strategy was implemented. Signal processing strategies were tested and improved, and results were correlated with the clinical scores given by the physicians to better understand them, initially test their reliability and select the most suitable ones. Results showed promise for this strategy in providing uniform, meaningful and new interpretations about different processes involved in the stages of the test, opening the possibility to develop custom user interfaces for clinical use.

1. Introduction

Balance is a complex phenomenon, central to the field of human biomechanics. No matter its cause, the loss of capabilities associated with motor control translates into significant reductions in the ability to perform daily tasks. In particular, the elderly population is subject to a higher risk of falls, which is associated to disabilities, the development of chronic degenerative diseases, and a consequent higher risk of death or a worsening quality of life. The fact that the global population is ageing *"at unprecedented rates"* [1], stresses the potential impact of these links between age and falls, so the possibility of assessing this risk results of great importance for public health [2]. There is a number of well-known clinical assessments in this field, such as the Tinetti scale or the Dynamic Gait Index

[4] Corresponding author. This work was carried out under and supported in part by the A216 UBACyT 2012-2014 project, Facultad de Arquitectura, Diseño y Urbanismo, Universidad de Buenos Aires (FADU-UBA), and by Facultad de Ciencias Fisicomatemáticas e Ingeniería, Pontificia Universidad Católica Argentina (UCA).

assessment of risk of falls (DGI) [3]. While these tests rely on the expertise of specialized physicians, which perform comprehensive evaluations and score subjects (factoring step frequency, symmetry, continuity, deviations and need of assistance, etc.), the possibility of *quantifying* balance patterns through objective indexes can help provide access to new, more uniform assessments. Recent technologies have opened the possibility to acquire and process balance patterns in innovative ways. Activity monitors involving accelerometers [4] are of special interest as low-cost alternatives to equipments like force platforms or high-speed video motion tracking systems, and the possibilities for complementing current clinical tests with information from these new sensors are considered of great value [5][6]. In order to approach this task, a wireless portable system was developed, which records linear accelerations in three axes [7]. This work is based on results from the evaluation of 42 subjects, over 60 years of age, through the DGI test, which will be detailed in following sections. Physicians performed the assessments while subjects were equipped with the new sensing device. The proposed line of work attempts to associate scores from the clinical examinations with experimental registers acquired through this system. This work presents and tests three new features and their evolution trough the stages of the DGI test, conceived as objective measures of factors that are taken into consideration in the clinical assessment (walking speed or changes in balance, etc.), aiming to give access to superior DGI evaluations, based on precise, uniform data with biomechanical meaning. Features were tested, and correlated to clinical scores, in order to prove and understand their meaning and scope. Finally, conclusions, further improvements and future lines of work are proposed.

2. Human Balance Assessment - Data acquisition

2.1. Portable and programmable wireless system for gait and balance assessment

In order to acquire relevant, objective measurements during classic clinical balance tests, a portable, wireless system for linear acceleration recording in three axes was developed [7]. This system (hardware shown in figure 1) includes a battery-powered *sensing device* that was conceived as a belt with an attached, small case, containing a Micro ElectroMechanical System, triaxial capacitive accelerometer, a *USB-powered terminal* connected to a personal computer, and a dedicated *graphical user interface,* through which the device is controlled. The sensing device is intended to be fitted tightly around the trunk of the subject, and its selected placement for its subsequent implementations is an approximation of the whole body's center of mass (CM), having the sensor fixed at the widest part of the hips and at the middle of the back, a common choice for similar applications [5][6]. The device will respond to the motion of the CM. Two alternative representations for the 3D-acceleration vector are shown in figure 2: three components -AnteroPosterior (AP), MedioLateral (ML) and Vertical (VE)-, or the magnitude of the vector (A_{MOD}) and two angles, θ_{AP} and θ_{ML}. The resulting system is a versatile and programmable wearable data logger and signal processor for 3D-acceleration signals, with a sampling rate of 100 Hz, sensing ranges of ±1.5 A_{unit}, ±2 A_{unit}, ±4 A_{unit} ±6 A_{unit} (A_{unit} = 9.8 m/s^2), and a top resolution of 0.001 A_{unit}. The high sampling rate and wide, programmable sensing range, combined with the good sensing resolution, are the main qualities of this system, suitable for both dynamic and static balance measurements, and with possible, further applications in sports, as well as its current implementation in normal and dysfunctional human balance analysis.

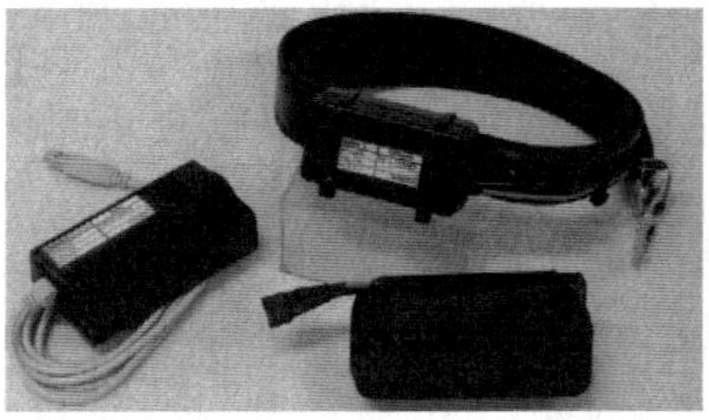

Figure 1. System components: Wearable sensing terminal (top), battery pack (bottom right), and USB base (left).

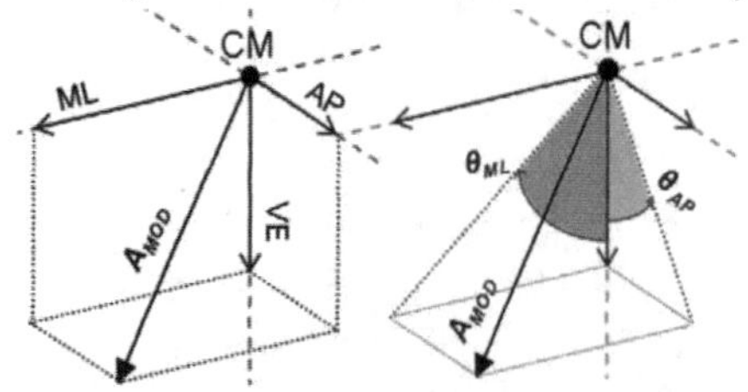

Figure 2. 3D vector components: (a) Linear coordninates (AP, ML and VE). (b) Vector magnitude A_{MOD} and angles θ_{AP} and θ_{ML}.

2.2. Dynamic Gait Index assessment of risk of falls

The Dynamic Gait Index (DGI) for gait assessment and risk of falls evaluation for the elderly consists of serveral excercises, mainly involving a normal, straight walk over relatively long distances (two 5 meters walks, in this case), with the addition of obstacles or commands given by a physician [3]. A detailed description for this particular implementation of this test is divided in stages as follows:

1. *Straight walk*
2. Straight walk with *speed changes* (slow down, normal or faster, when commanded)
3. Straight walk, with *horizontal head turns* (left or right, when commanded)
4. Straight walk, with *vertical head turns* (up or down, when commanded)
5. Straight walk, with *pivot* (180° turn, when commanded)
6. Straight walk, *stepping over a small obstacle*
7. Straight walk, *passing around an obstacle* and returning to a straight walk
8. *Stair climbing*

The aim of these stages is to evaluate each subject's ability to respond to slight changes during a normal walk, with the physician assigning a score between 0 and 3 points for each activity. A score of 0 indicates severe changes or the impossibility to perform the activity, and a score of 3 means little to no difficulty in making such changes. Each particular activity has some specific parameters that need to be accounted for in order to determine the score, for every subject. In particular, activities 1 through 4 have certain properties that make them better suited for a joint evaluation. The highest possible score is 24, and scores of 19 or less have been related to a greater incidence of falls in the elderly.

As described, these evaluations, performed by qualified physicians, result in a comprehensive, although qualitative estimation for each subjet's abilities, balance quality and risk of falling, so a unified standard would be needed, for example, if results from many subjects and/or evaluated by different physicians were to be compared (unless those physicians had previously discussed some basic guidelines for their scoring procedure). The actual description of the test gives some notions about what qualifies as each score, but the final decision as to what makes a "slight", "moderate" or "severe" change falls on each specialist. Another scenario in which a unified standard would be useful could be the study of one single subject over time, as this subject excercises in order to correct certain habits or patterns that affect its balance capabilities, therefore allowing to determine how much that subject has improved in that time. In any case, it could be noted that a quantitative, reliable and uniform source of information could be of value for the work of the physicians, not only for the analysis of each subject, but also for possible broader, statistical studies over larger populations. Furthermore, for cases where changes are not too evident or difficulties are not severe, a quantitative source of information could be able to precisely detrmine *how severe* these difficulties are or *how stable* certain features remain during the excercies, allowing for a more precise scoring system. On the other hand, this does not mean that such examinations should be completely replaced by new, automated methods. Given the complexity of the human balance process, a comprehensive, justified analysis covering many different factors is needed for these examinations to take place. The goal of adding new data to these factors would be to give an additional, reliable and unified source of information that physicians could take into account while evaluating patients, combining their knowledge with new sources of information.

2.3. Experimental setup

The studies presented in this work correspond to evaluations made by specialised physicians over 42 subjects (31 women and 11 men), with an average age of (77 ± 8) years, which manifested a level of risk of falls. Every evaluated subject gave informed, written consent, before performing the tests, which were previously authorized by the Ethics Committee of the medical institution where the evaluations took place (Unidad Asistencial Dr. César Milstein), and described to the subjects before taking place. Subjects are patients of the neurology service of the instiution, and they performed three classic evaluations of gait and balance: the Tinetti Test for gait and balance assessment, the DGI evaluation, and a Get Up and Go test, while wearing the new sensing device, placed at the desired position (estimation of the center of mass). This work focuses on the DGI assessment. The setting for the experiments involved marking the straight path that would be followed by the subjects (indicating start and stop marks, and with intermediate marks), calibrating the measuring devices and positioning cameras for image acquisition. High-speed videos were recorded with two Casio EXFH25 cameras, at 120 frames-per-second, with one camera recording the sagital plane and the other on the frontal plane. Regular videos were also recorded. Before the beginning of the tests, several anthropometric measurements were taken for each subject (height, weight, foot length, a stride record and leg lengths with and without footwear). Both the weight and single stride records were taken with a Vernier force platform (range: -800 N to 3500 N, accuracy: ±1.2 N), and acquired through a Logger Pro 3 data collector, by the Logger Pro software. A portable height meter (±1 mm), and tape measure (±1 mm) was used for height and leg measurements. While performing the tests, patients wore their usual footwear and viewing aids. Test scales were agreed upon and scored, simultaneously and independently, by a neurologist and a specialist in kinesiology, for a better control of the scoring procedure. A chronometer and mechanical step counter were also used for specific gait assessments. Subjects were finally equipped with the accelerometry sensing device (calibration strategy detailed in [7]), and its distance from the ground was also measured. From this point on, the clinical evaluations took their normal course, with the sensing device and the cameras recording the procedure.

3. Feature definition – Signal Processing

The stages for the DGI assessment involve a subject performing normal tasks, sometimes with the addition of obstacles or responding to commands. In general, a low-risk subject should be able to perform these activities making fast but steady changes as needed, without showing disturbances or great discomfort. In this section, a quantitative approach to the evaluation of such changes is presented, taking the form of a moving-window feature extraction process, so that its user is able to see how much these features change over time, during each stage.

3.1. Main factors involved in first DGI stages

When physicians evaluate a subject while performing the DGI assessment, they follow a series of general guidelines that point out what could be related to a specific score, from 0 to 3 [3]. Evaluations for stages 1 through 4 rely mainly on factors related to changes in speed or balance patterns, as well as to each subject's ability to continue performing the tasks. Quantifying these factors would be useful then for a joint analysis of all of these stages, since the scoring process relies heavily on the physicians defining what is to be considered severe, minor, good or bad, so their evaluation becomes subjective.

3.2. Feature definition

With the main factors involved in the evaluation of stages 1 through 4 in mind, the following quantitative features were proposed as a potential supplement for the clinical assessment:

I. Changes in speed Subjects can walk shorter distances over a constant period of time (resulting in lower speeds) for two reasons: a lower average step frequency at a constant stride length, or a reduced energy expenditure per step at a constant step frequency. The opposite would occur when increasing the speed. Considering the charactristics of the acquired acceleration signals, this translates into two features that can be calculated to factor speed changes, and other features: an average step frequency estimate ($\mathbf{F_{avg}}$), and an average energy expenditure per step factor ($\mathbf{E_{step}}$).

II. Changes in balance patterns, deviations, imbalance When the two previous features are removed from the acceleration signals, what is left is the actual "shape" of the acceleration pattern on each axis. The absolute value of a Fast Fourier Transform for the windowed signals (|FFT|), normalized by F_{avg} in the x-axis and by a factor related to E_{step} in the y-axis, would carry this information about the windowed signals. If this information is compared to an average pattern for a normal walk by the same subject, it would indicate how much that shape changed ($\mathbf{FFT_{evol}}$).

III. Subject's ability to continue walking Physicians can easily detect this event without the need for an automated source of information, so it will not be accounted for in the following sections.

3.3. Signal processing

3.3.1. Feature calculation Up to this point, three features have been selected as sources of useful information about performance on stages 1-4 of the test. These features are described in detail as follows:

- **Average step frequency estimate (F_{avg})** Previous work related to the sensing device and acquired balance and gait acceleration signals [8] has shown that information given by the sensor in the AP axis is the most reliable individual source for average step frequency estimation *on normal gait evaluations*. These signals are the most consistent, and their |FFT| have a peak at a fundamental frequency component ($F_{max}\{|FFT_{AP}|\}$) that can be directly linked to the actual mean frequency of the steps. Only when a very significant asymmetry between left and right steps occurs, that peak will take place at approximately one half of the actual mean step frequency, and those particular cases can be detected and compensated by comparing the three axes [8].Considering that the relevant information for these tests is the change in the average step frequency estimate, and not its absolute value, the proposed calculations of F_{avg} will be obtained by this method. When further analyzing this feature, this can be compensated, and relative indexes will be defined (F_{avg}/mean$\{F_{avg}\}$, for example) in order to overcome the possible difference between the real mean step frequency and the estimated F_{avg}.

$$F_{avg} = F_{max}\{|FFT_{AP}|\} \tag{1}$$

- **Average energy expenditure per step factor (E_{step})** An average energy expenditure factor E_{avg} for an acceleration signal x[n], from sample n_0 to n_0+ W, is defined in (2). Finally, the proposed average energy expenditure per step factor (E_{step}) is detailed in (3).

$$E_{avg} = \frac{1}{W}\sum_{n=n_0}^{n=n_0+W}(x[n] - mean(x[n]))^2 \tag{2}$$

$$E_{step} = \frac{E_{avg}}{F_{avg}} \tag{3}$$

- **Signal shape evolution factor (FFT_{evol})** The first step in order to obtain this feature is to calculate the FFT for the three signals provided by the accelerometer (mean values are subtracted in order to consider the dynamics, and a Hanning window is applied for better results). The results are then normalized by re-scaling both axes: the x-axis (f) is re-scaled dividing it by F_{avg}, ($f_N = f/F_{avg}$) and the y-axis is re-scaled by $\sqrt{E_{avg}}$. The normalized spectrum for a signal is defined as follows:

$$FFT_N(f_N) = \frac{|FFT(f/F_{avg})|}{\sqrt{E_{avg}}} \tag{4}$$

Once a normalized, new spectrum is obtained, it can be compared with other results. In particular, it is desired to compare the average $|FFT_N|$ of a normal walk with the $|FFT_N|$ of other stages, for the same subject, allowing to see how different the registers are. In order to determine this difference between two signals, named $|FFT1_N|(f_N)$ and $|FFT2_N|(f_N)$, a mean square error is obtained:

$$FFT_{evol} = \frac{1}{N_{bins}}\sum_{f_N=0}^{f_N=f_N\max}(|FFT1_N|[f_N] - |FFT2_N|[f_N])^2 \tag{5}$$

where N_{bins} is the number of bins for the normalized FFT (in this case 10000). FFT_{evol} will be calculated for the three acceleration signals, comparing each FFT_N spectrum to its mean counterpart for a normal walk. A higher FFT_{evol} implies a greater difference between signals. For a sampling rate of 100 Hz, fmax will be less than 50 Hz. As part of the process, bins over $f_N$max = 10 (fmax = 10 F_{avg}) are discarded, as they do not hold significant information compared to the lower frequencies.

3.3.2. Moving-window signal processing strategy Subjects were equipped with the accelerometer sensing device while performing the DGI assessment. Registers from stages 1-4 should be cut to remove initial and final transients. With the isolated relevant sections of the registers, three signals of variable length are left, as the duration of each stage depends on how fast the subjects can perform the activity. The goal of this work is to analyze the evolution of the proposed features while each stage is performed. In order to do so, a moving-window strategy was selected (figure 3). A window of size W_{size} (in seconds) was defined to select smaller segments of the signals, under which each feature would be calculated (for either one or the three signals). By moving the window one sample at a time (or 0.01s for a 100 Hz sampling rate), and recalculating the features within that new segment, the end result for each feature will be a plot showing its evolution over time, i.e. $F_{avg}(t)$. With this information, physicians will have access to accurate data indicating which changes have occurred, when they took place, and how severe they have been, adding relevant and objective information to their assessment.

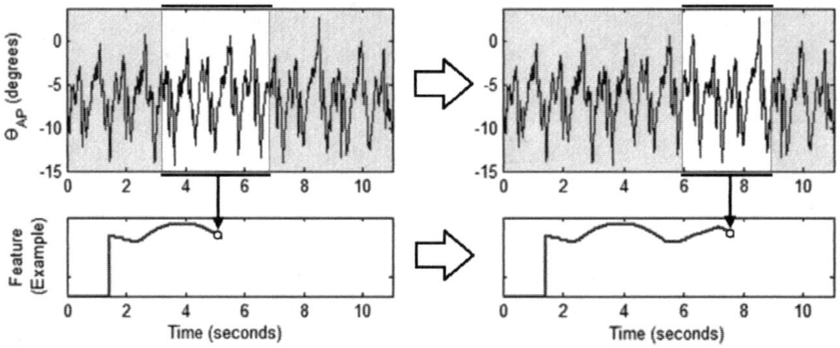

Figure 3. Moving-window feature extraction strategy for a signal. As the window is moved, a time chart for the feature (*Feature(t)*) is created.

The selected window size was defined as four times the average step period estimate of the normal walk (or $4/F_{avg0}$). Information obtained from the Tinetti test for the subjects involved in the DGI assessment showed that their average step frequency was above 1 Hz for a normal walk [8]. With this information in mind, this window size was selected because W_{size} needs to be large enough to contain at least a few steps (more than two consecutive steps) so that features preserve their meaning for the three axes, and also short enough to be sensitive to changes in the features. This W_{size} should also be compatible with the estimation of an average time between commands given by the physicians. The complete feature extraction process through this moving-window approach requires at least one normal walk register, as it is the basis against which the other stages are compared. The following are simplified signal processing stages required for the final feature extraction process:

1. Calculate mean F_{avg} and mean normalized spectrum $FFT_N(f_N)$ for a normal straight walk record (stage 1). Results are **1: F_{avg0}** and **2: $FFT_{N0}(f_N)$** for each signal.
2. Define **W_{size}**.
3. Use the moving window to calculate the evolution of F_{avg} and E_{avg} over time (for registers from stages 1-4), results are **1: $F_{avg}(t)$**, and **2: $E_{avg}(t)$** and **$E_{step}(t)$** for each signal
4. Normalize **|FFT|** on each sample with $F_{avg}(t)$ and $E_{avg}(t)$, to calculate **$FFT_{N1}(f_N)$**, for each signal
5. Get **$FFT_{evol}(t)$** for each component with $FFT_{N1}(f_N)$ and $FFT_{N0}(f_N)$.

It should be noted that step 1 actually requires to perform a variation of steps 2 through 5 on the normal walk signals, in order to get the three mean normalized |FFT| for that stage. Since E_{step}, as defined in (3), is not filtered, its evaluation through the moving-window process can lead to a feature that, while accurate, is sensitive to sudden peaks, which are expected for gait registers, especially for VE and ML or A_{MOD} and θ_{ML} signals. A filtered version of this feature (E_{step}') was developed, taking advantage of the FFT and the Hanning window relation to the signal energy, resulting in a smoother estimate (figure 5) that focuses on the energy expenditure in the central portion of the windowed signal. However, E_{avg} and E_{step} should not be discarded, and E_{avg} is needed to calculate FFT_{evol}.

4. Results

Figure 4 is a hystogram for the final scores of the DGI test. In can be seen that only 5 out of 42 subjects were diagnosed with a lower risk of falls. Considering the results of this test, most subjects would have a high risk of falling, and less than a 11.9% of them would be in better conditions. A more specific analysis, divided in stages, would point to particular conditions or difficulties for each subject. Some subjects could not perform all stages. A minimum of 40 registers per stage were obtained.

A total of 165 registers from stages 1-4 of the DGI assessment were pre-processed in order to satisfy the requirements set in section 3.3.2. Both linear coordinates and magnitude and angular components were obtained from each acceleration register, and the proposed features were calculated from these signals. Register length was variable, with a mean of (8.94 ± 3.79) seconds, a minimum of 4 seconds and a maximum of 26 seconds. W_{size} resulted in values of (2.63 ± 0.49) seconds. Window sizes remained adequately inferior to register lengths, as a lower F_{avg0} implies a larger W_{size}, but also that subjects will take longer to perform the tasks, resulting in longer registers.

- The proposed set of features is an attempt at isolating three characteristics of a gait signal: its fundamental frequency, and its intensity and shape on each axis. As described, these parameters should be independent, and this has been proved in some cases, (shown in figure 5 for θ_{AP}). However, this does not mean that features will be independent for every case and stage (i.e.: attempting to walk faster can lead to an increased energy expenditure, as well as a faster step rate). Subjects' ability to change these features independently is what will determine how related they are in practice.

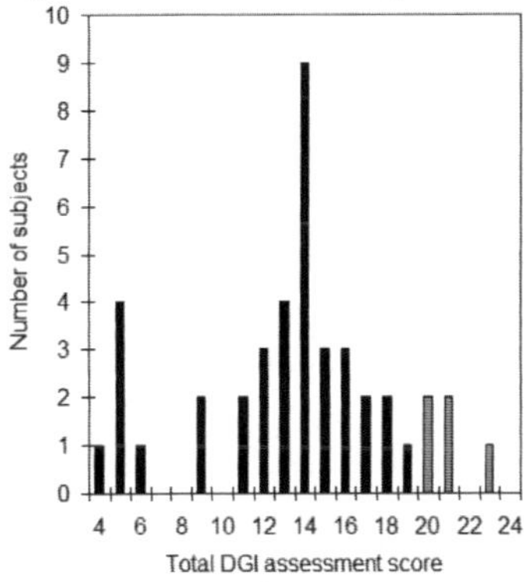

Figure 4. Results for DGI assessment of 42 subjects - Hystogram. Only 5 subjects received a positive evaluation.

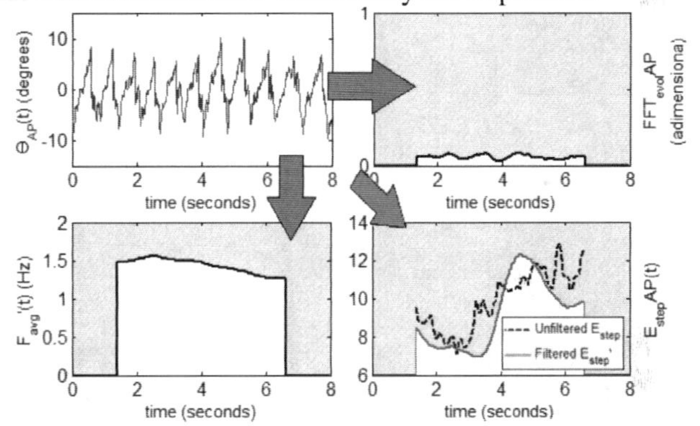

Figure 5. Independence between features for θ_{AP} signal (top-right): F_{avg}'(t) (bottom-left) decreases, while E_{stepAP}(t) (bottom-right) grows, and FFT_{evolAP}(t) (top-right) remains steady at a low value.

- $\mathbf{F_{avg0}}$ responded as expected for every normal walk assessment. One way to test this was to divide the number of steps in the registers (with previous, combined knowledge about the three normal walk signals for better accuracy) by the register time. Additionally, $\mathbf{F_{avg}(t)}$ has shown good response to speed changes in the other excercises. However, for some cases where subjects drastically changed their gait patterns, this feature has shown jumps to lower frequencies on those periods (22 cases, for stages 2-4). If this occurred in response to a sudden change in the actual step frequency, it would not be considered an issue. In these cases, however, it could be seen that the highest frequency component of the signal was lower than any detected or observable step rate, and that it no longer reflected the fundamental step frequency, but a lower drift, possibly due to changes in posture, or sudden imbalances. This implied that the biomechanical meaning of the parameter was lost for these cases, affecting the subsequent calculation of E_{step} and FFT_{evol}. Both an example of this and the proposed solution are shown in figure 6. *Two additional frequency estimators have been calculated to solve this*: one is a filtered version of F_{avg}(t), limiting its maximum change between samples; the other is the frequency maximum of the combined |FFT| of the AP and ML signals (considering that the ML-peak frequency for a normal walk should be half of the AP-peak frequency, and scaling the axes accordingly). With these three frequency values, the final, *new average step frequency estimate* (F_{avg}') was chosen as the one that most greatly reduced FFT_{evol} (the difference from a normal walk) for the three signals, on each sample. This estimation has shown improved resistance to these discontinuities. In result, the original $F_{avg}(t)$ will show jumps related to drastic changes in balance (which *is* relevant to the test), while F_{avg}'(t) will allow for the features to retain their biomechanical meaning (figure 6).

An analysis of F_{avg}'(t) and its general link to scores in the test was performed. Results from one subject showing great asymmetry between steps, as mentioned in section 3.3.1. (with video records supporting this) were adapted to allow the presented analysis, comprising all subjects. Table 1 shows results for F_{avg0} and F_{avg}'(t) (mean, maximum and relative standard deviation σ), grouped and averaged by score (which is specified for each stage). Results show consistency between F_{avg}' and the logic behind evaluations: more confident subjects perform tasks at higher speeds, with higher (speed changes) or lower (normal walk or with head turns) relative changes to F_{avg}'(t) than those with lower scores. Additionally, the last feature (σ{F_{avg}'(t)}/mean{F_{avg}'(t)}) is independent of the absolute value of F_{avg}', as proposed in section 3.3.1.

Table 1. F_{avg}'(t) sweep: Results from stages 1–4 of DGI test, grouped by score. Subjects with higher scores show a higher mean or peak step frequency on all stages. When asked to change speed, subjects with 2-3 points show higher relative standard deviations in F_{avg}', and when asked to walk normally while turning their heads, they show lower deviations than subjects with 0 or 1 point.

	Score	0 - 1	2	3	
Normal Walk	F_{avg0}	1.57	1.56	1.69	Hz
	σ{ F_{avg}'(t) } / mean{ F_{avg}'(t) }	0.04	0.03	0.02	adimensional
	Number of cases	7	25	10	Subjects
	Score	0 – 1	2	3	
Speed Changes	max{ F_{avg}'(t) }	1.80	1.87	1.95	Hz
	σ{ F_{avg}'(t) } / mean{ F_{avg}'(t) }	0.08	0.10	0.11	adimensional
	Number of cases	12	23	5	Subjects
	Score	0 - 1	2	3	
Horizontal Head turns	mean { F_{avg}'(t) }	1.46	1.58	-	Hz
	σ{ F_{avg}'(t) } / mean{ F_{avg}'(t) }	0.08	0.04	-	adimensional
	Number of cases	26	15	0	Subjects
	Score	0 – 1	2	3	
Vertical head turns	mean { F_{avg}'(t) }	1.39	1.58	1.65	Hz
	σ{ F_{avg}'(t) } / mean{ F_{avg}'(t) }	0.08	0.05	0.06	adimensional
	Number of cases	18	20	2	Subjects

It should not be expected, however, that every subject with a score of 0 should have a lower mean step frequency than every subject that received a score of 3. Physicians take many factors into consideration while scoring, but the analysis of general tendencies for this feature should put its reliability to test. Averaging between subjects was only implemented to allow for a general interpretation of the results, while taking this into account. Finally, the highest relative change in F_{avg}'(t) was found on subjects with a score of 3 in the walk with speed changes (0.11), meaning that they showed the most significant changes in speed, in relation to their average step frequency. Similar calculations were obtained from E_{step}'(t), for each acceleration coordinate, showing higher maximum values and relative standard deviations for stage 2 than for stage 1 for E_{step}' from θ_{AP} and A_{MOD}, as well as greater relative deviations for stages 3-4 than for stage 1 on all components, for example.

- E_{step}'(t) is a less noisy estimation than the unfiltered $E_{step}(t)$ (figure 5). In figure 7, E_{step}'(t) is variable, *and different,* between axes. Isolating E_{step}' values from the three signals is of interest, but *the ratio between AP and ML results* (indicating how much of the actual energy expenditure is dedicated to moving forward) is another feature that should be considered, and it can easily be obtained from these features, while it would be difficult for physicians to estimate it. Alternative filters (and energy estimators) can additionally isolate other important signal characteristics, like sudden peaks, etc.

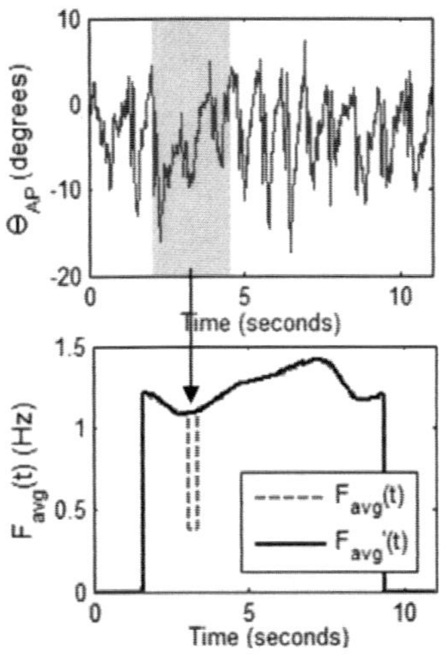

Figure 6. $\theta_{AP}(t)$ signal (top) and $F_{avg}(t)$ estimates (bottom): F_{avg}'(t) holds its meaning, while $F_{avg}(t)$ detects discontinuities.

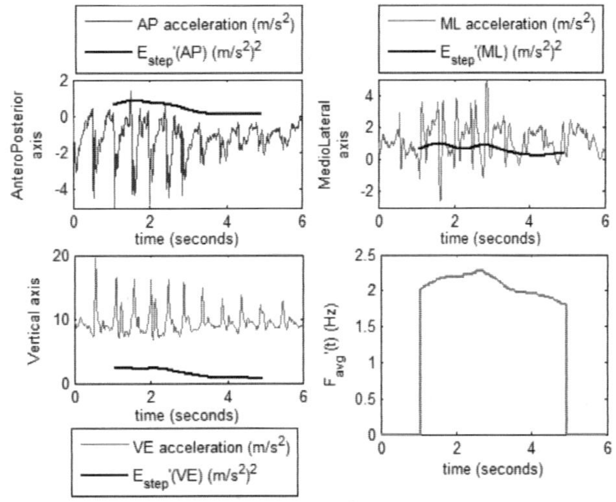

Figure 7. Filtered E_{step}'(t) results for subject in walk with speed changes (score: 2). Signals show different, significant changes in E_{step}'(t), while the average step frequency F_{avg}'(t) shows minor changes, stressing the relevance of this feature.

- As defined, $\mathbf{FFT_{evol}}$ is able to detect changes in signal shape, compared to the mean shape of a recent normal walk, but also independently. Since the mean shape of a normal walk is constant, changes in $FFT_{evol}(t)$ reveal changes in the actual signal, so FFT_{evol} can be used to assess how much the gait pattern changes over time, for all stages. Figure 8 shows the standard deviation of $FFT_{evol}(t)$ from two signals (θ_{AP} vs. A_{MOD}) during the normal walk (8(a)), and for the walk with horizontal head turns (8(b)), for all subjects. Subjects with higher scores tend to show lower values for both signals, especially in (8(b)). A similar tendency was found for the walk with vertical head turns. Results for θ_{ML} did not follow this trend. Additionally, the maximum standard deviation is significantly lower in the normal walk (0.07), compared to its equivalent in the walk with head turns in (8(b)) (0.30). This is consistent with the test, and can be used as a new feature, when evaluating a single subject.

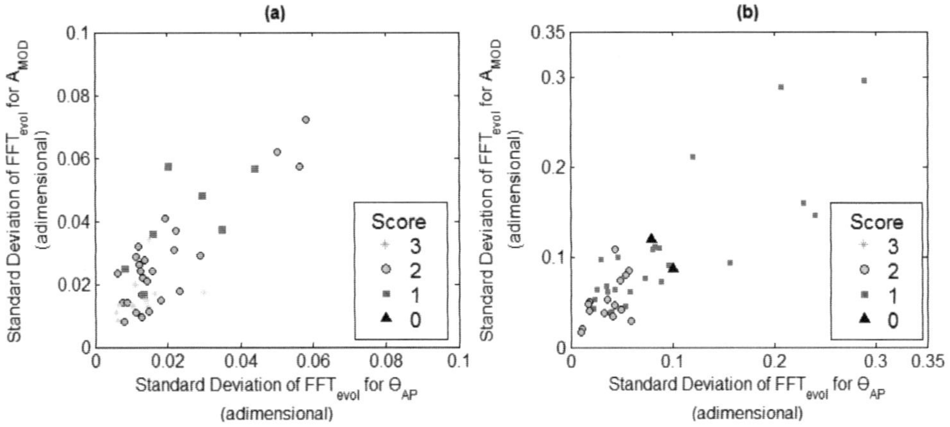

Figure 8. FFT_{evol} variations for (a) normal walk and (b) walk with horizontal head turns: Standard deviation of $FFT_{evol}(t)$ for θ_{AP} vs. A_{MOD}, for each subject. Overall values in (a) are lower than those encountered in (b). Additionally, subjects with lower scores show scattered, more significant values.

5. Conclusions and Future Work

In this work, a comprehensive analysis of human gait registers obtained from a 3D-accelerometer during the DGI assessment of risk of falls in the elderly has been proposed, based on three features, each of them holding a specific, uniform biomechanical meaning between subjects, laying the required foundations for an objective analysis. Information from the assessment of these features on 42 subjects was processed, and summarized through parameters that were separated by the scores received during the clinical evaluations. Results showed promise in the implementation of the features for the objective evaluation of gait registers, and also allowed to determine which ones are best suited for discerning good and poor general responses, with features from ML signals showing the most variable response, due to the many factors that affect their shape between subjects (ranging from almost square signals to less recognizable ones), which can be assessed in the future. While a direct indiviual relationship between any feature and the final score for a stage should not be expected, unless it showed extreme values (a combination of the three features and other assessments made by the physicians is what should determine the score), the presented results show the practical viability of this method as a uniform source of detailed, precisely graded information, as well as a verification of the tendencies that they are expected to follow in practice. More advanced, custom analyzes can be developed on this basis. A new graphical user interface, through which a user can process this information *during* the DGI assessment, obtain the features and show them to the physicians, either graphically or with simple, select indexes as shown in table 1, can be developed, providing access to relevant, objective information that can improve and broaden the scope of this assessment.

During the evaluation process, physicians remarked the possible effect of "training" on some of the tests. As subjects repeatedly perform these similar tasks, they get used to doing so, and this habit can affect how they respond in specific parts of the tests (stressing its effect on the timed, Get Up and Go evaluation). The proposed features can be used to actually measure this effect, by comparing two separate normal walks from the same subject at the beginning and end of the combined tests. Finally, it should be noted that certain variables from the original test can affect the proposed features: for example, the number of times and how often physicians command each subject to perform head turns or speed changes, which are not considered in the clinical assessment. For future work, commands can be emitted automatically, and saved with the acceleration signals, opening the possibility for an even more precise analysis (determining how much the features change, and how long it takes for the subject to perform the change).

6. References

[1] World Health Organization, *World Report on Disability*, 2011

[2] Baezner H et al, Association of gait and balance disorders with age-related white matter changes, *Neurology*, vol. 70, pp 935–942, 2008

[3] Shumway-Cook A and Wollacott M, *Motor Control: Theory and Practical Applications*, Williams and Wilkins, 1995

[4] Yang C and Hsu Y, A Review of Accelerometry-Based Wearable Motion Detectors for Physical Activity Monitoring, *Sensors*, vol 10, pp 7772-7788, 2010

[5] Mayagoitia R E, Lötters J C, Veltink P H and Hermens H, Standing balance evaluation using a triaxial accelerometer, *Gait and Posture*, vol 16, pp. 55-59, 2001

[6] Bugané F, Benedetti M G, Casadio G, Attala S, Biaggi F, Manca M and Leardini A, Estimation of spatial-temporal gait parameters in level walking based on a single accelerometer: validation on normal subjects by standard gait analysis, *Comput. Methods and Programs In Biomed,* XXX, Elsevier, 2012

[7] Vecchio R, Miralles M T, Ghersi I, Val C G, Capalbo I and Rodriguez H, Portable and programmable wíreless system for evaluation of balance and gait based on MEMS capacitive triaxial accelerometer, 2012, *Proc. Argencon 2012*, Córdoba, Argentina, 2012 [http://www.argencon.org.ar/sites/default/files/121.pdf]

[8] Miralles M, Vecchio R, Ghersi I, Paterson R, Pérez Akly M, Ferrando M, Paterson A and Álvarez F, Estudios de equilibrio en pacientes con riesgo de caída a partir de datos de acelerometría en tres ejes, *2012, Proc. XIV Jornadas Internacionales de Ingeniería Clínica y Tecnología Médica*, Paraná, Argentina

Contribution of arterial tree structure to the arterial pressure fractal behavior

Leandro J. Cymberknop[1,3], Ricardo L. Armentano[1,2,3], Walter Legnani[1,3], Franco M. Pessana[1,2,3], Damian Craiem[2], Sebastian Graf[2] and Juan G. Barra[2]

[1]Buenos Aires Regional Faculty, National Technological University, Buenos Aires, Argentina
[2]Favaloro University, Buenos Aires, Argentina
[3]School of Advanced Studies in Engineering Sciences, National Technological University, Buenos Aires, Argentina

Email: ljcymber@ieee.org

Abstract Arterial vascular beds can be characterized considering the arterial segments in terms of their physical properties. These and other trees have an open structure based on repeated bifurcations, following fractal rules. Fractal dimension (***FD***) quantifies the time series complexity defined by its geometrical representation. **Objective**: To evaluate the arterial pressure and diameter time series in order to assess the influence of arterial tree structure in arterial pressure fractal dimension (***FD***). **Methods**: Simultaneous aortic pressure and diameter were measured in 14 conscious dogs. A pair of ultrasonic crystals, a pressure microtransducer and a pneumatic cuff occluder were positioned in the upper third of the descending aorta. **Results**: Total reflection induced by the occlusion maneuver decreased ***FD*** concomitant to the aortic stiffening and early wave reflection. **Conclusion**: Arterial pressure fractality is highly dependent on the arterial tree structure.

1. Introduction

Arterial stiffening is a common but highly variable disorder that is associated with advancing age and exacerbated by many known cardiovascular disease risk factors, including genetic factors. The cardiac muscle provides blood flow to the arterial system, exerting hemodynamic forces on the vessel walls. About this, the main function of systemic circulation is to guarantee a continuous blood flow at capillary level. Previous works have stated that excessive arterial pulsatility is associated with various common diseases of aging and hypertension [1, 2]. In this sense, mechanic behavior of large arteries, denoted by its viscoelastic properties as well as the distributed nature of the terminations, play a fundamental role [3, 4]. Arterial vascular beds can be characterized either by determination of a global parameter or by considering the arterial segments in terms of their physical properties. One of the main features of arterial beds is the tree-like structure [4]. Accordingly, the term "arterial tree" refers to the branching structure from a main conduit artery up to the capillary beds, without including the latter. These and other trees in the arterial circulation have an open structure based on repeated bifurcations, following fractal rules [5]. A fractal time series behaves similarly at different degrees of magnification (observation scales), which may be deterministic or stochastic. In geometric terms, a fractal signal cannot be described or quantified by usual Euclidean measures, owing to its high irregularity [6]. Additionally, arterial wave reflection is recognized as an important phenomenon affecting pressure and flow contour from the ejecting ventricle [7]. Concerning the above, a time series complexity can be quantified in terms of its fractal characteristic. Part of the complexity arises from the basic structure of the cardiovascular system. The heart and the vasculature contain structures, which have a fractal-like appearance. The basic pattern of blood distribution is fractal, and this is imposed both by the anatomy of the vascular tree and by the local regulation of vascular tone [8]. In previous studies, a wavelet transform based technique was applied to arterial blood pressure and diameter time series, in order to reveal structures or patters that cannot be observed by usual procedures. Self-similar presence was confirmed, which is a necessary condition (but not sufficient) for a signal to be considered as a fractal [9]. As a consequence of that behavior, fractal dimension (***FD***) estimation was adopted, in order to quantify the loss of waveform complexity in blood pressure time series under pathological conditions [10]. In addition, these previous works suggested that stiffening is associated with a decrease in the fractal nature of the aortic pressure, but the mechanism remains unrevealed. Even more, ***FD*** can be seen as a measure of irregularity, roughness and variation [11].

To our knowledge, the effect of the arterial tree structure on arterial pressure fractality has not been previously reported. The goal of the present work was to analyze the fractal complexity of the arterial pressure wave in basal state and during total occlusion, where multiple branching reflections are avoided.

2. Material And Method

2.1. Experiments

Data of previous protocols performed in 14 male mongrel dogs (4.9±1.9 years, 22.2±2.9 kg) were recompiled to this study [12, 13]. In all cases, anesthesia was induced with intravenous thiopental sodium (20 mg/kg) and, after intubation, maintained with 2% enflurane carried in pure oxygen (4 L/min) through a Bain tube connected to a Bird Mark VIII respirator. Under sterile conditions, a left thoracotomy was made at the fifth intercostal space. A pressure microtransducer (Konigsberg P7, 1200 Hz frequency response) and a fluid-filled polyvinyl chloride catheter (2.8 mm OD, for later calibration of the microtransducer) were implanted in the descending thoracic aorta through a little incision in the left brachial artery. A pair of ultrasonic crystals (5 MHz, 4-mm diameter) was sutured on the adventitia of the aorta, after minimal dissection, to measure external aortic diameter. The transit time of the ultrasonic signal (1580 m/s) was converted into distance using a sonomicrometer (Triton Technology Inc., 100 Hz frequency response) and observed on the screen of an oscilloscope (Tektronix 465B) to confirm optimal signal quality. A polyvinyl chloride catheter (2.3-mm OD) was advanced through the left mammary vein to lie in the superior vena cava or right atrium for drug administration. A pneumatic cuff occluder made from silicon rubber was implanted around the descending thoracic aorta, proximally to the pressure transducer and the ultrasonic crystals, at a distance of 150 mm from the left ventricle (Figure 1). Aortic arc bifurcations (carotid, subclavian and brachiocephalic trunk arteries) were not subjected to occlusive maneuvers. Before repairing the thoracotomy, all cables and catheters were tunneled subcutaneously to emerge at the intercapsular space. At least one week after surgery, the aortic pressure was registered using the pressure microtransducer, which had been calibrated against a Statham P23-D transducer connected to the aortic fluid-filled catheter. The zero reference point was set at the level of the right atrium. The Statham transducer had been previously calibrated using a mercury manometer. The external aortic diameter signal was calibrated in millimeters using the 1-mm step calibration facility of the sonomicrometer. Aortic pressure and diameter signals were digitized every 4 ms on a computer using an analog to digital converter (National Instruments Lab PC) and digitally stored for later analysis. Instantaneous pressure-diameter loops were displayed on-line on the computer monitor. The acquisition started with approximately 20 consecutive beats in basal state (baseline state) and then the pneumatic occluder was inflated in order to obtain a total wave reflection (total occlusion state) [12, 13]. The instantaneous pressure-diameter loops were monitored and registered until stabilization was evidenced. After completion of the protocols, animals were euthanized with an intravenous overdose of thiopental sodium followed by potassium chloride; the correct position of the ultrasonic crystals was confirmed at necropsy in all cases.

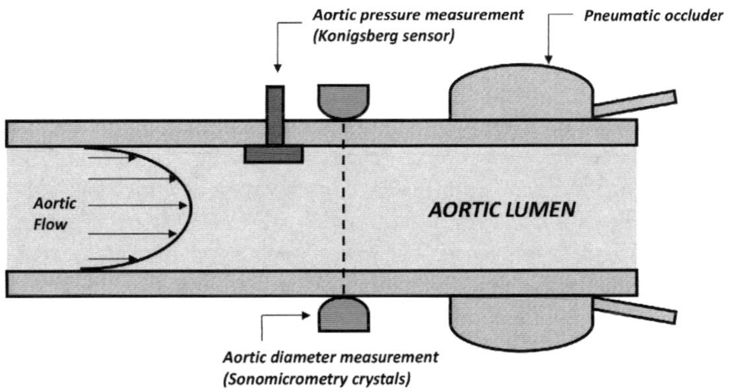

Figure 1. Descending thoracic aorta instrumentation. Blood pressure was measured through the implantation of a high fidelity Konigsberg® transducer. Aortic pulsatility was determined by means of the sonomicrometry technique. Reflections from periphery were avoided using a pneumatic occluder.

All protocols were approved by the Research and Development Council of the Favaloro University, and the study was conducted in accordance with the Guide for the Care and Use of Laboratory Animals published by the United States National Research Council (National Academy Press, Washington, DC, 1996).

2.2. Fractal Dimension. Higuchi's method

Fractal dimension quantifies how densely a metric space is occupied by the fractal set [14]. Moreover, ***FD*** determines the time series complexity measure defined by its geometrical representation [15]. From a theoretical point of view, a fractal can be defined as an affine self-similar set, whose Hausdorff dimension (a measure of the space "filled" by the set at its point's neighborhood) is strictly larger than its topological dimension [16]. Considering a time series of one time dependent variable, its ***FD*** value is included in the interval [1, 2]. While the Hausdorff dimension is the most relevant measure, on a practical level, Box counting (***BCD***) or Correlation dimensions (***CD***) are implemented more frequently [17]. The former has been selected for ***FD*** analysis, during the present study. Assessment of ***FD*** in this study was performed by applying the method proposed by Higuchi [18]. A number of subsets based on the original temporal series (***x(t)***, of length ***N***) are generated, considering an initial time value (***m***) and a temporal increment (***k***) as parameters, as follows:

$$x_k^m = \left\{ x(m); x(m+k); x(m+2k),.., x\left(m + \left[\frac{N-m}{k} \right] k \right) \right\} \tag{1}$$

The term [***(N-m)/k***] in (1) denotes the maximal time interval (Gauss notation) that can be considered for a selected m value. For each experimental time series, an averaged length is calculated (***Lm(k)***), as can be observed in the following expression:

$$L_m(k) = \frac{\sum_{i=1}^{\left[\frac{N-m}{k}\right]} \left| x(m+ik) - x(m+(i-1)k \right| \dfrac{N-1}{\left(\frac{N-m}{k}\right)k}}{k} \tag{2}$$

Then, the time series length function for each time increment (***L(k)***) is assessed, according to the expression:

$$L(k) = \frac{\sum_{m=1}^{k} L_m(k)}{m} \tag{3}$$

Finally, if $L(k) \propto FD$ is found, the time series morphology may be quantified by its ***FD*** value. The latter can be obtained by applying a linear regression method to a doubly logarithmic scale representation of ***L(k)*** against ***1/k***. In addition, maximal value of time interval ***k*** (k_{max}) should be emphasized, especially if adequate accuracy is required in the ***FD*** estimation process.

Higuchi proposed a method that may be applied to any kind of time series, stationary or not. However, the obtained value lacks of information related to the system involved (deterministic, chaotic or stochastic), which is responsible for the signal being analyzed. In consequence, the method should be applied in the evaluation of variations that have occurred in the same signal (before and after significant events or different physiological states, such as those considered for this procedure) [17]. During the present study, arterial pressure time intervals acquired at a sampling rate of 250 Hz were processed by means of Higuchi's method. A value of 16 was adopted for k_{max} and was estimated by applying linear regression analysis to consecutive groups of 5 points, belonging to the doubly logarithmic graph, which remained within an error band of 5% maximum variation.

2.3. Signal Processing algorithms development

Signal processing algorithms were developed on MatLab platform (MathWorks INC, Massachusetts, USA). Existing trends were eliminated (i.e. respiration induced fluctuations, electrical artifacts, etc.) by means of digital filtering (de-noising method) [19]. In order to characterize non-stationary events, Higuchi's method was applied to short segments of the time series (2 beats interval). A maximal window length was selected in order to ensure, simultaneously, maximal number of samples as well as a stationary behavior of the time interval (moving window method) [17].

2.4. Statistical analysis

Data were expressed as mean values ± standard deviation (SD). The presence of significant differences between baseline state and total occlusion state was assessed using a paired Student's t-test. For all statistical analyses, $P<0.05$ was adopted as statistically different.

2.5. Aortic mechanical behavior

Vascular mechanical properties were evaluated based on instantaneous recordings. In this sense, aortic stiffness was studied by means of pressure – diameter loops both in basal and occlusion states. The pure elastic relationship was obtained during diastolic phase. In terms of the contribution of the entire arterial tree (with the exception of the upper limb arterial conduits) to the pressure waveform morphology, augmentation index (AI_X) was calculated, which is considered as a measure of central pressure wave reflection.

2.5.1 Aortic stiffness

Changes in aortic wall stiffness were estimated assuming that the arterial wall is an isotropic homogeneous elastic material. Consequently, a linear elastic theory was applied, as follows:

$$E = \frac{dP}{d\varepsilon} \tag{4}$$

where $\boldsymbol{E}$ is the pressure-strain elastic modulus [20], $\boldsymbol{P}$ corresponds to the measured aortic pressure and $\boldsymbol{\varepsilon}$ constitutes the aortic strain. The variation of $\boldsymbol{\varepsilon}$ was obtained by referring the dynamic diameter to its non-stressed value. A biphasic model was adjusted, assuming two linear regions with different elastic modulus. According to this, low pressure slope is related to the elastin elastic response while the high pressure slope indicates the recruitment of collagen fibers. Both behaviors were quantified by means of linear regression analysis.

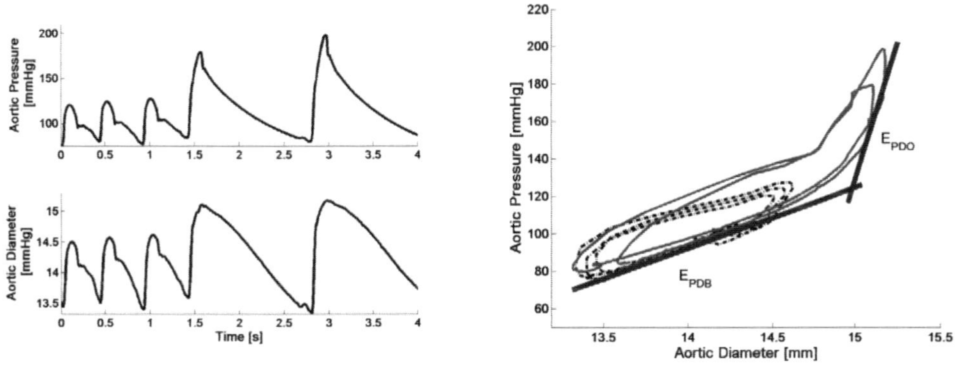

(a) ***(b)***

Figure 2. (*a*) Higher Panel: In-vivo measured aortic arterial pressure. **Lower Panel**: In-vivo measured aortic arterial diameter. Occlusion state is observed in the last two beats ***(b)*** Basal state (dashed line) and occlusion state (solid line) aortic pressure loops. $\boldsymbol{E_{PDB}}$ and $\boldsymbol{E_{PDO}}$ constitute the aortic elastic modulus for both physiological states, respectively.

2.5.2 Aortic Augmentation Index

Augmentation index is defined as the ratio of the peak above the *shoulder* of the pressure wave ($\boldsymbol{\Delta P}$) to the pulse pressure ($\boldsymbol{PP}$). In relation to this, assessment of $\boldsymbol{\Delta P}$ was performed from the fourth time derivative of the aortic pressure, according to [21]:

$$AI_X = \frac{\Delta P}{PP}\% \tag{5}$$

3. Results

Aortic pressure and diameter measured signals may be observed in Figure 2 (a), both at basal state (first three beats) as well as at total occlusion state (last two beats). Pressure-diameter loops behavior and $\boldsymbol{E}$ estimation for both states are shown in Figure 2 (b) for a representative dog.

Table 1. Aortic pressure aortic stiffness, augmentation index and fractal dimension variation between normal and occlusion states.

Parameter	Basal State	Occlusion State
Elastic Modulus [10^6dyn/cm^2]	0.70±0.17	1.79±0.85*
Augmentation Index [%]	21,38±5,19	49,40±7,41*
Fractal Dimension	1.07±0.02	1.02±0.01*

*P<0.05 was considered as significant different. Values are expressed as mean ± standard deviation

Concerning ***FD*** values continuous variation, a significant decrease in aortic pressure waveform complexity (pointed out by a ***FD*** diminution) is observed during the occlusion interval. Assessments obtained by applying Higuchi's method, both under baseline and after the occlusion maneuver, are detailed in Table I. Corresponding values of aortic stiffness and augmentation index are also included.

4. Discussion

The purpose of the present study was to assess the effect of arterial tree structure on arterial pressure fractal behavior. ***FD*** was applied, as a non-linear measure, in order to quantify the waveform morphology complexity (or roughness). To this end, Higuchi's method was implemented, which is widely utilized in non-linear signal processing literature. Furthermore, arterial wall stiffness was evaluated using the first derivative of the pressure-strain relationship while the effect of wave reflection was estimated from AI_X measurements. As a result, total reflection induced by pneumatic occluder decreased ***FD*** concomitant to the aortic stiffening and increment of AI_X.

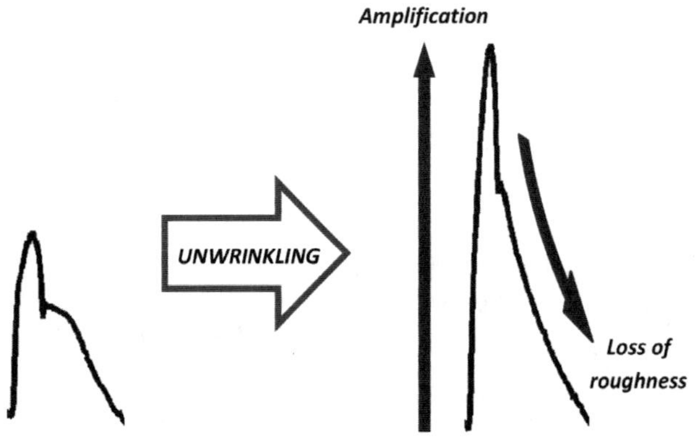

Figure 3. Aortic blood pressure unwrinkling phenomenon (loss of waveform complexity) as a consequence of the occlusive maneuver.

Regarding the above, it may be inferred that the higher ***FD*** values assessed in basal state are the result of the manifestation of reflected waves, emerging from the presence of vascular bifurcations. Therefore, this behavior suggests that under total reflection (only carotid, subclavian and brachiocephalic branches remained without being occluded) aortic blood pressure *unwrinkling* (pulsatility increase in conjunction with a loss of roughness) is the consequence of the absence of the fractal nature, induced by the multiple branching of the arterial tree (Figure 3).

It is noteworthy that ***FD*** analysis was performed over short time intervals, no longer than two heartbeats. This consideration allowed the insolation of the intrinsic mechanical response, preventing the contribution of reflex regulation mechanisms (which take place around the fifth heartbeat after occlusion) whose intervention might contaminate the waveform structure. Additionally, the heart rate decrease, induced during the occlusion maneuver, constituted another intriguing result. Considering that the development of the phenomenon occurred in less than a time constant of the system, the observation clearly deserves further investigation. Moreover, it is well known that physiological process originate complex, anisotropic fluctuations, which cannot be processed properly by usual measures. In this sense, fractal based techniques provide a nature based approach, in order to identify the presence of multi-scale interactions [8]. The present results suggest an evident trend between arterial wall reflection and the morphology of the acquired pressure time series. In previous studies [9, 10] it was demonstrated that loss of fractal complexity was related to an increase of arterial stiffness. In this study, the influence of wave reflection from the descending thoracic aorta and its posterior successive bifurcations (the upper vascular bed was not interrupted) was also evaluated. What remains to be investigated, is if arterial stiffness can be considered as a factor in FD diminution or may be induced by the presence of reflected waves.

In conclusion, arterial pressure fractality is highly dependent on the arterial tree structure. Further studies will be needed in order to demonstrate the role of arterial stiffness itself and its relation to wave reflection in the loss of arterial pressure fractal complexity.

5. Acknowledgments

This work was supported by National Technological University (NTU) – FONCyT – IP – PRH – 2007.

6. References

[1] Safar M. E., Levy B. I. and Struijker-Boudier H. Current Perspectives on Arterial Stiffness and Pulse Pressure in Hypertension and Cardiovascular Diseases 2003 *Circulation* 107(22): 2864–69.

[2] Armentano RL, Graf S, Barra JG, Velikovsky G, Baglivo H., Sanchez R, Simon A, Pichel RH and Lenvenson J Carotid wall visosity increase is related to intima media thickening in hypertensive patients 2004 *Stroke* 35.

[3] M. G. Taylor The input impedance of an assembly randomly branching elastic tubes 1966 *Biophys. J.* 6(1): 29-51.

[4] A. Avolio, Input impedance of distributed arterial structures as used in investigations of underlying concepts in arterial haemodynamics 2009 *Med. Biol. Eng. Comput.*, 47(2): 143-151, 2009.

[5] M. Zamir, Arterial branching within the confines of fractal L-system formalism 2001 *J. Gen. Physiol.* 118(3) 267-276.

[6] K. Falconer 2003 *Fractal Geometry: Mathematical Foundations and Applications* (2nd. ed. Wiley).

[7] R. Burattini and K. B. Campbell Comparative analysis of aortic impedance and wave reflection in ferrets and dogs 2002 *Am. J. Physiol. Heart Circ. Physiol.* 282(1): H244-255.

[8] V. Sharma Deterministic chaos and fractal complexity in the dynamics of cardiovascular behavior: perspectives on a new frontier 2009 *Open Cardiovasc. Med. J.* 3: 110-123.

[9] L. J. Cymberknop, W. Legnani, F. Pessana, D. Bia, Y. Zócalo and R. L. Armentano Stiffness indices and fractal dimension relationship in arterial pressure and diameter time series in-vitro 2001 *J. Phys.: Conference Series*, 332: 012024.

[10] L. J. Cymberknop, W. Legnani, F. M. Pessana, A. Crottogini and R. L. Armentano Coronary arterial stiffness is related with a loss of fractal complexity in the aortic pressure 2012 *Conf. Proc. IEEE Eng. Med. Biol. Soc.* 4200-03.

[11] J. B. Bassingthwaighte Physiological Heterogeneity: Fractals Link Determinism and Randomness in Structures and Functions 1988 *News Physiol. Sci.* 3(1): 5-10.

[12] R. L. Armentano, J. G. Barra, J. Levenson, A. Simon and R. H. Pichel Arterial wall mechanics in conscious dogs. Assessment of viscous, inertial, and elastic moduli to characterize aortic wall behavior 1995 *Circ. Res.* 76(3): 468-478.

[13] J. G. Barra, J. Levenson, R. L. Armentano, E. I. Cabrera Fischer, R. H. Pichel and A. Simon In vivo angiotensin II receptor blockade and converting enzyme inhibition on canine aortic viscoelasticity 1997 *Am. J. Physiol.* 272(2): H859-68.

[14] M. F. Barnsley 1983 *Fractals Everywhere* (2nd. ed., New York. Academic Press Professional).

[15] W. Klonowski From conformons to human brains: an informal overview of nonlinear dynamics and its applications in biomedicine 2007 *Nonlin. Biomed. Phys.* 1(1): 5.

[16] B. Mandelbrot 1983 *The Fractal Geometry of Nature* (Freeman, New York).

[17] B. S. Raghavendra and Narayana D. Computing fractal dimension of signals using multiresolution box counting method 2010 *Journal of Math. Sciences* 6(1).

Journal of Physics: Conference Series **477** (2013) 012030 doi:10.1088/1742-6596/477/1/012030

[18] T. Higuchi, Approach to an irregular time series on the basis of the fractal theory 1988 *Physica D.* 31: 277-283.
[19] D. L. Donoho De-noising by soft-thresholding 1995 *Information Theory IEEE Trans.* 41(3): 613-627.
[20] R. L. Armentano, E. I. Cabrera Fischer, J. G. Barra, J. A. Levenson, A. C. Simon and R. H. Pichel Single beat evaluation of circumferential aortic elastin elastic modulus in conscious dogs Potential application in non-invasive measurements 1994 *Med. Prog. Technol.*, 20(1-2): 91-99.
[21] R. Kelly, C. Hayward, A. Avolio and M. O'Rourke Noninvasive determination of age-related changes in the human arterial pulse 1989 *Circulation* 80(6) 1652-59.

Objective Response Detection of Multiple Auditory Steady-State Responses: Rice Detector vs Component Synchrony Measure

J J Silva[1], T Santos[2], C J Tierra-Criollo[1,2,3] and D B Melges[1]

[1]Graduate Program in Electrical Engineering - Federal University of Minas Gerais - Av. Antônio Carlos 6627, 31270-901, Belo Horizonte, MG, Brazil
[2] Graduate Program in Neuroscience - Federal University of Minas Gerais - Av. Antônio Carlos 6627, 31270-901, Belo Horizonte, MG, Brazil
[3]Biomedical Engineering Program - Federal University of Rio de Janeiro – Av. Horácio de Macedo 2030, 21941-914, Rio de Janeiro, RJ, Brazil

E-mail: jefersonjs2@yahoo.com.br, thamara.santos86@gmail.com, carjuliot@gmail.com, danilomelges@gmail.com

Abstract. Multiple auditory steady state responses (MASSR) is used to assess the integrity auditory pathways for different frequencies applied simultaneously without significant increases in exam time. Objective Response Detection (ORD) techniques have been employed for hearing screening since it conducts to response identification with statistical basis. This work aims at comparing time and detection rate of two ORD, Rice detector (RD) and Component Synchrony Measure (CSM). The binaural auditory stimulation was applied to 21 normal hearing infants and consisted of four amplitude-modulated (AM) tones presented simultaneously on each ear with the carriers 500, 1000, 2000 and 4000 Hz, and at 50 and 40 dB_{SPL} intensity levels. The CSM presented higher detection rates than RD for all carriers, both intensities and ears (statistical difference, assessed by proportion test with significance level of alfa=0.05, only for 500 Hz in both ears and for 1000 Hz in the left ear). Moreover, detection times were lower for CSM compared to RD (unpaired Wilcoxon rank-sum test with alfa=0.05 indicated statistical difference for 1000 Hz at 50 dB_{SPL} and 2000 Hz at 50 dB_{SPL} both for the left ear). Hence, the CSM could be considered more effective for hearing screening purposes.

1. Introduction

The detection of congenital hearing losses followed by the appropriate diagnosis and intervention allows the child development without impairment in their linguistic skills [1]. In this context, the improvement of hearing assessment techniques has been the motivation of several studies [2,3,4,5], which aimed to achieve more effective universal hearing screening programs. The employed techniques for the auditory system evaluation includes otoacoustics emission test [6], the Brainstem Evoked Response Audiometry (BERA) and, recently, the auditory response to amplitude-modulated sinusoidal stimulation [2,3,4].

The otoacoustics emission is a quick test performed to detect hearing loss for intensities higher than 30 dB_{HL} (HL - Hearing Level) [7]. However, it only evaluates the external ciliated cells, hence, it is not suitable to determine hearing threshold or identify whether the hearing loss is conductive or sensorineural [7]. On the other hand, BERA is able to determine the hearing threshold, although it becomes very time-consuming for evaluating specific frequencies hearing loss [2]. Moreover, BERA allows inferring about the hearing pathways integrity only until the inferior colliculus on the brainstem [8]. The assessment of cortical auditory information processing is possible through the Auditory Steady State Response (ASSR), which can be obtained by applying auditory clicks at high stimulation rates or modulated tones, leading to time overlapping evoked responses [9].

For the amplitude-modulated stimuli, the basilar membrane, located within the cochlea, vibrates in the specific region related with the carrier frequency. The basilar membrane vibration produces deflection on the ciliated cells that are in contact with the tectorial membrane. Since the ciliated cells depolarize for deflections in only one direction, a rectified stimulation response is obtained [2]. The spectrum of the ASSR presents a component in the modulation frequency, which is related to the hearing sensitivity to the corresponding carrier [2]. Therefore, employing several modulated tones at different frequencies simultaneously can result in a faster exam for hearing sensitivity evaluation [2]. This method is known as multiple auditory steady-state responses (MASSR) and employ objective response detection (ORD) techniques, which are based on statistical tests, for identifying the stimuli response with a maximum false positive rate, previously established.

Particularly, the frequency-domain ORD techniques have been studied in order to minimize the response detection time. These techniques may use only the magnitude of the Fourier Transformed EEG such as the Spectral F Test –SFT [6], only the phase as in the Component Synchrony Measure – CSM [10], or both such as in the Magnitude-Squared Coherence – MSC [11]. The SFT, CSM and MSC have been used to MASSR in order to minimize the detection time and maximize the detection probability [4,5]

Recently, FARINA et al. [12] introduced a frequency-domain ORD technique based on the Rice distribution (Rice Detector – RD), for somatosensory evoked potential detection, obtaining promising results. Therefore, this paper proposes comparing the performance of RD and CSM for MASSR detection.

2. Materials and methods

2.1. Component Synchrony Measure (CSM)

The CSM measures the degree of synchronism of each spectral component considering only the phase of Fourier Transform and can be estimated by [10]:

$$\hat{\rho}^2(f)=\left(\frac{1}{M}\sum_{i=1}^{M}\cos\theta_i(f)\right)^2+\left(\frac{1}{M}\sum_{i=1}^{M}\mathrm{sen}\,\theta_i(f)\right)^2 \tag{1}$$

where $\theta_i(f)$ is the phase of the ith EEG epoch of the spectral component f. If no response to the stimulation is expected (null hypothesis of response absence), the phase of EEG Fourier Transformed epochs is assumed to be randomly distributed between 0 and 2π and it can be showed that the CSM probability density function tends asymptotically to a chi-squared distribution with 2 degrees of freedom (χ_2^2) [10].

Hence, the critical value for the CSM estimates can be calculated by [10]:

$$\hat{\rho}_{crit}^2=\frac{\chi_{2,\alpha}^2}{2M} \tag{2}$$

Where M is the number of EEG epochs and α is the significance level of the statistical test.

The detection is obtained when the estimate exceeds the critical value ($\hat{\rho}^2(f)>\hat{\rho}_{crit}^2$).

Journal of Physics: Conference Series **477** (2013) 012031 doi:10.1088/1742-6596/477/1/012031

2.2. Rice Detector (RD)

The Rice Detector (RD) evaluates the presence of stimuli response based on the EEG Fourier Transformed epoch $X_i(f)$ and calculated as [12]:

$$\hat{r}_m = \frac{1}{M}\left[\left(\sum_{i=1}^{M} Re\left(X_i(f)\right)\right)^2 + \left(\sum_{i=1}^{M} Im\left(X_i(f)\right)\right)^2\right]^{1/2} \quad (3)$$

where M is the number of epochs used in the average calculation.

The $\hat{r}_m$ analytical critical value for a given significance level α can be calculated as [12]:

$$\hat{r}_{m_{crit}} = \left(\frac{2\sigma^2 \ln\left(\frac{1}{\alpha}\right)}{M}\right)^{1/2} \quad (4)$$

Since this value depends on the signal variance σ, and as this parameter is not known a priori, the statistics of interest can be redefined as suggested by [12]:

$$\hat{\xi} = \frac{\hat{r}_m}{\sigma} \quad (5)$$

Thus the critical value for the new metrics depends only on the significance level α and the number of epochs M [12]:

$$\hat{\xi}_{crit} = \left(\frac{2\ln\left(\frac{1}{\alpha}\right)}{M}\right)^{1/2} \quad (6)$$

Similarly, the detection is obtained when the estimate exceeds the critical value ($\hat{\xi}(f) > \hat{\xi}_{crit}$).

2.3. EEG acquisition

The electroencephalographic signals were collected from 21 individuals, aged between 9 and 11 years old, without history of neuropathies and normal hearing verified by tonal threshold audiometry. The signals were recorded, in a background noise controlled environment, by means of electrodes positioned in derivation [Cz] of the 10-20 International System. The ground electrode was located at [Fpz] and the signal reference about 3 cm below the inion.

The stimulus consisted of four AM tones presented simultaneously on each ear (binaural stimulation), at the intensities of 50 and 40 dB_{SPL} (SPL - Sound Pressure Level) by insert earphones 5A model (Aearo-Technologies). The modulation frequencies used for right ear were 77.15, 86.91, 98.63 and 104.49 Hz, respectively, for the carriers 500, 1000, 2000 and 4000 Hz. For the left ear, the modulation frequencies were 81.05 (carrier of 500 Hz), 94.73 (1000 Hz), 100.59 (2000 Hz) and 106.45 Hz (4000 Hz) [4].

The *AudioStim System* [13], developed in the Biomedical Engineering Laboratory of the Federal University of Minas Gerais (UFMG), was employed for both EEG acquisition and stimulation.

This research was approved by the Ethics Committee of UFMG (protocol n. 0369020300010) and the children legal sponsors signed written informed consent.

2.4. Pre-processing

The EEG signal was analogically band-filtered in the range from 30 to 300 Hz. Moreover, a notch filter at 60 Hz was also employed in order to minimize net noise influence. The signal was sampled at 1000 Hz, and then segmented into 1024-samples epochs. The epochs with more than 1% of samples exceeding 10 mV were discarded.

2.5. *Objective Response Detection*

The critical values for both techniques, $\hat{\rho}^2_{crit}$ and $\hat{\xi}_{crit}$, were calculated for the significance level $\alpha = 0.05$, using expression (2) and (6). Both the estimates and critical values were calculated by successively increasing the *M* number of EEG epochs. The detection time was based on five consecutive estimates exceeding the corresponding critical value.

2.6. *Statistical analyses*

The nonparametric unpaired Wilcoxon rank-sum test was applied for comparing the detection time between CSM and RD. The detection percentages were compared employing the proportion test. The significance level of 0.05 was used for both statistical tests.

3. Results

Figure 1 compares the detection performance of CSM and RD for each carrier frequency and the 21 individuals evaluated at intensity of 50 dB_{SPL}. The detection percentages were lower than 75% for CSM and RD at the carrier of 500 Hz in both ears. Detection response occurred for, 95.2% at 1000 and 4000 Hz, and 100% at 2000 Hz for right ear in CSM (Figure 1, top chart), while RD presented response only for 85.7% at 1000 and 2000 Hz, and 81.0% at 4000 Hz. For the left ear, CSM presented detection percentage equal to 85.7, 100 and 90.5%, for 1000, 2000 and 4000 Hz, respectively; while RD showed rates of 76.2, 90.5 and 76.2%. Hence, CSM showed superior performance for all cases, but without significant statistical difference.

The results for 40 dB_{SPL} are presented in Figure 2 for both ears. The detection rates for CSM and the right ear were of 85.7, 95.4 and 100%, respectively for the carriers 1000, 2000 and 4000 Hz. The RD showed 66.7% (1000 Hz), 90.5% (2000 Hz) and 81.0% (4000 Hz) for the same ear (Fig. 2, top chart). In the left ear, lower rates were found such as 66.7% (1000 and 4000 Hz) and 90.5% (2000 Hz) for CSM and 28.6% (1000 Hz), 85.7% (2000 Hz) and 57.1% (4000 Hz) for RD. The proportion test showed significant difference only for the carriers 500 Hz ($p<0.07$ for both ears) and 1000 Hz ($p<0.0005$ for the left ear), which presented detection rates lower than 77%.

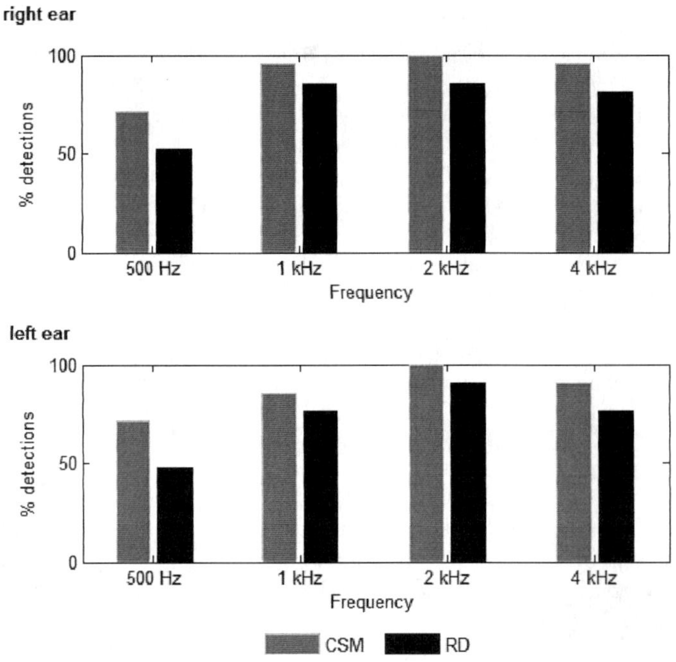

Figure 1. Detection rates for CSM and RD and each carrier at 40 dB_{SPL}.

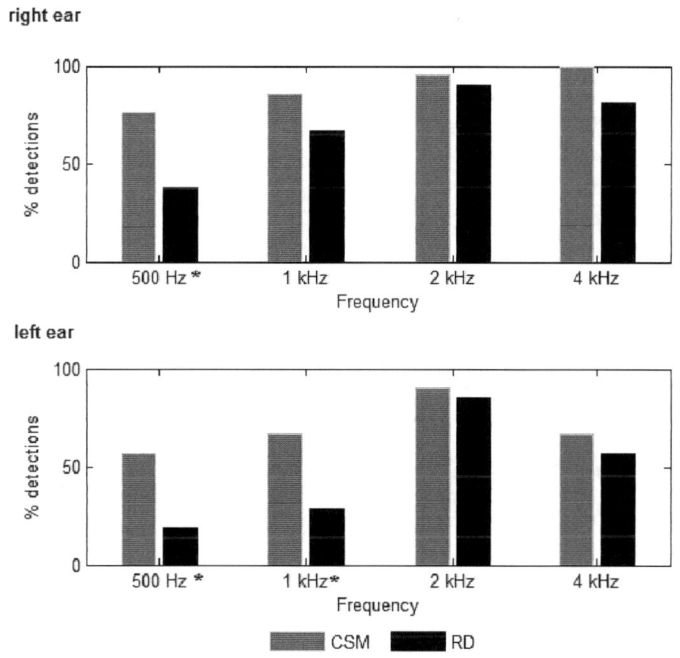

Figure 2. Detection rates for CSM and RD and each carrier at 50 dB_{SPL} (* indicates statistical difference).

The detection times for both techniques are presented in Fig. 3 for stimulation at 50 dB_{SPL} by means of boxplot. Table 1 summarizes the time medians for each carrier and both techniques and ears. It also presents the p-values for comparison of the medians by applying the unpaired Wilcoxon rank-sum test. Statistical significant difference was found only for 1000 Hz, left ear.

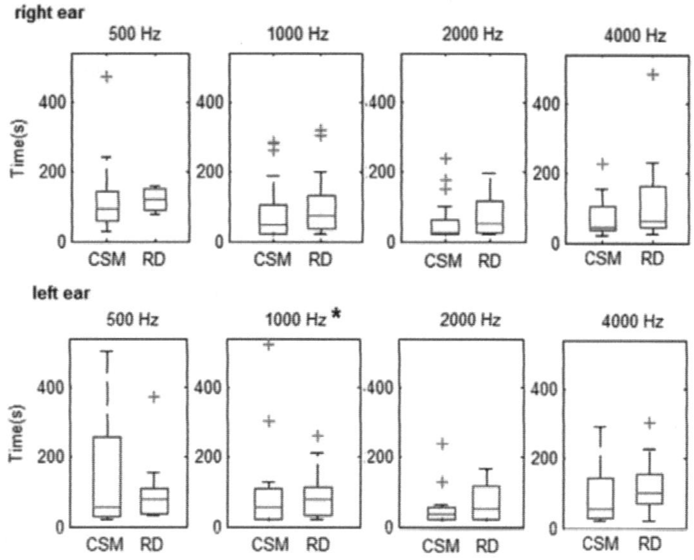

Figure 3. Boxplot of detection times for CSM and RD at 50 dB_{SPL} (* indicates statistical difference).

Table 1. Results of the unpaired Wilcoxon rank-sum test and medians of detection times (in seconds) for each carrier and both techniques and ears – stimulation at 50 dB_{SPL}.

	Carrier	CSM	RD	p-value
Right ear	500 Hz	95.2	119.3	0.3456
	1000 Hz	49.2	73.7	0.1596
	2000 Hz	26.6	50.2	0.1130
	4000 Hz	45.1	61.4	0.1749
Left ear	500 Hz	56.3	77.8	0.6024
	1000 Hz	56.8	80.4	0.0176
	2000 Hz	35.8	51.2	0.1181
	4000 Hz	57.3	102.4	0.1162

Finally, Fig 4 shows the detection times for CSM and RD and stimulation at 40 dB_{SPL}. The time medians are presented in Table 2 together with the result of the Wilcoxon test. Statistical significant difference was found only for 2000 Hz, left ear.

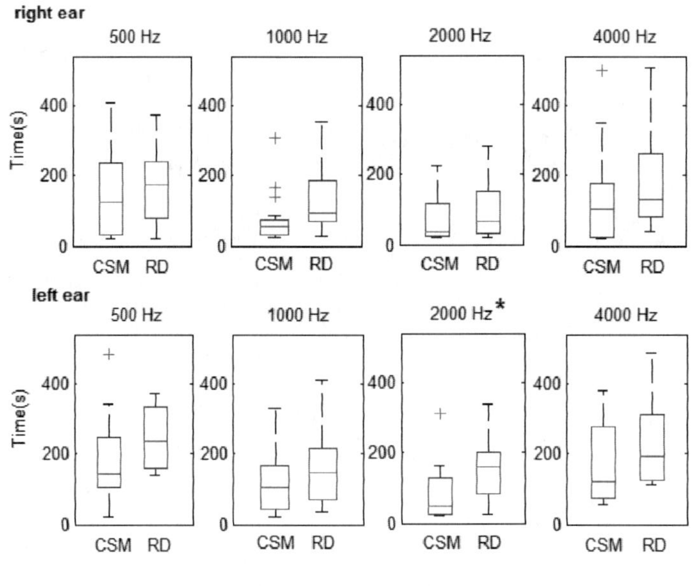

Figure 4. Boxplot of detection times for CSM and RD at 40 dB_{SPL} (* indicates statistical difference).

Table 2. Results of the unpaired Wilcoxon rank-sum test and medians of detection times (in seconds) for each carrier and both techniques and ears – stimulation at 40 dB_{SPL}.

	Carrier	CSM	RD	p-value
Right ear	500 Hz	123.9	172.5	0.7602
	1000 Hz	55.3	94.7	0.3596
	2000 Hz	36.4	68.6	0.0927
	4000 Hz	103.4	132.1	0.2602
Left ear	500 Hz	143.9	236.0	0.2615
	1000 Hz	106.0	146.9	0.3427
	2000 Hz	49.2	160.3	0.0121
	4000 Hz	119.3	193.0	0.1166

4. Discussion

This paper aimed at comparing the performance of CSM and RD for MASSR identification. To our known, this is the first study to employ RD in order to identify MASSR.

FARINA *et al.* [12] have introduced the use of RD applying to somatosensory evoked response detection, and this ORD technique presented performance equivalent to the Magnitude-Squared Coherence (MSC) [12]. The MSC uses magnitude and phase information of the Fourier Transformed EEG epochs, whilst the RD uses only magnitude information and the CSM only phase.

In our study RD showed detection rates lower than the CSM, regardless the frequency or intensity tested, leading to the interpretation that the phase is predominantly essential for ASSR detection. In fact, other studies reported the phase to be more important than magnitude in auditory evoked potential, particularly, for low signal-to-noise ratio [14,15]. Thus, using only magnitude information can lead to a relevant reduction in the detection performance, whilst there is a slight effect when only phase is employed [10,14]. The phase importance for response identification was also corroborated by [16], which reported the phase synchronism with the modulation frequency, even at threshold intensities, especially for amplitude-modulated stimuli, in experiments conducted with mammals.

Despite the low performance of RD, detection percentages difference between the ORD techniques was only significant for 500 Hz (both ears) and 1000 Hz (left ear). For the carrier of 500 Hz, the response identification was poor for both techniques, which agree with other authors results [3,17].

For audiology, both speed and consistency of response detection are important, respectively, as efficiency and stop criteria [1]. For this reason, in this work, the detection time was determined based on positive response identification in five consecutive ORD (CSM or RD) estimates. RD presented higher median detection times than CSM for all frequencies and both ears, although significant difference was only found for 1000 Hz at 50 dB_{SPL} and 2000 Hz at 40 dB_{SPL} on the left ear. This result agrees with other works, since the phase spectrum is important not only for achieving higher detection rates, but also for fastening the exam [10,14].

5. Conclusions

The CSM presented better performance than RD for both studied criteria: detection rate and detection time, with statistical significance for some of the employed carriers. These results point to the predominance of phase synchronism in the auditory response and indicate the CSM as preferable for hearing screening programs

Acknowledgments
To CAPES, FINEP, CNPq and FAPEMIG for financial support.

References

[1] Joint Committee on Infant Hearing - JCHT 2007 *Position statement: Principles and guidelines for early hearing detection and intervention programs* **120** 898-921.

[2] Lins O G and Picton T W 1995 Auditory steady-state responses to multiple simultaneous stimuli *Electroencephalogr Clin Neurophysiol* **96** 420–32.

[3] Picton T W, John S M, Dimitrijevic A and Purcell D 2003 Human auditory steady-state responses. *Int J Audiol* **42** 177-219.

[4] Santos T, Silva J J, Melges D B and Tierra-Criollo C J 2012 Component Synchrony Measure: Fase e Tempo de Detecção de Respostas Auditivas Múltiplas em Regime Permanente *XXIII Congr Bras Eng Bioméd* (Porto de Galinhas: XXIII CBEB) pp 1899–993.

[5] Zanotelli T, Santos T, Felix L B and Tierra-Criollo C J 2010 Deteção do Potencial Evocado Auditivo em Regime Permanente Utilizando Coerência e Coerência Múltipla *XXII Congr Bras Eng Bioméd* (Tiradentes: XXII CBEB) pp 1162-5.

[6] Dobie R A and Wilson M J 1996 A comparison of t-test, F test, and coherence methods of detecting steady-state auditory-evoked potentials, distortion product otoacoustic emissions, or other sinusoids *J. Acoust. Soc. Am.***100** 2236–46.

[7] Geal-Dor M, Adelman C, Levi H, Zentner G and Stein-Zamir C 2010 Comparison of two hearing screening programs in the same population: oto-acoustic emissions (OAE) screening in newborns and behavioral screening when infants *Int J Pediatr Otorhinolaryngol* **74** 1351-5.

[8] Zaeyen E J B 2005 *Aplicação da Coerência ao Eletroencefalograma para Investigar Características do Potencial Evocado Auditivo de Média Latência* (Federal University of Rio de Janeiro - UFRJ)

[9] Regan D 1989 *Human brain electrophysiology: Evoked potentials and evoked magnetic fields in science and medicine* (New York: Elsevier)

[10] Simpson D M, Tierra-Criollo C J, Leite R T and Infantosi A F C 2000 Objetive Response Detection in an Electroencephalogram During Somatosensory Stimulation *Annals of Biomedical Engineering* **28** 691-8

[11] Dobie R A and Wilson M J 1989 Analysis of Auditory Evoked Potentials by Magnitude-Squared Coherence *Ear and Hearing* **10** 2-13

[12] Farina Jr P D, Melges D B, Infantosi A F C and Miranda A M F L 2010 Técnica de detecção objetiva de resposta baseada na distribuição de rice *XXII Congr Bras Eng Bioméd* (Tiradentes: XXII CBEB) pp 915–8

[13] Martins H R , Romão M, Plácido D, Provenzano F and Tierra-Criollo C J 2007 Stimulator with arbitrary waveform for auditory evoked potentials *New. J. Phys* **90** 012031

[14] Picton, T W, Vajsar J, Rodriguez R and Campbell K B 1987 Reliability estimates for steady-state evoked potentials *Electroencephalogr Clin Neurophysiol* **68** 119-31.

[15] Cagy M, Infantosi A F C and Gemal A E 2000 Monitoring depth of anaesthesia by frequency-domain statistical techniques *Rev Bras Eng Bioméd* **16** 95-107.

[16] Joris P X, Schreiner C E, and Rees A 2004 Neural processing of amplitude-modulated sounds *Physiological reviews* **84** 541–77.

[17] A. Van Maanen and D. R. Stapells 2009 Normal multiple auditory steady-state response thresholds to air-conducted stimuli in infants *J Am Acad Audiol* **20** 196–207

Analysis of vectorcardiographic dynamic changes in patients with acute myocardial ischemia

R Correa[a1], P Arini[b,c], L Correa[a], M Valentinuzzi[c] and E Laciar[a]

[a] Gabinete de Tecnología Médica, Facultad de Ingeniería, Universidad Nacional de San Juan (UNSJ), San Juan, Argentina; rcorrea@gateme.unsj.edu.ar; lcorrea@gateme.unsj.edu.ar; laciar@gateme.unsj.edu.ar.

[b] Instituto Argentino de Matemática (IAM) "Alberto P. Calderón", Consejo Nacional de Investigaciones Científicas y Técnicas (CONICET), Buenos Aires, Argentina, pedro.arini@conicet.gov.ar

[c] Instituto de Ingeniería Biomédica (IIBM), Facultad de Ingeniería (FI), Universidad de Buenos Aires (UBA), Buenos Aires, Argentina, maxvalentinuzzi@arnet.com.ar

Abstract. This work evaluates the vectorcardiographic dynamic changes in ischemic patients before and during Percutaneous Transluminal Coronary Angioplasty (PTCA). Four vectorcardiographic parameters were computed in 51 ischemic and 52 healthy subjects with the objective of assessing the differences between both groups: ST Vector Magnitude Area ($\boldsymbol{ST}_{\mathrm{VMa}}$), T Vector Magnitude Area ($\boldsymbol{T}_{\mathrm{VMa}}$), ST Vector Difference ($\boldsymbol{ST}_{\mathrm{VD}}$), and T Vector Difference ($\boldsymbol{T}_{\mathrm{VD}}$). The conventional ST-Change Vector Magnitude (STC_{VM}) and Spatial Ventricular Gradient (*SVG*) were also calculated. Results indicate that the most of them show significant differences between healthy and ischemic subjects. Since, the statistical minute-by-minute PTCA comparison against a healthy population shows that ischemic patients monitoring reached values of Sensitivity = **99.5%** and Specificity = **99.4%**, when $\boldsymbol{ST}_{\mathrm{VD}}$, $\boldsymbol{T}_{\mathrm{VD}}$ and ***SVG*** were used in the classification. In conclusion the sensitivity and specificity for acute ischemia monitoring could be increase with the used of only three vectorcardiographic parameters.

1. Introduction

Myocardial ischemia is caused by a decompensation between the oxygen supply and demand; it is frequently associated with coronary atherosclerosis. The temporary occlusion of a coronary artery derives in a reversible ischemia, while a prolonged obstruction leads to myocardial infarction with serious consequences, such as malignant arrhythmias, heart failure and/or sudden cardiac death. Percutaneous Transluminal Coronary Angioplasty (PTCA) is a therapeutic procedure used to reestablish blood flow in narrowed arteries. Such procedure provides an attractive opportunity to study myocardial changes due to lack of coronary patency during its initial minutes [1]. Shortly after the beginning of balloon inflation, some changes can be detected in the electrocardiogram (ECG), such as ST-segment deviation and T-wave modifications, due to alterations in the ventricular repolarization process [2]. Several studies have demonstrated that ventricular depolarization is also modified during acute myocardial ischemia induced by PTCA [3-4]. In this context, other conventional index that

[1] Gabinete de Tecnología Médica - Facultad de Ingeniería-Universidad Nacional de San Juan-Av. Libertador General San Martín 1109 (O)- J5400ARL - San Juan, Argentina-Tel: + 54 - 264 – 4211700 (Ext. 313)- rcorrea@gateme.unsj.edu.ar.

provides information on the cardiac conduction system operation and on the ventricular action potential duration heterogeneity is the Spatial Ventricular Gradient (SVG) [5]. ter Haar *et al.* have reported that this gradient, in addition to ST analysis, has a potential role in detecting ischemia [6].
Within the vectorcardiography framework, the momentary cardiac electrical activity is representable by a single vector in the Euclidian space, i.e., the heart vector, and the VCG precisely describes both components, magnitude and direction, as time proceeds. Several studies have proposed its use for evaluating cardiac changes during myocardial ischemia or infarction. In a recent work, Hasan *et al.* concluded that vectorcardiographic analysis of beat-to-beat variability in ventricular depolarization and repolarization may provide markers of electrical instability in patients with myocardial infarction [7]. Eriksson used $\boldsymbol{ST}_{VM}$, ST-Change-vector magnitude ($\boldsymbol{STC}_{VM}$) and $\boldsymbol{QRS}_{VD}$ monitoring to give valuable prognostic information in cases of unstable angina and acute myocardial infarction, and concluded that vectorcardiography monitoring may identify myocardial reperfusion at an early stage [8]. Jensen *et al.* proposed $\boldsymbol{STC}_{VM}$= 0.05mV as the best criterion for the detection of myocardial ischemia in VCG studies that monitor coronary angioplasty [9].
Dellborg *et al.* demonstrated that monitoring myocardial ischemia with dynamic computerized continuous vectorcardiography (c-VCG) seems to be more efficient than Holter monitoring and may have a higher sensitivity [10]. Additionally, c-VCG has a full real-time capacity, so allowing monitoring over prolonged periods, while the results are immediately available without time-consuming analysis. Besides, Perez Riera *et al.* showed better specificity, sensitivity and accuracy of the computerized VCG when compared with conventional ECG in several cardiac pathologies [11]. Often, the advantage of the VCG is due to the constancy of the time relations between leads, while conversely, such relationships are lost in separate scalar lead analysis.
In a previous work, we analyzed a set of VCG parameters computed in the cardiac depolarization phase in ischemic patients [12]. The aim of this study was to differentiate a group of ischemic subjects from a population of healthy ones by VCG analysis of the ventricular depolarization-repolarization process. For that matter, four novel vectorcardiographic parameters (computed on repolarization segments and waves) and three conventional indexes (computed on depolarization-repolarization segments and waves) were evaluated.
As hypothesis, we state that the balloon occlusion modifies the morphology of the ST-T and, thereafter, its parameters can be used to characterize and monitoring the acute myocardial ischemia. To prove this, we evaluated the proposed and the conventional parameters in the classification scheme.

2. Materials

The Raw clinical records were extracted from the PTB diagnostic ECG and STAFF-III databases.
The first one contains the ECG records of 52 healthy subjects (39 men, mean age 42 +/- 14 yrs, and 13 women, mean age 48 +/- 19 yrs). The ECGs in this collection were obtained by Physikalisch-Technische Bundesanstalt (PTB); the National Metrology Institute of Germany. Each record includes 15 simultaneously measured signals: the conventional 12 leads (I, II, III, aVR, aVL, aVF, V1-V6) together with the 3-Frank lead-ECGs (X, Y, Z). Each signal was digitized at 1000 Hz, with 16 bits of amplitude resolution [13].
The second database consists of 51 ischemic patients (33 males, mean age 61 +/- 13 yrs and 18 women, mean age 60 +/- 10 yrs) admitted to the Charleston Area Medical Center in West Virginia, receiving elective PTCA in one of the major coronary arteries (STAFF-III study). The study was approved by the local investigation review board, and informed consent was obtained from each patient before enrollment [4]. The mean occlusion period was 5 min 7s. The locations of the 51 dilations were: Left Anterior Descending Artery in 11 patients, Right Coronary Artery in 14 patients and Left Circumflex Artery in 26 patients. We excluded patients with ECG signal loss during acquisition or with occlusion period less than 4 min 30 sec.
Nine standard leads (V1-V6, I, II, III) were recorded in the study using equipment by Siemens-Elena AB (Solna, Sweden), digitized at sampling rate of 1000 Hz and 0.6 μV amplitude resolution. Synthesized orthogonal X, Y and Z leads were obtained by the Kors transform [14]. A recent study

has demonstrated that Kors synthesis matrix provides a better estimation of Frank leads than the Inverse Dower transform in ischemic patients [15].
For each patient, two ECG records were obtained. One of them (denoted as before-PTCA Record) was acquired continuously at rest in the supine position prior to angioplasty, and the other (denoted as during-PTCA Record) was obtained during the PTCA procedure.

3. Methods

Figure 1 illustrates a block diagram of the different stages of the proposed analysis.

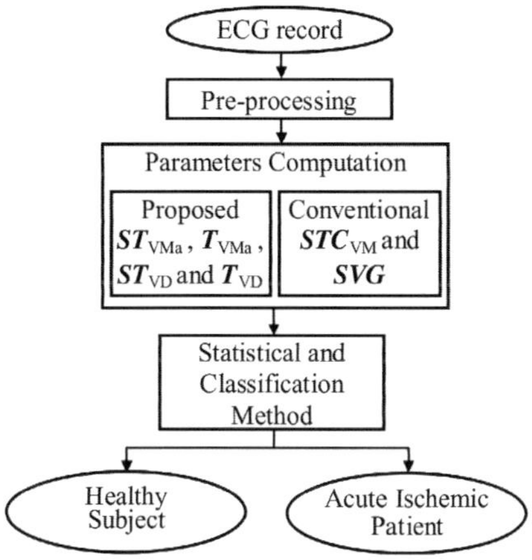

Figure 1. Proposed Analysis General Diagram.

3.1. Preprocessing

First, all ECG records were preprocessed with a band-pass filter (Butterworth, 4th order, 0.2–100 Hz, bidirectional) to reduce low and high frequency noise and a notch filter (Butterworth, 2th order, 50/60 Hz, bidirectional) to minimize the power line interference. A cubic spline interpolation filter was used to attenuate ECG baseline drifts and respiratory artifacts. Thereafter, the QRS complexes, T waves and their endpoints were detected in each ECG record using a wavelet-based technique [16]. Beats with a RMS noise level >40 μV (measured within a 40 ms window located at 2/3 of the RR interval) were excluded. In addition, ectopic beats were also eliminated by comparing incoming signals against a previously established template with the use of a cross-correlation technique. Additionally, with this technique, we also make the time alignment of beats, required to obtain an average beat, used in the parameters computation [17]. In this study, a visually low-noise normal beat extracted from the ECG record was selected as template (or reference) beat.

3.2. Parameters Computation

Four VCG parameters and two conventional indexes were computed for each detected beat.

- ST Vector Magnitude Area (ST_{VMa}): It is defined as the magnitude of the vector composed by the X, Y and Z Area between the amplitude of the ECG signal interval from each J-point to J-point+80ms (ST-segment) and the isoelectric level, that is,

$$ST_{VMa} = \left(aST_X^2 + aST_Y^2 + aST_Z^2\right)^{1/2} \tag{1}$$

  This parameters is, a modification of the STC_{VM} widely used in the monitoring of cardiac ischemia [7, 8], The principal differences is STC_{VM} is evaluated at the one point on de signal (usually j-point, or j-point+60ms), by the contrast the proposed ST_{VMa} is computed during all ST segment (figure 2a).

- T Vector Magnitude Area ($\boldsymbol{T}_{\text{VMa}}$): It is defined as the magnitude of the vector composed by the X, Y and Z Area between the amplitude of the ECG signal at interval of each T wave and the isoelectric level (figure 2b), that is,

$$\boldsymbol{T}_{\text{VMa}} = \left(aT_X^2 + aT_Y^2 + aT_Z^2\right)^{1/2} \tag{2}$$

- ST-T-Vector Difference ($\boldsymbol{ST\text{-}T}_{\text{VD}}$): It was defined as the difference area between the ECG signal at the current ST-T intervals (from each J-point to the T wave ends) and the reference of ECG signal (at ST-T intervals too) evaluated at the first 30s of each ECG record, that is,

$$\boldsymbol{ST} - \boldsymbol{T}_{\text{VD}} = \left(dA_X^2 + dA_Y^2 + dA_Z^2\right)^{1/2} \tag{3}$$

  The objective of this parameter is estimated all changes produced during left Ventricular repolarization (figure 2-c).

- T-Vector Difference ($\boldsymbol{T}_{\text{VD}}$): It was defined as the difference area between the ECG signal at the current T wave intervals and the reference of ECG signal (at T intervals too) evaluated at the first 30s of each ECG record, that is,

$$\boldsymbol{T}_{\text{VD}} = \left(dT_X^2 + dT_Y^2 + dT_Z^2\right)^{1/2} \tag{4}$$

  The objective of this parameter is estimated all changes produced in the T wave (figure 2-d).

- ST-Change Vector Magnitude ($\boldsymbol{STC}_{\text{VM}}$): The ST-vector is composed of the (X, Y and Z) ST-segment deviations from the isoelectric level, measured as the ECG signal amplitude at the J-point (Fig. 2-e); it is a widely used parameter when monitoring cardiac ischemia [7,8]. This vector is the difference between the ST-vector of the current beat and the averaged beat evaluated at the first 30s of each ECG record, that is,

$$\boldsymbol{STC}_{\text{VM}i} = \left[\left(ST_{Xi} - ST_{Xr}\right)^2 + \left(ST_{Yi} - ST_{Yr}\right)^2 + \left(ST_{Zi} - ST_{Zr}\right)^2\right]^{1/2} \tag{5}$$

  where r denotes the reference beat, i is the current beat, with i = 1, ..., N, and N stands for the total number of analysed beats(Fig. 2-b).

- Spatial Ventricular Gradient (SVG): It is defined as the vectorial QRS-T integral, that is:

$$\boldsymbol{SVG} = \left[\left(aQRS_X + aT_X\right)^2 + \left(aQRS_Y + aT_Y\right)^2 + \left(aQRS_Z + aT_Z\right)^2\right]^{1/2} \tag{6}$$

  where, $aQRS_X$, $aQRS_Y$, $aQRS_Z$ and aT_X, aT_Y, aT_Z are the QRS-complex and T-wave areas on the orthogonal leads, respectively, and all having units of mV·ms. Then, ***SVG*** has the same units. Unlike most other ECG parameters, the ***SVG*** is not influenced by changes in ventricular conduction pattern; it only changes if the distribution of the ventricular action potential morphology and/or duration is altered [6].

3.3. Statistical and Classification Methods

All parameters were computed for each detected sinus beat in every ECG record. First, we analysed the normality of these values using the D'Agostino-Pearson's test with the aim of quantifying the discrepancy between the parameters' distribution and the Gaussian distribution. It has been observed that the underlying variables' distribution is non-Gaussian. Afterwards, comparisons between healthy (mean of each parameter values of all ECG records) and ischemic subjects (parameters computed before and during PTCA, the last one grouped at 1 min intervals) were made using the non-parametric Mann-Whitney test.

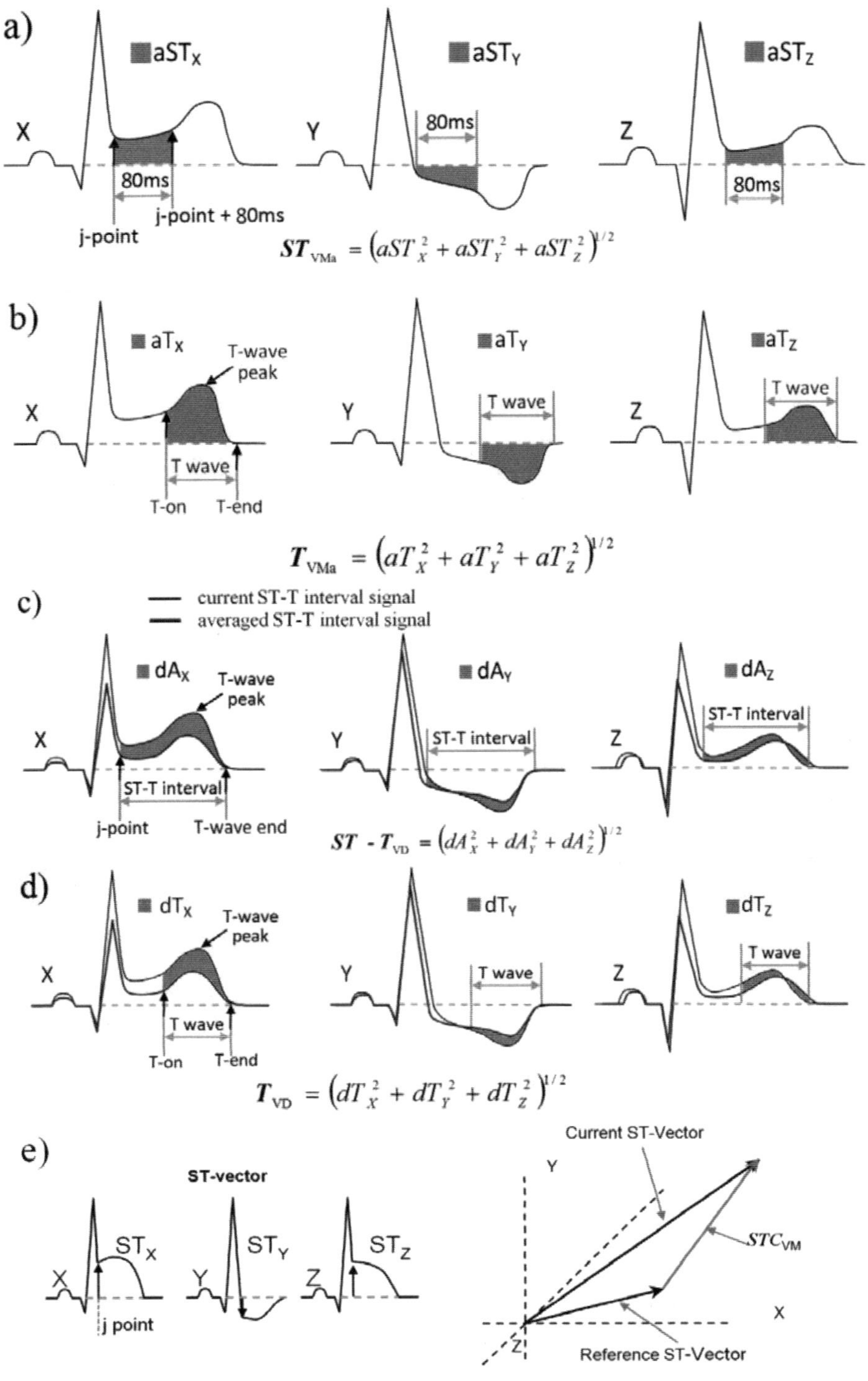

$$STC_{VM} = \left[(ST_{Xi} - ST_{Xr})^2 + (ST_{Yi} - ST_{Yr})^2 + (ST_{Zi} - ST_{Zr})^2\right]^{1/2};\ i = 1 - N$$

Figure 2. Vectorcardiographic Parameters computation.

The mean values of each parameter across the entire record for healthy subjects, and at 1 min interval for ischemic patients, were calculated. These values were used as inputs to a classifier based on *Linear Discriminant Analysis (LDA)* to distinguish (or separating out) ischemic patients from healthy subjects. Basically, the LDA classifier is a linear combination of variables, as follows. The LDA classifier is a linear combination of variables, as follows,

$$y = \mu_0 + \mu_1 X_1 + \mu_2 X_2 + ... + \mu_p X_p \quad (7)$$

where y is the output value of the discriminant function; μ_n (with $n=1,..., p$) stand for the coefficients of the discriminate function; X_n are the discriminate variables (QRS-loop and/or ST-T parameters) and p is the number of variables in the analysis.

The resulting discriminant function can be used to assign each ECG record to a particular class, ischemic patient or healthy subject, based on its values of discriminate variables. The model coefficients are estimated with a subset of ECG records for which the group is known. This subset of observations is sometimes referred to as the ***training subset*** (we used the 70% of ECG records of both populations). In order to validate the model, this discriminant function was used to predict the group of another different subset (referred to as ***validation subset***) of the ECG records (we used the remaining 30% of the ECG records).

To evaluate the performance of the LDA classifier, we computed the **Receiver Operating Characteristic** (ROC). It plots the Sensitivity (Se) against the 1-Specificity (Sp) values for the different possible cut-off points (cut-off values were swept between -5 and 5 in 0.01 steps) of the discriminant function. Then, the optimal cut-off point in the ROC curve was computed as the point nearest the top left-hand corner. This selection maximizes the Se and Sp sum, when it is assumed that the 'cost' of a false negative result is the same as that of a false positive one. Finally, the global performance of the classifier was evaluated with the Area Under the ROC Curve (AUC).

4. Results

Figures 3 depict the results of the statistical analysis described in section 3.3.The former, shows the mean and Standard Error of the Mean (SEM) of each computed parameter in healthy and ischemic populations before and during the PTCA procedure. The values marked with * indicate the statistical significance (p-value < 0.05).

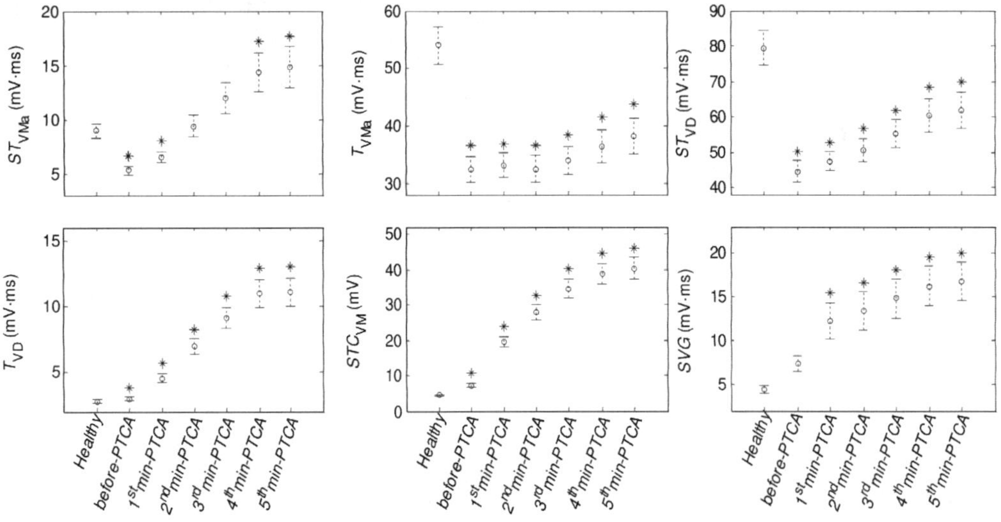

Figure 3. Mean and standard error of the mean of each parameter computed in healthy and acute ischemic subjects (the parameters for the acute ischemic patients were grouped at intervals of 1 min, thereby obtaining 5 groups for each record). *Denoted statistical significance, p_value< 0.05 (see section 3.3).

Table I show the classification results (Sens and Spec mean values) for each vectorcardiographic parameters studied and for the best combination of them (***ST-T***$_{VD}$, ***T***$_{VDa}$ and ***SVG***), before and during PTCA procedure.

Table I: Classification outcomes for each computed index and for the best combination of them

	ST$_{VMa}$		***T***$_{VMa}$		***ST-T***$_{VD}$		***T***$_{VDa}$		***STC***$_{VM}$		***SVG***		***ST-T***$_{VD}$, ***T***$_{VDa}$ and ***SVG***	
	Sens	Spec	Sens	Spec	Sens	Spec	Sens	Spec	Sens	Spec	Sens	Spec	Sens	Spec
before-PTCA	63.93	69.25	82.80	69.75	67.07	69.88	58.07	79.50	71.33	54.50	69.53	81.25	90.07	84.31
during-PTCA 1st min	75.00	50.88	82.67	67.38	88.40	94.19	80.80	83.38	75.07	76.94	68.80	73.75	92.40	93.31
during-PTCA 2nd min	44.53	46.88	75.07	72.69	92.53	96.38	85.07	84.00	83.87	78.63	60.87	81.88	94.60	96.88
during-PTCA 3rd min	53.40	55.06	79.93	70.69	95.07	97.56	86.27	88.88	84.47	81.88	57.27	76.56	96.87	97.56
during-PTCA 4th min	51.87	63.25	78.20	68.69	95.00	97.63	89.00	88.81	88.87	82.00	66.67	61.19	97.60	98.19
during-PTCA 5th min	46.20	75.81	72.60	70.56	96.53	98.25	91.13	90.50	87.07	83.50	62.20	63.00	99.53	99.38

Table II shows the Area Under the ROC Curve (AUC), for classification schemes using only one index at a time and using the parameters ***ST-T***$_{VD}$, ***T***$_{VDa}$ and ***SVG*** combined.

Table II Area Under the ROC Curve (AUC) for different classification schemes

	ST$_{VMa}$	***T***$_{VMa}$	***ST-T***$_{VD}$	***T***$_{VDa}$	***STC***$_{VM}$	***SVG***	***ST-T***$_{VD}$, ***T***$_{VDa}$ and ***SVG***
before-PTCA	0.73	0.78	0.76	0.68	0.63	0.83	0.91
during-PTCA 1st min	0.64	0.77	0.95	0.85	0.82	0.81	0.98
during-PTCA 2nd min	0.43	0.78	0.98	0.89	0.89	0.77	0.99
during-PTCA 3rd min	0.55	0.75	0.98	0.91	0.92	0.72	1.00
during-PTCA 4th min	0.62	0.73	0.99	0.92	0.94	0.67	1.00
during-PTCA 5th min	0.61	0.70	0.99	0.93	0.94	0.66	1.00

5. Discussion and Conclusions

Several studies have demonstrated the potential automated vectorcardiographic analysis usefulness to evaluate cardiac changes during ischemia or infarction [5-11]. Furthermore, in [9], Jensen *et al.* compared on-line computerized VCG's derived from 12-lead ECG's, and concluded that the first one is a more sensitive method for detecting myocardial ischemia during coronary angioplasty. However, most of the studies based on VCG use the ***STC***$_{VM}$ and ***SVG*** parameters.

Also these two conventional indexes, here in we examined 4 vectorcardiographic parameters computed in the ST-T interval to further describe cardiac dynamic changes during an episode of acute ischemia induced by PTCA.

On the basis of the statistical analysis, it can be observed in figure 3 that all the vectorcardiographic parameters (with same exception: the ***ST***$_{VMa}$ for 2nd and 3rd min during PTCA and ***SVG*** before-PTCA) produced significant differences (p-value < 0.05) between healthy and ischemic populations, before and during PTCA. Besides, the four proposed parameters, similar to conventional indexes, tend to gradually increase its mean value during angioplasty. This similar behaviour could be used in a combined analysis to improve monitoring of acute myocardial ischemia.

The discriminant analysis (see Table I) indicates that the vectorcardiographic parameter with the best global performance was ***ST-T***$_{VD}$, which reached Sen=96.5%, Spec=98.2%, and AUC=0.99.

Moreover, minute-by-minute statistical comparison of the PTCA against a healthy population shows that acute ischemic patients monitoring is greatly enhanced when the parameters ***ST-T***$_{VD}$, ***T***$_{VDa}$ and ***SVG*** are used in the classification, reaching **Sens = 99.5% and Spec = 99.4%**, at the 5th minute of the PTCA. Meanwhile, the best classification using only the conventional ***STC***$_{VM}$ or ***SVG*** indexes, reaching values of Sens = 88.9 % and Spec = 82.0% at the 4th minute the occlusion, which are lower than those reported above. Additionally, the best AUC value (see Table II), using the parameters $\boldsymbol{ST}$-$\boldsymbol{T}_{\mathrm{VD}}$, $\boldsymbol{T}_{\mathrm{VDa}}$ and $\boldsymbol{SVG}$ is **AUC = 1.00**, after the 3th minute of the balloon inflation. Thus, it indicates high effectiveness for the proposed classification technique. This AUC value is considered of high accuracy in diagnostic tests [18].

In conclusion: The proposed technique based on vectorcardiographic study could be used in addition to the conventional ST-T analysis for a better monitoring of ischemic patients. From the clinical point of view, the most important future application would be the ambulatory monitoring of ischemic patients in Holter or stress tests studies.

Acknowledgments
This study was supported by grants from *Consejo Nacional de Investigaciones Científicas y Técnicas* (CONICET – PIP538), *Agencia Nacional de Promoción Científica y Tecnológica* (ANPCYT – PICT-O-UNSJ 0027) and *Universidad Nacional de San Juan* (CICICTCA-UNSJ I972).

Reference

[1] Engblom H and Strauss DG 2011 Electrocardiography of ischemic heart disease. In: *Multimodal cardiovascular imaging: Principles and clinical applications,* Edited by Pahlm O and Wagner G, 1st ed, (New York, USA: McGraw-Hill Medical, c2011) 128-145.

[2] Bayés de Luna A 1999 Electrocardiografía Clínica, 1st Ed. (Barcelona, Spain: ESPAXS S.A.)

[3] Surawicz B, Orr CM, Hermiller JB, Bell KD, Pinto RP 1997 QRS changes during percutaneous transluminal coronary angioplasty and their possible mechanisms. *J Am CollCardiol* **30**:452-458.

[4] Ringborn M, Pettersson J, Persson E, Warren SG, Platonov P, Pahlm O, Wagner GS 2010 Comparison of high-frequency QRS components and ST-segment elevation to detect and quantify acute myocardial ischemia. *J. Electrocardiol* **43**:113–120.

[5] Draisma HHM, Schalij MJ, van der Wall EE, Swenne CA. Elucidation of the spatial ventricular gradient and its link with dispersion of repolarization. *Heart Rhythm* 2006; 3:1092.

[6] ter Haar CC, Maan AC, Schalij MJ, Swenne CA 2012 ST and ventricular gradient dynamics during percutaneous transluminal coronary angioplasty. *Computing in Cardiology* **39**:341-345.

[7] Hasan MA, Abbott D, Baumert M. 2012 Beat-to-beat vectorcardiographic analysis of ventricular depolarization and repolarization in myocardial infarction. *PLoS ONE* **7**(11) :e49489.doi:10.1371/journal.pone.0049489

[8] Eriksson SV 1999 Vectorcardiography: a tool for non-invasive detection of reperfusion and reocclusion. *Thromb Haemost* **82**:64-67.

[9] Jensen SM, Johansson G, Osterman G, Reiz S, Näslund U 1994 On-line computerized vectorcardiography monitoring of myocardial ischemia during coronary angioplasty: comparison with 12-lead electrocardiography. *Coronary Artery Dis* **5**:507-514.

[10] Dellborg M, Malmberg K, Ryden L, Svensson AM, Swedberg K 1995 Dynamic on-line vectorcardiography improves and simplifies in-hospital ischemia monitoring of patients with unstable angina. *J Am Coll Cardiol* **26**:1501-1507.

[11] Pérez Riera AR, Uchida AH, Ferreira Filho C, Meneghini A, Ferreira C, Schapacknik E, Dubner S, Moffa P 2007 Significance of vectorcardiogram in the cardiological diagnosis of the 21st century. *Clin Cardiol* **30**:319-323.

[12] Correa R, Arini P, Valentinuzzi M, Laciar E 2013 Novel set of vectorcardiographic parameters for the identification of ischemic patients. *Med EngPhys* **35**:16-22.

[13] Bousseljot R, Kreiseler D, Schnabel A 1995 Nutzung der EKG-Signal daten bank CARDIODAT der PTB über das Internet. *Biomedizinische Technik, Band 40, Ergänzungsband* **1**:317.

[14] Kors JA, van Herpen G, Sittig AC, van Bemmel JH 1990 Reconstruction of the Frank vectorcardiogram from standard electrocardiographic leads: diagnostic comparison of different methods. *Eur Heart J* **11**:1083-1092.

[15] Man S, Algra AM, Schreurs CA, Borleffs CJ, Scherptong RW, van Erven L, van der Wall E, Cannegieter SC, Schalij MJ, Swenne CA 2011 Influence of the vectorcardiogram synthesis matrix on the power of the electrocardiogram-derived spatial QRS-T angle to predict arrhythmias in patients with ischemic heart disease and systolic left ventricular dysfunction. *J of Electrocardiol* **44**:410-415.

[16] Martinez JP, Almeida R, Olmos S, Rocha AP, Laguna P 2004 A wavelet-based ECG delineator: Evaluation on standard databases. *IEEE Trans Biomed Eng* **51**(4):570-581.

[17] Laciar E, Jané R, Brooks DH 2003 Improved Alignment Method for Noisy High Resolution ECG and Holter Records using Multi-Scale Cross-Correlation. *IEEE Trans Biomed Eng* **50**(3):344-353.

[18] Swets JA 1988 Measuring the accuracy of diagnostic systems. *Science* 240:1285-1293.

Journal of Physics: Conference Series **477** (2013) 012033 doi:10.1088/1742-6596/477/1/012033

New method to obtain spontaneous baroreflex gain curve as a function of the systolic pressure

Juan Carlos Perfetto 1, Aurora Ruiz 2, Carlos Dattellis 3
1 Biomedical Engineering Institute, Buenos Aires University, Buenos Aires
Email: jperfet@fi.uba.ar
2 Fernandez Hospital, Buenos Aires
3 Favaloro University
ARGENTINA

Abstract. We present a non invasive new method based on the analysis of blood pressure and heart rate variability[1] to obtain baroreflex gain as a function of the systolic pressure. This method can explain certain results obtained in orthostatic tolerance studies, such as great variations of the baroreflex sensitivity[2, 3]. This approach help to understand the behavior of the baroreflex and to place some results in a more complete context.

1. Introduction

The baroreflex is a well-known system to stabilize arterial blood pressure by means of negative feedback. The baroreceptors are placed in the carotid sinus and the aortic arch.
In a laboratory environment the stimulation pressure can be controlled, and the RR interval changes response registered. In this way the entire baroreflex function can be computed.
This is an open loop estimation because we control the pressure instead of the natural control, which is the autonomic system. This invasive method can be very risky if we whish to reach curve boundaries.

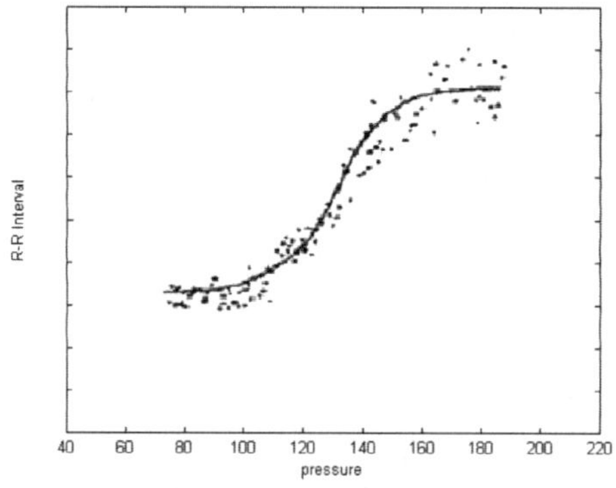

Fig. 1 - Baroreflex curve

Journal of Physics: Conference Series **477** (2013) 012033 doi:10.1088/1742-6596/477/1/012033

This has been done in early 1930, and as many other biologic functions, its shape is sigmoid. We can therefore distinguish tree important points, the threshold, the saturation and the mid level.

The baroreflex has a minimal value of pressure to work named threshold, and beyond a determined value (saturation) the response is not increased. In the neighborhood of the mid level the sigmoid is almost linear and its derivative is constant. For this reason it is accepted to use that value as a unique characteristic of the baroreflex function. This value is called baroreflex sensitivity (BRS) and is measured in ms / mmHg.

Gain can be estimated by computing the derivative of the sigmoid function. We can also assume that real baroreflex gains should have similar shapes. If these hypotesis were thrue we can expect that points extracted from baroreflex sequences could have good correlation with these curves. Different mid values and pressure ranges before saturation had been simulated to mimic variability between subjects, this is shown in Fig. 2.

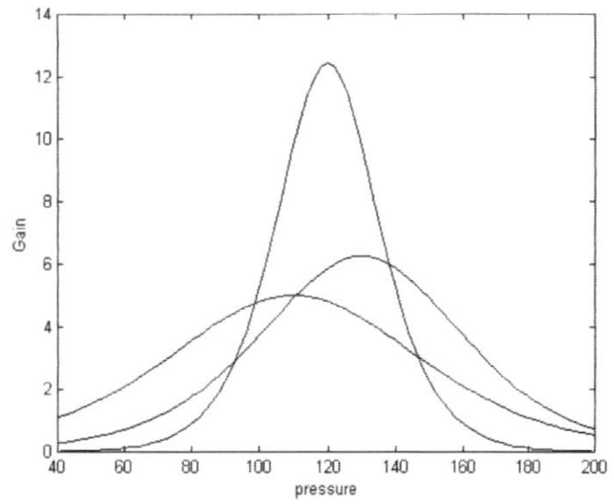

Fig. 2 The parameters of each subject make the curve more flat or acute

The baroreflex gain remains within certain boundaries in a certain subject in different physiological situations, and changes will be result of distinct modulation mechanism of the Auntonomic Nervious System. Therefore the BRS is function of the operation point having it's maximum at a certain arterial pressure wich can be assumed as the reference value to be stabilized. With the subject at rest operation points are near the mid values and we don't commit great errors assuming linear relationship, but if our interest is to study the behavior of the system in situations where the work points are far from the mid level we have to abandon this simplification. This is the case of tilt test, a widely used technique in orthostatic intolerance studies

The tilt test push patient from equilibrium and usually produces an arterial pressure fall, and in case of inadequate compensation leads the patient to syncope. Subject is registered in a tilt bed at rest and next at 70 degrees. In this case new points from de curve are detected. These points have lower values of BRS and now have no sense average them. We are presenting a new method to reconstruct this curve from spontaneous techniques. We cannot obviously reconstruct the whole curve, but instead we can reconstruct it in the zone of work.

2. Data Acquisition

All of the subjects participating in this study were selected from patients of the Fernandez Hospital. We used a commercial isolated amplifier with acquiring software, a Colin non-invasive blood pressure monitor, and a computer, which records the signals. Both channels are digitized at 1200 samples/sec.

3. Preprocessing

Signal acquisition during tilt test includes electrocardiogram (ECG) and continuous blood pressure curve (CBPC).
Baroreflex action is evaluated by studying correlations between heart rate (HR) and blood pressure (BP) changes. Two signals must be available as digital data for computer processing: beat to beat interval and beat to beat systolic and diastolic BP. They are obtained respectively from ECG and CBPC. This implies detection of R wave on ECG to compute intervals between them and determination of maximum and minimum BP for every heart beat on CBPC.

4. Method description

The tilt test has two stages, the first lasts twenty minutes with the patient at rest, this bring us a cloud of spots near the maximum of the baroreflex curve, each of these spots corresponds to different gains of baroreflex sequences. The second stage has a maximun duration of forty five minutes, and is associated with lower arterial pressure values, in some cases very far from mid level. Another test like to sit and stand up may produce similar effects in pressure. These changes in pressure induces compensation by the baroreflex
The first part of the method we propose consist in detect baroreflex sequences from time series of both channels. To do this task we adopt the sequential method [4] and also identify three or more consecutive cycles of blood pressure increments coupled with RR interval concordant changes. Only values of correlation coeficient greater than 0.85 between both point series are included to assure linear relationship between point serial pairs. The quotient between segments is the gain. The minimun increment to be considered is 1 mmHg for the arterial pressure. For each ramp we register not only the gain but also the corresponding arterial pressure. The method also allows a lag of one beat between ramps.

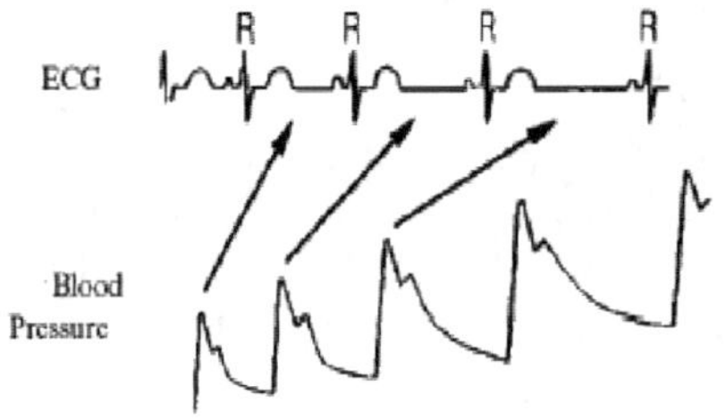

Fig. 3 -Concordant changes between arterial pressure and RR Interval

The sequential method then averages all the gains from the selected ramps. The value obtained is the BRS. In our method instead of computing the average from all systolic arterial pressure points, pressure space is divided in non-overlapping windows of 10 mm of Hg wide. Then gain values corresponding to pressure values within each window are averaged. The abcisa will be the central pressure of the corresponding window. Each ramp of baroreflex detected contributes to aproximate real points of the gain curve, for this reason we can only reconstruct part of the curve.
The philosophy upon this is to characterize the same phenomena by means of averaging the slopes obtained from each selected sequence. The biological signals have great variability and we will have different gain values at the same pressure range. To partially reconstruct the baroreflex curve covered we must interpolate between points, each points separated the following by 10 mm Hg. Cubic splines are used for this task.

Journal of Physics: Conference Series **477** (2013) 012033 doi:10.1088/1742-6596/477/1/012033

5. Results

In Fig 3 we can observe results of the sequential method, were baroreflex sequences are detected and their slope shown. The segments are placed at a distance from the left margin coincident with the sistolic arterial pressure at the begining of the sequence. If we observe their slopes are very close when sequences have same or similar arterial systolic pressure.

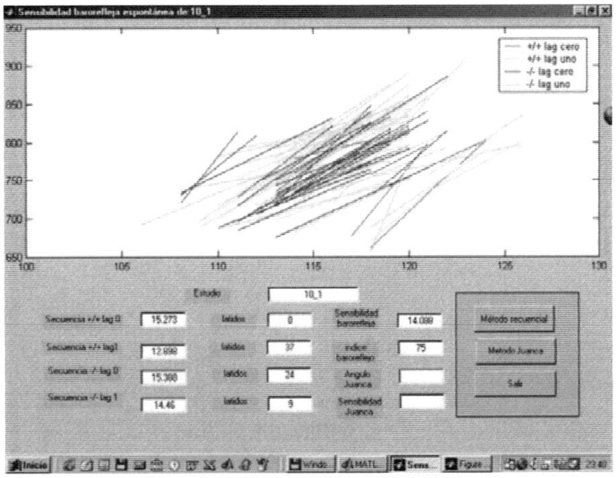

Fig. 2 - Sequential method from one subject

We are able to reconstruct important parts of the whole curve of baroreflex function of several subjects with only rearranging the data collected during tilt test studies. In the figure 4 we can see the results of curve fitting process of five of them. In many occassions the partially reconstructed curves are enough to give us an idea of how the entire curve could be. In some cases we cannot reconstruct the curve from impaired baroreflex subjects, but in this case others methods also fails. In other ocasions only two points are extracted which represent a part of the curve. We have also tried others sizes for the systolic pressure windows to obtain better curve resolution, but due to the spontaneous characteristic of the method, when reordering the samples it can be noted that they are not evenly spaced. Therefore results are similar, but with higher distortions, and the conclusion is that is not worth increasing the complexity. On the other hand when the window size is greater the gain is computed using more gain points and its value more confident and the reconstructed curve more smooth.

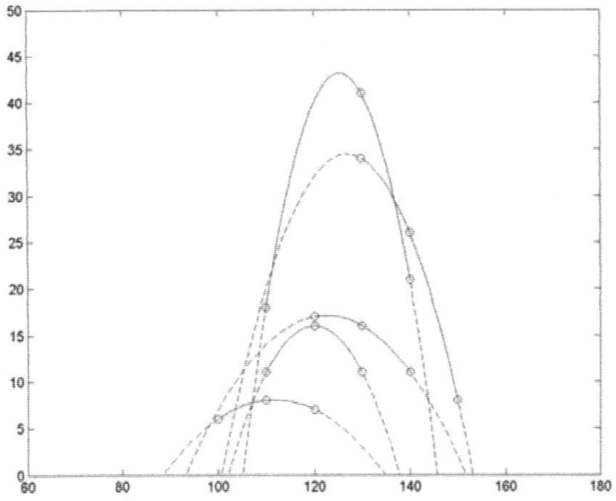

Fig. 3 – Baroreflex gain curves from different subjects . Solid lines joins computed points while dashed lines indicate extrapolation zones.

6. Conclusion

By rearranging the data collected doing standard procedures of assessing baroreflex gain in the time domain, the analysis of the results can be enhanced, incorporating new features like the central pressure to be stabilized, the width of the curve, and the part of the curve the subject has activated.

Those features are not obvious in a traditional analysis, but here are exposed and can be very useful in another studies of cardiovascular diseases.

We do not present numerical results, but we are trying to introduce this method in the analysis and discussions on this subject

Recently it is viewing a growing interest in studying the baroreflex, since it has been proved its utility to assess various cardiological diseases. It is known that others diseases modify the baroreflex curves, for example diabetes and others neuropathies, so this method can also be applied

In this context non-invasive methods like this are receiving great importance, and clinical acceptance because there is no risk to patient.

References:

1. Task Force of the European Society of Cardiology and North American Society of Pacing and Electrophysiology, 1996 Heart rate variability: standards of measurement, physiological interpretation and clinical use. Task Force of the European Society of Cardiology and the North American Society of Pacing and Electrophysiology. *Circulation*,. **93**(5): p. 1043-65.

2. Bechir, M., et al., 2003 Dysfunctional baroreflex regulation of sympathetic nerve activity in patients with vasovagal syncope. *Circulation*, **107**(12): p. 1620-5.

3. Pitzalis, M., et al., 2003 Enhanced reflex response to baroreceptor deactivation in subjects with tilt-Induced syncope. *Journal of the American College of Cardiology, 2003.* **41**(7): p. 1167-1173.

4. Parati, G., M. Di Rienzo, and G. Mancia, 2000 How to measure baroreflex sensitivity: from the cardiovascular laboratory to daily life. *J Hypertens*, **18**(1): p. 7-19.

Comparison of central systolic pressure estimates obtained from ultrasound images and applanation tonometry

Graf[1,2,3], L. Bortolin[1], A. Pascaner[1], D. Craiem[1,2]

[1] Fac. de Ingeniería y Ciencias Exactas y Naturales, Universidad Favaloro
[2] CONICET, Buenos Aires, Argentina
[3] Fac. Regional Buenos Aires, UTN

Abstract. Current non invasive assessment methods to estimate central pressure are based on the use of arterial tonometry: directly on the carotid artery or using a mathematical transfer function applied to the radial artery waveform. The first one is not recommended in obese patients or in the presence of carotid plaques, whereas the use of a unique transfer function to different population has been questioned. In this work we evaluated two alternative methods to assess systolic local arterial blood pressure: 1) from the analysis of B-mode carotid diameter waveforms (SBP_Dia) and 2) from the radial artery pressure waveform using an n-point moving average filter (SBP_Rad), and we estimated the accuracy of both when compared to direct carotid tonometry (SBP_Ton). In 20 asymptomatic subjects (49±11 years, range: 38-73; pulse pressure: 53±9 mmHg, range: 36-70), SBP_Ton was 132±13 mmHg (range 115-154) and correlated significantly with SBP_Dia ($R=0.96$, $p<0.05$) and with SBP_Rad ($R=0.93$, $p<0.05$). Mean difference between SBP_Dia and SBP_Ton was 0.85±4.0 mmHg, and 0.18±5.0 mmHg between SBP_Rad and SBP_Ton, independent of pressure levels. In conclusion, both alternative methods were found to allow an accurate and precise estimation of systolic local arterial pressure, when compared to direct carotid tonometry.

1. Introduction

Estimation of central blood pressure has gained increased attention over the past few years, based on evidence that central pressure may more accurately predict cardiovascular damage when compared to brachial pressures [1,2,3]. Current non invasive assessment methods are based on the use of arterial tonometry: directly on the carotid artery [4] or using a mathematical transfer function applied to the radial artery waveform to transform it into a central waveform [5,6]. In both cases, a calibration procedure is required, assuming that mean arterial pressure (MAP) is nearly constant throughout the large artery tree and that diastolic pressure does not change substantially [7]. Direct carotid tonometry requires a trained operator and is not recommended in obese patients or in the presence of carotid plaques. By other side, the use of a unique transfer function to be applied to different population has been questioned [8].

A recent study, propose to derive central systolic pressure from the radial artery pressure waveform using an n-point moving average filter [9]. This method, which is simple and avoid the use of a transfer function, showed an excellent correlation and agreement when compared to directly measured aortic root systolic pressures at cardiac catheterization [9]. Another approach, is to derive pressure waveforms from arterial distension waves obtained from echographic images [10,11]. In a recent study, we evaluated the validity to assess systolic local arterial blood pressure using B-mode ultrasound image analysis and the implementation of an iterative calibration model. This method was found to allow an accurate and precise estimation of systolic local arterial pressure, with an underestimation error of ~ 2% independent of pressure levels, when compared to carotid and femoral arterial tonometry [12]. Therefore, the aim of this study was to evaluate in a general population these two alternative methods to assess systolic local arterial blood pressure, 1) from the analysis of B-mode carotid diameter waveforms, and 2) from the radial artery pressure waveform using an n-point moving average filter, and to estimate the accuracy of both when compared to direct carotid tonometry.

2. Material and methods

2.1. Study subjects

Non invasive study was performed in 20 (12 men, 8 women) asymptomatic subjects ranging in age from 38 to 72 years old (mean 49±11 years). Measurements were performed after 15 min rest period in a temperature-controlled environment. In order to ensure similar haemodynamic conditions, for each arterial segment, echographic and applanation tonometry recordings were performed consecutively, with a separation time of about 3 minutes. Systolic (SBP_{bra}) and diastolic (DBP_{bra}) brachial blood pressure values were determined in the arm by sphygmomanometer procedure (supine position), between echographic and tonometer recordings. Hypertension was defined by blood pressure of 140 and/or 90 mmHg or above and/or presence of hypertensive medication.

Left carotid arterial waveforms were assessed by experienced sonographer physicians with high-resolution B-mode echography (ATL HDI 5000, Miami Lakes, EE.UU) using a 7.5-MHz probe. Briefly, a sequence of B-mode images of at least 10 seconds were transferred to a computer, and digitized for offline analysis using an automated computerized system (Hemodyn4M®, Oxitech, Argentina). In order to obtain the diameter waveform, the sequences of images were analyzed automatically frame by frame. The anterior and posterior walls were detected. After analyzing the overall sequence, the software yielded as output, the internal diameter waveform, which was calculated as the difference between the far and near wall movement.

Arterial pressure waveforms were obtained at the left radial and left carotid arteries by applanation tonometry using a Miller pen-type tonometer (SPT 301, Millar Instruments, Houston, Texas, USA) and a dedicated hardware and software system [12]. The instantaneous pressure waveforms of at least 10 cardiac cycles were digitized every 2 ms and stored on the hard disk of a PC for off-line post processing.

2.2. Data analysis

MAP was estimated at the upper arm using SBP_{bra} and DBP_{bra} values, as 38% of pulse pressure above diastolic pressure [13,14,15]. For each tonometry signal an average waveform was constructed of at least five beats.

The carotid waveforms were calibrated by assigning the corresponding DBP_{bra} to the minimum value and the resulting brachial MAP to the average value. The peak of the calibrated carotid (SBP_Car_Tonometer) artery waveforms were considered as the reference systolic value.

The radial waveforms were calibrated first by assigning the corresponding DBP_{bra} to the minimum value and the resulting brachial MAP to the average value. Then a 125 moving average point filter was applied [9], and the peak of the resulting filtered signal was considered as the systolic estimates (SBP_Rad_tonometer)

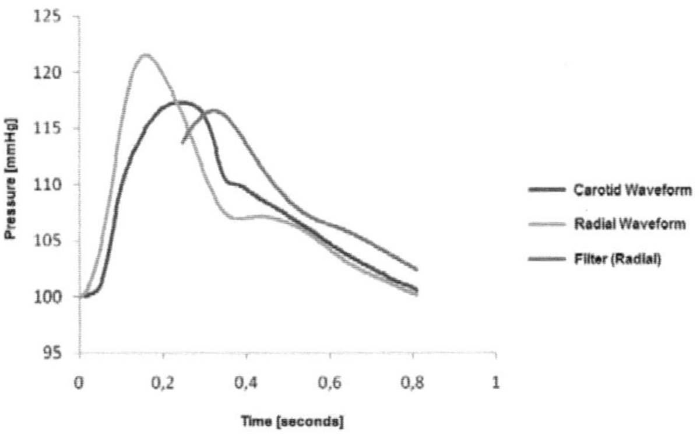

Figure 1. Typical example of carotid tonometer waveform (in black) and radial tonometer waveform (in gray) after calibration procedure. Radial waveform after applying a 125 point moving average filter appears superimposed.

The diameter waveforms (D) were calibrated using an exponential approach proposed by Vermeersch et al [11]. Briefly, it is assumed that pressure (P) and arterial cross-section (A) are related by an empirically derived exponential function. The derived pressure waveform is calibrated to DBP_{bra} and MAP by iteratively changing a wall rigidity coefficient (α). The peak of the calibrated carotid waveform (SBP_Car_Diameter) was considered as systolic value. See figure 2

$$A(t) = \frac{\pi \cdot D(t)^2}{4} \quad , \quad \alpha = \frac{A_d \cdot \ln\left(\frac{SBP_{bra}}{DBP_{bra}}\right)}{A_s - A_d}$$

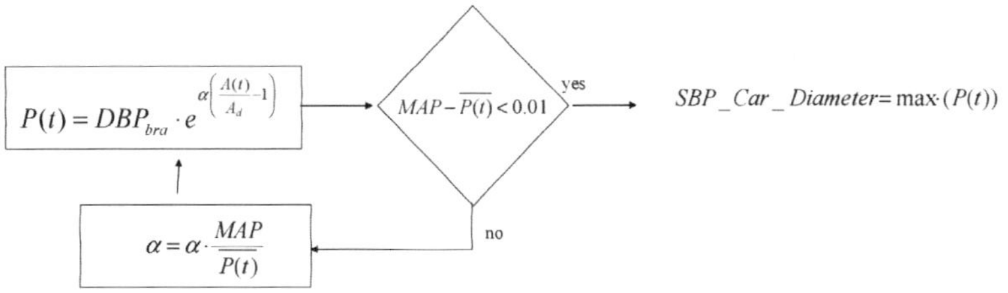

Figure 2. Iterative exponential calibration approach. A(t): carotid cross-section, A_s and A_d: systolic and diastolic carotid cross-section, P(t): estimated carotid pressure waveform, $\overline{P(t)}$: average value, α: wall rigidity coefficient

Values are presented as mean ± standard deviation. Differences were compared with paired t-test and the relationship between variables was evaluated by linear regression and Bland and Altman analysis. A value of $p<0.05$ was considered statistically significant

3. Results

Subject's characteristics are shown in table 1.

Table 1. Subjects characteristics

(n=20)	Mean±SD	Min	Max
Age (years)	49 ± 11	38	72
Body Mass Index (Kg. m^{-2})	27 ± 3	23	33
Sex (male,%)	58		
SBP_{bra}, mmHg	139 ± 12	117	155
DBP_{bra}, mmHg	86 ± 12	61	108
PP_{bra}, mmHg	53 ± 9	36	70
Hypertensión, %	63		

Mean carotid systolic pressure value obtained with direct carotid tonometry was 132±14 mmHg. There were no significant differences between SBP_Ca_Diameter and SBP_Ca_Tonometer values (p=ns, paired t-test), and between SBP_Ca_Tonometer and SBP_Rad_Tonometer values (p=ns, paired t-test).

Figure 4 (left panel) shows carotid SBP estimated by exponential B-mode calibration method plotted against carotid SBP estimated by tonometry. The two methods correlated significantly ($R=0.96$, $p<0.05$). In figure 4 (right panel), the difference between direct carotid tonometry and echocardiography-derived pressure estimates is plotted against the average of the two methods. The mean difference between the two methods was 0.85±4.0 mmHg, whereas the range of the limits of agreement (mean ± 2 standard deviation) was -7 to 9 mmHg, showing no trend with increasing pressure levels.

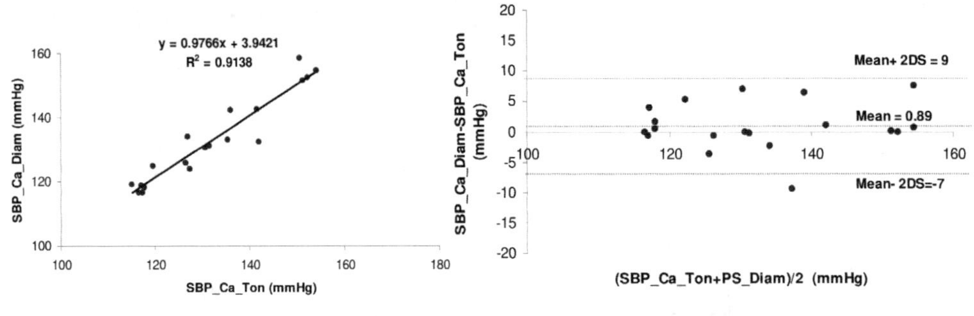

Figure 4. Left: carotid SBP by the exponential diameter calibration method plotted against carotid SBP by the direct tonometer method (n=20). Right: Bland and Altman plots. The dotted line is the mean difference and the dashed lines represent the mean ± two times SD

Similarly, figure 5 (left panel) shows SBP estimated by the radial calibration method plotted against carotid SBP estimated by tonometry. The mean difference between direct carotid tonometry and radial-derived pressure estimates was 0.18±5.0 mmHg, whereas the range of the limits of agreement (mean ± 2 standard deviation) was -10 to 10 mmHg, showing no trend with increasing pressure levels.

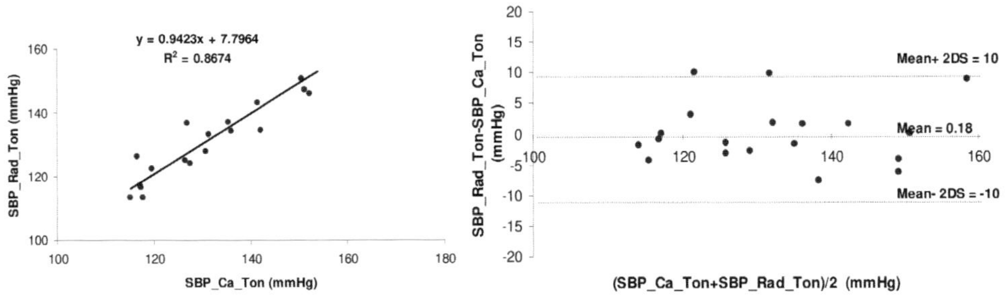

Figure 5 Left: SBP derived from calibrated radial waveform after moving average filter plotted against carotid SBP by the direct tonometer method (n=20). Right: Bland and Altman plots The dotted line is the mean difference and the dashed lines represent the mean ± two times SD

4. Discussion

In this study, we investigated the validity of using two approaches to estimate systolic local arterial blood pressure: from the analysis of B-mode carotid diameter waveforms and from the radial artery pressure waveform using a 125 point moving average filter. Due to the impossibility to obtain a direct measurement of aortic root pressure using an invasive procedure, direct carotid tonometry was used as reference method. Carotid tonometer waveforms were calibrated according to the procedure proposed by Kelly and Fitchet [7], using brachial diastolic and mean arterial pressures. Instead of deriving MAP from brachial tonometry, in this study MAP was computed as 38% of pulse pressure above diastolic pressure. Recently we showed that such value in the formula introduced an error of only 0.1% in mean brachial pressure estimation resulting in an error of 0.2% in systolic carotid estimates, independent of pulse pressure levels [13,14].

Calibrated radial pressure waveforms were filtered using a 125 point moving average filter. This number was established according to Williams et al, who determined that one quarter of the tonometer sampling frequency (in this case 500 Hz) most accurately defined the value of central systolic pressure [9]. In general, this alternative method was relatively simple to implement. We found no significant differences between SBP_Rad and SBP_Ton methods. The mean difference was only 0.18 mmHg, and 95% of the differences were less than two standard deviations, showing no trend.

Using calibrated diameter distension waveforms instead of tonometric waveforms avoids the technical difficulties and required operator skills associated with direct carotid applanation tonometry. Moreover, it use is not limited to superficial arteries. Several studies showed a very good agreement in systolic values, when comparing direct carotid tonometry with linearly or exponentially calibrated diameter distension waveforms [4,10,11,12]. However, compared to linear calibration, exponential calibration significantly improves the pressure estimation [10,11,12]. We found no significant differences between SBP_Dia and SBP_Ton methods. The mean difference was only 0.85 mmHg, and 95% of the differences were less than two standard deviations, showing no trend.

In summary, our results showed that the two alternative methods presented good accuracy within the acceptability limits of the AAMI criteria (differences <= 5 ± 8 mmHg). However, the results were obtained in a small sample and without severely hypertensive patients. It is therefore necessary to perform more studies, including a higher number of patients to confirm the results here found.

5. Conclusion

The two alternative methods used based on exponentially calibrated B-mode carotid diameter and on radial tonometry after application of a moving average filter, were found to allow an accurate and precise estimation of systolic local arterial pressure, when compared to direct carotid tonometry.

Acknowledgments
This research was supported in part by grant from PIP No 112-200901-00734 (CONICET)

References

[1] The CAFE Investigators for the ASCOT Investigators. Differential impact of blood pressure lowering drugs on central aortic pressure and clinical outcomes—principal results of the Conduit Artery Function Evaluation study: the CAFE study. Circulation 2006;113:1213–25.
[2] Roman MJ, Devereux RB, Kizer JR, et al. Central pressure more strongly relates to vascular disease and outcome than does brachial pressure: the Strong Heart Study. Hypertension 2007;50:197–203.
[3] Wang K-L, Cheng H-M, Chuang SY, et al. Central or peripheral systolic or pulse pressure: which one best relates to target organs and future mortality. J Hypertens 2009;27:461–7.
[4] Van Bortel LM, Balkestein EJ, et al. Non-invasive assessment of local arterial pulse pressure: comparison of applanation tonometry and echo-tracking. J Hypertens 2001; 19:1037-1044
[5] Chen CH, Nevo E, Fetics B, et al. Estimation of central aortic pressure waveform by mathematical transformation of radial tonometry pressure. Validation of generalized transfer function. Circulation 1997;95:1827–36.
[6] Fetics B, Nevo E, Chen CH, Kass DA. Parametric model derivation of transfer function for noninvasive estimation of aortic pressure by radial tonometry. IEEE Trans Biomed Eng 1999;46:698 –706.
[7] Kelly R, Hayward C, Ganis J et al. Non-invasive registration of arterial pressure pulse waveform using high-fidelity applanation tonometry. J Vasc Med Biol 1989; 3:142-9
[8] Hope SA, Meredith IT, Cameron JD. Arterial transfer functions and the reconstruction of central aortic waveforms: myths, controversies and misconceptions. J Hypertens 2008;26:4 –7.
[9] Williams B, Lacy PS,Yan P, Hwee C-N, Liang C, Ting C-M. Development and Validation of a ovel Method to Derive Central Aortic ystolic Pressure From the Radial Pressure aveform Using an N-Point Moving Average Method. JACC 2011; 57(8):951–6
[10] Kips J, Vanmolkot F, Mahieu D, Vermeersch S, Fabry I, Hoon J, Van Bortel L, Segers P. The use of diameter distension waveforms as an alternative for tonometric pressure to assess carotid blood pressure. Physiol. Meas. 31 (2010) 543–553
[11] Vermeersch SJ, Rietzschel ER, De Buyzere ML, De Bacquer D, De Backer G, Van Bortel LM, Gillebert TC, Verdonck PR and Segers P. Determining carotid artery pressure from scaled diameter waveforms: comparison and validation of calibration techniques in 2026 subjects. Physiol. Meas. 29 (2008) 1267–1280

[12] Graf S, Craiem D, Armentano R. Non invasive assessment of carotid and femoral arterial pressure using b-mode ultrasound diameter waveforms. Conf Proc IEEE Eng Med Biol Soc. 2012;1:5610-13
[13] Graf S, Craiem D, Armentano R. Validity of a new method to estimate mean arterial pressure at brachial level. Conf Proc IEEE Eng Med Biol Soc. 2010;1:2861-4
[14] Graf S, Armentano R. Non Invasive Assessment of Carotid and Femoral Arterial Pressure: differences in calibration using measured and calculated mean brachial pressure. IFMBE Proceedings 2009; 25/IV, pp. 793–796
[15] Verrij E, Vincent H, Bos W. Rule of thumb to calculate mean arterial pressure at the brachial artery level. J Hypertens 2008; 26:1043-1048.

Wound healing assay in a low-cost microfluidic platform

A J Conde[1], E Salvatierra[2], O Podhajcer[2], L Fraigi[3] and R E Madrid[1]

[1]Laboratorio de Medios e Interfases, FACEyT-Universidad Nacional de Tucumán/INSIBIO-CONICET
Av. Independencia 1800, 4000, S. M. de Tucumán, ARGENTINA

[2]Fundación Instituto Leloir, Buenos Aires, ARGENTINA

[3]Instituto Nacional de Tecnología Industrial, San Martín, ARGENTINA

E-mail: aconde@herrera.unt.edu.ar

Abstract. Cell migration is a very important stage in different pathological processes such as wound healing, vascularisation, immune response, tissue regeneration and cancer. Traditional wound healing assays are done by scratching a monolayer on standard culture surfaces. However this method can lead to data misinterpretation and poor comparability in results from different experiments. The microfluidic wound healing assay is a mature alternative to the traditional method, but it is expensive and bulky due to the need of syringe pumps and can be difficult to implement by the end user. Here we show an alternative low-cost microfluidic platform based on the MainSTREAM platform. We also demonstrate the feasibility of a microperistaltic pump to carry out these kind of assays. A laminar flow of PBS-Trypsin-PBS confined in a microfluidic channel with a previously cultured monolayer of IIB-Mel LES melanoma cells was used to generate the wound. Cell migration rate could be quantified successfully.

1. Introduction

Cell migration is the movement of cells from one location to another and it can involve single cells or groups of them and it is regulated by different phenomena such as physicochemical gradients [1], cell-cell contacts [2] and substrate topology [3] among others. Cell migration is an essential step in different pathologic and physiological processes such as wound healing, cancer, inflammation, cell growth and differentiation [4]. There are several in-vitro cell migration assays [5]. The most common are the wound healing assay and the Boyden chamber assay. The first is performed by introducing a wound by scratching the cell monolayer with a pipette tip, a blade or similar. The migration rate can be quantified by monitoring cell movement towards the artificially generated wound. The Boyden chamber assay is performed by seeding cells on top of a permeable membrane that separates two culture chambers. Cell migration can be quantified by observing the number of cells that migrate through the membrane.

Although the traditional wound-healing assay is simple, economical and can result in very high throughput assays [6], the inherent nature of the scratching can damage the cells on the monolayer which can cause leaking of the intracellular contents towards the wound complicating the interpretation of data [7]. Also the artificially generated wound size in this method can vary up to 33% [8]. With these drawbacks in mind, data analysis and direct comparison of experimental conditions can be misleading. Furthermore, in the traditional assay, cell culture media and reagents volumes are not insignificant.

Some of the aforementioned shortcomings have been already addressed utilizing microfluidics [9] and it has been demonstrated that this new assay is a mature alternative for the conventional wound-healing assay [10]. In this assay the cells are cultured inside a microfluidic channel and the wound is generated by parallel flows of trypsin and buffer/media that mix only by diffusion (due to the laminar flow condition). The trypsin flow detaches the exposed cells, while the cells exposed to the buffer remain attached. Migration rate of the cells towards the generated wound can be quantified like the standard assay. Another advantage of microfluidic systems is that physicochemical gradients and shear stress can be easily generated [10, 11]. Furthermore, it offers a much better control over the cell microenvironment being possible, for example, to reduce dilution of secreted cell signaling molecules in the culture media (due to the small volume of the microchannels) [12].

Although such devices offer an alternative solution to the traditional wound healing assay, they are quite difficult to implement by the end user, i.e. biologists, and have a high cost associated with support equipment, i.e. syringe pumps can cost more than 4.000 US dollars each and two of them are needed at least. Moreover, they might be difficult to place near or within detection instruments, incubators and so on limiting system portability. Furthermore, long tubing is required for interfacing the microfluidic chips to macro world, resulting in large dead volumes and poor liquid control due to pressure buildups [13].

Here, we propose a low-cost microfluidic system for the implementation of the wound healing assay based on the MainSTREAM platform developed by Sabourin et. al. [13]. This platform incorporates all components required for the flow of sample to waste through the Lab On a Chip system (i.e., pumps, valves, microfluidic reaction chips, reagent vials, waste containers, component interconnections, and computer controlled actuation). We demonstrate for the first time the use and feasibility of a low-cost microperistaltic pump for the generation of the multiple parallel laminar flows needed for the wound generation. IIB-Mel LES melanoma cells with well-characterized migratory phenotype [14] were used for this study.

2. Materials and Methods

2.1. Microfluidic chip and microfluidic platform

Microfluidic channels were milled in polymethyl methacrylate (Clarex, Japan) layers by a CNC milling machine (Minuth, Argentina). For best optical results, the cell culture channels were all-through milled in a 0.5 mm thick PMMA layer and other two 1.0 mm thick layers were used for sealing and connection purposes according to the MainSTREAM platform standards [13]. The PMMA layers were then bonded via a UV-assisted bonding process [15]. Bonding faces of the PMMA layers were exposed to a 250W mercury UV light for 2 minutes, aligned in a CNC milled aluminum frame and subsequently bonded in a bonding press (Shimeq, Argentina) at 80°C and 2MPa for 15 minutes. The bonding press was then fan cooled until the plates' temperature reached 40°C in order to avoid deformations and cracking of the chips.

PDMS (Sylgard 184; Dow Corning, USA) used for elastomeric components was mixed in a 10:1 mass ratio of elastomer to curing agent and placed under vacuum to remove air bubbles. PDMS components were cured overnight at 65°C.

The PMMA base plate, pumps, connectors, vial holders, tubing interfaces were also milled with a CNC machine, designs are described elsewhere [13].

Inlets and outlets of the chips were connected to 10 ml glass vials (Sigma Aldrich, USA) via 1.6mm OD and 0.2mm ID PTFE tubing (Bola, Germany). Vials were held in position by custom made PMMA vial holders attached to the base plate. In order to avoid bubbles the inlet and the outlet vials were pressurized with 0.5 bar filtered air.

LEGO Mindstorms NXT 2.0 servomotors and control unit were used to drive the pumps.

A photograph of the complete system can be seen in Figure 1.

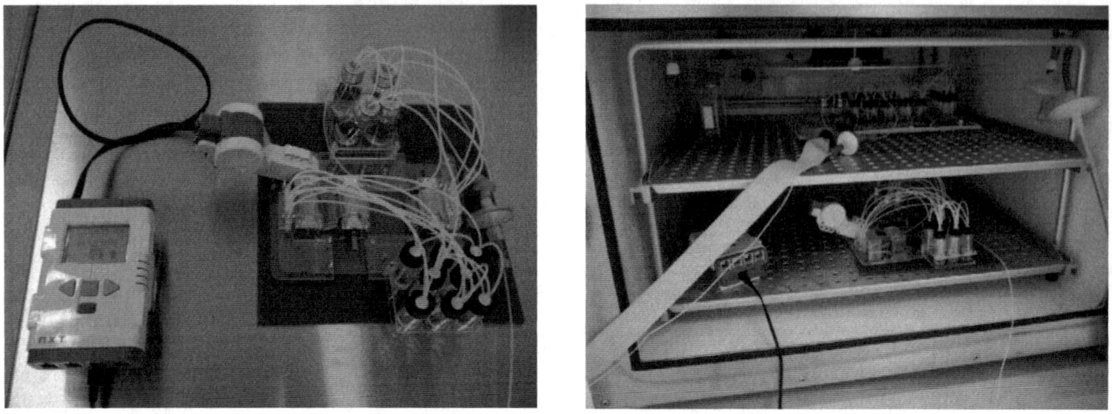

Figure 1. The complete microfluidic system for the wound-healing assay is shown on the left. On the right, it is shown that the complete system can fit easily within an incubator, only a power cable and gas connection for pressurization are needed. Note the small footprint and portability.

2.2. Assembly, sterilization and preparation of the chip and microfluidic platform

The complete system was assembled in a laminar hood under sterile conditions. Glass vials, screw caps and PTFE tubing were previously autoclaved. When the system was fully assembled, sterilization of pumps, connectors and the chip was done by flushing a 0.5 M NaOH solution at 5μl/min for 30 min. This was followed by 30 min washing with sterile H_2O to remove all the remaining NaOH and finally coating the chip by flushing it with complete medium (10% FBS (Natocor, Argentina) in DMEM (Invitrogen, USA)) at 5μl/min for 1 hour at 37°C.

2.3. Cell loading and culture

Subconfluent IIB-Mel LES cells were detached with 0,05% trypsin-EDTA, counted in a haemocytometer and resuspended at 2x10e6 per ml in complete medium.

Cell loading was done from the outlet connector chip (previously removing the outlet tubing) of the system and with the pump in reversal mode, i.e. sucking. LEGO motors power was set to 90% in order to have a good cell seeding distribution [16].

Cells were left to attach with no flow for 2 hours. Then, the cells were flushed with a 50 nl/min flow for 5 hours and finally were left in culture with a 500 nl/min flow overnight or until the monolayer was formed.

2.4. Laminar flow confirmation and microfluidic wound assay

In order to confirm the laminar flow inside the cell culture channel, the outer vials were loaded with miliQ water and the center vial with green food dye. Pumps were started and observation of the laminar flow was carried out with a stereoscopic microscope coupled with a CCD camera.

For the wound assay, the middle vial was replaced with 0,05% trypsin-EDTA (Invitrogen, USA) and the outer vials with PBS or media with no serum in order to avoid deactivation of the trypsin. In this way, only the center portion of the cells are to be removed. The pumps were set to 50 μl/min at 37°C for 20 min. Confirmation of the wound was observed in a contrast phase inverted microscope. Then, the cells were left at 500 nl/min flow with fresh media and pictures were taken at 1 hour, 2 hours, 3 hours, 5 hours and 18 hours.

3. Results

3.1. Visualization of laminar flow

We were able to effectively obtain a clearly defined laminar flow in the centre of the culture channel. A photo of the assay can be observed in Figure 2. In this experiment, the Reynolds number was 2.12 (D=500 μm, ν=1.10e-6 m^2/s, V=0.00424 m/s), confirming that the laminar flow was stably formed.

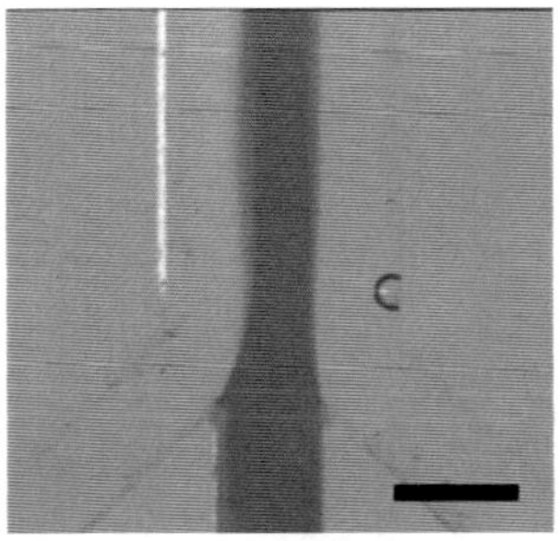

Figure 2. Laminar flow confirmation inside the microfludic channel. Scale bar is 500 μm.

3.2. Visualization of the wound

As expected, cell detachment was observed in the centre of the channel confirming also the laminar flow inside the microfluidic channel. Photographs of the cells cultured inside the channel before and after trypsin flushing can be seen in Figure 3. We show the beginning/merging point of the channel in order to be able to compare with the laminar flow seen on Figure 2.

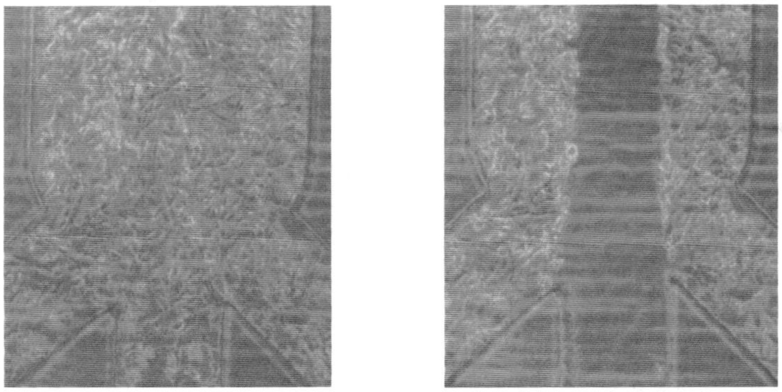

a b

Figure 3. Cell monolayer inside the microfluidic channel after 24 hours of culture a) before and b) after the trypsin flushing.

3.3. Cell migration after the wound assay

Photos of the cell migration towards the wound at different times can be observed in Figure 4. Pictures were taken at the center region (related to the length) of the microfluidic channel instead of the beginning/merging point as showed in Figure 3. This was done in order to ensure that the flow was fully developed. Note the invasion and pseudopodia extension of some of the cells in the centre of the photos pointed out by the red arrow.

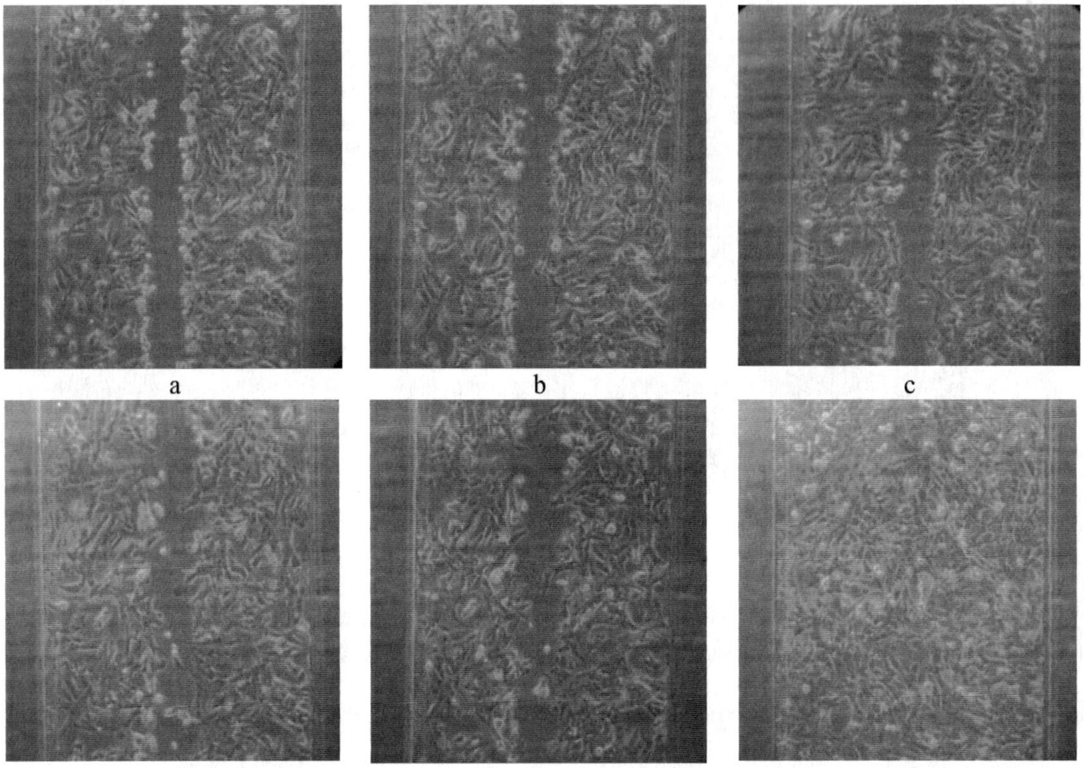

Figure 4. Cell migration observation. a) Just after the trypsin flush, b) 1 hour after the wound, c) 2 hours after the wound, d) 3 hours after the wound, e) 5 hours after the wound and f) 18 hours after the wound. The red arrow shows migration and pseudopodia expansion of a particular cell at different times.

Quantification of the migration process can bee seen in the graph of Figure 5. The migration distance was calculated as the short side of the rectangle with 500 μm of lenght without cell in the wound. Such area was measured in the segmented image using ImageJ software.

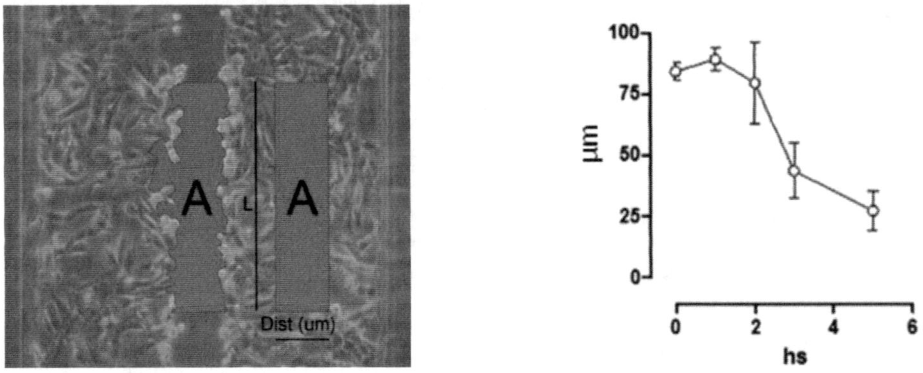

Figure 5. Quantification of the cells' migration towards the wound. a) Representation of migratory distance calculus. Wounds areas were measured in segmented images and l (Dist.) were calculated as $l=A*L^{-1}$. t The average of three areas of the wound was calculated. Values represent the mean ±SD in every time measured.

4. Discussion
We have demonstrated that it is possible to carry out microfluidic wound healing assays in a low-cost and easy to use microfluidic platform. A complete microfluidic system (including the control units and motors) can cost down to 20 times less than the systems based on syringe pumps. We have also demonstrated the feasibility of the microperistaltic pump for performing this kind of assays. Moreover, multiple migration assays can be carried out at the same time and in the same chip because of the pump's 8-channels design. Just for numbers, 4 simultaneous migration assays can be carried out with one pump, eight with two pumps and 16 with four pumps. It is very easy and straightforward to put these 4 pumps in a single platform because of the innate modular design of the MainSTREAM platform. Thus, considering the cost per migration assay, using the four pumps at a time, we could roughly estimate the initial cost per assay in our platform as 35 USD.

5. Conclusion
A low cost and easy to use microfluidic system with a microperistaltic pump was developed and tested. Wound edges of confluent cell monolayers for cell migration assay were obtained using a combination of laminar flows with media/PBS and trypsin in such a way that the central portion of the channel is in contact with the trypsin flow. This central channel can be coated with different matrix proteins as well as with proteins and factors or combinations to be tested. In addition blood flow can be simulated for endothelial cell assays and dynamic adhesion assays with tumor cells on endothelial monolayer. Although the preferred method is the Boyden chamber, it has the disadvantage that it cannot be monitored in real time, and gradients through pores could be disrupted by cell migration itself.

6. Acknowledgements
This work was supported by grants from the Agencia Nacional de Promoción Científica y Tecnológica (ANPCyT), Consejo Nacional de Investigaciones Científicas y Técnicas (CONICET), and institucional funds from the Instituto Superior de Investigaciones Biológicas (INSIBIO) and Consejo de Investigaciones de la Universidad Nacional de Tucumán (CIUNT). Mirofabrication was made in INTI, Bs.As. and the biological cultures in the Fundación Instituto Leloir.

References
[1] Janetopoulos C and Firtel R A 2008. *FEBS Lett.* **582** 2075-2085
[2] Hidalgo-Carcedo C, Hooper S, Chaudhry S I, Williamson P, Harrington K, Leitinger B, Sahai E 2011. *Nat Cell Biol.* **13** 49-58
[3] Pot S A, Liliensiek S J, Myrna K E, Bentley E, Jester J V, Nealey P F and Murphy C J 2010. *Invest. Ophthalmol. Vis. Sci.* **51** 1373–1381
[4] Hulkower K I and Herber R L 2011. *Pharmaceutics* **3** 107-124
[5] Liang C C, Park A Y, Guan J L 2007. *Nat. Protoc.* **2** 329–333
[6] Yarrow J C, Perlman Z E, Westwood N J and Mitchison T J 2004. *BMC Biotechnol.* **4** 21
[7] Nikolic D L, Boettiger A N, Bar-Sagi D, Carbeck J D and Shvartsman S Y 2006. *Am. J. Physiol. Cell Physiol.* **291** 68-75
[8] Pinco K A, He W, Yang J T 2002. *Mol. Biol. Cell.* **13** 3203-3217
[9] Nie FQ, Yamada M, Kobayashi J, Yamato M, Kikuchi A and Okano T 2007. *Biomaterials* **28** 4017–4022
[10] van der Meer AD, Vermeul K, Poot AA, Feijen J and Vermes I 2009. *Am. J. Physiol. Heart. Circ. Physiol.* **298** 719-725
[11] Shih H, Liu M, Weng T, Chen Y, Liao W, Tung Y 2012. *Proc. 16th Int. Conf. Min. Syst. Chem. Life Sci.* Okinawa, Japan
[12] Felder M, Sallin P, Barbe L, Haenni B, Gazdhar A, Geiser T and Guenat O 2012. *Lab Chip* **12** 640-646
[13] Sabourin D, Skafte-Pedersen P, Jensen Søe M, Hemmingsen M, Alberti M, Coman V, Petersen

J, Emnéus J, Kutter J P, Snakenborg D, Jørgensen F, Clausen C, Holmstrøm K and Dufva M 2012. *J. Lab. Autom.* Sept. 26

[14] Ledda F, Adris S, Bravo I, Kairiyama C, Bover L, Mordoh J, Chernajovsky Y and Podhajcer O L 1997. *Nat. Med.* **3** 171 - 176

[15] Truckenmuller R, Henzi P, Herrmann D, Saile V, Schomburg W K 2001. *Microsystem. Technol.* **10** 372–374

[16] Hemmingsen M 2012. PhD Thesis. DTU Nanotech. Technical University of Denmark

Monitoring heart rate variability on-line used mobile telephone 3G e-health service oriented

Goroso, D. G[†*]; Silva[*], R. R. da, Battistella[†], L. R.; Odstrcil[£], M.; Paolini[£], M.

† Physical Medicine and Rehabilitation Institute of Clinical Hospital, SP, Brazil.
* Research Technological Center of the University of Mogi das Cruzes, SP, Brazil.
£ Center of Information Technology of National University of Tucumán

Corresponding author's e-mail: daniel.goroso@hc.fm.usp.br

Abstract. We are living an epidemic of cardiovascular and cerebrovascular diseases that have led to many researchers worldwide increase their technical studies and methods of identification, diagnosis and treatment. Moreover, risk estimate based on factors such as age, sex, blood pressure, smoking, and lipid levels is still incomplete and depend on interest of subject by going to the hospital to do lab tests. The Brazilian experience that recorded each year over 90.000 deaths caused by cardiovascular and cerebrovascular diseases, only in 2008 the Unified Health System log about 200.000 hospitalizations, which resulted in an approximate cost of $ 180 million public resource. In addition, this is responsible for approximately 40% of early retirements and leading cause of disability in the age group above 50 years. Following the premise that "The best cure is prevention", this work seeks to provide a mobile tool to prevent early cardiac arrhythmia by means of monitoring on-line and assessing heart rate variability. Preliminary results proved effective in Sao Paulo for study of heart rate variability. This new service uses mobile phone to record, evaluate and transmit information, which currently is in clinical validation.

Keywords: Tool, Prevention, Cardiac Arrhythmia, FLEEM, mobile phone.

1. Introduction

Heart rate variability (HRV) is the physiological phenomenon of variation in the time interval between heartbeats. It is measured by the variation in the beat-to-beat interval. Other terms used include: "cycle length variability", "RR variability" (where R is a point corresponding to the peak of the QRS complex of the ECG wave; and RR is the interval between successive Rs), and "heart period variability".

Methods used to detect beats include: ECG, blood pressure, ballistocardiograms[1,2], and the pulse wave signal derived from a photoplethysmograph (PPG). ECG is considered superior because it provides a clear waveform, which makes it easier to exclude heartbeats not originating in the sinoatrial node. The term "NN" is used in place of RR to emphasize the fact that the processed beats are "normal" beats.

The HRV analysis has been proven to be a powerful tool for predicting mortality after a myocardial infarction[1], for measuring the risk stratification of diabetic patients, for evaluating the fetal cardiovascular system maturation[3], or for obstructive apnea detection[4,5]. The Task Force of the European Society of Cardiology and the North American Society of Pacing and Electrophysiology[1] propose several techniques for measuring the HRV. Two different types or methods are proposed[6]: a) statistics in the time domain over the RR series[7]. Examples are the standard deviations or the calculation of the percentage of values that differ more than 50 ms from the previous one; and b) analysis in the frequency domain, or spectral analysis, which allows for decomposing the full variability in its frequency components: both low- and high frequency components[7].

In this study, a novel system of monitoring physiological signals takes advantage of advances in information and communication technology to acquire, process and visualize physiologic information of people in their day to day. The system measured heart rate to perform arrhythmia diagnostics. The system that allows the "Control of your health in your hands" was developed between the University of Sao Paulo, the Research Technological Center of the University of Mogi das Cruzes (SP, Brazil) and Center of Information Technology of National University of Tucumán. This product called FLEEM (Free Living Energy Expenditure Monitoring System®)[8,9,10]. FLEEM can monitor HR 24h/day, 7day/week with a device that combines technology, comfort and reliability. The sensors are noninvasive, low power consumption and high sensitivity. The information is pre-processed on a mobile device and centralized on a server that does the monitoring and accompaniment for each user, allowing that data always travels with the user[8]. Different means are used to transmit information of evaluations (Wi-Fi, Bluetooth and GSM), which ensures connectivity and extends coverage. Physiological data are processed in a telemedicine centre and accompanied by a specialized team[9]. The monitoring of the system is non-invasive, non-intrusive, 24 x 7. As the HRV recorder will detect a "non-standard" heart rate (HR) signal, it has been necessary to compare those signals to "standard" ECG recording signals in order to evaluate the arrhythmia detection ability of the new system. Simultaneous recording of HR signals from both the mobile telephone recorder and a conventional Holter recorder was compared by two independent cardiology specialists with regard to signal quality for performing arrhythmia diagnosis. In addition, calculated R–R intervals from the two systems were correlated. A total number of 16 patients participated in the study, but in this paper is shows the monitoring the only one patient. The goal is to test the loyalty of the FLEEM System® to detect with acceptable quality the arrhythmia diagnosis.

2. Materials and Methods

This section contains four parts. First is the description of FLEEM Systems®. Second is the data acquisition. Third is the methods study protocol and HRV analysis. Finally, the whole data process is described.

2.1. FLEEM Systems® - Telemedicine

The FLEEM System® (Free Living Energy Expenditure Monitoring System) [8,9,10] is a system service that allows the monitoring of medical variables (heart rate, temperature, motion, etc.) and transmits them to a cell phone operating system Android via sensor Zephyr HxM BT® that measured the HR. This mobile phone has a set of others applications that, in addition to processing the measured data, can record data from the accelerometer for assessing energy expenditure or if the patient has or not a fall (fall control). The android app (Google Play Signs Physiological on-line®) also allows recording the amount of ingested food and calculate the calorie consumption. All these pre-processed data, are transmitted to an application server via (TCPIP) via 3G or Wi-Fi synchronously (on line) or asynchronous, while it does not have connection, stores the data, and when you have it, transmits them[8]. The phone's GPS helps to determine the geographical position of the patient. Once measurements arrive at the server, doctor who is appointed to the benefit of the patient, can display this information in different graphic formats so that they help you make decisions. These graphics are calculated based on measurements and typical data of the patient (sex, biotype, weight, height, etc.) and can be customized by the doctor. The application architecture is LAMP and is based on free tools without licensing costs, generating a runtime environment flexible, scalable, global, regime 24 × 7, safe, reliable coverage and massive capacity to care for patients at the same time.

The Figure 1 shown the outline of FLEEM System®.

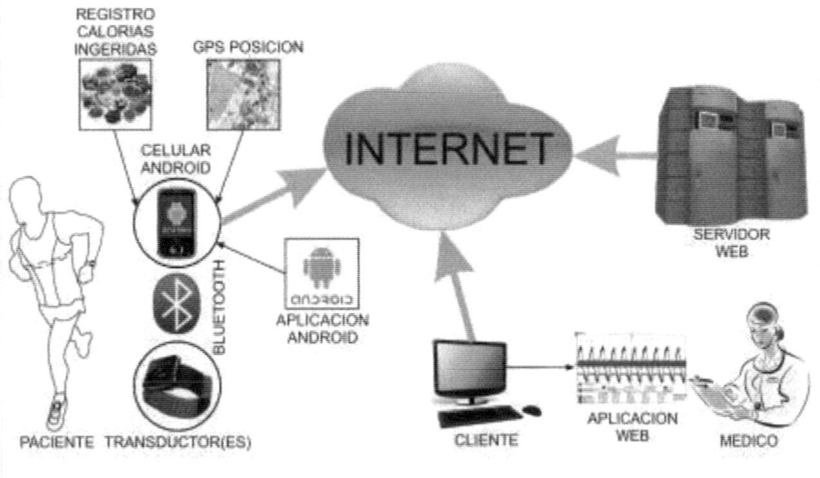

Figure 1. Outline of operation of the integrated system of monitoring physiological signals.

2.2. Data Acquisition

Figure 2 shows the physical layout of the sensors in the body of the subject[9]. Heart rate monitor sensor is installed through an adjustable elastic band directly in the chest of the subject and connected with mobile phone via Bluetooth connection. From the cellular mobile device detected and indirectly generated registered information are transmitted by the radio service (GPRS) general package of the 3G network of the cell for a large distance mobile or wireless technology for short distances or via USB connection[9].

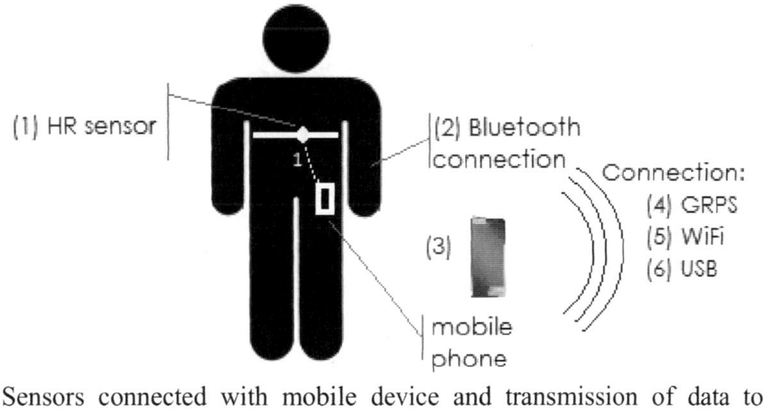

Figure 2. HR Sensors connected with mobile device and transmission of data to the central of telemedicine.(1) Heart rate monitor. (2) Connection between sensor and mobile phone. (3) Mobile phone. (4), (5) and (6) Connection 3G network between mobile phone and server of telemedicine.

2.3. Methods

The tests were conducted initially in laboratory environment and then under various free-living environments in order to record the actual data without interfering with routine activities of any subject.

2.3.1. Protocol

Ethical approval for the study was granted by the Mogi das Cruzes University Research Ethics Committee protocol FR-281644. This paper presents the monitoring the heart rate of a patient of male gender, clinically diagnosed with cardiac arrhythmia. The acquisition of the physiological signal was carried out through the sensor was connected via Bluetooth with mobile device. Monitoring and calculations was carried out through the FLEEM System® web platform over a period of 2 weeks. The protocol was that any given day within those two weeks of study the patient do not swallow medication for regulation of the arrhythmia. The patient must not inform that day was within those two weeks. This was done to test the loyalty of the system of monitoring FLEEM System® designed for detection of cardiac arrhythmias, among other analyses tools[8,9,10].

2.3.2. Data Analysis.

To analysis the HRV is used pNNx family index because it is a widely used measure of heart rate variability statistics. The software application was done in Java J2ME based in pNNx software package [xx][1]. For each RR interval series (24 hours, each 5 min), the absolute values of the differences in consecutive NN intervals were obtained. The heart rate variability is recorded during dream states (baseline), light activities (rest) and moderate activities (active), at different times of day, but during the rest are enrolled in pre-and postprandial breakfast and lunch. The data analysis separates differences distributed in the following intervals in ms: [0, 50], (50, 80], (80, 120] (120, 140] and (140, 160][6,7], classifying total sample data. Thus, with this classification is possible a distribution histogram and calculated probability occurrence, which allows calculation of the correlation time intervals for each activity state. For analysis is necessary assess the discrimination between a variability under normal and pathological conditions for each record into different states[11] (sleep, rest, activity) by using pNNx thresholds from less than 30 ms to more than 150 ms (each 20 ms interval)[7,12]. In addition, we have analysis with special care to find out the changes in the autonomic tone before and after meals during normal days. The heart rate variability represents adaptive responses of autonomic nervous system to challenges for circulation as much as by breathing that which to some extent reflects the sympathovagal balance[1,4,5]. Figure 3 shows record of heart rate variability from one case monitoring FLEEM System ®.

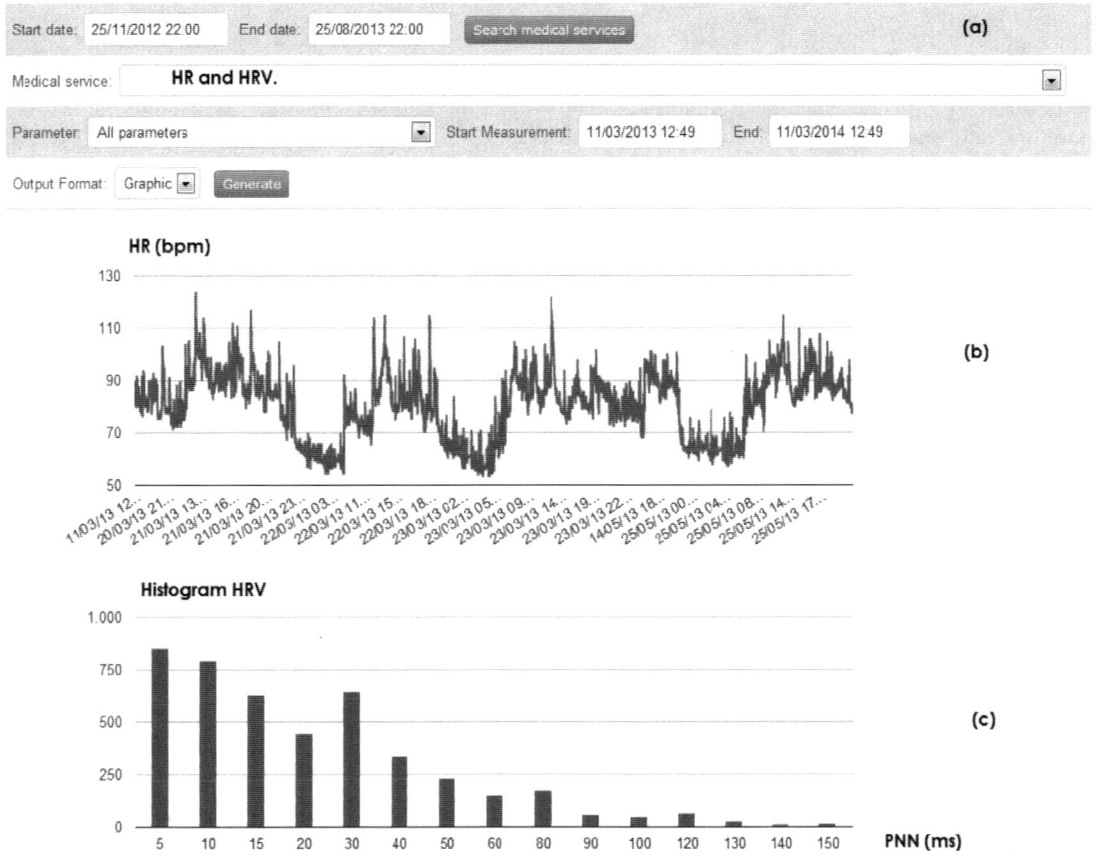

Figure 3. Recording of the heart rate monitoring FLEEM System® web platform (www.fleem.com.ar). (a) data concerning the patient and service of monitoring. (b) On-line monitoring the cardiac frequency of a patient during March/11/2013 to May/25/2013. (c) General histogram of heart rate variability.

The time domain methods are computationally simple, but they lack the ability to distinguish between sympathetic and parasympathetic contributions of the HRV. The power spectrum estimation techniques avoid the problem of leakage and provide better frequency resolution. In this way, spectral analysis of HRV recorded via mobile phone can be a powerful method for studying the pathophysiologic processes in certain diseases and it can also be used in the daily clinical practice. The standard frequency bands used for power calculation[1,7] were the very low frequency (VLF) band, ranging from 0 to 0.04 Hz, the low frequency (LF) region, from 0.04 to 0.15, and the high frequency range, form 0.15 to 0.40 Hz.

The HF reflects the fast changes in the beat-to-beat variability due to the parasympathetic or vagal stimulation. The LF region represents a mixture of sympathetic stimulation of the heart.

The spectrogram was obtained and, after that, the power spectra were calculated for each region, as well, as the total power spectrum. The LF/HF ratio, of interest for the clinical diagnosis, was also calculated[1,7].

3. Results

In this work, we have focused on the implementation of a computerized tool that used mobile telephone 3G e-health service-oriented for analyzing the HRV recording. The FLEEM System monitored on-line, for each HR, the following measures were obtained: mean VLF power, mean LF power, mean HF power, total power, and LF/HF ratio. The Figures 4a and 4b shows the output of the HRV software package, which represents HR and HRV for one subject monitored with/out arrythimia medication (case 1 and case 2, respectively). The period analyzed was six hour from 02:00hs to 02:30h for each protocol.

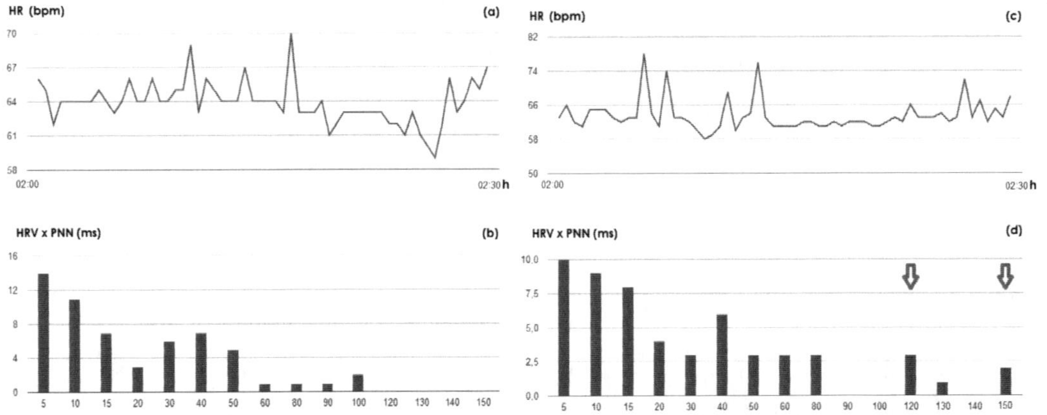

Figure 4. Recording of the heart rate (HR) and heart rate variability (HRV) monitoring FLEEM System® web platform (www.fleem.com.ar). (a) HR subject case 1. (b) HRV histogram subject case 1. (c) HR subject case 2. (b) HRV histogram subject case 2.

The Figure 4d shows variability greater than 100 ms intervals, being that these peaks are not registered in the situation in which the patient ingested his medication to control cardiac arrhythmia (Figure 4b). Observe that in the situation of control (Figure 4a and 4b) the cardiac frequency is much more uniform, fact that is shown in the histogram presenting peaks just for a variability in the PNN [5, 15] ms. The Table 1 shows the HR summary for each case.

	Min	Max	Range	Mean	Median	Mode	Var	Std
Case 1	56.5	69.71	22.21	63.59	63.26	63.75	8.80	2.96
Case 2	56.5	79.93	43.43	73.32	73.23	63.36	126.71	11.25

Table 1. HR summary obtained during 30 min in different day with the patient with arrhythmia medication (case 1) and without arrhythmia medication (case 2).

Figure 5 shows the a plot of the Fast Fourier Transform (FFT) and the power spectrum for the HR register presented in Figure 4.

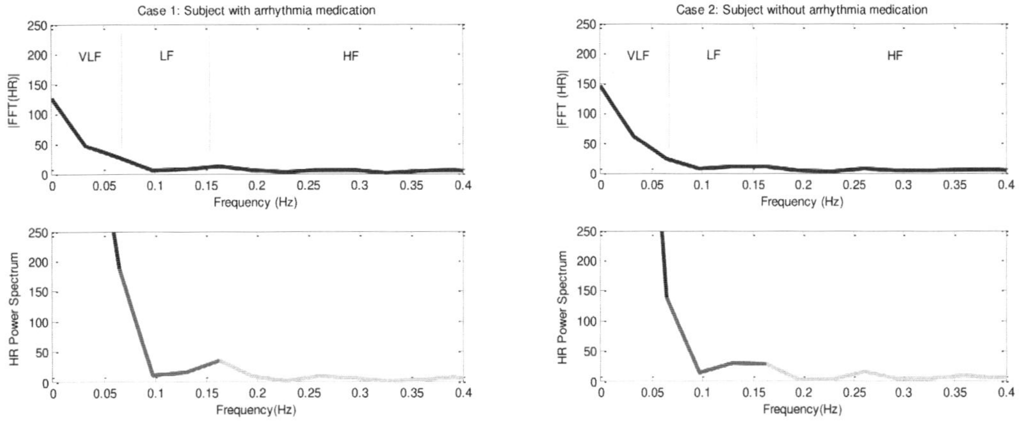

Figure 5. Spectrum of cardiac frecuency signal. Case 1 and Case 2. VLF: Very Low Frequency. LF: Low Frequency. HF: High Frequency.

Table 2 shows the VLF, LF, HF power, and for LF/HF ratio calculated from the plot shows plotted in the Figure 5 for each case.

patient	mean VLF Power	mean LF Power	mean HF Power	Total Power	LF/HF
Case 1	$1.59\text{x}10^{+3}$	62.01	8.04	$4.86.10^{+3}$	7.71
Case 2	$2.12\text{x}10^{+3}$	53.90	7.80	$6.50\text{x}10^{+3}$	6.90

Table 2. Summary obtained during 30 min in different day with the patient with arrhythmia medication (case 1) and without arrhythmia medication (case 2). Measures in [ms^2].

The results obtained in the analysis of heart variability in the domain of time (Figure 4 and table 1)[12], and in the domain of the frequency[13] (Figure 5 and Table 2) show that there is alteration in cardiac autonomic control system. Note is that Std reported in Table 1 in Case 2 is nearly 4 times greater than in the Case 1, being that the variance is 10 times greater. This means that the patient, during the period that this supposedly relaxed (02:00h-02:30h) presents a wide variability in relation to your heart rate average (see it values means and max, Table 1).

The analysis of heart variability in the frequency domain shows absolute values of power[13] altered of the Case 2 in relation to the control, Case 1. There is predominance of power average and total in case 2, this last almost 1.5 greater than in case 1[12]. These results are indicative of the presence of cardiac frequency components that alter the normal heart rate[12,13].

The patients used the wireless sensor while doing physical sport activities and the quality of the recorded HRV signals made it possible to perform arrhythmia diagnostics even under such conditions. In this study, the monitoring was realized daily and the analyzed was during nocturne period. Consequently, this makes possible improvements in correlating arrhythmias to physical activities. The integration of traditional risk factors with monitoring daily of heart rate variability, and the self-care, represent an effective approach for identification of stroke patients at risk for early mortality.

4. Conclusions

This tool enables the monitoring and detecting arrhythmia cardiac based on the variability of RR records analyzed. According to the table 1 and 2 and Figure 4 and 5 notes are mainly the effect of the medication for the regulation of cardiac arrhythmia, which was detected in the online of heart rate monitoring. The analysis carried out in the space of time and frequency space confirmed present in the patient cardiac arrhythmia on the day in which it not ingested medication, concluding in this way that the FLEEM system is a useful and appropriate tool for the monitoring of HR, HRV and detection of cardiac arrhythmia via online using 3G technology.

References

[1] Malik, M. et al. (Chairman of Writing Committee of Task Force of the European Society of Cardiology and the North American Society of Pacing Electrophysiology): Heart Rate Variability: Standards of Measurement, Physiological Interpretation, and Clinical Use. Circulation, 93,1043-1065. (1996).

[2] A. J. Camm, M. Malik et al., “Heart rate variability: standards of measurement, physiological interpretation, and clinical use,” Circulation, vol. 93, no. 5, pp. 1043–1065, 1996.

[3] M. R. Ortiz, S. D. Aguilar,J. Alvarez-Ramirez, A. Martínez, C. Vargas-Garcia, R. González-Camarena, J. C. Echeverría. Prenatal RR fluctuations dynamics: detecting fetal short-range fractal correlations. vol. 26, no13, 1241–1247, 2006.

[4] Gula, L.J., Krahn, A.D., Skanes, A., Ferguson, K.A., George, C., Yee, R. and Klein, G.J., Heart rate variability in obstructive sleep apnea: a prospective study and frequency domain analysis. Annals of Non-invasive Electrocardiology (ANE). v8. 144-149. 2006.

[5] Penzel, T., Kantelhardt, J.W., Grote, L., Peter, J.H. and Bunde, A., Comparison of detrended fluctuations and spectral analysis for heart rate variability in sleep and sleep apnea. IEEE Transactions on Biomedical Engineering. v50. 1143-1151.

[6] R. Balocchi, F. Cantini, M. Varanini, G. Raimondi, J. M. Legramante, and A. Macerata, “Revisiting the potential of timedomain indexes in short-term HRV analysis,” Biomedizinische Technik, vol. 51, no. 4, pp. 190–193, 2006.

[7] J. J. Sollers, T. W. Buchanan, S. M. Mowrer, L. K. Hill, and J. F. Thayer, “Comparison of the ratio of the standard deviation of the r-r interval and the root mean squared successivedifferences (SD/rMSSD) to the low frequency-to-high frequency (LF/HF) ratio in a patient population and normal healthy controls,” Biomedical Sciences Instrumentation, vol. 43, pp. 158–163, 2007.

[8] Lopez, L.J.R. and Goroso D.G. Monitoring System of Assessment of Energy Expenditure to Human in Free-living Condition by Mean of Information and Communication Technologies. in Proc. Med-e-Tel, Luxembourg (Lu), 2010, pp. 550-555. ISBN 2179-3220

[9] L.J.R. Lopez , D.G. Goroso e L.M. Battistella. Sensor Network for Assessment of Energy Expenditure design based on Android. In IFMBE Proc. 2011; 3:130-134.

[10] L.J.R. Lopez , D.G. Goroso e L.M. Battistella. Relationship between triaxial stand-alone and embedded accelerometers in a mobile phone for Healthcare Applications. Medical Engnnering & Physics, 2012.

[11] Akselrod, S., Gordon, D., Ubel, F.A., Shannon, D.C., Berger, A.C. and Cohen, R.J., Power spectrum analysis of heart rate fluctuations: a quantitative probe of beat-to-beat cardiovascular control. Science. v213. 220-222. 1981.

[12] Malliani A, Montano N. Heart rate variability as a clinical tool. Ital Heart J.;3(8):439-45. 2002.

[13] Montano N. et al. Heart rate variability explored in the frequency domain: a tool to investigate the link between heart and behavior. Neurosci Biobehav Rev. 33(2):71-80. 2009.

Assessing effective connectivity in epileptogenic networks. A model-based simulation approach.

Florencia Jacobacci*[1,2], Martin Sapir*[1,2], Santiago Collavini*[1,2], Silvia Kochen[2], Alejandro Blenkmann[1,2]

1 Facultad de Ingeniería y Ciencias Exactas y Naturales, Universidad Favaloro, Buenos Aires, Argentina
2 IBCN "Prof. E. De Robertis", Facultad de Medicina, Universidad de Buenos Aires (UBA) - CONICET, Buenos Aires, Argentina.

e-mail: florenciajacobacci@gmail.com, martinsapir@gmail.com, santiagocollavini@gmail.com, ablenkmann@gmail.com

Abstract. Different connectivity configurations were simulated using epileptogenic and non epileptogenic neuronal populations. Connectivity between them was measured using Partial Directed Coherence and Directed Transfer Function. The results were satisfactory and in some cases of clinical utility. The methodology that was used is discussed in comparison with previous works.

*** Jacobacci, Sapir and Collavini contributed equally to this work.**

1. Introduction

Epilepsy is a neurological pathology characterized by the recurrence of seizures which is the result of abnormal electrical activity in groups of neural populations. The *epileptogenic zone* (EZ) is the area of cortex indispensable for the generation of seizures. It may include different overlapped areas: the *seizure onset zone* (SOZ), an abnormal region of the brain which becomes hyperexcitable and synchronized and triggers the epileptic seizure; the *irritative zone* (IZ), an area of cortex involved in epileptic interictal activity; and other areas, e.g. lesion zones [1]. To keep seizures under control, the most common treatment is drug-based. Unfortunately, 30% of these patients do not respond to pharmacological treatment [2] and the surgical treatment, i.e. tissue resection, may be the best option in 50% of these cases [3]. In order to eradicate seizures after cortical resection the EZ has to be correctly defined.

Intracranial electroencephalography (iEEG) recording of brain activity is the gold standard method of assessing the location and extent of the SOZ and IZ since it provides electrical signals directly recorded from the cortical surface (*electrocorticography*, ECoG) or areas deep inside the brain (*StereoElectroencephalograpy*, SEEG). The incorrect identification and resection of the pathological region would result in the recurrence of seizures after surgery. As a consequence, it is crucial to determine its anatomical location, its functional interactions, and its connectivity pattern. The EZ is organized as a network, commonly known as *epileptogenic network*, and is formed by different neuronal populations connected to each other. Understanding this organization may have implications in the surgery prognosis.

It has long been recognized that neuronal interactions are directional [4]. Being able to assess the directionality of neuronal interactions in the EZ is highly desirable in order to have a deeper understanding of the behavior and extension of pathological network.

Different kinds of connections have been defined in neuroscience in the last decades. Anatomical connectivity refers to the physical trajectory of the nervous fibers that connect different regions of the brain. This type of connectivity is almost stable over time and can be assessed using diffusion tensor imaging. Functional connectivity is a concept that reflects the functional interaction of different temporal correlations between remote neurophysiological events. It is simply a statement about the observed correlations; it does not provide any direct insight into how these correlations are mediated. Functional connectivity is studied using functional neuroimaging techniques such as functional magnetic resonance imaging (fMRI). Effective connectivity, on the other hand, is closer to the intuitive notion of a connection and can be defined as the influence on neural system exerts over another. This connectivity may vary over different situations, conditions or tasks, denoting the dynamics of information processing between different areas in the brain. In pathological situations, as in epilepsy, evidence indicates that effective connectivity may be affected [5][6]. A significantly different connectivity pattern distinguishes the epileptogenic zone from other cortical regions not only during the ictal event, but also during the inter- and pre-ictal periods. This indicates that the lesional nodes play a leading role in generating and propagating ictal EEG activity by acting as the hubs of the epileptic network originating and sustaining seizures.

The aim of this work is to be able to test tools that estimate the effective connectivity between neuronal populations within the epileptogenic network. The knowledge of this characteristic would offer better insight on the pathology and a better planning of resection surgery, resulting in a better quality of life for the patient. In order to study the reliability of the connectivity estimator, we propose the use of non-linear physiologically plausible models of neuronal populations. These models have proven to be useful in resembling real iEEG signals [7]. Besides, they allow variation of the parameters so as to set the connectivity between the different groups of neurons.

2. Method

To test the reliability of the connectivity estimators, we generated a series of data representing pairs of populations with different characteristics and different connectivity. These populations consisted of epileptic and/or non-epileptic neurons with a one-way connectivity gain. We defined three different levels of coupling based on the activity induced by an epileptogenic population over a non-epileptogenic one. These levels were denoted as low, intermediate and high. Four basic cases were defined in which all the possible combinations of said populations were represented. Each case was simulated four times, according to high, medium and low connectivity or none at all, thus defining sixteen different series of data.

Lastly, two final simulations were carried out with three populations which represented a more relevant and realistic case. The connectivity was measured using Partial Directed Coherence and Directed Transfer Function [8][9][10]. Statistical significance of connectivity measurements was tested using a non-parametric surrogate statistical method. Phase randomized surrogate data was used for the tests [11]. Finally, to account for multiple comparisons statistics, the false discovery rate (FDR) method was applied [12].

All simulations and data analysis were done in Matlab 7.12.0 (The MathWorks, Inc). Our own scripts were developed and functions from ARFIT, BIOSIG, EEGLAB and SIFT toolboxes for Matlab were also used for this purpose [13][14][15][16][17].

2.1. Physiologically plausible model

The model used is a reversion of a neurophysiologically relevant model which represented the generation of evoked potentials in the visual cortex [18]. Later on, this model was extended to generate spontaneous interictal spikes from multiple coupled neuronal populations by altering the model parameters [7]. The simulation of this kind of pathological electrical activity has successfully been used to evaluate sLORETA algorithm for the resolution of the inverse problem on ECoG [19].

The diagram of the model used to simulate normal and epileptic brain activity in a realistic way can be seen on Figure 1. On this model, each population represents thousands of neurons and is composed of three subpopulations: pyramidal neurons, excitatory inter-neurons, and inhibitory inter-neurons. In turn, each subpopulation is characterized by: a) a non-linear sigmoid function S(v) that relates the average membrane potential of the subpopulation with the density of action potentials and b) a second order linear system which relates the density of pre-synaptic impulses to the average membrane potential; excitatory post-synaptic potential (EPSP) or inhibitory post-synaptic potential (IPSP), depending if the generating subpopulation is excitatory or inhibitory. Besides, there are four constants C_i (i=1, 2, 3, 4) that represent the amount of synapses between the subpopulations.

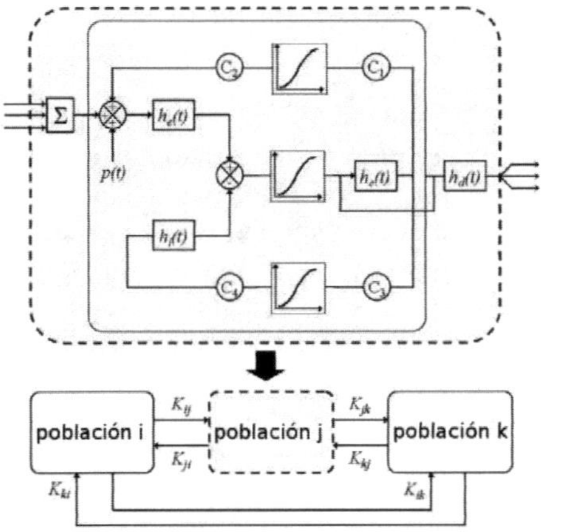

Figure 1. Physiologically plausible model of coupled neural populations. The top of the figure shows the diagram for a cortical population where the upper loop represents the subpopulation of excitatory inter-neurons, the middle loop represents pyramidal neurons and the lower loop represents inhibitory neurons. On the bottom of the figure, connection with other populations can be seen.

The impulse response of the second order systems represents the EPSP and IPSP of the modeled neurons. In the case of pyramidal neurons and excitatory inter-neurons this is $h_e(t)=Aate^{-at}$ whereas for inhibitory inter-neurons it is $h_i(t)=Bbte^{-bt}$, where A and B are the average synaptic gains for each model, and a and b are linked to the membrane average time constant and the average distributed delays in the dendritic tree. In normal conditions, A=3,25mV, B=22mV, a=100s–1 and b=50s–1. The populations modeled using these parameters are called *passive* and the electrical activity they generate is "background activity". On the other hand, increasing the average excitatory synaptic gain to values between 3,4mV and 3,6mV leads these populations to a state of rhythmic generation of interictal spikes. These populations are called *active*.

Finally, to define the coupling between the populations, the matrix K is used, where each element K_{ij} represents the coupling intensity between the origin population *i* with the destiny population *j*.

The mean membrane potentials in pyramidal neurons are a measure of the neural activity in a relatively small area of cortex and are proportional to the potentials recorded with intracranial electrodes [19]. Therefore, these signals are used as output of the system. The coupling parameters define the connectivity between populations and can be adjusted to simulate different network configurations.

A coupling criterion was defined based on the behavior of a *passive* neuron when receiving information from an *active* one. The passive neuron would start to show spikes with a firing rate related to the coupling parameter. At a certain level of coupling the passive neuron would have a firing rate equivalent to that of the active neuron (at least 30 spikes per minute); this level was defined as high connectivity. Low connectivity was defined as the lowest level that produces a firing rate different than zero (about 5 spikes per minute). A third, intermediate level was defined in between these two.

The assessed connectivity estimators presented later in this work may predict these configurations based on the information of membrane potential in the pyramidal neuron populations. These potentials were selected as an appropriate measure since real iEEG recordings are proportional to the mean membrane potential of pyramidal neurons populations. The reader is referred to [7] and [18] for more details of the model used.

2.2. *Multivariate causality based measures of connectivity*

The methods for studying effective connectivity between neuronal populations were based on mathematical analysis of the membrane potential in the pyramidal neuron populations. These data were presented in the form of multiple simultaneous time series.

The basic concept of bivariate causality introduced by Granger is the causal influence in the context of linear regression models. If the variance of the autoregressive prediction error of one time series at the present time is reduced by inclusion of past measurements from a second time series, then the second is said to have causal influence on the first i.e. the prediction of one time series could be improved by incorporating the knowledge of a second one [4][20].

2.2.1. Multivariate autoregressive (MVAR) model Let $X(t)=[X_1(t), X_2(t),...,X_k(t)]^T$ denote the measurement from k channels at time t and T *means* matrix transposition. We can describe $X(t)$ in the form of a MVAR model [9]. In equation 1, p is the model order, that is to say, how many previous values of X are needed to estimate the current value. $A(i)$ is a matrix of dimension kxk containing the autoregressive coefficients A_{mn}, m and $n=1,...,k$ which relate channel m with channel n for the time lag i. $E(t)$ is a vector of length k containing the error term for each channel, i.e. part of the signal X at time t that is not explained by the previous values of X.

$$X(t)=\sum_{i=1}^{p} A(i)X(t-i)+E(t) \tag{1}$$

2.2.2. Directed Transfer Function In the definition of Granger causality given at the beginning of section 2.2, only two time series are considered. Directed Transfer Function (DTF) is a multivariate method that gives pairwise directional information from one autoregressive model fit to a larger number of channels. It has been shown that multivariate DTF function can be interpreted within the framework of Granger causality [9] . Transforming equation 1 to the frequency domain [8], we obtain:

$$A(f)=I-\sum_{j=1}^{p} A(j)exp(-i2\pi f\Delta tj) \tag{2}$$

Where I is the identity matrix of dimension kxk and Δt is the sampling interval.

Equation 2 can be rewritten as:

$$X(f)=A^{-1}(f)E(f)=H(f)E(f) \tag{3}$$

In equation 3, $H(f)$ is the transfer matrix of the system for each time frequency point.

DTF from channel j to channel i, representing the causal influence from j to i at frequency f, is defined as:

$$\theta_{ij}^2(f)=|H_{ij}(f)|^2 \tag{4}$$

The normalized DTF expresses the ratio of influence of channel j to channel i to the joint influences from all the other channels to the channel i, and has a value between 0 and 1.

$$\gamma_{ij}^2(f)=\frac{|H_{ij}(f)|^2}{\sum |H_{im}(f)|^2} \tag{5}$$

It is important to highlight that a zero value of DTF may not mean a total absence of causal influence between the two channels [9].

2.2.3. Partial Directed Coherence Partial Directed Coherence (PDC) was introduced as a modification/generalization of directed coherence and DTF. While DTF represents a balance of signal power that spreads from one structure to another over many possible alternative pathways, PDC portrays the relative strength of direct pairwise interactions. As a consequence, DTF marks the existence of signal pathways connecting structures either directly or indirectly while PDC resolves the existence of direct connections[23]. Thus, PDC provides clearer and more immediate frequency domain connectivity picture of Granger causality for the simultaneous analysis of more than two time series [10]. Besides, PDC has the computational advantage of avoiding the matrix inversion in equation 3 that the calculation of DTF requires.

The normalized squared PDC expresses the ratio of influence of channel j to channel i to the joint influences from channel j into all other channels, and has a value between 0 and 1.

$$\pi_{ij}^2(f) = \frac{|A_{ij}(f)|^2}{\sum |A_{im}(f)|^2} \tag{6}$$

2.3. *Simulation of data, adjustment of MVAR model and calculation of DTF and PDC*

Sixty seconds simulations were considered a good balance between the number of data points to synthesize and analyze and sufficient interictal record with abundant epileptic spikes. Based on previous studies sampling frequency was set to 200 Hz [23]. Average excitatory synaptic gain was set to 3.25mV to generate passive populations and to 3.52mV to generate active ones. Null, low, medium and high one-way connectivity between the populations were used.

Two coupled poupulations

Four groups were defined according to the populations and direction of connectivity possibly involved in a simulation consisting of two populations. The name of each group consists of two characters that identify the type of neurons that take part of the coupling as passive (P) or active (A). The first character refers to the population where the information flow is originated and the second one refers to the population which is the destiny of this flow.

These groups were simulated for each connectivity level: null (N), low (L), intermediate (M) and high (H). Simulations will be quoted using this code (e.g.: simulation PA-M refers to a passive population coupled to an active one with intermediate connectivity with the information flowing from the passive to the active population). The result was 15 simulations (due to AP-N and PA-N being equivalent) which cover all the possible combinations of unidirectional and null connectivity between active and/or passive populations.

Many coupled populations

The two final cases represented an active population (A) connected to a passive one (P1), which in turn was connected to another passive population (P2). The coupling scheme used can be seen in figure 2. Connectivity between the active population A and the passive population P1 (A→P1) was set to be medium. Two different connectivity patterns were assumed. In one of these cases the connectivity between the passive populations (P1→P2) was set to be low (this case will be quoted as APP-ML), while in the other it was set to be medium (which will be quoted further in this work as APP-MM).

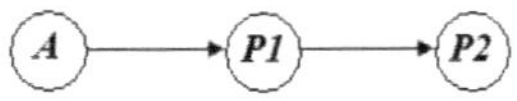

Figure 2. Coupling scheme for APP-ML and APP-MM simulations.

Since spike duration is 200-300 ms long, and based on physiological previous works [6] we constrained the MVAR model to use information from the previous 100ms to estimate the actual signal value. Therefore, the time series data were adjusted using an order 20 MVAR. To test this premise, the Schartz Bayes Criterion (SBC) was used to define an optimal MVAR model order. The Matlab ARFIT Toolbox was used for this purpose [13][14].

The connectivity is most likely to be time variant in this case due to the mutual influence of brain regions [21]. As a consequence, a Kalman filter was used to estimate the MVAR model coefficients on every time sample using Matlab's Biosig toolbox [15]. Then, PDC and DTF were calculated for every time point and every frequency going from 0 to Nyquist frequency using our own Matlab scripts. PDC and DTF were chosen over the multivariate version of Granger causality [22] due to the fact that PDC and DTF consider the frequency domain and they are normalized estimators of connectivity [23].

2.4. Non-parametric statistical test for significance
An empirical distribution technique using surrogate data was used [9],[11]. Surrogate data were constructed by performing an independent phase randomization in each of the time series data. This method has no assumptions about connectivity distributions and allows testing a no connection hypothesis. The data was transformed using Matlab fast Fourier transform command. The phase was then randomized independently for each channel, making sure that the phase spectrum remained odd. Afterwards, the signal is transformed to the time domain using inverse fast Fourier transform. Phase randomization algorithm maintains the power spectrum of the signal while destroying any information flow in the time domain.

In order to have good statistical significance, we created an empirical distribution for the causal measures out of 500 surrogates. Since phase randomization method assures that there is no interaction among the channels, this distribution gives the estimator behavior for the null hypothesis case. P-value of every time-frequency was obtained in comparison with the surrogate distribution.

2.5. Multiple comparison statistics
The approach on assessing statistical significance of the results that was discussed in the previous section bears a probability that there will be false positive detections. Considering the fact that the connectivity matrix has a time-frequency dimension of 600000 points (50 frequency by 12000 time points), approximately 3000 points may show a statistically significant connectivity due to the multiple comparisons (assuming that the null hypothesis is true, alpha value of 0.05 and a normal distribution of the connectivity estimator). In order to correct for multiple comparisons a false discovery rate (FDR) method was applied to the connectivity matrix which eliminated false results with a 5% error rate. The implementation for the FDR method belongs to the SIFT toolbox for EEGlab [16][17].

3. Results
The simulated model of the neural populations proved to be effective, as it resulted in a physiologically plausible signal in which the algorithms could be tested. Active and passive populations were simulated in different cases so as to represent the different combinations of populations that could be present in the neural cortex.

To fulfill the criteria described in section 2, connectivity coupling parameters K were defined as 100, 170 and 275 for low, medium and high simulations respectively.

The order 20 MVAR showed little to no difference compared to lower orders despite the fact that the SBC is known to penalize for the use of a large order model. As a matter of fact, it is more acceptable to err due to overfitting rather than subfitting, hence the preference over the largest order. On figure 3, SBC for simulation AP-L can be seen. Similar results were obtained for the rest of the simulations. According to SBC, MVAR model order should be between 3 and 25 depending on the considered simulation.

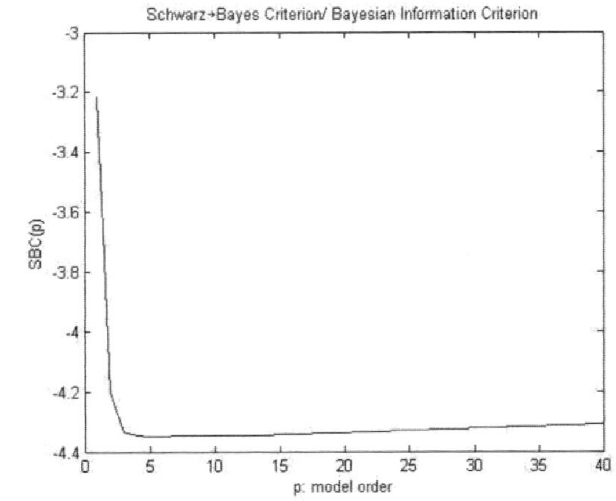

Figure 3. Bayesian criterion shows that the chosen model order is acceptable

Connectivity was measured in these simulations, and the effectiveness of the PDC and DTF algorithms turned out to be determined by the neuronal populations involved. Since the simulations involving two coupled populations can not include indirect connectivity between them, results of PDC and DTF did not differ in those cases. As expected, in the many coupled populations, some differences were found in the estimation of connectivity using PDC and DTF [10],[23].

Non-coupled populations

Simulations PP-N, AA-N and PA-N (equivalent to AP-N) were estimated correctly, as the PDC and DTF plots ended up being void of all information.

Groups PA, PP and AA

In most cases, the effective connectivity measurements properly showed the coupling direction among populations until the surrogate data method and the statistical test for significance were applied, resulting in the equivalent of no connectivity at all between the populations. This would suggest that the method has low sensitivity for this kind of application.

Group AP

Connectivity measures from simulations AP-H and AP-M are showed in figures 4 and 5 respectively. Different frequency connectivity patterns can be observed. Connectivity results in simulation AP-L were not statistically significant when contrasted with the no connectivity hypothesis.

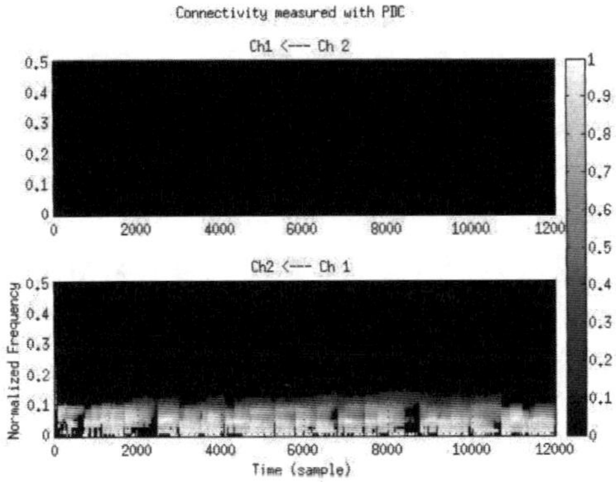

Figure 4. PDC of simulation AP-H. The connectivity is mainly in the lower frequency band. Ch1: A population, Ch2: P population.

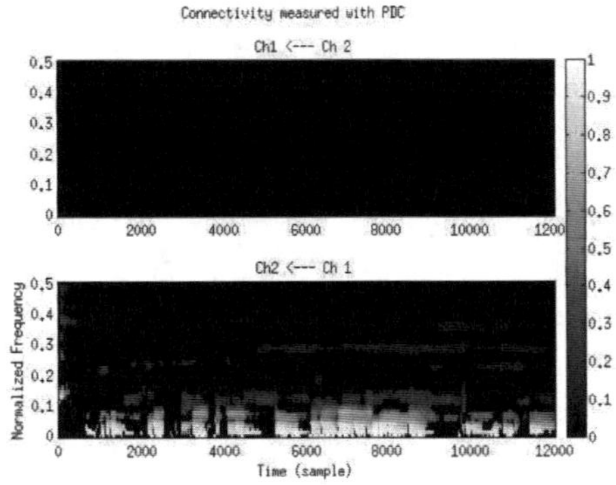

Figure 5. PDC of simulation AP-M. The connectivity is mainly in the lower frequencies and also spread over medium frequencies. Ch1: A population, Ch2: P population.

Journal of Physics: Conference Series **477** (2013) 012037 doi:10.1088/1742-6596/477/1/012037

Many coupled populations

PDC and DTF results for simulation APP-MM can be seen on figures 6 and 7, respectively. Differences between PDC and DTF were visible in these cases. Significant connectivity was measured for A→P1 and P1→P2. DTF in this case did not depict the indirect connectivity between A and P2, as it could be expected [9].

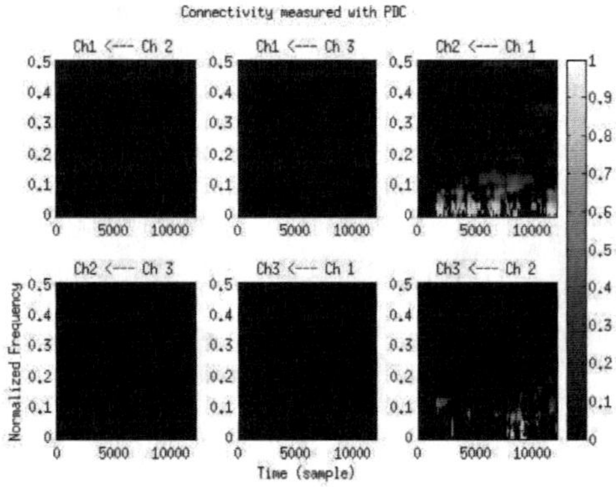

Figure 6. PDC for APP-MM simulation. A→P1 and P1→P2 show effective connectivity in the lower frequency band. P1→P2 connectivity shows lower strength. Ch 1: A population, Ch 2: P1 population, Ch 3: P2 population.

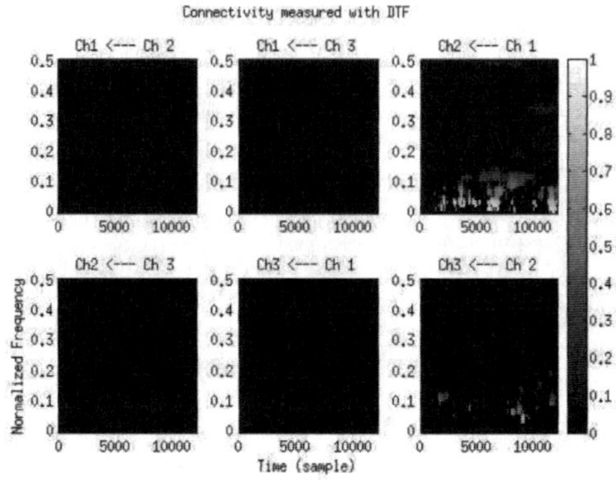

Figure 7. DTF for APP-MM simulation. A→P1 and P1→P2 show effective connectivity in the lower frequency band. P1→P2 connectivity shows lower strength. Ch 1: A population, Ch 2: P1 population, Ch 3: P2 population.

PDC and DTF results for simulation APP-ML can be seen on figures 8 and 9, respectively. Connectivity was measured successfully for A→P1 but a non significant measure was found for P1→P2. Again, the difference between DTF and PDC measures can be seen. DTF for APP-ML depicts a very low indirect connection between A and P2.

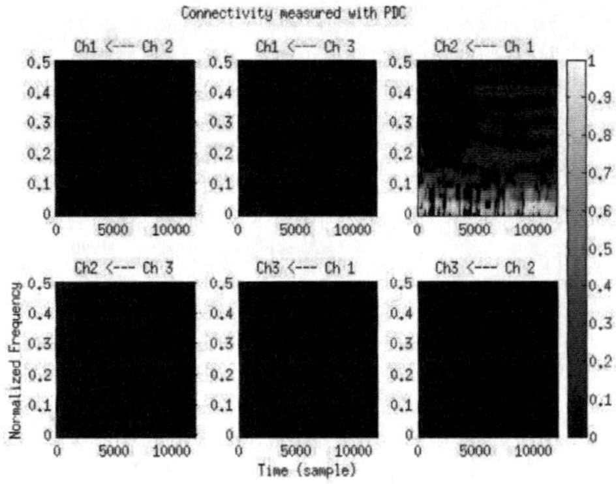

Figure 8. PDC for APP-ML simulation. A→P1 show effective connectivity mainly in the lower frequency band. P1→P2 connectivity is not detected. Ch 1: A population, Ch 2: P1 population, Ch 3: P2 population.

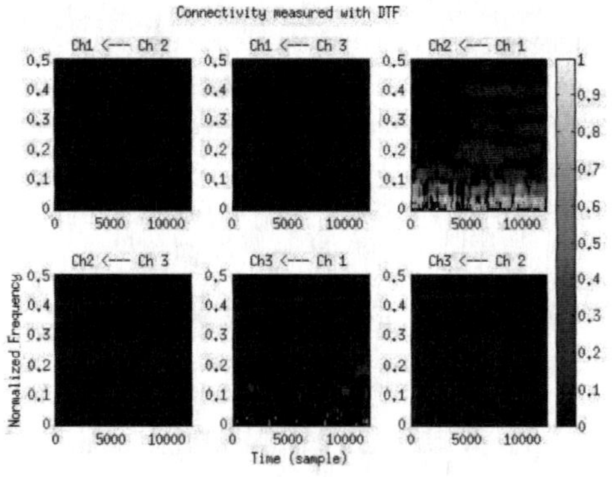

Figure 9. DTF for APP-ML simulation. A→P1 show effective connectivity mainly in the lower frequency band. P1→P2 connectivity not detected. Indirect connectivity is measured between A→P2 Ch 1: A population, Ch 2: P1 population, Ch 3: P2 population.

4. Discussion and conclusion

Results show that PDC and DTF methods, corrected by FDR method, describe correctly the connectivity for simulations AP-M, AP-H, PP-N, AA-N and PA-N (equivalent to AP-N). On simulation AP-L, the measured connectivity was not significant. This could be due to the fact that low connectivity gain was defined as the least possible to alter the firing rate of the passive population, thus providing little information flow. As a consequence, causal influence resolution achieved by the method may be below what we expected to resolve.

The estimation of connectivity in the simulations belonging to group PA and PP was not significant. This might be caused by the fact that information coming from passive populations does not modify the firing rate of the destiny populations. As for simulations of group AA, results were not significant either. This could be due to the non-linearity of the signals, which would make it difficult for the MVAR model to get a good estimation of the temporal dynamics of this signal.

The fact that the best outcome came from simulations AP-M and AP-H shows that the estimation of connectivity between two populations using this method might be only possible when the firing rate of the destiny population is noticeably altered.

Three population simulations offer a more realistic approach of the connectivity on an epileptogenic network. Connectivity from the active population A to the passive P1 was measured for simulations APP-MM and APP-ML with both PDC and DTF. These results are of importance in the clinical interpretation of epilepsy network configuration, since this method allows the determination of how epileptic activity is spread from its origin. Meanwhile, connectivity from P1 to P2 could only be measured in simulation APP-MM. These simulations use the passive population in the middle of the connectivity path as an information pathway. Again, the case is properly represented by PDC and DTF once the firing rate in the destiny populations is altered in a noticeable way. Since P1 is a passive population "activated" by A, the coupling P1→P2 behaves more like an AP coupling than a PP coupling. PDC and DTF for AP-L simulation proved to be not significant while AP-M did. As a consequence, the results for many coupled populations are consistent with what was found for simulations of two coupled populations.

We noticed differences in the frequency response of PDC and DTF according to the strength of couplings. These differences will be studied in future works.

As PDC and DTF methods are linear estimators it is evident that the connectivity estimation will be more effective in signals simulated by the use of linear models. Previous works have already proven the success of DTF and PDC for estimation of signals generated linearly via autoregressive simulation [23]. Recent work has been done on estimating connectivity based on simulated data of physiological significance [28]. However this work is based on a two-dipole model with linear dynamics which is far from representing the non-linear dynamics of real brain signals.

Astolfi et al. [26] simulated a non-linear population model signal [7] and then other population signals to simulate a neural network. However, the other populations with connections to the non-linear one were generated based on delays and escalations of it [26]. This type of connections were well described by PDC, however physiological plausibility of this signals is not clear.

It is important to remark that, in this work, we applied the model proposed by Wendling et al. [7] which is representative of the physiological reality. This means PDC and DTF's effectiveness was tested with a model that generates physiologically plausible neuronal signals. However, PDC and DTF are linear estimators of causal influence. Using non-linear connectivity estimators may provide better results, since this model includes non-linear elements and it has been shown that subtle relations between signals are not accessible by linear methods [27].

In [7] the authors proposed the use of a nonlinear method to analyze the connectivity of simulated signals, the nonlinear correlation coefficient h2 to measure the degree of coupling between two signals. Although h2 proved to be effective at estimating the connectivity between two nonlinear signals, it does not give information about the flow direction of the information.

DTF has already been used to study the propagation of epilepsy real neural information using intracranial recordings [8]. However, it is to be noted that no statistical tests for significance were applied. In our work we have observed that the use of FDR reduces considerably the number of connections that are noticed.

It should also be noted we did not attempt any pre processing of the data, which may help improve the statistical significance in the results. Also, the use of other methods for multiple comparisons corrections, such as voxel-cluster approach, may provide more consistent results [1]

For future works, attempting to pre-process the data may produce more sensible results. Some possibilities include a normalization based on the variance of the signals. It remains unclear, yet, how to consider the influence of interictal spikes in this procedure. Moreover, an approach analyzing only interictal spikes is interesting. Besides, the use of information theory approach will avoid the linearity constrains that are used in PDC and DTF.

Concluding, we evaluated the use of PDC and DTF as estimators of effective connectivity between philologically plausible populations in epilepsy. The methods show sensitivity to distinguish the coupling from active to passive populations at intermediate and high levels. This tool may be useful in the description of epileptogenic networks. However other techniques would be necessary to achieve more sensible results.

References

[1] Rosenow F and Lüders H S 2001 *Brain* **124** 1683-700
[2] Kwan P et al. 2010 Definition *Epilepsia* **51** 1069-77
[3] Engel J 1998 *Epilepsy Res.* **32** 1-11
[4] Ding M, Chen Y and Bressler S L 2006 in Handbook of Time Series Analysis
[5] Van Mierlo P et al. 2013 *Epilepsia,* In press
[6] Varotto G, Tassi L, Franceschetti S, Spreafico R, Panzicca F 2012 *NeuroImage* **61** 591-8
[7] Wendling F, Bellanger J J, Bartolomei F and Chauvel P 2000 *Biol. Cybern.* ***83*** 367–78
[8] Ge M, Jiang X, Bai Q, Yang S, Gusphyl J and Yan W 2007 *Conference proceedings: Annual International Conference of the IEEE Engineering in Medicine and Biology Society*
[9] Kaminski M, Ding M, Truccolo W A and Bressler S L 2001 *Biol. Cybern.* ***85*** 145–57
[10] Sameshima K and Baccalá L A 1999 *Journal of neuroscience methods* **94** 93–103
[11] Theiler J, Eubank S, Longtin A, Galdrikian B, and Doyne Farmer J 1992 *Physica D* **58** 77–94
[12] Genovese C R, Lazar N A and Nichols T 2002 *NeuroImage* **15** 870–8
[13] Neumaier A and Schneider T 2001 *ACM Trans. Math. Softw.* **27** 27-57
[14] Schneider T and Neumaier A 2001 *ACM Trans. Math. Softw.* **27** 58-65
[15] Available at: http://biosig.sourceforge.net/
[16] Mullen T, Delorme A, Kothe C and Makeig S 2010 An Electrophysiological Information Flow Toolbox for EEGLAB Society for Neuroscience Conference, San Diego, CA, USA, 2010
[17] Delorme A, Mullen T, Kothe C et al 2011 EEGLAB, SIFT, NFT, BCILAB, and ERICA: New Tools for Advanced EEG Processing *Computational Intelligence and Neuroscience* **2011**
[18] Jansen B H and Rit V G 1995 *Biol. Cybern.* **366** 357–66
[19] Blenkmann A, Beltrachini L, von Ellenrieder N, Kochen S, Muravchik C H 2011 *XIV Reunión de trabajo en Procesamiento de la Información y Control*
[20] Granger C W J 1969 *Econometrica* **37** 424-38
[21] Omidvarnia A, Mesbah M, O'Toole J M, Colditz P, Boashash B 2011 *7th International Workshop on Systems, Signal Processing and their Applications (WOSSPA)*
[22] Barrett A B, Barnett L, Seth A K 2010 *Physical Review E* **81** 041907
[23] Baccalá L A and Sameshima K 2001 *Biol. Cybern.* **84** 463-74
[24] Blenkmann A 2012. PhD Thesis UNLP. Available at: http://sedici.unlp.edu.ar
[25] Haufe S, Nikulin V V, Müller K R and Nolte G 2013 *NeuroImage* **64** 120-33
[26] Astolfi L et al. 2005 *IJBEM* **7** 62-5
[27] Le Van Quyen M, Adam C, Baulac M, Martinerie J and Varela F J 1998 *Brain Res.* **792** 24-40
[28] Lopes da Silva F H, Pijn J P and Wadman W J 1994 *Prog. Brain Res.* **102** 359-70
[29] Bullmore E T, Suckling J, Overmeyer S, Rabe-Hesketh S, Taylor E, Brammer M J **1999** *IEEE Transactions on Medical Imaging* 18 32-42

9th Argentinean Bioengineering Society Congress (SABI 2013) IOP Publishing
Journal of Physics: Conference Series **477** (2013) 012038 doi:10.1088/1742-6596/477/1/012038

Playing with your mind

Mauro Rodríguez, Ramiro Giménez, Pablo Diez, Enrique Avila, Eric Laciar, Lorena Orosco, Agustina Garcés Correa

Gabinete de Tecnología Médica (GATEME) – Universidad Nacional de San Juan
San Juan, Argentina.

pdiez@gateme.unsj.edu.ar

Abstract. A Brain-Computer Interface (BCI) is a communication system between the brain and a machine like a computer. Some BCI systems have been used to help people with disabilities and sometimes, with entertainment purposes. In this paper, a BCI-game system is developed. It allows controlling the altitude of a ball inside of a glass pipe according to mental concentration level, which is measured on EEG signals of the user. The system is automatically adjusted to each user, hence, it is not needed any calibration step. Ten subjects participated in the experiments. They achieved effective control of the ball in a few minutes, demonstrating the feasibility of the BCI-game system.

1. Introduction

A Brain-Computer Interface (BCI) is a non-muscular communication and control channel [1]. Communication can be defined as a process to express and share experiences among people, where the machine can be used as an accessory tool. A new trend for man-machine interaction is using the brain signals to promote a natural interface [2]. In such sense, a BCI has been used for helping people with disabilities to command a robotic wheelchair [3], to spell on a computer screen [4] or to navigate through virtual worlds [5].

On the other hand, a BCI system was used for entertainment purposes. For example, a race car was controlled on a computer screen with a BCI based on steady-state visual evoked potentials [6]. A BCI was used to gain the control on the balance of an animated character on a tightrope in an immersive 3D game [7]. Moreover, some commercial BCI systems for entertainment purposes have been developed. For example, Mattel Inc. presented a toy which uses brain waves to steer a ball through an obstacle course, the MindFlex [8]. A summary of commercial BCI systems can be found in a Wikipedia webpage named: Comparison of consumer brain-computer interface devices [9].

In this paper, a method to calculate the attention (or concentration) level of the user is presented. Moreover, a BCI-game for controlling the altitude of a ball inside a glass pipe (Figure 1) is proposed. The elevation of the ball is correlated with the attention level, i.e., higher altitude corresponds to higher concentration of the user.

The remainder of this paper is organized in two parts. The first one is a description of the preliminary study which was performed in order to evaluate the attention level of the subjects and then, designing the BCI-game. In the second part, the developed BCI-game system is described.

2. Preliminary study

The preliminary study was performed to determine the best EEG channel configuration, as well as to detect changes in the EEG signals, which are related to the concentration of the user. Hence, the

Journal of Physics: Conference Series **477** (2013) 012038 doi:10.1088/1742-6596/477/1/012038

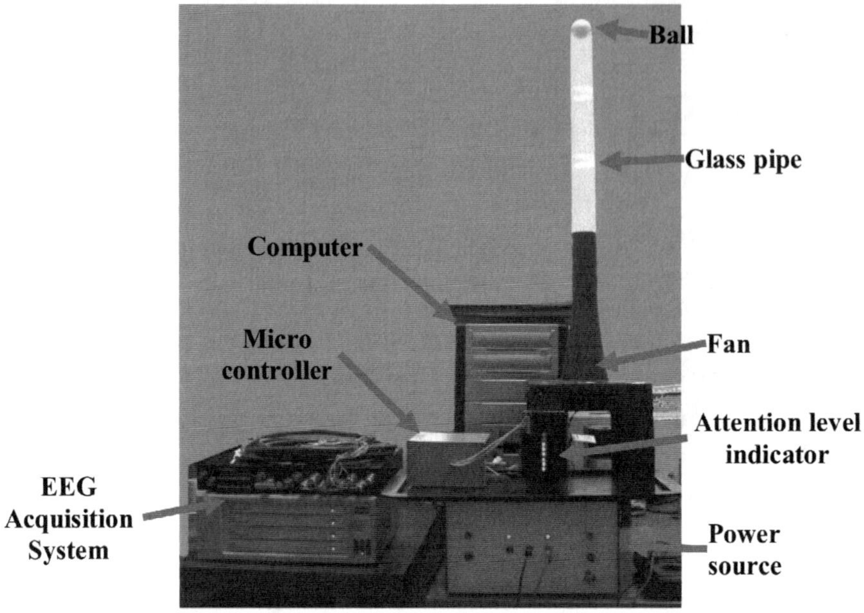

Figure 1: The BCI-game system

attention level is measured over the pre-frontal and frontal cortex where complex cognitive processes are located [10]. The EEG was measured with five channels as depicted in Figure 2.

Four subjects (33.5±1.9 years old) participated in the preliminary study. Each subject performed three different mental tasks of 2 minutes duration each, namely:

- **Relaxation:** The subject was told to simply relax and try to think of nothing in particular.
- **Math operation:** The user performs a countdown in steps of 3 beginning with a random number of 4 digits.
- **Geometric task:** the user was instructed to visualize a three-dimension block figure being rotated about an axis.

Note that the last two mental tasks leads to the high-attention level, whereas the first one corresponds to the low-attention level. The EEG signals were acquired using a Grass 15LT amplifier system and digitalized at 256 Hz with a NI-DAQPad6015. The analogical pass-band filters were set at 3 and 100Hz and a notch 50 Hz filter was applied as well.

The EEG spectral content was divided into several rhythms or bands [11]:

- Delta rhythm (<4Hz): it is encountered during deep sleep stages.
- Theta rhythm (4-7Hz): it occurs during drowsiness.
- Alpha rhythm (8-13Hz): it is related to relaxation stages.
- Beta rhythm (14-30Hz): it is associated with an activated cortex.
- Gamma rhythm (>30Hz): it is related to a state of active information processing of the cortex.

Hence, the alpha, beta and gamma rhythms were considered in the current work for analyze the attention level. The power spectral density of the EEG was estimated with the periodogram method (see section 3.1) and then, the power of each band was calculated. The periodogram was calculated using windows of 2, 3, 4 and 5 s duration.

Journal of Physics: Conference Series **477** (2013) 012038 doi:10.1088/1742-6596/477/1/012038

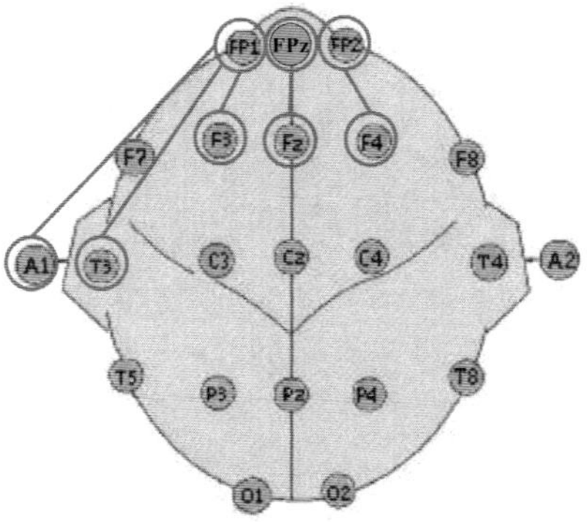

Channel	Electrode position
1	Fp1- F3
2	Fpz - Fz
3	Fp2 - F4
4	Fp1 - T3
5	Fp1 - A1
GND	A2

Figure 2: EEG channel measured in the preliminary tests.

2.1. Results of the preliminary study

The figure 3 depicts the EEG power bands during the *relaxation* and *math operation* tasks. The four subjects presented a similar behaviour on the alpha and beta rhythms, as showed in figure 3. In other words, the alpha and beta powers are lower as higher the attention level is. The gamma rhythm also presented differences for each mental task; however those variations were different to each subject.

Those changes in the EEG rhythms were observed with more amplitude in two EEG channels: Fp1-A1 and FP1-T3. Consequently, the configuration Fp1-A1 is adopted, because those positions are free of hair, which allows an easy electrodes placement.

Finally, a one-way ANOVA test was applied to the EEG power bands in order to evaluate the statistical significance of the differences between relaxed and attention states. The EEG power bands were calculated using the periodogram of 5 s EEG segments; thus, 47 segments were extracted from each 2 minutes mental task. Theta rhythm do not present differences between both states ($p=0.269$). However, alpha, beta and gamma rhythms present significant differences ($p<0.05$) between low- and high-attention level.

With the results obtained from this preliminary study the BCI-game system was developed. The following section describes the BCI-game system.

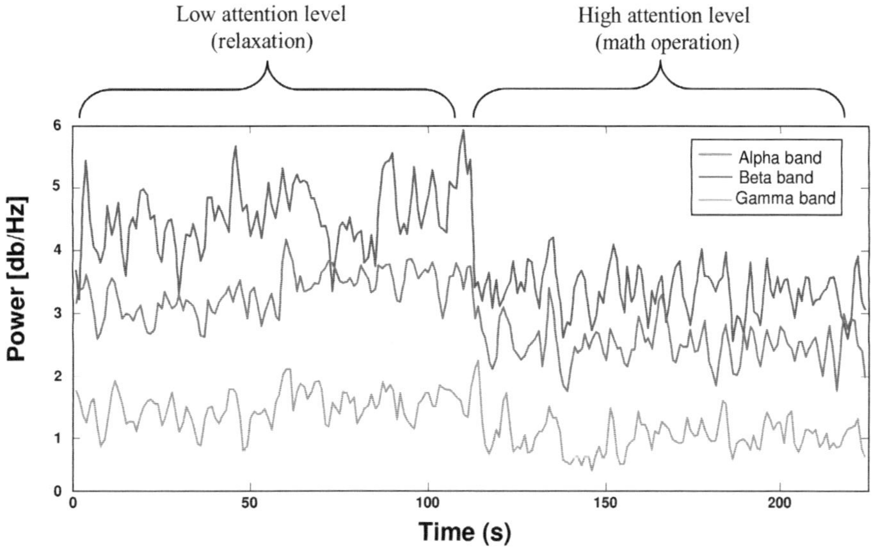

Figure 3: Power EEG bands when the *relaxation* and the *math operation* tasks were performed. The EEG was measured at Fp1-A1 channel.

3. BCI-attention-game

Figure 4 presents a scheme of the BCI-game system. It is composed of an EEG acquisition system, which measures the brain activity. Then, the EEG signal is fed into the computer and the user's attention level is calculated. According to this attention level, the computer sends a command to the microcontroller. The microcontroller has two functions, the first function is controlling the rotation speed of the fan, and the second one is commanding the attention level indicator. Finally, the air expelled by the fan moves up/down the ball inside the glass pipe.

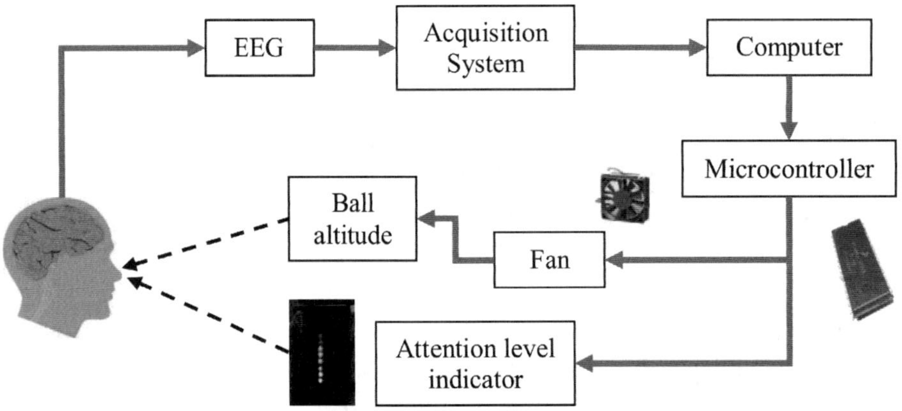

Figure 4: The different BCI-game components.

3.1. EEG signal processing method

First, a Butterworth band-pass digital filter, order 6, with 6 and 30 Hz cut-off frequencies was utilized. Afterwards, the periodogram was computed. It is an estimation of the power spectral density based on the Discrete Time Fourier Transform (DTFT) of the signal x[n]:

$$P_{xx}(f) = \frac{T_s}{N}\left|\sum_{n=1}^{N} x[n]e^{-j2\pi f n T_s}\right|^2 \tag{1}$$

where $P_{xx}(f)$ is the periodogram, T_S is the sampling period, N is the number of samples of the signal and f is the frequency. To compute the periodogram, the Fast Fourier Transform (FFT) with 5 s length Hann window was used.

Then, a value representing the attention level, named Attention Power (AP), is proposed. It is based on the power of alpha and beta rhythms and is calculated as:

$$AP = \sum_{f=8}^{25} P_{xx}(f) \tag{2}$$

The AP value is lower as higher the attention level of the user is and vice versa, i.e., they are inversely related. This is in according with the results obtained from the preliminary study, i.e., the alpha and beta powers are lower as higher the attention level is. Finally, the AP values were filtered with a moving average filter of 5 points length and thus, a smoothed version of AP was obtained. Hence, artifacts that could affect the power of the EEG rhythms (and the AP value) were attenuated or eliminated. Consequently, smoothed movements of the ball into the glass pipe were achieved. In figure 5, the AP values for different attention levels are depicted.

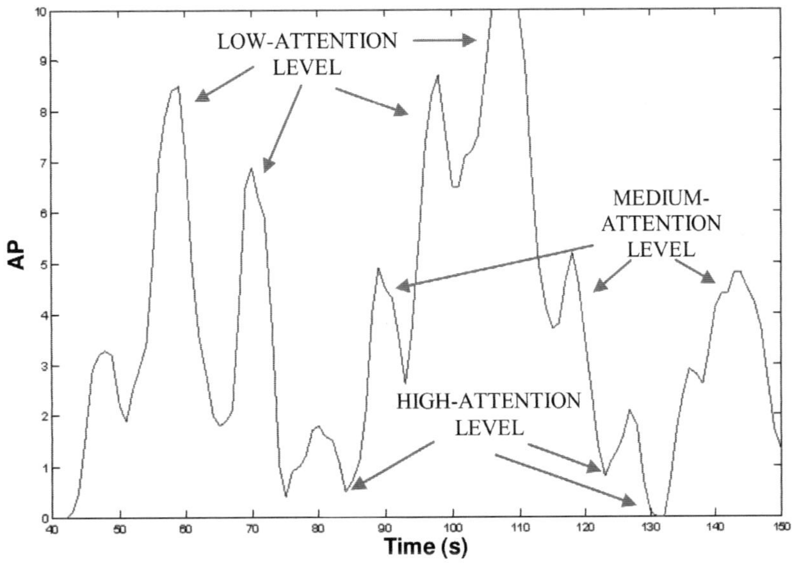

Figure 5: Smoothed version of the AP values for different attention levels.

3.2. *Quantification of AP values*

The AP value is a continuous variable (as depicted in figure 5), but the speed of the fan is a discrete variable. Hence, in order to translate AP values to different speeds of the fan a quantification process is needed. These quantification levels were determined empirically from the preliminary study and were defined as percentages of the maximum AP value.

The maximum and minimum AP values were automatically adjusted by the algorithm (figure 6), thus avoiding any calibration/configuration step. On the other hand, sometimes the user achieves a very high-attention level which is hard to reach again. Consequently, the user cannot reach the quantification levels based on that maximum AP value. Therefore, the user cannot moves up the ball again. The same effect was observed for minimal AP values. Hence, a modification of the maximum and minimum values is proposed. This modification is a slowly decrease of the maximum value across the time. The minimum AP value is slowly increased across the time (see figure 6).

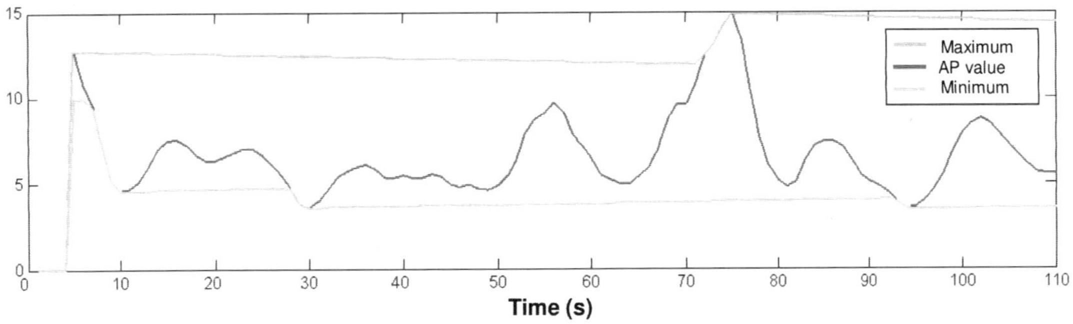

Figure 6: The AP values for different attention levels and the automatic adjustment of the maximum and the minimum.

3.3. Ball altitude control system

In order to conduct the air expelled by the fan into the glass pipe a structure was designed (Figure 1 and 7). Consequently, a light ball levitates inside the glass pipe. The altitude of the ball is controlled by the air expelled by the fan. A microcontroller Motorola 68HC908JL3 was programmed to control the rotation speed of the fan through a pulse width modulation (PWM) scheme. Therefore, when the user is concentrated (corresponding with low-AP value) the microcontroller increases the rotation speed of the fan.

On the other hand, the microcontroller commands the attention level indicator (figure 4). It is an arrangement of light-emitting diodes (leds), which indicates the changes in the attention level of the user. Due to air and ball dynamics, the movement of the ball inside the glass pipe presents a delay in reflecting the concentration changes. On the contrary, the attention level indicator produces a faster feedback to the user of his/her attention level (figure 1 and 7). Each led represents a specific attention level. Thus, green leds represent low-attention levels, yellow leds correspond to medium-levels and red leds represent high-concentration levels.

Therefore, the attention level indicator and the rotation speed of the fan are simultaneously adjusted by the microcontroller according to the attention level of the user. This is translated into different velocities of the ball inside the glass pipe. For example, when green leds are on, the ball descends inside the glass pipe (as fast as less green leds are on). When red leds are on, the ball ascends inside the glass pipe (as fast as more red leds are on). When yellow leds are on, the ball is still or moves very slowly.

In the glass pipe are limited three zones with white stripes (figure 1 and 7). These zones are used to evaluate the user performance (see Results section).

4. Results

Ten subjects (2f and 8m; 39.5± 15.5y) played with the BCI-game. They were randomly asked to levitate the ball to one of the three zones in the glass pipe. The task was repeated three times per position and the results are presented in table 1.

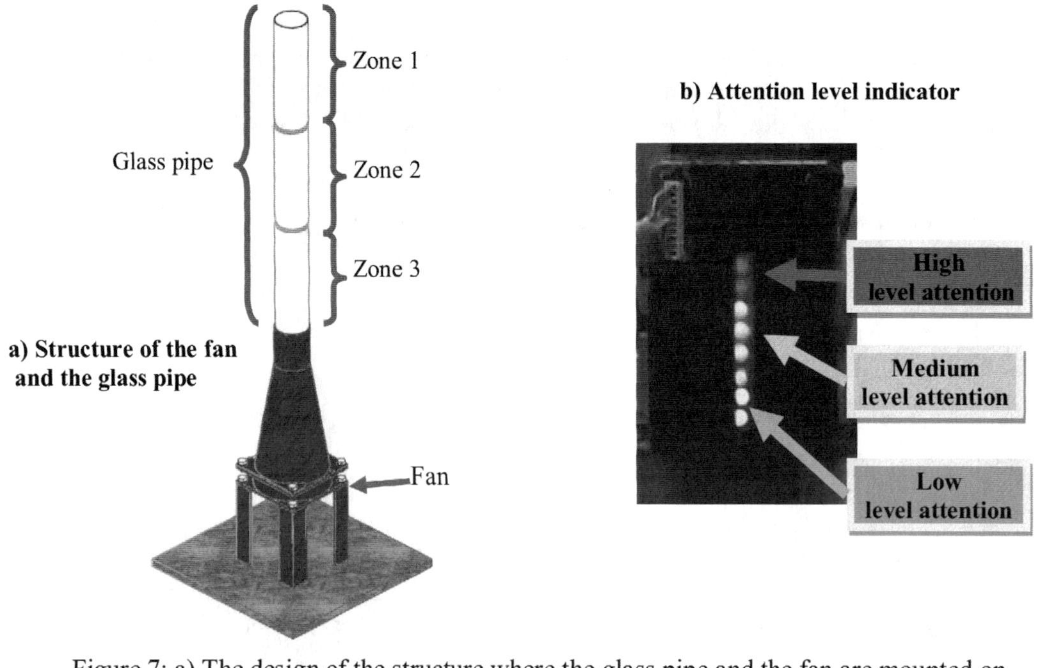

Figure 7: a) The design of the structure where the glass pipe and the fan are mounted on.
b) The box and the leds of the attention level indicator

Table 1: Results of the ball levitation

Subject	Altitude: Low (zone 1)	Altitude: Medium (zone 2)	Altitude: High (zone 3)	Average
1	3	2	3	88.9 %
2	3	3	3	100 %
3	3	3	3	100 %
4	2	2	3	77.8 %
5	3	3	3	100 %
6	3	2	3	88.9 %
7	3	3	3	100 %
8	3	0	3	66.7 %
9	3	3	3	100 %
10	3	3	3	100 %
Total	**29**	**24**	**30**	
Average	**96.7 %**	**80 %**	**100 %**	

Finally, the subjects answered a questionnaire about their own perception of the BCI-game.The questions were:

A) Was it hard to conduct the ball to the top of the pipe (zone 1)?
B) Was it hard to conduct the ball to the middle of the pipe (zone 2)?
C) Was it hard to drive the ball to the bottom of the pipe (zone 3)?
D) Do you think that your attention level was correlated with the altitude of the ball?
E) While the experiment advances, did you achieve a better control of the ball?
F) With training, do you think that you could perform it better?

The subjects could answer as 1) Yes; 2) No or 3) Medium to each question. The answers of the subjects are resumed in the table 2.

Table 2: Answers of the subjects expressed as percentages

Question	Yes	Medium	No
A: the ball to the top	--	--	100%
B: the ball to the middle	50%	20%	30%
C: the ball to the bottom	10%	70%	20%
D: correlation of attention level and the altitude	80%	20%	--
E: better control with the time	60%	40%	--
F: performing better with training	100%	--	--

5. Discussion

From the preliminary study some useful insights were obtained. For instance, when a subject modifies his/her attention level, it is possible to detect the corresponding modifications in the EEG signals. These modifications were observed in the spectral content of the EEG signals measured from the pre-frontal and frontal cortex. Moreover, those changes are reflected in the alpha, beta and gamma rhythms. This fact was statistically confirmed using a one-way ANOVA test.

During the game, the subjects were not specifically instructed how to perform the relaxation or concentration mental tasks. Hence, each subject chooses a different way to change his/her attention level. For example, some subjects close his/her eyes to relax, other ones breathe deeply trying to relax. A subject achieves low-attention levels speaking about anything in particular and another subject concentrates in environmental sounds in order to increase his attention level.

Achieving high-attention levels is a relatively easy task since everybody can concentrates in something. Once the subject is concentrated and the ball is in the top of the glass pipe, relaxing is a harder task, which takes several seconds driving the ball to the bottom of the glass pipe. However, conducting the ball to the zone 2 is more difficult, i.e., it is hard to maintain a medium-attention level. This is in according with the answer to the questionnaire, to drive the ball to the middle zone (zone 2) was the hardest task, while to conduct the ball to the top of the glass pipe was the easiest task.

The 80% of the subjects (question D) found correlation between his/her attention level and the altitude of the ball, while 20% of the subjects found a medium correlation. The ability to control de ball inside the glass pipe increases with the time (question E), which can be improved even more with training (question F). This shows that controlling the ball requires some training.

Subjects with different ages and gender tested the BCI-game, no differences in the performance were observed among them. The system is based on electrodes collocated at Fp1-A1 position. This is very important because only one EEG channel is acquired; besides those positions are free of hair allowing the easy electrodes placement. The proposed EEG signal processing method is a simple way to estimate the attention level of the user. The AP value, along with the automatic adjustment of the maximum and minimum AP values, allows a suitable controlling of the ball inside the glass pipe.

6. Conclusions

In this paper, an automatic method to detect the attention level of a person was presented. Moreover, the system does not need any calibration or adjustment. Then, a system to control the altitude of a ball inside a glass pipe was designed. This allows that people play with their own brain activity.

The proposed BCI-game is simple, although, it allows evaluating how people change their own attention level. Besides, the system could be used in some diseases like deficit-attention disorders. Moreover, the attention level BCI could be used to operate some systems by people with motor disabilities.

References

[1] Wolpaw JR, Birbaumer N, McFarland DJ, Pfurtscheller G, Vaughan TM. Brain-computer interfaces for communication and control. *Electroen. Clin. Neuro.*, vol. 113, no. 6, pp.767-791, (2002).

[2] Torres Müller SM, Diez PF, Bastos-Filho TF, Sarcinelli-Filho M, Mut V, Laciar E. SSVEP-BCI Implementation for 37-40 Hz Frequency Range. *33rd Annual International Conference of the IEEE EMBS,* Boston, Massachusetts USA, August 30 - September 3, 2011

[3] Diez PF, Torres Müller SM, Mut VA, Laciar E, Avila E, Bastos-Filho T, Sarcinelli-Filho M. Commanding a robotic wheelchair with a high-frequency steady-state visual evoked potential based brain–computer interface. *Med Eng Phys* , vol. 5, pp. 1155–1164 (2013).

[4] Hwang HJ, Lim JH, Jung YJ, Choi H, Lee SW, Im CH. Development of an SSVEP-based BCI spelling system adopting a QWERTY-style LED keyboard. *J Neurosci Methods*. Vol. 208(1), pp. 59-65 (2012)

[5] Scherer R , Lee F, Schlögl A, Leeb R, Bischof H, Pfurtscheller G. Toward Self-Paced Brain–Computer Communication: Navigation Through Virtual Worlds. *IEEE Transactions on Biomedical engineering*, vol. 55, no. 2, (2008)

[6] Martinez P, Bakardjian H, Cichocki A. Fully Online Multicommand Brain-Computer Interface with Visual Neurofeedback Using SSVEP Paradigm. *Computational Intelligence and Neuroscience* Vol. 2007, Article ID 94561, 9 pages (2007)

[7] Lalor EC, Kelly SP, Finucane C, Burke R, Smith R, Reilly RB, McDarby G. Steady-State VEP-Based Brain-Computer Interface Control in an Immersive 3D Gaming Environment. *EURASIP Journal on Applied Signal Processing*. Vol 2005:19, pp. 3156–3164 (2005)

[8] ©2013 Mattel, Inc., http://mindflexgames.com/

[9] http://en.wikipedia.org/wiki/Comparison_of_consumer_brain-computer_interface_devices

[10] Ortega Caballero P., Design of interactive processing methods for brain signals analyzis (in Spanish) Universidad de Chile, Facultad de Ciencias Físicas y Matemáticas, Departamento de Ciencias de la Computación, 2005.

[11] Sörnmo L, Laguna P. Bioelectrical Signal Processing in Cardiac and Neurological Applications, Elsevier Academic Press (USA); ISBN 0-12-437552-9, pp.1-668. (2005)

Position control with PID regulation for a FES system: preliminary results.

L Schiaffino[1,2] and C B Tabernig[1]

[1] Laboratory of Rehabilitation Engineering and Neuromuscular and Sensorial Research (LIRINS) of the National University of Entre Ríos, Route 11, Km 10, Oro Verde city, CP(3101), Argentina.
[2] Electronic Department of the National University of Entre Ríos, Route 11, Km 10, Oro Verde city, CP(3101), Argentina.

E-mail: lschiaffino@bioingenieria.edu.ar

Abstract. For people with quadriplegia at C3 and C7 levels functional electric stimulations (FES) could be a solution to make precision movements using the wrist. Closed loop control for FES systems enables to improve precision, stability and diminishes the impact of external perturbations. The goal of this work is to present a control position of the wrist articulation whenever the movement is produced by FES. In this article, a control design is developed along with the results of its first evaluation process on three voluntary individuals without neurological damage. A PID controller is used and it is adjusted experimentally for each test subject. A Constant reference for the wrist extension angle at sagittal plane and time evaluation of 10 minute was used to probe de PID. For two subjects the regulators responded reasonably with low mean errors and low standard deviations. For subject 3 the designed regulator did not succeed in stabilizing the position as from minute 6 in the second test. Preliminary studies of this work suggest that it could be possible to continue with the study of the closed loop position control with PID strategy in some people with spinal cord injuries whose movements are assisted by FES.

1. Introduction

People with quadriplegia at C3 and C7 levels possess mobility of their arms but are not capable of clutching their fists strong enough to grasp, hold, and release objects. Therefore, they are unable to perform daily activities that are needed for their own survival, which leads them to depend on either other people or auxiliary systems to manage such activities. It is estimated that in Argentina, 50% of spinal cord injuries are at cervical level [7]. Functional electric stimulations (FES) have been widely used in such patients in order to increase the strength of their palms and thus achieve the flexion/extension of the wrist both in therapeutic treatments and neuroprosthetics. These systems can be configured individually for each user given that the type of control or residual movement varies from patient to patient.

In order to improve their performance, neuroprosthetic systems require a specific control of the articular position so that the movement trajectories can be achieved satisfactorily. At the same time, it is necessary to develop control strategies so that the FES equipment can give autonomy to the patients in their daily activities [1].

Initially, most of the FES systems adopted open loop control strategies [2, 3]. Presently, some therapists prefer this equipment because of the simplicity in its use and because they are able to adjust the levels of stimulation according to each patient´s needs in a simple way. However, the open loop result is not as satisfactory because the external load noise and the muscular fatigue cannot be eliminated [1].

Later on, closed loop controllers were designed to adjust FES systems parameters through feedback [4,6]. In these systems, an error signal is developed which results out of the difference between the real output and the desired one. This type of control improves precision and stability, and diminishes the impact of the external perturbations on the system. At the same time, the development of automatic controllers allows patients and those who assist them not to press switches to activate and control FES equipment. The proportional integrative derivate controller (PID) is a feedback control strategy that is widely used in industrial control systems. It is through this controller that the error between the measured variable and the desired point of adjustment can be corrected and minimized through the proportional (P), integral (I) and derivative (D) action. Therefore, in FES systems, for which precision and stability are essential, it is very important to correctly define the P, I, and D values. Traditional estimation of P, I and D is based on industrial experience through a manual tuning called Ziegler-Nichols method. The use of this method has not been fully successful for FES control systems [8], thus making other forms of tuning control systems necessary.

The goal of this work is to present a position control of the wrist articulation whenever the movement is produced by FES. A PID controller is designed and it is adjusted experimentally for each test subject at constant reference. An adjusted PID produces an accurate relative stability and a robust response time. In this article, the design of the controller and the results of its first evaluation process on voluntary individuals without neurological damage are presented.

2. Design methodology

In order to achieve the proposed goal, three volunteers with no neuromuscular damage and whose peripheral nerves were intact were tested. All subjects agreed to participate in the experimental protocol procedures. The methodological steps to design the PID controller were:

- Experimental determination of the model to be controlled: for the PID controllers design it was necessary to calculate the estimated model system to which the control was applied. According to the transfer function of the model it is possible to determine the parameters of the controller. In this case, the model is the muscle osteoarticular of the wrist whenever the extension of the right hand is produced by the action of the FES. This model has as the stimulating voltage as its input and the angular position of the articulation under study in the sagittal plane as its output.
- Design of the PID controller: Once the transfer function of the step before was estimated for each subject, the design of the controller was made and the parameters were calculated according to functioning premises of the complete closed loop system. A good relative stability and a robust response time according to the system to be controlled were established as requirements.
- Preliminary implementation of the closed loop regulators: In this step, the PID controllers designed on step before were implemented in a digital form. The control loop in real time was executed in each subject through a portable computer and an AD/DA data acquisition module. The control action inferred on the stimulating level of the FES equipment in order to keep the angular position of the articulation of the wrist as a reference constant value.

3. Design implementation

In order to make the design of the PID regulators and to evaluate their preliminary functions in real time an experimental platform was mounted and made out of (see figure 1):

- A Dorsiflex FES stimulator to achieve the articulation of the wrist movement. This equipment works with stimulation frequency of 23 [pps], biphasic pulses with positive rectangular and negative excursion and pulse width of 0.23 [msec]. This stimulator has an output from 0 to 130 volt (measured with no connection to a subject) regulated by a low voltage level selector. This selector may vary from 0 to 8 [V]. Stimulating pulses were applied through superficial adhesive conducting rubber electrodes. For the implementation of the closed loop, the Dorsiflex was adapted so that it could be commanded from a digital computer.
- An electro goniometer for the measurement of the sagittal angular position of the articulation of the wrist. This equipment uses the rotational position sensor Murata brand Model SV01A103AEA101R00 .
- A USB AD/DA data acquisition module Data Translation model DT9804. This module uses 16 bits conversors. The data acquisition was established at 5000 [Hz] for all cases.
- A portable Toshiba Satellite PSAG8U computer with MATLAB 7.1 software.

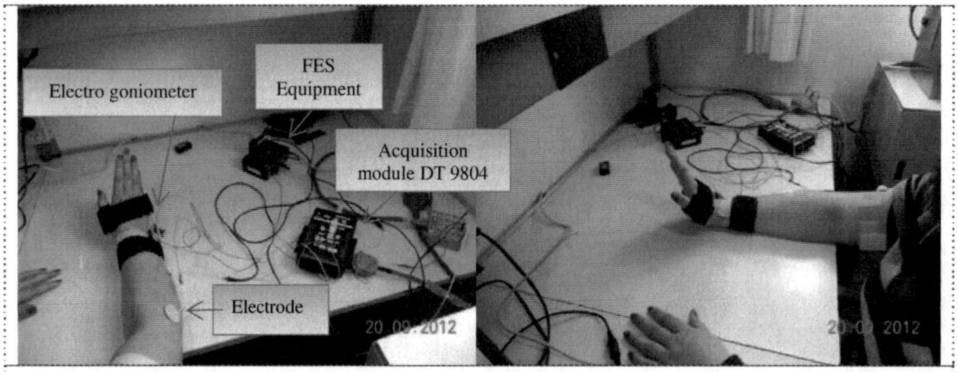

Figure 1. Photos of the experimental platform for the design and test of the PID controllers with subject 2.

3.1. Experimental determination of the model to be controlled

In order to estimate the transfer function of the model for each subject, the angular position of the right hand´s wrist was registered (extension movement) when the extensor carpi radialis longus and brevis muscles and extensor carpi ulnaris were electronically stimulated by the Dorsiflex system. The previously described electro goniometer was used by applying its centre on the biestiloideal line. The angular variation of the wrist was measured with this equipment in its sagittal plane.

The maximum stimulation voltage tolerated was established for each subject. They were requested to be relaxed without inducing any voluntary movements on the studied articulation.

The stimulation level along with the angular position was acquired by using the acquisition Data Translation module DT9804. The data was stored in the portable computer and later on processed by the Matlab software.

A ten-minute-long register was perform on each subject by separate in which the level of stimulation was raised up to the maximum tolerated level, in step-like stimulus and in random increases and decreases of the stimulating voltage. The obtained data was processed in the system identification toolbox of the Matlab software in order to obtain an estimated the model of the system. This system is represented by the stimulation level as the input and the angular variation of the wrist (extension movement) in its sagittal angle as the output. In the utilized toolbox the estimation by error prediction, with Gauss-Newton search method and determinant minimization criteria (maximum 20 iterations) were selected. Part of the data of each registry was used to calculate the model and the rest of the data was used to validate it. In figures 2 and 3 the described method can be observed for subject number 1.

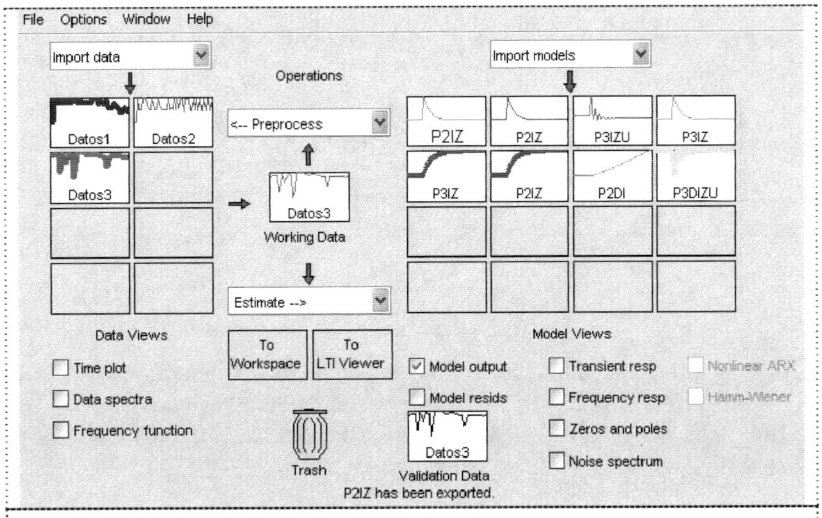

Figure 1. Obtaining the models for subject 1 using the system identification toolbox of Matlab.

The model was approached by a four order transfer function G(s) as in equation (1) with one pole in the origin, dead time (Td), one real zero (-1/Tz) and gain (Kp).

$$G(s) = \frac{K_p.(T_z.s+1).\exp(-T_d.s)}{s.(T_{p1}.s+1).(T_{p2}.s+1).(T_{p3}.s+1)} \qquad (1)$$

The model given by the toolbox that caused the least number of errors using the validation data was chosen for each subject. On Table 1 the time constants, gain and delay chosen for each subject are shown based on the generic function G(s).

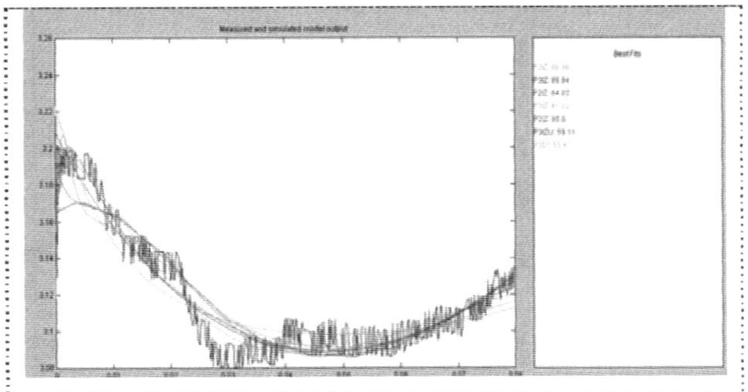

Figure 2. Validation of the models for subject 1 using the system identification toolbox of Matlab.

Table 1. Time constants, gains and delays of the transfer functions chosen for each subject.

	Kp	Tp1	Tp2	Tp3	Tz	Td
Subject 1	0.0126	1.476	6929.5	261.71	11403	0
Subject 2	0.076	0.234	7.35	$4.58\ 10^7$	$1.41\ 10^8$	0
Subject 3	0.098	0.0016	64.08	11312	2863.8	0

3.2. PID controller design

Based on the estimated model for each subject a parallel form PID regulator was designed with the generic transfer function as in equation (2).

$$C_{PID}(s) = K_p + \frac{K_i}{s} + \frac{K_d.s}{(1+s/N)} = K\frac{(Tz_1.s+1).(Tz_2.s+1)}{s.(1+s/N)} \qquad (2)$$

In this controller a low pass filter was included in the derivative action in order to avoid high frequency noises. Each PID was designed so that the system would have a good relative stability and a robust response time. The design criteria adopted was 50degrees and a bandwidth of 4 [rad/sec]. The parameters K, Tz1 and Tz2 for each PID are shown on table 2.

Table 2. PID parameters (K, Tz1, Tz2 and N) for each subject.

	K	Tz1	Tz2	N
Subject 1	136.26	14	28	100
Subject 2	19.9	14	28	100
Subject 3	21.315	85	85	100

3.3. Closed loop implementation.

A real time close loop was implemented with negative feedback. This diagram block is shown in figure 3. A PESAG8U Toshiba Satellite portable computer was utilized, a USB data acquisition module data translation model DT9804 with AD/DA converters of 16bits, and a MATLAB acquisition toolbox. The sample frequency was chosen at 5000 [Hz].

The PID regulators described on the previous step were implemented in digital way using the Tusting transformation. The angular measure of the extension of the wrist in the sagittal plane and the electric stimulation were executed in the same way as in section 3.1.

The adapted Dorsiflex system was used. The level of stimulation of the Dorsiflex was changed by the digital value of the PID controller output. This digital value was transformed into a low-tension voltage by a D/A converter of the DT9804 module. This low voltage value replaced the Dorsiflex equipment level selector. In order to avoid abrupt changes in the levels of stimulation which could cause damage or pain in the subjects of study, before activating the close loop automatic regulation a voltage level increase ramp was implemented from 0 to reference value (0.5 V/sec increase). Subsequently automatic regulation was initiated through which the referential position was kept constant throughout the duration of the test.

4. Design evaluation methodology

The PID performance for each subject was evaluated as follows:

a) through the calculation of the relative stability of the system (PID design with estimated experimental model) for each subject . This stability, which is defined as each system´s margin to become unstable, was measured by the gain margins and phase margins in each case.

b) through the calculation of error and its standard deviation. These variables were measured by executing real time close loops in two different days for 10 minutes each. The two tests for each subject were performed under the same experimental conditions in different days, so that it could be observed whether the designed regulators at fixed parameters could keep the control in spite of the variability of the system. In each case the angular reference, real angular position, low stimulation voltage value and the error (the difference between the proposed reference and real angular position of the articulation of the wrist) was registered. Figure 1 shows two photos of the experimental set up of the control system applied to subject 2.

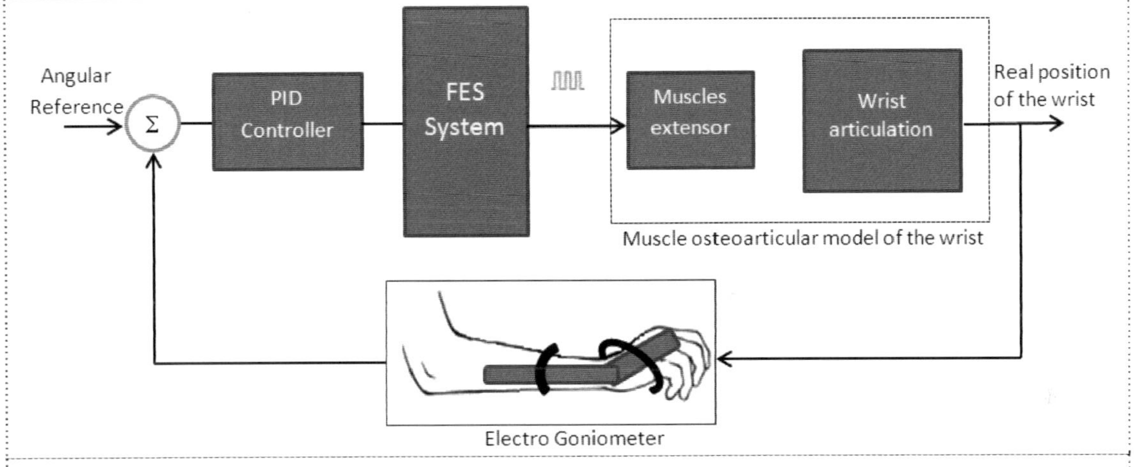

Figure 3. Scheme of the control setup in real time.

5. Results

The calculated relative stability for each system (PID designed with estimated experimental model) for each subject is shown on Table 3. The parameters on this table are gain margin (GM), phase margin (PM) and critical gain (Kc). Figure 4 shows the measured values for subject 3.

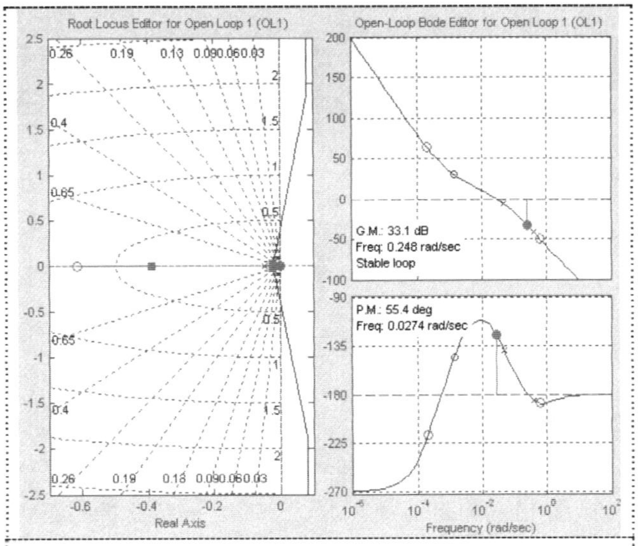

Figure 4. Measured phase gain and phase margin for subject 3.

Table 3. Calculated relative stability for each system
(PID designed with estimated experimental model for each subject).

	GM [dB]	PM [deg]	Kc
System 1	41.9	45.1	124
System 2	33.1	55.4	24.9
System 3	21.4	57.1	11.7

Figures 5 to 7 show the signals of the closed loop system at real time for all subjects. Table 4 shows the mean error and their standard deviation for each test in real time.

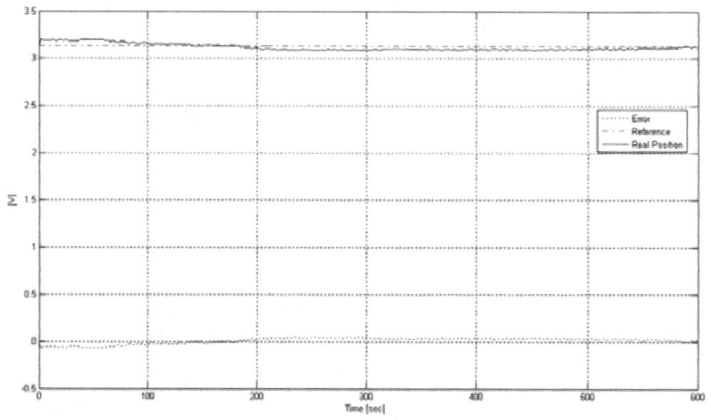

Figure 5. First record for subject 1 of the control system working in real time.

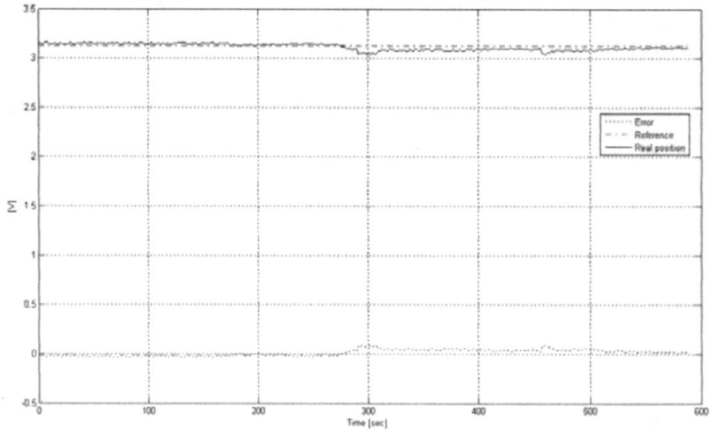

Figure 6. Second record for subject 2 of the control system working in real time.

Table 4. Mean error and their standard deviation for each test in real time closed loop.

	Subject 1 Test 1	Subject 1 Test 2	Subject 2 Test 1	Subject 2 Test 2	Subject 3 Test 1	Subject 3 Test 2
Mean Error	0.013	0.045	0.079	0.023	0.012	-0.064
Standard deviation	0.0321	0.0648	0.139	0.0275	0.0285	0.17

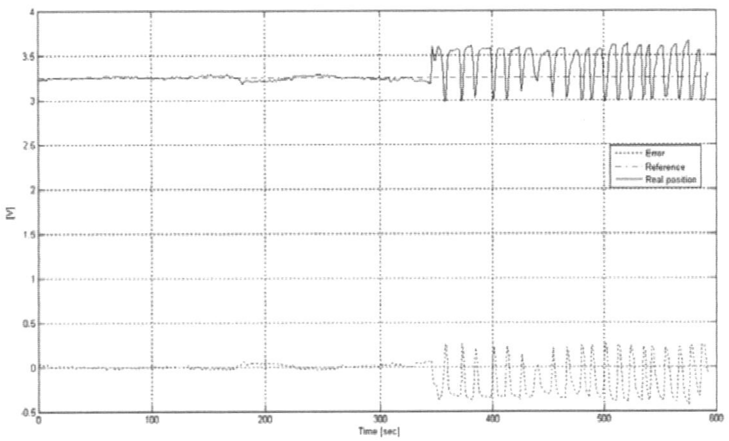

Figure 7. Second record for subject 3 of the control system working in real time.

6. Conclusion and discussion

The closed loops simulations with the experimental model and the designed controllers for each subject, showed a good relative stability was shown according to the values of gain margin, phase margin, and critical gain as detailed on Table 3. At the same time, subject 1 shows the highest relative stability whereas subject 3 has the lowest.

The muscle osteoarticular system of the wrist is a non-linear model, which varies with time because of the various causes that affect it such as muscular fatigue [4]. For subjects 1 and 2, at constant reference for the wrist extension at sagittal angle and during the period in which real time control was executed, the regulators responded reasonably in both tests with low mean errors and low standard deviations.

This suggests that by designing regulators with good margin of relative stability it is possible to control the proposed position in these two subjects despite the variability of the system to be controlled. In the case of subject 3, the first test can be included within the above conclusions for subjects 1 and 2. However, for test 2 (figure 7), the designed regulator did not succeed in stabilizing the position as from minute 6. The position remains oscillating around the reference value. In the simulation and according to the values on table 3, the PID for subject 3 had the lowest relative stability. This suggests the possibility of redesigning the controller so that a better relative stability can be achieved and repeat the experimental tests in real time. Another alternative solution for this subject, and in accordance to what other authors have presented before [4,5,6], is the application of non-linear control strategies such as fuzzy logic or neural network. In future, closed loop control strategies will be included in portable systems for home treatments of quadriplegic patients. These control systems will be implemented in micro controllers as it is simpler and less expensive to implement technologically PID regulators in them rather than non-linear control strategies. The preliminary studies of this work suggest that it should be possible to continue with the study of the closed loop position control with PID strategy in some people with spinal cord injuries whose movements are assisted by functional electrical stimulation.

7. References

[1] Cheng L, Zhang G, Wan B, Hao L, Qi H and Ming D 2009 Radial Basis Function NeuralNetwork-based PID Model for Functional Electrical Stimulation System Control *Proceedings of 31st Annual International Conference of the IEEE EMBS* 3481-84

[2] Hausdorff J M and Durfee W K 1992 Open-loop position control of the knee joint using electrical stimulation of the quadriceps and hamstrings *Medical and Biological Engineering and Computing* 29 269-80

[3] McNeal D R, Nakai R J and Meadows 1989 Open-loop control of the freely-swinging paralyzed leg Biomedical Engineering *IEEE Transactions on.* **36** 895-905

[4] Kurosawa K, Futami R, Watanabe T and Hoshimiya N 2005 Joint Angle Control by FES Using a Feedback Error Learning Controller *IEEE Transactions on Neural Systems and Rehabilitation Engineering* **13** 359-71

[5] Davoodi R and Andrews B J 2004 Fuzzy logic control of FES rowing exercise in paraplegia *IEEE Trans. Biomed. Eng.* **51** 541-3

[6] Hussain Z, Tokhi M O, Jailani R and Ahmad F 2010 Parameter optimization of FES-assisted indoor rowing exercise using MOGA *Proceedings of Fourth Asia International Conference on Mathematical/Analytical Modelling and Computer Simulation* 77-80.

[7] www.msal.gov.ar

[8] Watanabe T, Matsudaira T., Hoshimiya N. and Handa Y. 2005 A test of multichannel closed-loop FES control on the wrist joint of a hemiplegic patient *Proceedings of 10th Annual Conference of the International FES Society*

Acknowledgments

Authors wish to acknowledge the assistance of Carolina Carrere, Rodolfo Ramirez, Gustavo Zajac, Renata Gietz and Gabriela Merino for their collaboration with the present work. We want to acknowledge the National University of Entre Rios for its economic support.

Characterization of a Pneumatic Artificial Muscle for its application in an Active Ankle-Foot Orthosis

C Lescano[1,3], C Herrera[1], R Mirabal[1], R Rodrigo[2] and S Rodrigo[1]
[1] Gabinete de Tecnología Médica
[2] Laboratorio de Metrología, Departamento de Electromecánica
Facultad de Ingeniería, Universidad Nacional de San Juan, Av. Libertador San Martín 1109 Oeste, San Juan, Argentina.
[3] CONICET: Consejo Nacional de Investigaciones Científicas y Técnicas

E-mail: clescano@gateme.unsj.edu.ar

Abstract. The training of lower limbs with robotic exoskeletons or active orthoses has proven to be an effective tool to help patients with various types of neurological movement disorders, to recover their functional ambulation capacity. A key component to achieve this purpose is the type of active element used to compensate for muscle weakness, typically observed in such locomotor disorders. Based on these considerations, this paper presents the results of an experimental test developed for calibration, tuning and behavioral identification of a commercial pneumatic artificial muscle *Festo*®, with view to its application as active element of an active ankle - foot orthosis prototype for human gait rehabilitation. It is expected that such prototype would enable to recover the ankle joint function in patients with Myelomeningocele. The actuator performance analysis from the achieved tests, as well as the study of gait patterns in persons with Myelomeningocele, evidence the potentiality of this actuator as an active element of a robotic exoskeleton applied to human gait rehabilitation.

1. Introduction

Today many medical centres worldwide are implementing therapeutic treatments in an attempt to restore locomotion in patients with neuromotor disorders. The principle on which these treatments are based is neuronal plasticity or neuroplasticity, defined by the World Health Organization as the ability of nerve cells to regenerate anatomically and functionally, after being under pathological environmental or development influences, including trauma and diseases. The application of this principle specifically to the case of locomotor disorders of neurological origin allows from the practice and experience of rhythmic and repetitive movements of the human body in three-dimensional space, that the nervous system can be restructured to generate a series of proprioceptive and exteroceptive stimulus that neutralize posture patterns and abnormal movements patterns and create new ones, favourable for proper locomotion [1].

In this sense, the locomotor training with robotic exoskeletons or active orthoses (an external structural mechanism that interacts in parallel with the human operator actively assisting the lower limbs for automated practice walking in a treadmill, combined with partial support of the weight of the body), has proven to be an effective tool to help patients with various types of neurological movement disorders, to recover their functional ambulation capacity, by providing a steady and controlled movement of their lower limbs body, while releasing the therapist to physical stress imposed by this task [2, 3, 4, 5, 6].

For the design and development of such gait rehabilitation systems it is necessary to consider three fundamental components: structure, mechanisms of articular actuation and control strategies of the device. A review of the state of the art in terms of the structural characteristics of locomotive rehabilitation exoskeletons indicates that many of these systems lack of an active control of the ankle joint, determining an unnatural and unrestricted foot movement, which further prevents downloading the additional weight imposed by the use of this device. [3]. It is important to mention that under normal conditions, the ankle joint plays an essential role for shock absorption and progression during stance phase of the gait cycle, besides facilitating the advancement of the lower extremity during the swing phase [7, 8, 9].

On the other hand, the joint actuation of robotic exoskeleton for gait rehabilitation developed so far is made by motorized actuators only controling hip and knee joints in the sagittal plane, which results in certain degree of inconsistency with their adequate spatial biological motion. Therefore, the current trend is focused on the use of alternative actuators to electric motors that mimic the real joint mechanical behavior [10], as in the case of series elastic [11], magneto-rheological [12], the robotic tendons [13] and pneumatic muscles actuators [14, 15]. Particularly the latter ones show significant comparative advantages, such as its high power to weight ratio that significantly reduces power consumption, and its variable compliance allowing a better adaptation and interaction with the human operator.

In reference to the various control schemes proposed for these robotic rehabilitation systems, it appears that the developed algorithms until now contemplated little or null patient involvement in the rehabilitation process, besides being based on normal gait patterns rather than adapting to assistance degree needed for each considered neuropathological gait condition. In this regard, Yang *et al* [16] have noted that a key issue to achieve optimal results through the use of these rehabilitation systems is the definition of efficient control strategies and joint actuation, based on the human operator involvement decisions and in kinematic and kinetic characteristics of his or her ambulation.

Taking into account this general background of robotic exoskeletons for gait rehabilitation and particularly, those related to its joint actuation mechanism, this paper is oriented to identify the mechanical behavior of a commercial *Festo*® pneumatic artificial muscle, PAM [17]. To achieve this experimental test bench for PAM was developed, which allowed characterizing its behavior in static conditions for a range of loads equivalent to those recorded during normal human locomotion. The purpose is to apply it as an active element for an exoskeleton prototype that will command specifically the ankle joint, through an active ankle-foot orthosis (AAFO).

A literature review indicates that the function of the main plantar flexor muscles (in this case, *Gastrocnemius* (*Medial* and *Lateral* heads) and *Soleus*) are severely impaired in patients with low level spinal cord injuries (SCI), such as those from Myelomeningocele (MMC) [18, 19]. In this situation, it is expected that the AAFO offsets the plantar flexor weakness (characteristic of this level of SCI) at the final stage of stance phase, in where this muscle group generates the main active power demands required to forefoot takeoff, previous to swing phase [7, 8, 9].

The following sections will first describe the materials and methods used to develop the mentioned experimental test bench. Secondly, the criteria and procedure used for the PAM characterization in static conditions is explicated. Subsequently the results for measurement instrument calibration, as well as of the static characterization of PAM and a comparison with the results obtained from estimations of skeletal muscle force developed during a normal gait cycle will be shown. Finally, the conclusions of this work will be presented, also carrying a brief discussion on the feasibility of using PAM as active elements of robotic exoskeletons for human gait rehabilitation.

2. Materials and Methods

2.1. *Experimental system*

In order to model the PAM behavior and to analyze which control strategies is adequate for the intended application, numerous experimental test were performed. In figure 1 an outline of the implemented test system is displayed, composed by PAM, load cell, linear encoder, pneumatic circuit, pressure cell, signal conditioning circuit and data acquisition board. Such elements are now described in detail.

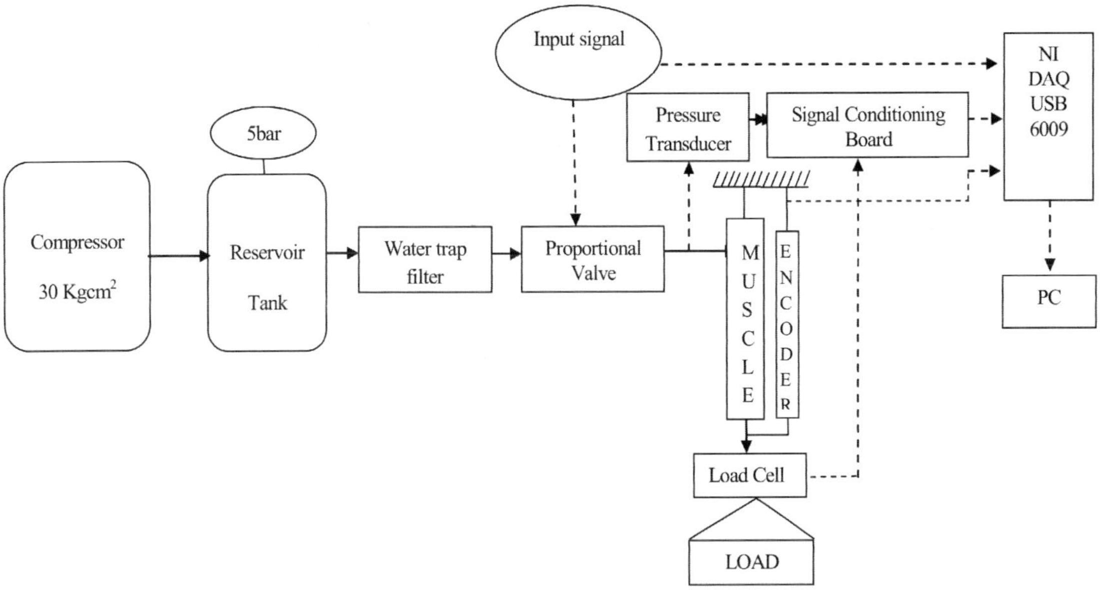

Figure 1. Setup for the implemented experimental system.

-Pneumatic Artificial Muscle: a DMSP-40-400N-RM-CM *Festo*® model with press-fitted connections was used (figure 2), 400mm nominal length and 40mm inside diameter [18]. The contraction system is formed by a pressure-tight length of rubber hose, sheathed in high-strength fibers. The fibers create a rhomboidal pattern with a three-dimensional grid structure. When internal pressure is applied, the hose expands in its peripheral direction, thus creating a tensile force and a contraction motion in the muscle's longitudinal direction. According to manufacturer's specifications, is capable of exerting a theoretical maximum force of 6000N for a 1 to 6 bar range of operating pressures, and developing a maximal permissible contraction of 25% of its nominal length.

Figure 2. Views of experimental setup for pneumatic muscle actuator.

-Load cell: The pneumatic movable end actuator was connected in serial with a MTS-1 'S' type strain load cell *BSL*® model, 1000 lb capacity, used to measure the force exerted by the PAM, conditioning then the output signal through an AD623 amplifier.

-Rectilinear Displacement Transducer: PZ34-S-250 model fabricated by *Gefran*® was arranged in parallel with the pneumatic actuator, which delivers a proportional voltage to the longitudinal displacement experienced, thus recording the contraction of the actuator when pressurized.

-*Pneumatic circuit*: a compressor with 30 $Kgcm^2$ capacity provides compressed air that is accumulated in a reservoir tank to stabilize the pressure supplied to the actuator. The working pressure was set at 5 bar maximum, taking into account the operating conditions of the PAM (Figure 3). After the accumulator tank, the air flows through a water trap filter and a VPPM-6L-L-1-G18-0L6H-V1N-S1 proportional valve *Festo®* model, which regulates the pressure inside the artificial muscle based on an analog voltage variable between 0 to 10V. Different types of reference signals were provided to the valve by a signal generator.
-*Pressure transducer*: AKS 330 *Danfoss®* model allows measure and verifies if the pressure supplied to the actuator corresponds to the pressure value supplied by the regulator valve.
-*Signal conditioning board*: to properly record the signal from the load cell, the circuit displayed in figure 4, based on AD623 amplifier, was used. Also, the output signal of the pressure cell (4 to 20mA) was translated in a voltage on a 220Ω resistor.

Figure 3. Pneumatic circuit.

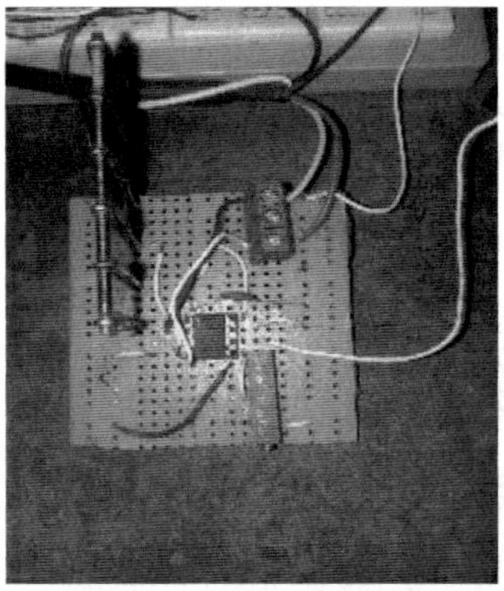

Figure 4. Conditioning signal board

-*Data Acquisition Board*: The NI USB-6009 *National Instrument®* model was used, considering its sampling rate, data communication by USB port and its adaptability to various programming languages. Four of its analog inputs were used to acquire signals from the rectilinear displacement transducer, load and pressure cells and the set point of the pressure regulating valve. Signals from each of the measuring instruments were captured and later sent to a PC for processing and visualization of results. In addition, for the capture and storage of the data a Virtual Instrument (VI) developed in *Labview®* was used.

2.2. *Measurement calibration instruments*

After implementing the physical system for experimental test, the static calibration of the used measurement instruments was achieved for two conditions: without load and for a 250-3000N load range. For each load value, a range from 0 to 5Vcc was applied to the pressure proportional valve. As well, the calibration curve for the proportional pressure valve was obtained applying as consign signal different *Vcc* values at 1V steps and measuring then the output pressure through a Serie 700 P27 pressure module *Fluke®* model, 0-20 bar range, associated with a *Fluke®* 725 multifunction process calibrator.

On the other hand, the calibration for displacement transducer was achieved by registering the *Vcc* values obtained for each longitudinal displacement of its axis. Also, in order to know the displacement for PAM and transducer, PAM nominal length without pressurization was measured, registering then its final length under pressure.

Furthermore, the pressure transducer calibration was based on the applied pressure to the PAM, registered by pressure module and multifunction process calibrator instruments, and by measured voltage at a 220 Ω electrical resistance. Finally, the load cell calibration was achieved applying different load values and then measuring the output voltage. Table 1 shows the equations corresponding to the obtained calibration curves.

Table 1. Calibration equations for measurement instruments.

Proporcional valve	*Presure [bar]=1.002*Voltage [V] + 0.026*
Rectilinear Displacement Trasducer	*Voltage [V]=-0.823*Displacement [cm] + 24.35*
Pressure cell	*Voltage [V]=0.117*Presure [bar] + 0.9979*
Load cell	*Voltage [V]=0.00077*Force [N] + 0.1508*

3. Pneumatic artificial muscle characterization

A comparison of PAM to biological skeletal muscle helps to visualize the importance of exploring this alternative as an active element for an AAFO for gait rehabilitation, in order to compensate the weakness of plantar flexors muscles. While the microscopic features of biological muscles cannot be emulated in technology yet, a general replication of the macroscopic functionality is possible in the form of biomimetic actuators, one of which is a DMSP from *Festo*® [17]. An analogy can be made between biological skeletal muscle and PAMs because in both cases, force is generated through activation resulting in dynamic behavior of force-length change relation highly nonlinear. Additional similarities between them include a displacement of 35% of its resting length and capability to operate as part of an antagonist pair [20].

In addition, the force-generating capacity of the PAM is about 6 times the body weight, allowing replicating the developed forces during human activity [7-9]. On the contrary, biological skeletal muscle operates with greater efficiency, responds more quickly and has more accurate control than that of a PAM, which implies the need to find an accurate PAM model representing its dynamic behavior, and then to define control strategies that attempt to emulate the regulation exerted by central nervous system during human locomotion.

Taking into account such similarities and in order to evaluate the possible application of PAM as an active element for an AAFO for gait rehabilitation, its static behavior for different load conditions was registered, then representing from experimentally obtained data the isobar curves relating the PAM load capacity and the actuator longitudinal contraction.

Also, to compare these results with those obtained from a Hill muscle model of the skeletal muscle (which incorporates elastic and contractile elements to represent the skeletal muscle contraction dynamics), previously obtained database was used. This database corresponds to estimation of muscle forces developed during a gait cycle of an adult subject (25- year-old male with a height of 1.70 m and a weight of 70 kg) with normal cadence, obtained then of combining inverse dynamic and global static optimization techniques applied to a multibody model of human body. Details of the applied procedure can be shown in [21].

In particular, the force-length relation estimated for the plantar flexor muscle group was evaluated for 50 -60% of gait cycle, based on considerations related to its crucial role for a powerful plantar flexion during forefoot takeoff phase, which is precisely the function expected to be performed by the PAM in the AAFO to be developed. A literature review indicates that in normal conditions, the plantar flexor muscle group is capable of exerting a force of 3-4 times body weight to ensure forefoot takeoff during walking, whereas during running this force increases to 6-8 times body weight [22].

With respect to the variation of skeletal muscle length during a gait cycle, Lieber [23] has signalled that *muscle length changes are relatively small during the time when muscles are active*, while other researchers suggest that during the stance phase of gait when the cat soleus first actively lengthens and then actively shortens, the total length excursion is only about 10% of its maximum length (i.e, muscle rest length), and is probably even less when one considers the fact that the Achiles tendon can actually absorb some of the length change [24]. Given the invasiveness of practice, unfortunately it is not possible to know if these shortening conditions are similar in humans. However, these results are a reference to visualize the possible operating conditions that would occur in humans during forefoot takeoff.

Finally, to make the PAM results comparable with the corresponding to plantar flexor muscle group, the ratio muscle length with respect to resting muscle length (l/l_0) was represented at the axis of abscissa, and the ratio developed muscle force referred to maximum muscle force (F/F_{max}) in the ordinate axis.

4. Results

In figures 5 to 8 the calibration curve achieved for the measurement instruments are shown. The result presented for each measurement instrument corresponds to a linear adjustment of the data.

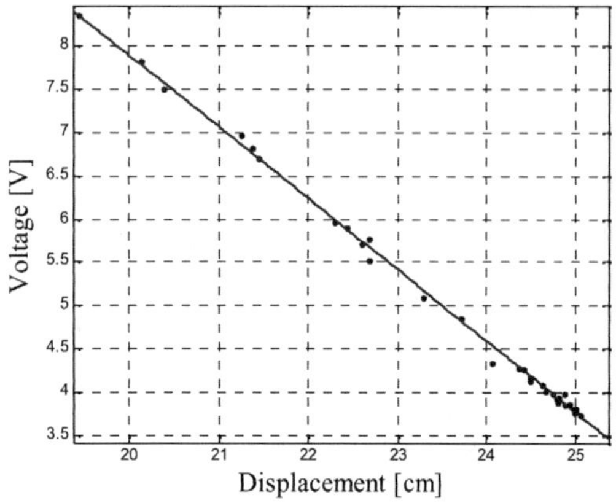

Figure 5. Calibration curve for encoder.

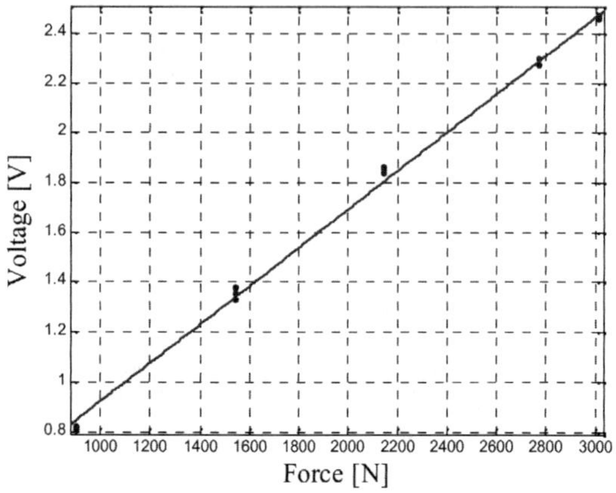

Figure 6. Calibration curve for load cell.

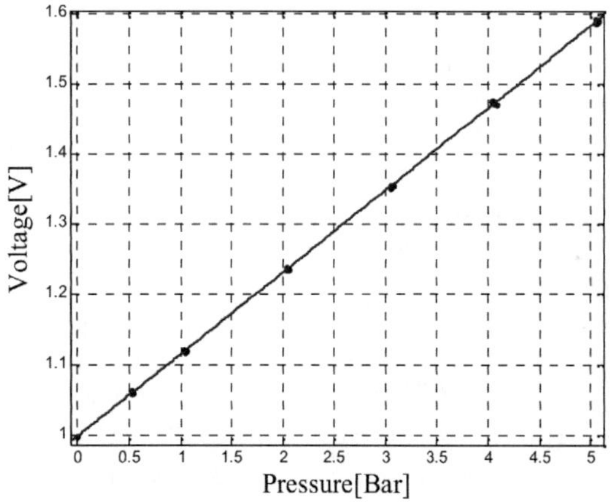

Figure 7. Calibration curve for pressure cell.

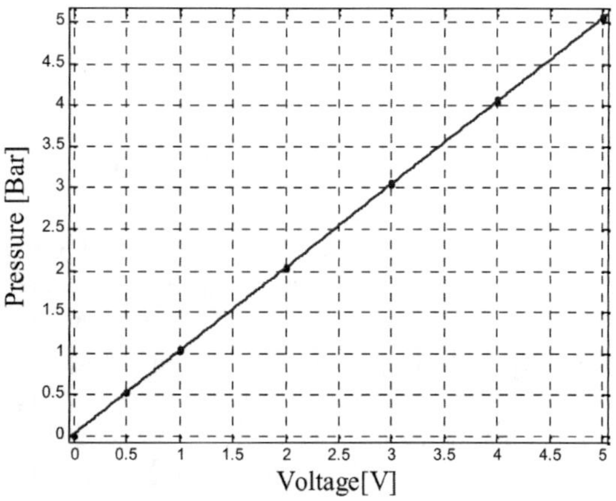

Figure 8. Calibration curve for proportional valve.

In addition, the statistical variables obtained for each calibration instrument as a result of linear adjustment is presented in table 2. It is remarkable the significant linear relationship among the involved variables, evidenced by a regression coefficient of 99%.

Table 2: Statistical parameters for linear adjustment of the calibration instruments.

Instrument	R^2	*SSE* (Sum of Squares Due to Error)	*RMSE* (Root Mean Squared Error)
Encoder	0.9985	0.090	0.0549
Load Cell	0.9952	0.02681	0.04376
Pressure Cell	0.9998	0.0002349	0.002668
Proportional Valve	0.9999	0.005458	0.01372

As well, the isobar curves obtained for the experimentally evaluated PAM are displayed in figure 9, relating permissible force F [N] as a function of the contraction h [%] of its nominal length, determined as $(l-l_0)/l_0*100\%$.

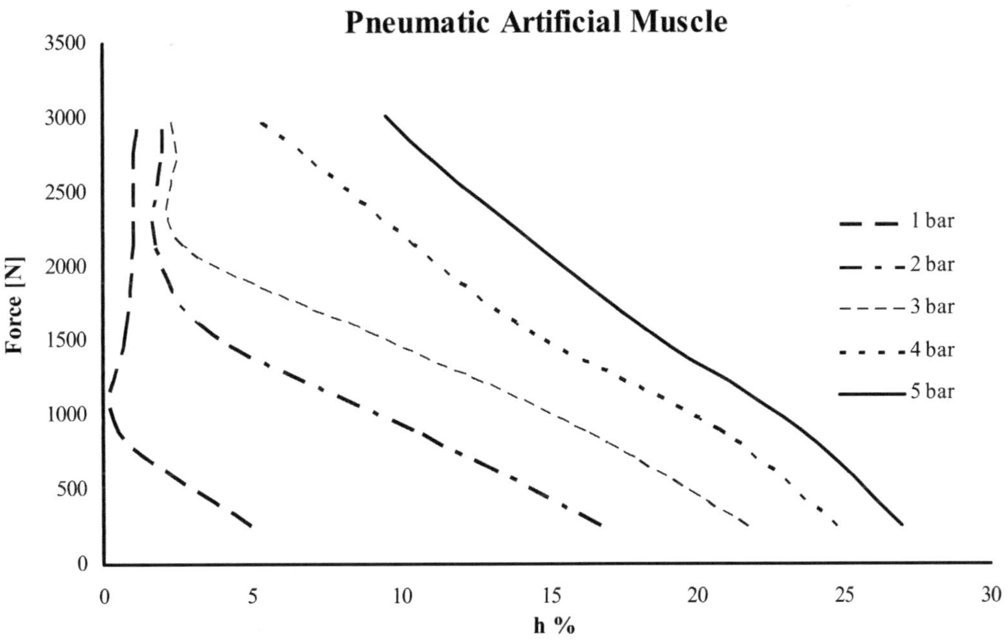

Figure 9. Isobar curves for evaluated PAM.

Furthermore, the force-length relations for the tested PAM and estimated activity of the plantar flexor muscle group during gait are represented in figures 10 and 11 in order to visualize if there exist similarities between both types of muscles. In the case of PAM, figure 11 displays for each inlet pressure value, the relation between PAM force/maximal PAM force and PAM length/PAM rest length. Note that although the evaluated PAM is capable to support a theoretical maximal load of 6000 N, the relative load values have been measured for a maximal load of 3000 N.

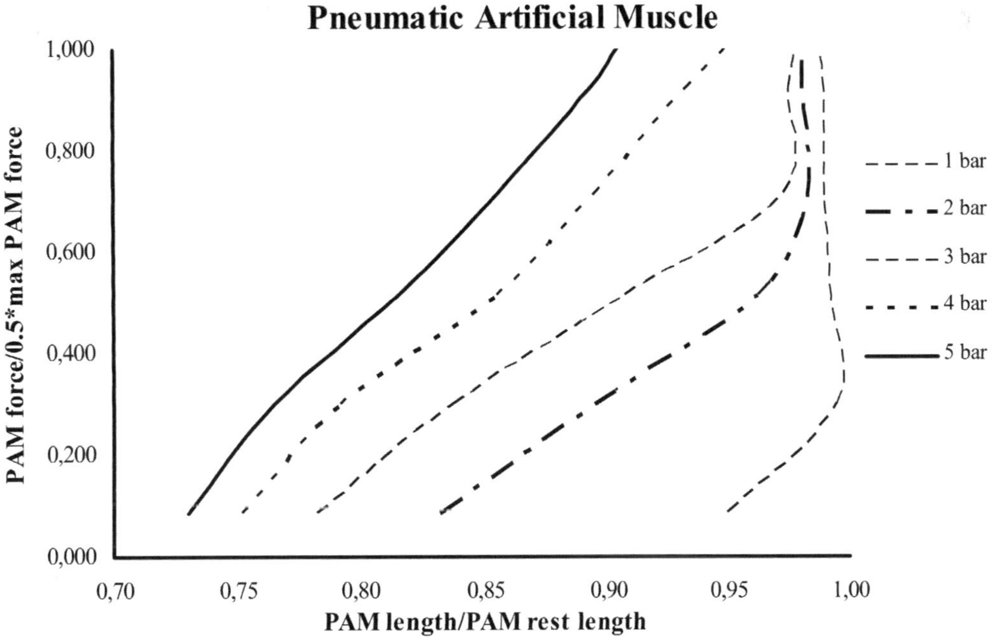

Figure 10. Relative force – length curves for pneumatic artificial muscle experimentally evaluated.

As well, figure 11 graphs the relation between muscle force/maximal isometric muscle force and muscle length/muscle rest length for the plantar flexor muscle group during a gait cycle and estimated by the solution of indeterminate biomechanical problem through optimization techniques [21].

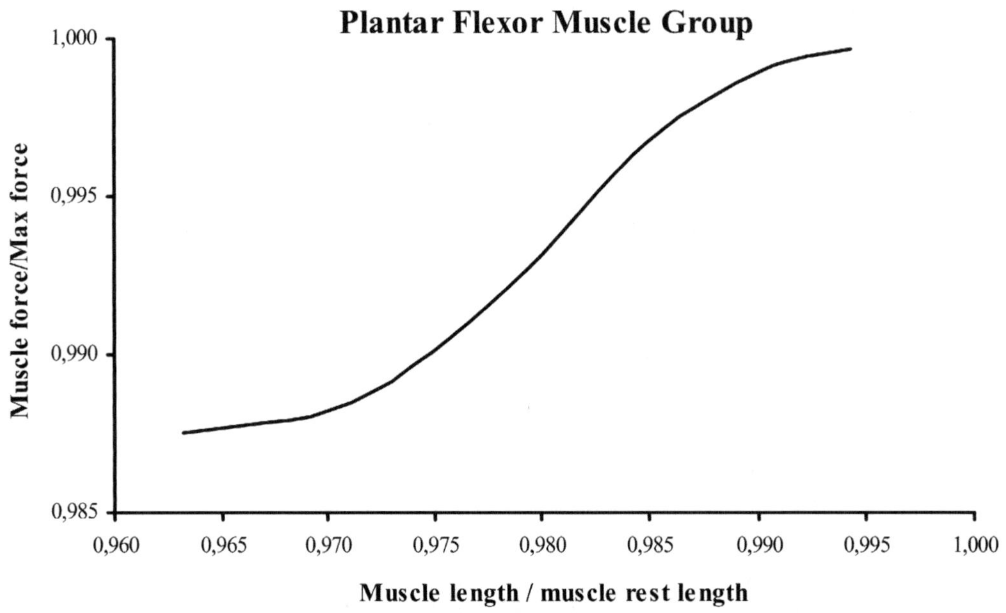

Figure 11. Relative force – muscle length relation estimated for plantar flexor muscle group.

5. Discussion and conclusions

Numerous efforts have been made searching a joint actuator for motor rehabilitation exoskeletons, which properly emulate the behavior of human skeletal muscles [2-6, 10-15]. In this sense, research shows advantageous features of PAM in comparison with other types of actuators for such applications [14-15], related with its high power to weight ratio and variable compliance allowing a better adaptation and interaction with the human operator.

With the aim of identifying the mechanical behavior of a commercial PAM *Festo*® with view to its application as an active element of an AAFO for gait rehabilitation, an experimental test bench was developed. This experimentation allowed the calibration of measuring instruments and evaluation of PAM performance, which characterizes its behavior in static conditions for a range of loads equivalent to those recorded during normal human locomotion.

At this stage, the behavior of PAM in static conditions was analyzed vertically placing it while underwent various inlet pressures and variable loads, set according to operating conditions which shall be subjected in the provided for active orthotic device. Such conditions are referred to attend during the final stage of the stance phase in the gait cycle, the function of the main plantar flexor muscles (*Gastrocnemius (Medial* and *Lateral* heads) and *Soleus*) for forefoot takeoff, seriously impaired in patients with SCI at the lumbosacral region, such as those from MMC [18, 19]. Note that in normal conditions, the plantar flexor muscle group is capable of exerting a force of 3-4 times body weight to ensure forefoot takeoff during walking, whereas during running this force increases to 6-8 times body weight [22]. Given this information and considering the use of active orthotic device by a person body weight 70kg (686.7N), the evaluated load range was from 250 to 3000 N, corresponding to the required level of force for the development of normal locomotion.

Regarding the calibration of instruments achieved the results displayed in Figures 5 to 8 show a highly linear behavior, further demonstrated by the statistical parameters presented in Table 2 displaying in all cases a linear fit with a regression coefficient more than 99%. In the case of the encoder, new measurements will improve the measurement error, observed increased from the rest of evaluated instruments.

Analysis of PAM isobar curves displayed in figure 9 and obtained from the relationship between the force exerted and the rate of shortening developing by PAM indicate that at 1 bar inlet pressure and loads less than 1000N, contraction achieved is less than 5%, while for higher load values PAM does not experience effective contraction. Meanwhile, for 2-3 bar input pressures, this contractible behavior of PAM (until 20%) occurs for loads lower than 2000N. For higher input pressure (4-5 bar), the contraction percentage reaches up to 27%. Furthermore, taking into account these results and comparing them with PAM operating curves provided by the manufacturer can be said that the conditions tested (maximum load of 3000 N) are within the operating range specified.

On the other hand, the representation in figure 10 of the PAM static characteristics based on the relative ratio PAM force/maximal PAM force and PAM length/PAM rest length, shows an isometric behavior at 1 bar inlet pressure and loads less than 1000N, while at increasing input pressure PAM is able to develop greater contractions, which resembles the concentric contraction properties exhibited by human skeletal muscle. It is also noted that the PAM is able to support loads greater than 4 times the body weight and develop shortening of about 10% of its nominal length for this level of loading.

Furthermore, linking these PAM operating conditions with the results presented in figure 11, which displays on the one hand, that at 50% of gait cycle the maximum force developed by the main plantar flexor muscles is about 3000 N (equivalent to 4.44 times the body weight of the subject analysis) and on the other hand, that the shortening experienced by such muscle group is around 3% in the period 50-60% of gait cycle, it can be said that the PAM behavior could satisfy the conditions of work required for the intended application. As well, these results are consistent with the above mentioned regarding the level muscle force required for forefoot takeoff [22] and the relatively small muscle length changes during its activation time along the gait cycle [23, 24].

As a conclusion, analysis presented in this paper is part of the evidence required to assess whether the characteristics of the analyzed PAM, meet the specifications to be performed by an active element for ankle foot orthosis, which compensates the weakness of the plantar flexors in patients with MMC. The results of experimental test for the evaluated PAM indicate a biomimetic behavior at macroscopic level, similar to that manifested by the plantar flexor muscles group in the analyzed static conditions. Particularly, two characteristics of PAM emulate its biologic equivalent:

1. Static load is proportional to inlet pressure, which can so play the part of the neural excitation required for muscle tissue activation,
2. Static load continuously decreases when contraction ratio h increases, making the actuator a true contractile actuator [25].

In future works experimental test for the PAM in dynamical conditions will be done. Then, is predicted using a neural network for modeling the nonlinear PAM behavior. Such identification will allow characterizing the real dynamical behavior of PAM, previous to the definition of control scheme required to its operation as an active element for the intended AAFO, to be applied for gait rehabilitation in MMC patients. Such control scheme must be able to regulate PAM in order to emulate the function of human plantar flexor muscle group at the forefoot takeoff during the gait cycle, a question that requires a careful analysis at the time of implementing the AAFO for gait rehabilitation.

Acknowledgment

The authors would like to thank to Universidad Nacional de San Juan and to Gobierno de la Provincia de San Juan for financial support, as well as to the staff of Instituto de Materiales y Suelos for his technical assistance.

6. References

[1] Purves D, Augustine GJ, Fitzpatrick D, Katz L, LaMantia A and McNamara J 2004 *Invitación a la Neurociencia* (Buenos Aires: Editorial Médica Panamericana)

[2] Jezernik S, Colombo G, Keller T, Frueh H and Morari M 2003 Robotic Orthosis Lokomat: a rehabilitation and research tool *Neuromodulation* **6** 108-115
[3] Moreno J, Brunetti F, Rocon E and Pons J 2007 Immediate effects of a controllable knee ankle foot orthosis for functional compensation of gait in patients with proximal leg weakness *Med Biol Eng Comput.* **46** 43-53
[4] Colombo G, Wirz M and Dietz V 2001 Driven gait orthosis for improvement of locomotor training in paraplegic patients *Spinal Cord* **39** 252-5
[5] Wirz M, Colombo G and Dietz V 2001 Long term effects of locomotor training in spinal humans *J Neurol Neurosur Ps* **71** 93-96
[6] Beyl P, Naudet J, Van Ham R and Lefeber D 2007 Mechanical design of an active knee orthosis for gait rehabilitation *Proc. IEEE 10th Int. Conf. on Rehabilitation Robotics* **1** 100-5
[7] Sánchez J, Prat J, Hoyos J, Viosca E, Soler C, Comín M, Lafuente R, Fabregat A and Vera P. 1999 *Biomecánica de la marcha humana normal y patológica* (Valencia: Instituto Biomecánico de Valencia) p 31-120
[8] Perry J 1992 *Gait Analysis: Normal and Pathological Function* (USA: Slack Inc.)
[9] Winter D 2009 *Biomechanics and Motor Control of Human Movement* (Canada: John Wiley and Sons Inc.)
[10] Dollar A and Herr H 2008 Lower extremity exoskeletons and active orthoses: challenges and state-of-the-art *IEEE T Robotic Autom* **24** 144-158
[11] Blaya J and Herr H 2004 Adaptive control of a variable-impedance ankle-foot orthosis to assist drop foot gait *IEEE Trans Neural Syst Rehabil Eng,* **12** 24-31
[12] Naito H, Akazawa Y, Tagaya K, Matsumoto T and Tanaka M 2009 An ankle–foot orthosis with a variable-resistance ankle joint using a magnetorheological-fluid rotary damper *J Biomed Eng* **4** 182-91
[13] Boehler A, Hollander K, Sugar T and Shin D 2008 Design, implementation and test results of a robust control method for a powered ankle foot orthosis (AFO) *Proc. IEEE Int. Conf. Robot. Autom.* 2025-30
[14] Ferris D, Czerniecki J and Hannaford B 2005 An Ankle-Foot Orthosis Powered by Artificial Pneumatic Muscles *J Appl Biomech* **21** 189-197
[15] Ferris D, Gordon K, Sawicki G and Peethambaran A 2006 An improved powered ankle foot orthosis using proportional myoelectric control *Gait Posture* **23** 425-8
[16] Yang CJ, Niu B and Chen Y 2005 Adaptive Neuro-Fuzzy Control Based Development of a Wearable Exoskeleton Leg for Human Walking Power Augmentation *Proc. IEEE/ASME Int. Conf. on Adv. Intelligent Mechatronics (California: USA)* 467-472
[17] http://www.festo.com/rep/en_corp/assets/pdf/info_501_en.pdf
[18] Gutierrez E, Bartonek Å, Haglund-Åkerlind Y and Saraste H 2005 Kinetics of compensatory gait in persons with myelomeningocele *Gait Posture* **21** 12-23
[19] Öunpuu S, Davis RB, DeLuca PA 1996 Joint kinetics: methods, interpretation and treatment decision-making in children with cerebral palsy and myelomeningocele *Gait and Posture* **4** 62-78
[20] Caldwell DG, Tsagarakis N and Medrano-Cerda GA 2000 Bio-mimetic actuators: polymeric pseudo muscular actuators and pneumatic muscle actuators for biological emulation *Mechatronics* **10**, 499-530
[21] Rodrigo SE, Ambrósio JA, Tavares da Silva MP, Penisi OH 2008 Analysis of human gait based on multibody formulations and optimization tools, *Intl J Mechanics based Design of Structures and Machines* **36** 446-477
[22] Giddings VL, Beaupre GS, Whalen RT, Carter DR 2000 Calcaneal loading during walking and running *Med Sci Sports Exerc* **32** 627–634
[23] Lieber RL Skeletal 1992 *Muscle Structure and Function. Implications for Rehabilitation and Sports Medicine* (Baltimore: Williams and Wilkins)
[24] Gregor RJ, Roy RR, Whiting WC, Lovely RG, Hodgson JA and Edgerton VR 1988 Mechanical output of the cat soleus during treadmill locomotion: in vivo vs. in situ characteristic *J Biomech* **21** 721-732
[25] Tondu B 2007 *Humanoid Robots: Human-like Machines* (Vienna: ITech).

VIRTUAL REALITY UPPER LIMB MODEL CONTROLLED BY EMG SIGNALS

Fabricio Muri, Celina Carbajal, Ana M. Echenique, Hugo Fernández, Natalia M. López.

Gabinete de Tecnología Médica, Facultad de Ingeniería, Universidad Nacional de San Juan.

nlopez@gateme.unsj.edu.ar

Abstract. This work presents the design and development of a six-channel system for acquisition and conditioning of electromyographic signals collected in the upper limb. The main objective of the work is to create a system that can be used as rehabilitation and training instrument for potential users of myoelectric prostheses. Using software rendering, feature extractions, classifier training and design of the mechanical model of the human arm, with the running movements of flexion, extension, pronation and supination of the forearm and the grasp in a reality environment virtual, providing rehabilitation therapy to different patients.

1. Introduction

Assistive technologies have emerged in recent years in order to provide rehabilitation devices for people with varying degrees of disability. Especially for upper limb amputees, or patients with reduced mobility and control, several devices have been developed and find clinical application [1]. Nowadays, the construction of prosthetic systems as extensions of any human body limb has huge improvements, which present essential features such as: flexibility, aesthetic, lightweight and multi-functionality.

Many robot systems for upper limbs rehabilitation have been developed, such as MANUS system, which has two degrees of freedom (DOF) for active control of force-feedback, and conducted clinical tests with successful results [2]. Specifically designed for neurorehabilitation, the ARMin device [3] is an exoskeleton with 6 DOF, in which the patient is positioned in a wheelchair and the arm is attached to the structure to complete the exercise sequence. It allows a complete position control and preserves the natural biomechanics of the elbow and shoulder joints, assuring the patient comfort. The ArMin is equipped with force and position sensors and has three functioning modes: movement therapy mode, game therapy mode and ADL training mode. Another alternative for upper limb rehabilitation is the use of Virtual Reality (VR) tools, providing visual feedback to the user, by the repetitive exercises or in game like tasks. In this area is important to cite the works of Adamovich et al [4] and Morrow et al [5], between others.

Similarly, science and engineering have made significant advances in the area of digital signal processing, especially surface electromyographic signals (sEMG) and its applications in the control of active devices. Moreover, sEMG are used as control input in devices [6] designed for post-stroke rehabilitation programs, or Virtual Reality (VR), using the information contained in the signal about the intentionality of the user and the muscular force involved.

The main objective of this work is to design a system capable of acquire sEMG signals, digitalise and process them, to be used as a control input to a virtual object that represents a model of the upper limb. This VR object emulates the user's movements, providing a training system for potential prosthesis users or to patients with reduced mobility in the upper limb, by repetitive exercises or tasks performed in the VR environment.

A customized circuit was designed to acquire the six channels sEMG of the upper limb, constructing a data base of ten normally limbed subjects. Software was developed under Matlab®/Simulink® that performs the A/D conversion, processing and classification of user intent,

Journal of Physics: Conference Series **477** (2013) 012041 doi:10.1088/1742-6596/477/1/012041

done by an artificial neural network (ANN). Finally a control strategy for real-time simulation is performed and the virtual reality animation allows the visualization of upper limb movements, to validate the results.

The paper is organized as follows: Section II explains the design and developing of the acquisition stage, the database, the sEMG signal processing, feature extraction and simulation of the VR object. In Section III, the results of sEMG signals classification using different combinations of features parameters are showed and finally, the conclusions are exposed.

2. Materials and Methods

2.1. Design and developing of the acquisition circuit

The block diagram of Figure 1 provides an overview of the building blocks of the front-end circuit that amplify the information contained in the sEMG signals and that way separate them from any noise.
The sEMG signal is captured by disposable surface electrodes (Ag-AgCl), which are highly stable and the electrolytic gel junction provide a very low noise level. They are connected to the amplification stage using shielded cables, supporting mobility. An embedded instrumentation amplifier, INA 129, of Texas Instruments® was used, setting the gain in 950V/V. For electrical safety conditions, an optical isolation stage was interposed between the subject and the devices connected to the (electric) voltage line, feeding the circuit with two batteries of 9 Volts. Then, signals are filtered by a 4th order bandpass Butterworth filter, in Sallen-Key structure, with cutoff frequencies between 10Hz and 500Hz.
The circuit was designed with six acquisition channels. Only four of them were used for the classification of movements, leaving the possibility of acquiring sEMG signals from other muscles with other purposes. The A/D conversion is performed by a PCI 6024-E from National Instruments® board, with a sample frequency of 1KHz [7].

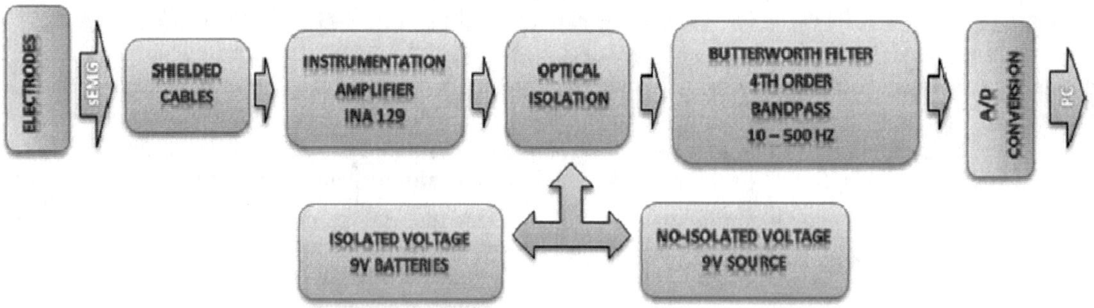

Figure 1: Block diagram of the designed system.

2.2. Database

After previous informed consent, sEMG data corresponding to six classes of motion were collected from 4 volunteers, all normally limbed subjects, (3 male, 1 female, 27 ± 3 years old).
Electrodes were placed in *brachial biceps*, *brachial triceps*, *brachioradialis* and *pronator teres*, during flexion, extension, pronation and supination of the forearm, and grasp, all them interspersed with rest. The users were prompted to perform the movements by a video, which assures the task execution in the programmed time. The test is divided into two sequences; one is used for training the classifier and the other one for validation. Also, an extra session is done by the subjects to control the virtual arm in real time.

2.3. EMG Processing, Feature extraction and Classification.

All digital processing is performed using the Matlab/Simulink® software. In order to avoid the electrical line interference a digital notch filter was used, with a cutoff frequency of 50 Hz and a Q factor of 250. Due to the accumulative properties of the sEMG descriptors, it is very important to remove the noise of baseline, in order to avoid erroneous control commands (Fig. 2). The data was segmented into non overlapped windows of 50 ms (N samples) to obtain a feature vector of each segment. Several sEMG features are described in bibliography, in temporal or frequency domain [8]. However, the choice of the best feature vector was made selecting those descriptors that better represent the information contained in the signal, without detriment of the real-time performance. The features selected are good estimators of the muscular force developed during contraction. These are:

Energy:

$$E = a\sum_{i=1}^{N} x_i$$

Integral of Absolute Value:

$$IAV = \frac{1}{N}\sum_{i=1}^{N}|x_i|$$

Willison Amplitude:

$$WAMP = \sum_{i=1}^{N} f(|x_i - x_{i+1}|) \quad f(x) = \begin{cases} 1 & if \;\; x > \varepsilon \\ 0 & otherwise \end{cases}$$

Waveform Length:

$$wl = \sum_{i=1}^{N}|\Delta x_i| \qquad \Delta x_i = x_i - x_{i-1}$$

The continuous feature vector resulting (Fig. 3) is suitable for the use as control reference.The classification was implemented by an ANN, using a multilayered backpropagation learning algorithm. We implemented a network of three hidden layers with 16, 12 and 8 neurons in each and 6 neurons in the output layer. Figure 4 shows the training vectors corresponding to a sequence of the six described movements.

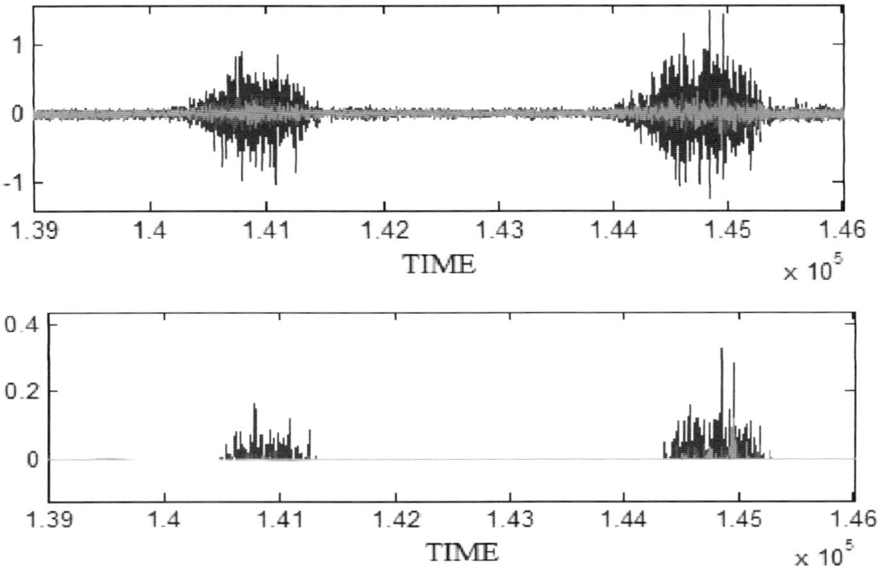

Figure 2: Up: Raw sEMG signals. Down: sEMG filtered and without noise baseline.

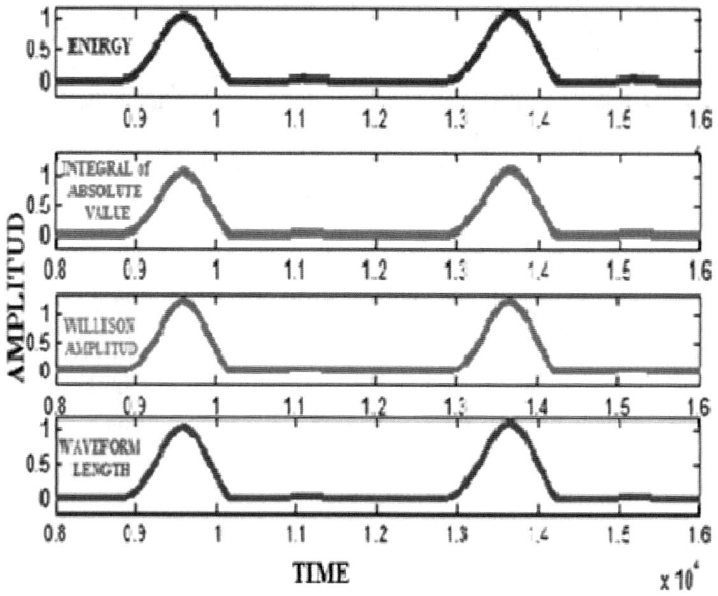

Figure 3: Feature vector corresponding to the signal of Figure 2.

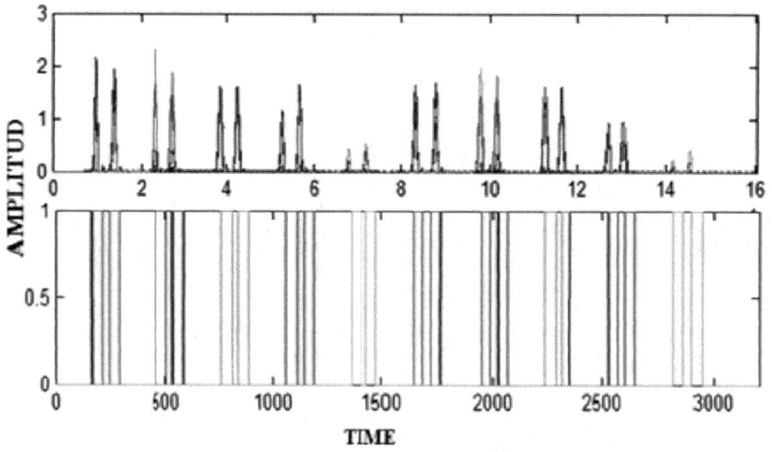

Figure 4: Top: Feature vector corresponding to 4 channels of acquisition. Bottom: Output vector corresponding to the six classes of movements.

2.4. Virtual Model of the Upper Limb.

The next stage is the upper limb modeling, represented by cylindrical and spherical rigid bodies, with the SimMechanics® and Virtual Reality® toolboxes [9]. Firstly, the tasks, movements and dimensions of the upper limb need to be defined. Inertial parameters, mass and center of gravity were calculated according with the average of the measurements made in the volunteers. Also, bodies´ moment of inertia, degrees of freedom and coordinate axis must be specified in the toolbox, to use the virtual reality environment. The signals obtained in the classification stage are translated to control commands by a PID control system, controlling the VR model actuators that emulate the behavior of the upper limb. These actuators include rotational joints in the elbow, wrist and fingers.

The elbow joint was used as reference to the forearm segment and hand, which are represented by a cylinder and a sphere. Fingers were represented only by two segments. Finally, sensors and actuators blocks inserted in the model provide the interaction tool with the others Simulink blocks (Figure 5). The control is implemented by a PID controller.

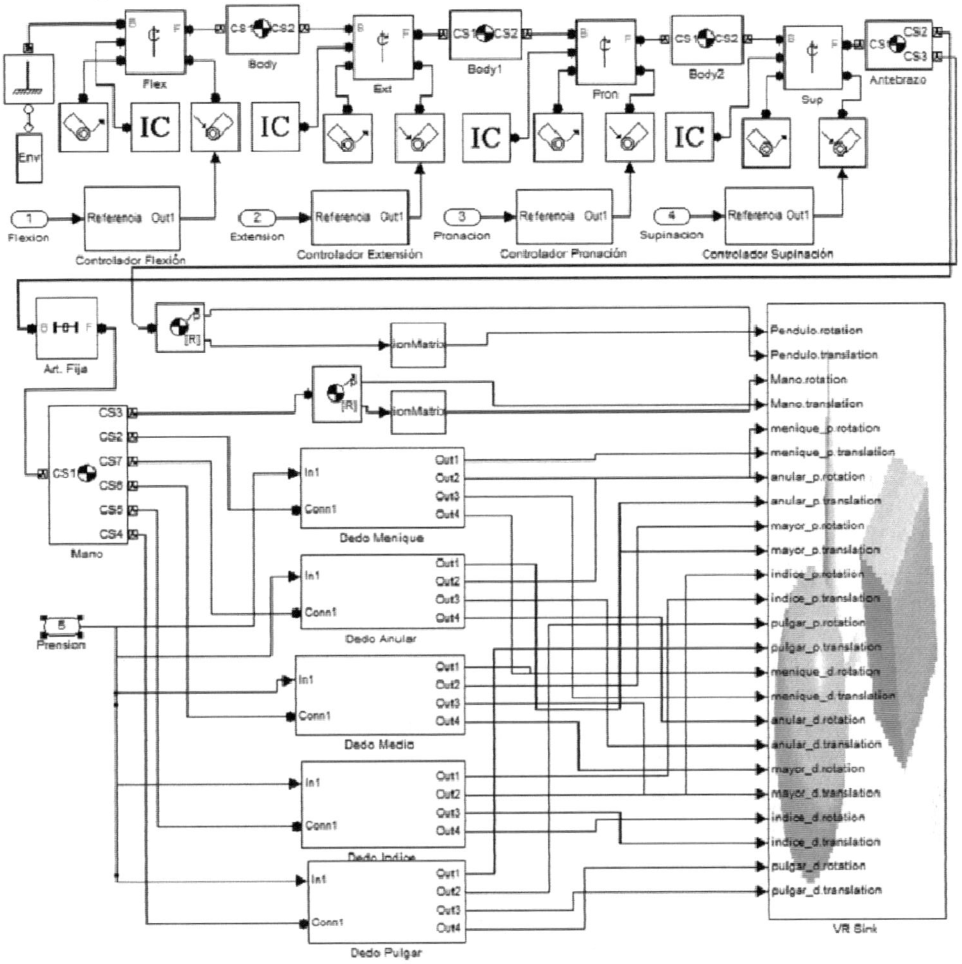

Figure 5: Block diagram of the Upper Limb Model in SimMechanics®

3. Results

The ANN validation was done offline with the data for that purpose. Table 1 shows the average percentage of positive ratings (APPR) per volunteer, calculated by the porcentual relation between the number of successful movements and the total number of movements, by one hundred.

	VOLUNTEER 1	VOLUNTEER 2	VOLUNTEER 3	VOLUNTEER 4
APPR [%]	**69,75**	**92,5**	**72,75**	**82,5**

Table 1: APPR considering E, IAV, WAMP and WL.

It is noted that the average ranges from 69.75% to over 92%. The variability among volunteers is attributed to sEMG signals which differ from causes such as the location of electrodes, layout and individual physiology, among others.

In order to reduce the computational time, only WAMP and IAV features were taken in to account, and Table 2 presents these results. It is possible to see that the volunteer 2 increase slightly the average score, while the remainders suffer moderate decrease.

	VOLUNTEER 1	VOLUNTEER 2	VOLUNTEER 3	VOLUNTEER 4
APPR [%]	**49,5**	**93,75**	**56**	**76,25**

Table 2: APPR considering WAMP and WL.

Table 3 presents the effect occurred when only the WAMP is extracted from the signals, showing a significant decrease of the APPR, especially in a couple of volunteers that are under the range of accepted values.

	VOLUNTEER 1	VOLUNTEER 2	VOLUNTEER 3	VOLUNTEER 4
APPR [%]	**31,25**	**84,5**	**50,25**	**70,75**

Table 3: APPR considering WAMP

From the above, is possible to conclude that efficiency and good performance of the classifier depends more of the feature extraction stage, than for the classifier in self.

The virtual model was tested online with all subjects, demonstrating a good performance and acceptance. The users do not report any lack of controllability, nevertheless the system is not able to correctly identify the grasp state in all situations. (To see an experimental session, please visit http://www.youtube.com/watch?v=bGsZUq_8mYA)

4. Conclusions and Discussion

The system proposed can be used as a complete training rehabilitation device, due to its portability, lightweight and embedded signal conditioning. The algorithms and model developed in Matlab has the potentiality to be compiled as executable software, and used in any personal computer. The control objective of this work was accomplished and also, a versatile and documented sEMG signals database was obtained.

The different stages of signal processing (pre-processing, feature extraction and classification) provide a control signal that was used in the actuators of the virtual model. By the results presented, it was demonstrated that the use of a feature vector, consisting of E, IAV, WAMP and WL, generates the best results in the classification process. The use of ANN is one of the best alternatives to the sEMG signal classification, by their ability to learn and solve problems that are not linearly separable. This allowed us to get offline classification rates higher than 70%. The virtual model of the upper limb was built with SimMechanics®, providing a VR environment in which is possible to perform online experiments and check the results obtained offline.

We conclude that this system can be used as wearable robot simulator, for upper limb rehabilitation and also as training stage for potential users of myoelectric prostheses. Future works include the use of microcontrollers, inclusion of programmed tasks in the VR environment and shoulder simulation, in order to add more degrees of freedom to the model. Also, the system must be tested with people with upper limb disabilities.

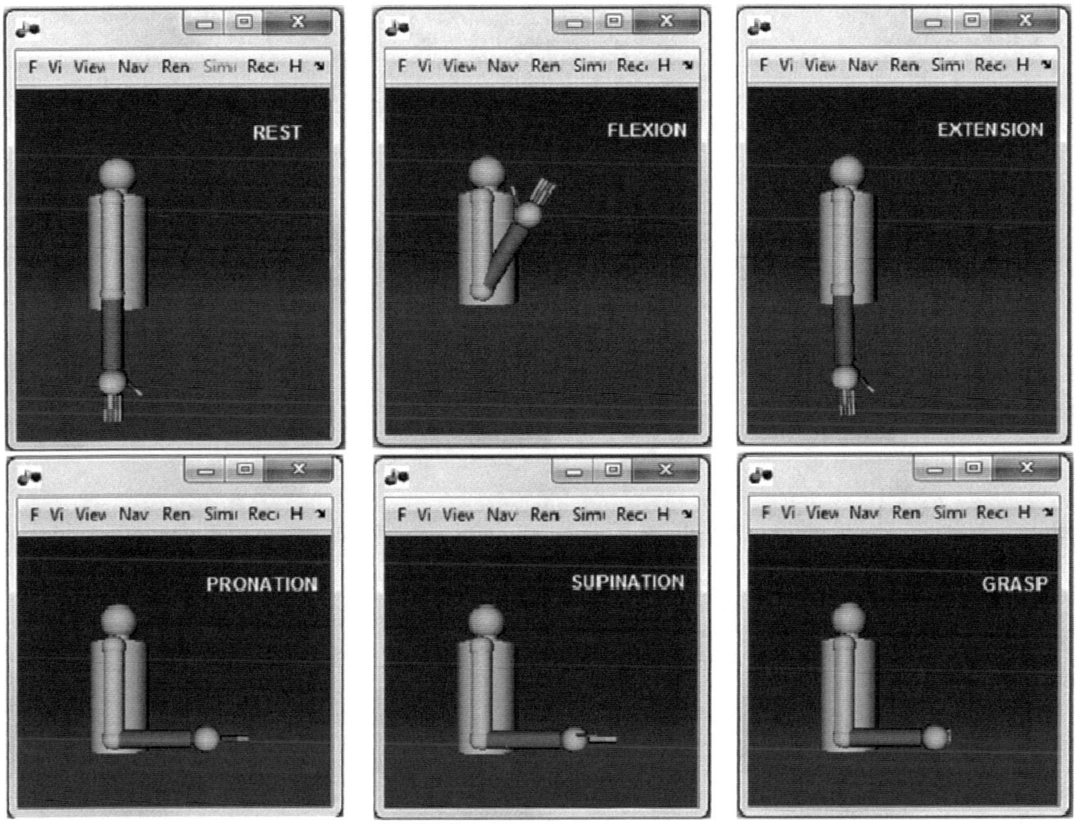

Figure 6: The virtual model performed with the V-RealmBuilder Toolbox®, which simulates online the movements made by the volunteers

References

[1] M. Zecca, S. Micera, M.C. Carrozza, P. Dario, "Control of multifunctional prosthetic hands by processing the electromyographic signal", Critical Reviews in Biomedical Engineering, Begell House, 2002, pp. 459– 485.

[2] Krebs H.I., Volpe B.T., et al. "Increasing productivity and quality of care: Robot-aided neuron rehabilitation", Journal of Rehabilitation Research and Development, 2000, Vol.37, No.6, pp.639-652.

[3] Mihelj, M.; Nef, T. & Riener, R. "ARMin – Toward a Six DoF Upper Limb Rehabilitation Robot", IEEE International Conference on Biorobotics and Biomechatronics, pp. 1154-1159, Italy, February 2006, IEEE, Pisa

[4] Adamovich S., Merians, A., Boian, R., Tremaine, M., Burdea, G., Recce, M., & Poizner, H.(2005). A Virtual Reality Based Exercise System for Hand Rehabilitation Post-Stroke. Presence, Special Issue on Virtual Rehabilitation, Vol.14, No.2, April 2005 pp.161-174, ISSN:1054-7460

[5] Morrow, K., Docan, C., Burdea, G. & Merians, A. "Low-cost Virtual Rehabilitation of the Hand for Patients Post-Stroke", Proceedings International Workshop on Virtual Rehabilitation, pp 6-10, ISBN: 1-4244-0280-8, August 2006, New York, NY

[6] Merletti R, Botter A, Cescon C, Minetto MA, Vieira TM. "Advances in surface EMG: recent progress in clinical research applications". Crit Rev Biomed Eng. 2010; 38(4):347-79

[7] http://www.ni.com/pdf/products/us/4daqsc202204_ETC_212-213.pdf

[8] Zardoshti-Kermani M.: "EMG Feature Evaluation for Movement Control of Upper Extremity Prostheses". IEEE TransactionsonRehab. Engineering, vol. 3, pp.324-333, 1995.

[9] "V-Realm™ Builder. User´s Guide and Reference". MatLab Toolbok.

9th Argentinean Bioengineering Society Congress (SABI 2013) IOP Publishing
Journal of Physics: Conference Series **477** (2013) 012042 doi:10.1088/1742-6596/477/1/012042

Prototype of a Communications System for Blind People in a City

J.C. Fuentes, J.P. Graffigna, J.Dipane and M.Segura

Gabinete de Tecnología Médica. Facultad de Ingeniería.
Universidad Nacional de San Juan, San Juan, Argentina.

email: jgraffig@gateme.unsj.edu.ar

Abstract. Assistive Technology (AT) is an interdisciplinary research area that allows finding solutions for individuals having a given disability[1] by easing or improving skills and functions in order to perform daily activities. A technology is regarded to be assistive whenever it fits well to the needs, skills and capabilities of the user, specifically by taking into account the intended activity and limitations arising in the context and environment where the person performs such activity. This work intends to solve the problems of vision-impaired persons in order to help them move autonomously through a city. To this aim, a prototype based as an FM device to transmit traffic lightsinformation to blind people is presented. Both the prototype design and its development are described, altogether withsteps and preliminary experiences. Results and conclusions show the perception levels and evaluation responsesstated by blind people, as well as thedrawbacks and advantages of the prototype.

1. Introduction

A main objective of Assistive Technology (AT) is to ease or allow for accomplishing a certain task within a given setting. The professionals that work with AT are in charge of primarily pondering the various factors involved, so as to recommend the proper technical assistance that best fits to the user's needs and capabilities in his/her specific environment[2]. Therefore, it is possible to state that AT is not aimed at solving the impairing condition but, rather, it tries to attain acceptable functional results that the user seeks to do his activities.

When designing AT systems, devices are engineered to let the user be a functionally independent individual. In the case of sensorial impairment, ATs may offer the necessary assisting means to allow the user acquire such sensorial information. To decide what kind of AT has to be developed for a given case calls for a prior analysis on the impairment state itself, its residual function and level. With this information, it is then possible to respond to whether the sensorial enhancement devices can be used or not. For example, if the residual sensorial function is insufficient, alternative sensorial ways have to be used[3]. Therefore, for persons whose vision cannot be used as the interface with the intended technology, other senses, like hearing and tactile functions may be studied as alternate ways.

For blinds, specifically, ATs offer help in various areas, such as obtaining visual information from the context, manipulating and controlling objects, bringing mobility and orientation for AT users, and other applications.

Mobility and orientation are significant problems for the blind people. Normally all the senses are sharpened to detect smells, sounds, air drafts, textures, temperature differences, and the like, in order to alert or bring awareness to the person about the environment where he/she is located.

There are several solutions to the problem of mobility [5], starting from canes and ending at electronic devices for travel assistance (Electronic Travel Aids or ETAs). With all its inherent limitations, the cane is undoubtedly the best device in use so far for its lower cost and greater simplicity. The ETAs use different physical principles and, at present, there is not any totally general or complete. Generally, these are based on physical principles of ultrasound, laser, vision, etc.[3][4][11].Canes or glasses with headphones incorporate ultrasonic sound emitting audio signals according to the characteristics of the objects that are close to the user.

The spatial orientation and recognition of objects beyond the reach of the cane is the goal of many experimental systems. These seek to convert environment variables (information from images, three-dimensional reconstructions, ultrasonic signals, laser emitters, and similar sources) into useful information for auditory or tactile stimulators. After training, the user may interpret patterns, contrast objects located in front of the sensor, or experience visual phenomena as a perspective, a parallax, etc.

In order to let the user know his/her position and allow for safe mobility in a city, it is requiredto overcome many cultural, physical and information barriers. The first constrains are related to the scarce importance given by architects, businesses, employers and the general public to the problem of disability. On this regard, it is ubiquitous to find buildings built offside the construction line, which is the referential line for the mobility of blind people. Stores and businesses do not properly care about the risks of signs placed on sidewalks or other artifacts impeding pedestrian traffic. Motorists may also cause annoying hurdles when parking their cars on the sidewalk, obstructing the crosswalk or downplaying traffic regulations. Thesebehavioral attitudes call for an awareness policy for the society which should include not onlyurban planners and municipal entities but all pertaining agencies and entities involved in such issue. The most relevant physical barriers for blind people in the public environment, excluding the above mentioned ones, are the uneven, broken sidewalks, the inability to be oriented in open spaces or the hurdles of crossing normal- to high-traffic streets. In this case, the lack of physical information is an obstacle to overcome.

While there are systems based on GPS, with speakers voicing information on the local surrounds, such systems cannot report dynamic events that change in minutes or seconds. For example: intersections having or lacking street lights; the state of traffic lights; recognition of the bus line;the transient variations or changes of the environment (barriers, curbs, ditches, culverts, poles), etc.

In many countries, the assisting technology to bring accessibility for blinds and visually impaired persons for crossing streets featuring traffic lights involvesusing an audible signal or some variation of the sort. However, in large cities, where these devices are implemented, they present some limitations, which create confusion, annoyance and distrust.

The traffic lights with sound signals emit a beep sound to indicate the pedestrians when to cross and, depending on the time left to complete the crossing, it speeds up its beep frequency. These traffic light arrangements may present some problems like, for example, off-squared corners or highly-noisy areas where this disturbance may lead to confusion [7]. In addition, the distress experienced by neighbors may impel them to destroy such devices.

Other developed tools use other technology types that resort to remote devices activated by the user. Such a solution requires that all blind people count with these devices. [6]

The present work aims at designing a Communications System and developing a prototype based on FM signal emissions that reportboth the static and the dynamic information on the surrounding environment to the blind person. To this aim, short-range radio stations are placedthroughout the city,from which they receive information for different events and thus can be informed audibly. This system orients not only blind people, butalso provides guiding information to the general public.

Moreover, the proposed solution is designed on the idea of a universal design and does not require employing complex elements by the user.

2. Methodology

Upon studying and analyzing previous works, the design and construction of the device is carried out in a series of steps. They include a needs survey of the local blind population, the technological aspects related to the design, the development and testing of the specific prototype for further improvements, or, as an approved release.

The survey was conducted in cooperation with the Association for the Blind of the Province of San Juan. The survey trialsinvolved blind people helping assess the most probably needed capabilities and activities, particularly those relating to mobility in the city, and in their daily lives. Other related inquires implied defining some aspects of the prototype.

The prototype design has to meet three aspects. First, it should be a device meeting the universal design paradigm that regards'a product within itsusage environment, with easy affordability and access to the largest number of people as possible, including all aspects possible for accessibility, by simplifying daily lifeand by building products, services and environments of simple makes and usage for all people'. The secondconsideration was to use ready, market available, low-cost technology components. Finally, the design was based on a model proposed by Assistive Technology Cook[5].This model is proposed to define an AT system by analyzing the solution as a whole; that is, by considering the activity to be done, the functional characteristics of the person and the context of use (see Fig. 1).

The ATs for blind people require an interface capable ofcapturing/receiving information from the context (Machine-context interface); it feeds such data to a processor which, in turn, produces usable information transmitted to the person via another interface. Specifically for this current work, the proposed design is based on atraffic lightsinterface to receive information fromthe context, a processor that convert the events into voice messages transmitted to uservia an FM receiver (the processor itself) and renders the information through the speakers to the user (Man-machine interface).

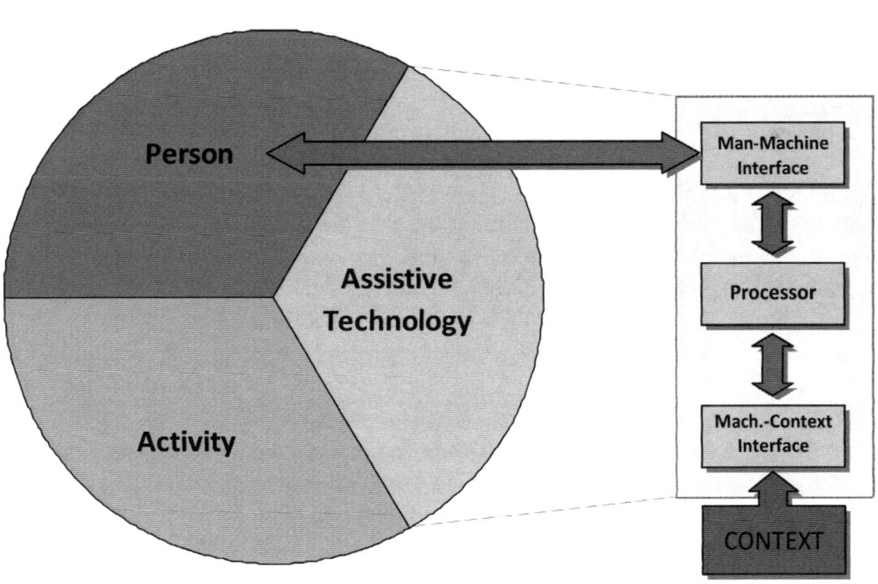

Figure 1. Interaction of the AT with the other components of the HAAT model.

The communication system for the blind can provide static and dynamic information with three basic elements. The first one is an FM radio transmitter with a band operating in the commercial spectra that allows any user to receive the signal. The second element is a voice synthesizer capable of converting information of text format into an audio signal to be transmitted. Finally, it features a microprocessor system that receives information on dynamic events; then, it constructs messages with the correct information and releases the text message. A scheme is presented in Figure 2.

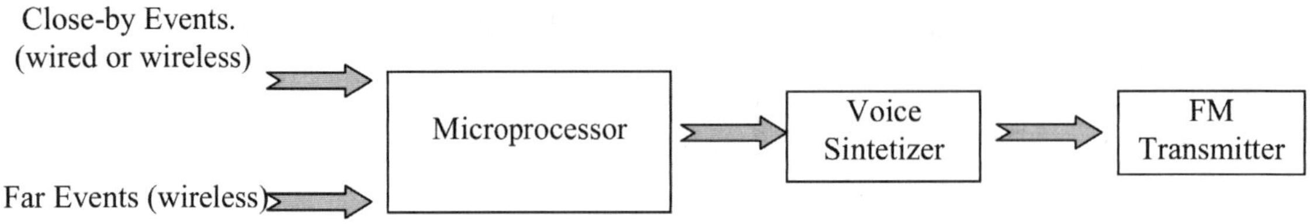

Figure 2. Schematic Diagram of the Prototype

The general framework of this system is presented in Figure2. Each device is placed at the street corners, or intersections, toemit information about the nearest events. All devices transmit an FM signal in a same frequency and,this way,the users receive information about the closest transmitter. The prototype is developed using wired events, but is possible to extend this scheme to wireless events as well.

3. Preliminary Results

3.1. Survey on the Needs of Blind People.

The purpose of this survey stage was to study the specific problems of the blind in the city of San Juan and to consider the information (messages) rendered by the prototype in accordance to the needs stated by the participant subjects.

This section also presents the results and some conclusions stemming from a studymade with a group of 12 blind people. This research part was undertaken in the context of an oral survey.

As a preliminary stage, the group was informed on the proceedings of the planned experience and the general concept of the project. Subsequent meetings were held with small groups, where they answered a brief questionnaire based on the criteria of the International Classification of Functioning, Disability and Health (ICF).

We sought people who do their daily mobility autonomously, and who have severe visual impairment. Almost everyone in the group had acquired the disability in adulthood. None of them has shown any discordance in mental functions. Regarding hearing and balance, only 16% of the test group had a slight deficiency. None showed any loss of motor functions.

As for mobility, on average each person visitsthe city 7.75 times a week, being this rate greater for individuals aged 20-35 years (10 times). We also took account of the transport means they used. About 80% of the individuals take mainly the bus and, to a lesser extent, taxis.In addition, approximately 33.3% of the sample cannot read or write in Braille.

Regarding the use of TA, all have stated they use the cane, and they know about computers with screen readers, recorders, calculators and tactile floors.

Then, the group was consulted about the prototype to implement. First, we gave a general description of the idea and the proposed technology. When asked which information was more important for them, the following priority list was specified: the current position of the individual; the states/changesof traffic lights;the bus stops and bus numbers; the most challenging circuits in the downtown; landmarks; ways to stop a taxi; important public landmarks and buildings (Hospitals, Social Services Offices, etc..).

3.2. Voice Synthesizer.

Twelve different data sheets for speech synthesizers were analyzed. A comparative study was made considering the following features: method of synthesis, input/output, evaluation board, Spanish language support, etc. We selected a model developed by Digital Acoustic Corporation [8], that uses a method based on concatenation phonemes, has a serial input and <standard analogic output as well as support in Spanish (Figure 3).

Figure 3. Voice Synthesizer.

The synthesizer was installed and configured using acomputer. We had no stability problem that had to be solved by the manufacturer. The synthesizer was blocked and we had to define a strategy to solve this problem.

3.3. Transmitter and Transmission Range Test.

Under the premises of universal design, it was sought for a transmission mode ofcommon, widespread use, which could be inexpensive to install and operate, and be capable to communicate the human voice easily within the 50 meter range. Some analog and digital alternatives were evaluated. From the above considerations, it was opted for FM transmission, which is a common means for cellular phones, portable MP3 devices and conventional radios, while being inexpensive as well.

We used a commercial transmitter featuring an analog input, a single-pole antenna and maximum power of 1dBW, as shown in Figure 4.

Figure 4.Commercial FM Transmitter.

For experimentation, a spectral scan was carried out using a Rhode& Schwartzspectra analyzer, choosing a transmission frequency that may not interfere with commercial frequencies. The frequency was set at 94.3MHz. In the tests, we sense a strong signal in path-loss ranges, but the signal-to-noise ratio dropped in presence of many obstacles and background noise.

On open space having low noise (-90dBm), and by adjusting the power to -10dBm, the signal could be received from up to 150m, when the signal power was similar to the base power noise (Figure 5).

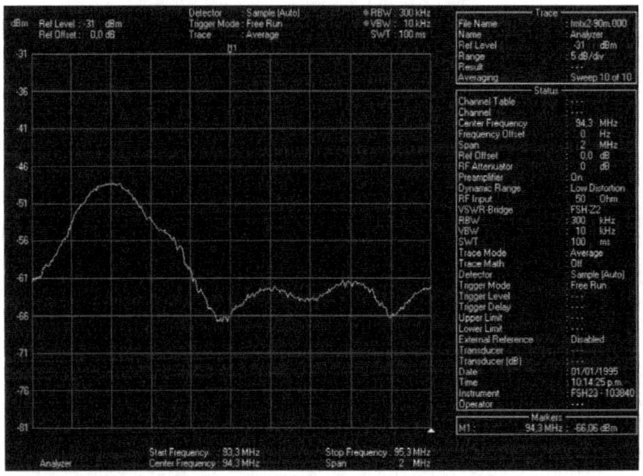

Figure 5.Power Spectra. Distance to transmitter: 150m.

Upon completing this stage, a cell-commutation test is done. This refers to the ability of receiver to switch between two different stations emitting on the same frequency, when it commutes or switches from one cell/station to another. For this test, we used two transmitters T1 and T2 calibrated on a same frequency and power level. Gap distances were tested at 300, 200 and 180 meters. Both devices were tested in open space and with little background noise. For all instances, a receiver was shift from T1 to T2, where we can distinguish three areas: T1 reception area, switching area and T2 reception area. The switching area was noisy and was reduced when the distance between T1 and T2 is reduced. This phenomenon can be seen in Figure 6, where the signal must be greater than the noise in order to reduce the switching zone.

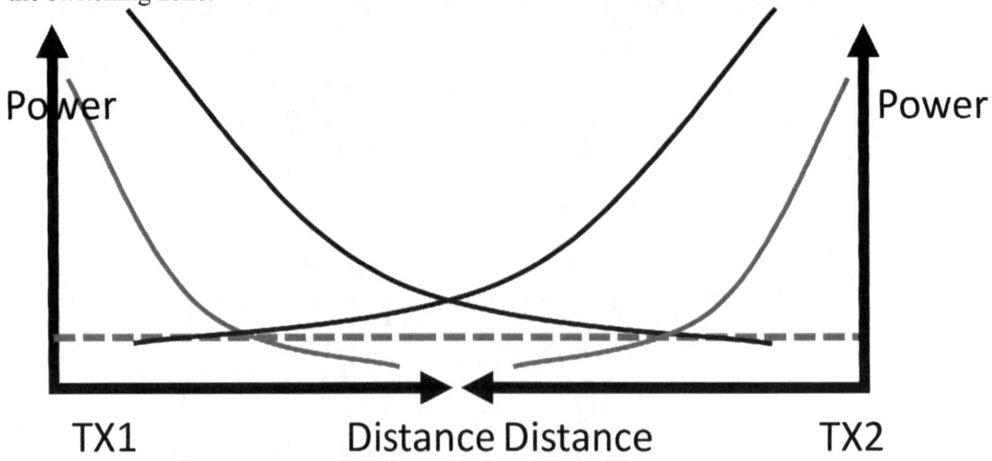

Figure 6.Noise effect on the switching zone.

4. Development

After performing a preliminary study of the most sensitive components, we used a basic microcontroller to operate the control logic with the synthesizer. The microcontroller was placed on a board to receive opto-isolated traffic lights signal and to send messages to synthesizer.

The software was laboratory tested, and all elements for electrical safety were verified. Figure 7 shows a view of the device.

Figure 7.Developed Prototype.

5. Results of tests with blind users.

The test with four blind people was performed at the corner of Mitre and Caseros Streets, San Juan-Argentina. The transmission frequency firstly chosen was 93.3MHz. Previous power measurements with the spectrum analyzer indicated that this frequency attained the best conditions to test the module for this location. The antenna's length is 80cm, telescopic type,and maximum input power. Cell-phones were used as audio receivers for testing. The participants had cell phones of different makes and models.

As a first activity, they were given a brief instructional introduction on device operation, as well as all features relevant to spatial orientation and cardinal points. Each individual had to be assisted in the earlier trials. As a final test, each individual was permitted to cross the intersection with silent supervision. Then, they were left alone to operate by themselves in said zone. Figure 8 shows a prototype installed at the intersection and a blind participant crossing the street.

Figure 8.Evaluation with blinds.Installed Device and Blind Person using the FM receiver.

After tests,the participants were inquired through a short questionnaire: The general responsesfor each question were as follow:

-Did you feel safe with the information provided by the messages of the device?

'The participants were confident with the information given by the device, so that they needed no further than a brief explanation about the delivered messages'.

-Have you had a hard time understanding the operation of the prototype?

'The learning period needed was very short due to the simplicity of the method. Besides, the transmission was crisp and clear'.

-Are the messages clear?

'The messages are clear, simple and the transmission came inwithout any electrical or electromagnetic noise'.

-What message would you add to the above ones?

'Messages with information about wheelchair ramps were not taken into account'. Due to the lack of tactile floors, these messages are very important for blind pedestrians. It must be emphasized that missing or broken drain lids are very dangerous hazards for blinds.

-What message would you change?

It was proposed to change a word in Spanish that could cause confusion: "cruce" (crossing/to cross) for "intersección" (intersection)

6. Conclusions

The main goal of this job isto transmit static and dynamic information to blind people to let them walk and cross safely at intersections using traffic lights information. The dynamic information transmitted is the current state of traffic lights and the static information corresponds to the environment of the corner in a 50 meter range.

At first, we analysed the main needs expressed by the local blind community in order to discern the particular characteristics of the problem. Then, we looked through and studied different existing alternatives but could not find any similar solution to the here proposed one.

As regards device technology, we analysed various alternatives in the design stage, and then we tested the selected components. The tests consisted in the evaluation of the synthesizer and the cell switching of the FM transmitters.

Mounting the device was a simple task, because of the separate modules that worked correctly at the laboratory. A main problem during the development of the assembly was the timing for the messages, from the microcontroller to the synthesizer module.

Participants indicated that the prototype is very easy to use. The messages are clear and simple to understand excepting for some minor corrections. The device is very useful for their daily commutes.

It is expected that this prototype, in combination with tactile floors at the sidewalk-curb border, will produce excellent results.

As an extension into a future project, plans areto use the prototype with other events, and to have a server in the city linked to the devices.

References

[1] OMS, ORGANIZACIÓN MUNDIAL DE LA SALUD: Clasificación Internacional del Funcionamiento, de la Discapacidad y de la Salud (CIF) 2001.

[2] GRAFFIGNA, Juan Pablo: "Sustitución Parcial de la Visión mediante Realimentación Táctil". 2009. Universidad Nacional de San Juan. ISBN: 987-05-7233-6.

[3] ZELEK, Jhon: The E. (Ben)& Mary Hochhausen Fund for Research in Adaptative Technology For Blind and Vissually Impaired Persons. 2002.

[4] MARCIN R. STRAKOWSKI, BOGDAN B. KOSMOWSKI, Ryszard KOWALIK, and Paweł WIERZBA: An Ultrasonic Obstacle Detector Based on Phase Beamforming Principles. IEEE Sensors Journal, VOL. 6, NO. 1, FEBRUARY 2006.

[5] COOK A. HUSSEY .S.: Assistive Technologies, Principles and Practice. 2da Ed.Mosby. 2002

[6] MARTINEZ, Alfonso : "Nuevos Semáforos sonoros en la ciudad de León", Ciudad de León, ONCE

[7] DIAZ, Eduardo: "Sistema de orientación para invidentes y débiles visuales aplicados para el primer cuadro de la ciudad" Universidad de Antioquia, Mayo 2008.

[8] Digital Acoustic Coorporation:User manual TTS-EM-HD2.

[9] Martínez Evelio, "El ABC de las Telecomunicaciones", la Revista RED en la edición especial Noviembre 2002.

[10] CIDAT-ONCE: Catálogo de Material Tiflotécnico. Organización Nacional de Ciegos de España. 2012.

[11] DEBNATH Niranjan, Jaya Bharati THANGIAH, Shamala PARARASAINGAM, Dr. Syed ABDUL KADER ALJUNID: A Mobility Aid For The Blind With Discrete Distance Indicator And Hanging Object Detection. TENCON 2004. IEEE.

[12] DEWHURST, D: Electronic Aids for Visually Impaired People. Vis-Ability DK. 2005

AUTHOR INDEX

AUTHOR INDEX

Institute of Physics
Dirac House, Temple Back
Bristol BS1 6BE UK

CURRAN ASSOCIATES INC
proceedings.com